Tuberculosis in Animals: An African Perspective

Asseged B. Dibaba • Nicolaas P. J. Kriek •
Charles O. Thoen
Editors

Tuberculosis in Animals: An African Perspective

Springer

Editors
Asseged B. Dibaba
Department of Pathobiology, College of Veterinary Medicince
Tuskegee University
Tuskegee, AL, USA

Nicolaas P. J. Kriek
Department of Paraclinical Sciences, Faculty of Veterinary Science
University of Pretoria
Onderstepoort, South Africa

Charles O. Thoen (deceased)
Department of Veterinary Microbiology and Preventive Medicine
Iowa State University
College of Veterinary Medicine
Ames, Iowa, USA

ISBN 978-3-030-18688-3 ISBN 978-3-030-18690-6 (eBook)
https://doi.org/10.1007/978-3-030-18690-6

This Springer imprint is published by the registered company Springer Nature Switzerland AG.
The registered company address is: Gewerbestrasse 11, 6330 Cham, Switzerland

This book is dedicated to the memory of Charles Thoen, coeditor of this book who died unexpectedly before its publication. We will remember him for his enthusiasm, commitment, and the contributions that he made until his death to the knowledge base of bovine tuberculosis in all its forms.

Preface

Africa, considered to be the cradle of humankind, is a vast continent characterized by the richness of and variation in its ecosystems and human and animal populations. It faces major challenges as part of the developing world, and many of the countries on the continent are of the poorest on the globe. The lack of financial resources, inadequate medical and veterinary services, levels of poverty, underdevelopment, lack of service delivery, political instability, ethnic conflict, and serious diseases of humans and animals complicate these issues in most, if not all, of its countries.

This book deals specifically with tuberculosis of animals in Africa and its zoonotic implications for human populations in the various countries. It is known that human tuberculosis was rife in Egypt at the time of the Pharaohs, and it is thought that tuberculosis in cattle existed in North Africa since ancient times. At that time, and even now, nomadic human–animal population movements seem to facilitate the spread of the disease throughout various regions in Africa.

Tuberculosis is one of the major economically important global diseases of humans and animals. It is of as great importance in Africa as it is elsewhere and it is endemic in African countries that report their disease status to the WHO and the OIE. In many of the countries, confirmation of the causal agent and determination of species and strains are not practiced because of the lack of suitable laboratories and of inadequate financial, human, and material resources. In most of the African countries, available funds are allocated to dealing with more immediate devastating animal diseases such as rinderpest (now eradicated), CBPP, and foot-and-mouth disease. Bovine tuberculosis, however, remains important as it limits the ability of countries to participate in international trade in animals and animal products to countries that are free from the disease, or are in the process of eradicating it.

For many years, it has been assumed that colonization and the importation of foreign cattle resulted in the introduction and spread of bovine tuberculosis in Africa. However, recent research suggests that unique African strains of *Mycobacterium bovis* occur in Western and Eastern Africa, and it appears that the disease in cattle in Africa was present before the importation of bovine tuberculosis-infected foreign cattle from Europe and Australia. It does appear though that the introduction of the foreign strains of *M. bovis* contributed to the large diversity of *M. bovis* strains isolated from cattle across Africa.

Tuberculosis in animals is caused by a number of bacteria of the genus *Mycobacterium*. It is a chronic, contagious disease and, dependent on the mycobacterial species, has a wide host range and various routes of transmission.

The epidemiology of tuberculosis in animals is complex, and is even more so in Africa, given the extreme variation in ecosystems, poor border control, and farming practices that vary from extensive to intensive and, in many instances, the movement of livestock over vast distances because of nomadism and transhumance.

In parts of Africa, more recently, spread of *M. bovis* to wildlife has become common, and in some instances, wildlife species became maintenance hosts of the disease. In this respect, the role of lechwe (*Kobus leche*) in Zambia and that of African buffaloes (*Syncerus caffer*) in South Africa, Zimbabwe, Tanzania, and Uganda are good examples. This development has serious consequences for countries attempting control and eradication of the disease in domestic stock as it has been shown that when bovine tuberculosis occurs in a number of species in a complex ecological system, eradication by applying current control measures becomes impossible. Cross-border control of the movement of wildlife is even more difficult than that of livestock, and this movement may become a bone of contention in the development of the large number of trans-frontier parks that straddle the border of a number of southern African countries.

The increasing use of contemporary molecular techniques, such as VNTR, in African countries as presented in this book, provides greater insights not only into the status of bovine tuberculosis in African countries, but, moreover, the interrelatedness of outbreaks across international borders.

The information provided by current African research presented by the different contributors to the book and the reports of different African countries provide a better understanding of the epidemiology of *M. bovis, M. africanum*, and other closely related pathogenic mycobacteria in cattle, wildlife, and zoonotic tuberculosis in Africa. This improved understanding should be utilized by policy makers and animal and human health authorities to improve their decision-making and chances of success when attempting to control and eradicate the disease in their respective countries. It is also clear that future activities aiming to control and eliminate tuberculosis in animals and humans should employ interdisciplinary collaboration between medical and veterinary medical professionals as embodied by the philosophy of the "One Health" approach.

Tuskegee, AL — Asseged B. Dibaba
Onderstepoort, South Africa — Nicolaas P. J. Kriek
Ames, IA — Charles O. Thoen

Contents

Part I Human and Animal Tuberculosis in Africa

1 Introduction 3
Nicolaas P. J. Kriek

2 The Current Status of Bovine Tuberculosis in Africa 15
Asseged B. Dibaba, C. J. Daborn, S. Cadmus, and A. Michel

3 Bovine TB Zoonosis in Africa 31
Paul D. van Helden and Anita Michel

4 The Control of *Mycobacterium bovis* Infections in Africa: A One Health Approach 41
S. I. B. Cadmus, P. I. Fujiwara, J. A. Shere, B. Kaplan, and C. O. Thoen

5 Tuberculosis in African Wildlife 57
Anita L. Michel and Paul D. van Helden

6 The *Mycobacterium tuberculosis* Complex in Africa 73
Sven D. C. Parsons, Michele A. Miller, and Paul D. van Helden

Part II Epidemiology of Bovine Tuberculosis in Africa

7 Epidemiology of Bovine Tuberculosis in Africa 89
Asseged B. Dibaba and C. J. Daborn

8 Molecular Epidemiology of *Mycobacterium bovis* in Africa 127
Adrian Muwonge, Franklyn Egbe, Mark Bronsvoort, Demelash B. Areda, Tiny Hlokwe, and Anita Michel

9 The Diagnosis of Bovine Tuberculosis 171
Nicolaas P. J. Kriek, Demelash B. Areda, and Asseged B. Dibaba

10 The Control of Bovine Tuberculosis in Africa 237
Asseged B. Dibaba and Nicolaas P. J. Kriek

Part III Country Reports

11 Bovine Tuberculosis: Status, Epidemiology, and Public Health Implications in Burkina Faso . . . 273
Adama Sanou

12 The Status of Bovine Tuberculosis in Cameroon . . . 283
Julius Awah-Ndukum, Nkongho Franklyn Egbe, and Victor Ngu-Ngwa

13 Bovine Tuberculosis in Egypt . . . 305
Aziza Amin

14 Status of Bovine Tuberculosis in Ethiopia: Challenges and Opportunities for Future Control and Prevention . . . 317
Demelash B. Areda, Adrian Muwonge, and Asseged B. Dibaba

15 Bovine Tuberculosis in Ghana . . . 339
Dorothy Yeboah-Manu and Adwoa Asante-Poku

16 The Status of Bovine Tuberculosis in Malawi . . . 351
Poya E. C. Njoka and Asseged B. Dibaba

17 Bovine Tuberculosis in Nigeria: Historical Perspective, Burden, Risk Factors, and Challenges for Its Diagnosis and Control . . . 363
Simeon Idowu Babalola Cadmus

18 Bovine Tuberculosis in Rwanda . . . 379
Gervais Habarugira, Joseph Rukelibuga, and Manassé Nzayirambaho

19 BTB Control Strategies in Livestock and Wildlife in South Africa . . . 387
Anita L. Michel, Donald R. Sibanda, and Lin-Mari de Klerk-Lorist

20 Bovine Tuberculosis in the Republic of Sudan: A Critical Review . . . 403
Z. A. Ishag, El Tigani Asil, Ali Parsaeimehr, and Guo-Qing Shao

21 The Changing Landscape of Bovine Tuberculosis in Tanzania . . . 415
Bugwesa Z. Katale, Hezron E. Nonga, and Rudovick R. Kazwala

22 Holes and Patches: An Account of Tuberculosis Caused by *Mycobacterium bovis* in Uganda . . . 425
A. Muwonge, L. Nyakarahuka, W. Ssengooba, J. Oloya, F. Olea-Popelka, and C. Kankya

23 Bovine Tuberculosis in Zambia . . . 445
Sydney Malama, Musso Munyeme, and John B. Muma

Part I
Human and Animal Tuberculosis in Africa

Chapter 1
Introduction

Nicolaas P. J. Kriek

1.1 Introduction

Africa is a vast continent, and it is the second largest on the planet. It is subdivided into 54 countries with a large diversity of ethnic groups and languages, an enormous livestock population, and a unique diversity of wildlife. Its human population in 2013 exceeded 1.1 billion; and it is expected to reach 2.5 billion by 2050, and 4.4 billion by 2100. The African countries have a tumultuous history, and currently they remain some of the least developed and poorest of all countries globally. The impact of incursions and colonization by other nations, and human migration shaped the current distribution of the main ethnic groups, and the contour and size of the individual countries that now exist in Africa.

Livestock numbers on the continent are large, and this form of agriculture is critical in sustaining many of the communities, of which in Western and Eastern Africa, large numbers live nomadic lives. Nomadic communities participate in one or other form of transhumance, characterized by the seasonal migration of people and their livestock across international boundaries, making disease control in livestock almost impossible. This diversity of people, livestock husbandry practices, and wildlife present major challenges to policy makers and regulatory authorities, including those managing and controlling diseases. Many of the epidemic diseases of humans and animals of global importance are prevalent on the continent and affect the wellbeing of humans and the distribution of its livestock and wildlife. A large number of these diseases are zoonoses, and livestock, but particularly wildlife, play a major role in sustaining these infections. The complexity of the demographics of Africa and the lack of technical personnel and financial resources have a profound effect on the ability of the continent to manage these important diseases.

N. P. J. Kriek (✉)
Department of Paraclinical Sciences, Faculty of Veterinary Science, University of Pretoria, Onderstepoort, South Africa

A. B. Dibaba et al. (eds.), *Tuberculosis in Animals: An African Perspective*,
https://doi.org/10.1007/978-3-030-18690-6_1

Tuberculosis is of major importance in livestock, wildlife, and humans and has a global distribution. With its many manifestations and numerous species susceptible to infection by the mycobacteria that cause the disease, it is endemic on the continent and is present in its humans, livestock, and wildlife. The content of this book deals with the various aspects of *Mycobacterium bovis* infection in Africa in livestock (cattle in particular), wildlife, and humans that are prone to contract this serious infection.

1.2 A Brief Overview of the Important Aspects

The chapters in this book deal with specific aspects of the disease caused by *M. bovis* in humans, livestock, and wildlife with the focus on Africa and the challenges associated with the detection, epidemiology, control, and eventual eradication of the disease from the continent. It also contains a number of country reports that focus on the specific situation in each of the countries. As only a few of the countries of Africa have the ability, and committed to report on the status of tuberculosis in their human and animal populations, these may be assumed to reflect the status in most, if not all, countries in the various regions of Africa: those on the Mediterranean rim, Western Africa, and the Sahel, and in Central, Eastern, and Southern Africa.

The second chapter provides a broad overview of the status of the disease and its control in Africa. Bovine TB (BTB) is globally a notifiable disease and is classified by the World Animal Health Organization (the OIE) as being of socioeconomic and of public health importance. It is now also classified as one of the important neglected zoonoses. Against this background and the known presence of the disease in humans and animals in Africa since antiquity, little is known about its current status even though the first diagnosis of BTB in livestock was made in South Africa in 1880 and in wildlife during the course of the 1920s, and it is now known to occur throughout the continent.

For long, it has been assumed that the European settlers introduced BTB into Africa during the time of colonization, but it subsequently transpired that there are a number of indigenous strains, the Af clonal complexes of *M. bovis*, causing BTB in Africa. Control of the disease is complicated by its characteristically slow, insidious spread in cattle herds and populations, the lack of specific clinical signs, and the difficulty of eradicating the disease from livestock even in the developed countries in the world. The limited knowledge about the disease by African farmers, consumers of products, veterinary authorities, and policy makers, and limited human and financial resources are the major reasons why little attention is still given to the control and eradication of the disease from African countries.

The complexity of the epidemiology of BTB is a major reason for the difficulties experienced in controlling it. Bovine TB is a multihost disease that also occurs in wildlife, some of which are maintenance hosts. Given the experience in some of the developed countries, such as the UK, New Zealand, and the USA, it appears to be impossible to eradicate the infection once one of the wildlife species in the same ecosystem becomes a maintenance host of the infection. Initially, BTB was considered to be a disease of intensively farmed cattle housed during winter, and that it would not be a problem in extensively farmed cattle. This still appears to be the situation in Africa where there remains a perception, shown since to be false, that local breeds were less susceptible to the infection, and that the extensive management systems practiced in many parts of the continent, will limit the extent of the disease, and thus its importance. Based on these assumptions, the lack of information about the distribution and extent of the infection in African livestock and wildlife, and the limited financial and human resources, policy makers and veterinary authorities mainly focus on the control of what they consider to be a more serious threat, the acute epidemic and endemic infectious diseases.

The marked increase in the size of the human population on the African continent and the rapid rate of urbanization are causing an increasing change in animal husbandry practices. To augment their income and to satisfy the burgeoning demand for milk of the expanding urban populations, there is now a trend to establish small-scale dairy farming in urban and periurban areas, and these practices result in a marked increase in the prevalence of BTB in many of those herds. As the disease remains uncontrolled in most of the countries on the continent, increasing numbers of people consume unprocessed, infected milk, given the ethnic cultural habit in many parts of Africa to consume raw, unprocessed, or soured milk, and meat.

Internationally, the control of BTB is increasingly complicated by the presence and persistence of the disease in free-ranging wildlife and its bidirectional spread at the wildlife–livestock interface. It is anticipated that Africa, with its rich diversity of wildlife species, will also be plagued by this problem. Examples of maintenance hosts that have already been identified in Africa include African (Cape) buffaloes (*Syncerus caffer*), greater kudus (*Tragelaphus strepsiceros*), and Kafue lechwe (*Kobus leche kafuensis*), and there may be others. The role of these maintenance hosts is largely poorly researched, and they may in future prove to be a major obstacle in controlling and eradicating BTB from Africa.

The most pressing problems in dealing with the issues are the lack of application, and the absence of statutory control policies including surveillance for the presence of BTB in most of the African countries. These events result in a lack of information required to develop sound control policies and programs. In most African countries, there is a total lack of information about the presence of the disease, its impact on the livestock sector, and its role in humans as a serious, but neglected, zoonosis.

It is critical for African countries to address the challenges and dangers of the presence of the disease, given the rapid expansion in the numbers of humans and animals on the continent that will enhance the impact of the disease, and if it remains uncontrolled, it may reach catastrophic proportions. Additionally, the presence of uncontrolled BTB across the continent will inevitably have a major negative impact on the international drive to eradicate TB in humans within the next few decades, irrespective of the specific cause and the presence of zoonotic TB.

Chapter 3, dealing in more detail with zoonotic TB caused by *M. bovis*, also emphasizes the lack of information about the prevalence of the disease in humans in Africa, primarily because of the lack of sufficient laboratory infrastructure and funding to distinguish between *M. tuberculosis* and *M. bovis* isolates. Humans contract the disease in different ways: by drinking unpasteurized milk, by inhaling infected droplets when in close contact with infected animals shedding mycobacteria, and by eating meat contaminated with *M. bovis*. The prevalence of zoonotic TB in Africa is unknown but it is known to be higher in countries in which the prevalence of the disease in cattle is high and in countries where BTB is common, and in which 10–15% of humans may suffer from zoonotic TB.

There is a general attitude, based on the universal lack of information about zoonotic TB and anecdotal evidence, that zoonotic TB in Africa is not a problem, and thus does not merit monitory expenditure to create the necessary diagnostic infrastructure. As a result, the role that it may play in the epidemic occurrence of TB in humans in Africa is largely ignored. This is against the background of the extent to which it occurred in some of the European countries before BTB was controlled in cattle and milk was pasteurized, where 50% of TB cases in children with lymphadenitis, and 65% of intestinal TB cases, were due to an *M. bovis* infection. Zoonotic TB is the main reason for introducing pasteurization of milk and for justifying the massive financial expenditure over many decades by developed countries to control BTB in their national cattle herds. Large sections of the African population do not have access to pasteurized milk and prefer to drink raw milk, or partially fermented milk. Ethnic practices also cause people to eat high-risk meat and meat products, and contrary to common belief, certain sectors of the nomadic people live in close association with their animals, thus enhancing the likelihood of contracting the disease caused by *M. bovis* by inhalation.

The likelihood that HIV-AIDS may enhance human infection with *M. bovis* in Africa has been raised as an additional reason why more attention should be given to this infection, but it appears, according to the limited available information, that this is not the case. One further important complication of not identifying *M. bovis* in human TB cases is the likelihood of the development of drug-resistant *M. bovis* strains as it is already resistant to rifampicin, one of the front-line drugs used for treatment of TB in humans. In this context, XDR *M. bovis* strains have been identified, and if they are to extend into Africa, their presence may develop into a health threat that would be most difficult to deal with.

The current *liaises faire* approach by policy makers and the human and veterinary regulatory authorities may yet prove to be a risky one, given the general lack of information about the disease. The occurrence of hotspots of *M. bovis* infection has been reported in Tanzania where, in areas, the prevalence of zoonotic TB in humans reached 10–38%. There may be many more such hotspots, and with the expected increase in the human population and urbanization, the general lack of control of the disease in cattle and the ignorance of farmers and consumers about the disease, zoonotic TB may have disastrous effects in times to come. As an example, an average herd prevalence of 10.5% and an in-herd BTB prevalence of up to 90% have been reported in intensive dairy farms in Ethiopia, and it is only reasonable to expect that this would eventually also be reflected in the number of zoonotic TB cases in those areas.

Chapter 4 deals with the control of the disease by applying the multidisciplinary approaches of One Health to facilitate the control and eradication of human TB, thus implying also the control and eradication of BTB. The complexity of BTB, because of the involvement of multiple susceptible species including humans, and the presence of wildlife maintenance hosts, makes it imperative that a broad multidisciplinary approach be adopted to engender some hope of success with the attempted control and eradication of both BTB and zoonotic TB in Africa. The WHO and other international agencies recently emphasized the importance of controlling BTB and of zoonotic TB in an attempt to eradicate human tuberculosis during the course of the next few decades. To do this, a number of issues will have to be addressed including determining the extent of BTB in Africa, the burden of the disease in the human population, the role of human-to-human transmission, and understanding the molecular epidemiology of the disease to allow the development of appropriate control strategies. The extent, cost to control, and complexity of the disease are such that to have any hope of successfully dealing with it will require a multidisciplinary approach. The One Health approach would enable authorities to focus on improved surveillance, the development of novel diagnostic methods, coordinated research, controlling the disease in targeted populations, and emphasizing food safety by raising the awareness of all stakeholders about the risks involved. It too will aid in developing effective policies and focusing on joint human and animal health interventions with committed funding by the respective governments and international donors. These activities require the development of a comprehensive plan for the continent, which can conceivably be coordinated by the AU, and aligning the activities to those included in the WHO roadmap for the eradication of zoonotic TB.

Although it has been known since the 1920s that various African wildlife species may contract BTB, for many years little attention has been given to the disease in wildlife in Africa. The occurrence of BTB in African wildlife and its impact on the epidemiology of the disease is dealt with in Chap. 5. Currently, at least, 29 different free-ranging wildlife species have been diagnosed with BTB. Few of them, such as African buffaloes, greater kudus, warthogs, and Kafue lechwe are maintenance hosts, sustaining the infection and transmitting the disease to cattle and other species at the interface. Most of the others are spillover hosts that may, or may not, play a role in transmitting the disease to other wildlife, livestock, and humans.

Globally, the role of wildlife as maintenance hosts of *M. bovis* infections, and their role in the epidemiology of the disease increasingly became apparent during the last few decades. This situation creates a major challenge for Africa against the background of the diversity of its wildlife and the extensive intermingling at the interface between wildlife, livestock, and humans in many parts of the continent. Bovine TB in wildlife may also impact negatively on the species infected, particularly those that are endangered and close to extinction, on conservation efforts, and on the income generated by the lucrative wildlife ranching enterprises in Southern Africa. Fundamentally, the only way in which the infection in wildlife could eventually be controlled would be by vaccination, but no effective vaccine has yet been developed. For African countries that want to attempt to control and eradicate

BTB, the extent of the infection and the epidemiology of the disease in wildlife will have to be determined, to allow the development of effective control programs that will have the potential to guide activities to eradicate the disease.

Wildlife, livestock, and humans are also infected by other species of the *Mycobacterium tuberculosis* complex (MTC) that evolved between 40,000 and 70,000 years ago from a common ancestral human pathogen of African origin. Many of these mycobacteria are host-specific but others cross species barriers. Some of the mycobacterial species are primarily found in Europe and Asia, but others, the African sublineage of the RD9-deleted clade, occur mostly in Africa. *Mycobacterium africanum*, a common cause of TB in humans in Western Africa, is rarely isolated from domesticated animals. The details of other wildlife-related species, including the chimpanzee bacillus, the dassie bacillus, *M. mungi*, *M. surricattae*, and *M. orygis*, are dealt with in Chap. 6. These organisms appear to be particularly host-specific and transmission to humans and livestock seems to be unlikely. However, being able to identify these mycobacteria may be important in the understanding of the epidemiology of TB in humans and animals, and the consequences of these infections for free-living wildlife and domesticated species. It is important, though, to keep in mind that the growth characteristics of these bacteria are poorly defined, that they are usually slow growers, and that they are difficult to culture. In addition, the genetic markers used to define lineages of the MTC may be ambiguous, and may result in the misclassification of some of the species in the MTC.

To adequately design control strategies for BTB, knowledge of its epidemiology that varies from country to country and even within countries is important. The data available for Africa are largely incomplete, fragmented, historical, often contradictory, and difficult to process. That which is known is collated in Chaps. 7 and 8, respectively, presenting the data based on the Epidemiologic Problem Oriented Approach (EPOA) and its molecular epidemiology. The lack of formal control measures for BTB, prevailing animal husbandry practices such as transhumance (leading to the uncontrolled movement of cattle and other livestock within and between countries), increasing urbanization and burgeoning smallholder dairy farming in urban and periurban areas, and the presence of the infection in a large number of wildlife species are distinctly different from the situation on other continents. Because of these differences, the epidemiology of BTB and zoonotic TB is anticipated to be different from those on other continents and unique for the African continent. Diagnosing the infection remains one of the most critical issues and is an additional complicating factor because of the lack of sensitivity and specificity of the various currently available diagnostic techniques.

The primary mode of transmission of the disease is dependent on the individual species, and often varies, depending on the local animal husbandry practices, within countries. In cattle, the main mode of transmission is by inhalation, but in certain settings, *per os* infection is important. Aerosol transmission is particularly important in cattle in a confined airspace and with crowding, but it also occurs when they are exposed to droplets of contaminated water, droplets expelled during eructation when grazing, and by inhaling contaminated dust particles. The presence of lesions in the palatine tonsils indicates that *per os* infection may be more important in cattle than previously anticipated. However, *per os* infection appears to play a lesser role in the transmission of the infection, as much higher doses than with aerosol transmission are required for cattle to contract the disease. This mode of infection, however, is more commonly seen in goats and camels. Few calves, unless fed pooled milk containing *M. bovis,* contract the disease by consuming infected milk because of the low occurrence of the infection in the udder of cattle. Predators, such as lions, too are prone to contract the disease *per os* while eating *M. bovis*-infected organs and tissues. Licking, grooming, and suckling also may result in *per os* transmission. Other forms of transmission, such as the percutaneous route, appear to be important in lions and in kudus.

Several host-specific risk factors play a role in cattle contracting the disease, and these are often related to specific management practices. Commonly, higher prevalences are encountered in urban and periurban smallholdings following uncontrolled movement of cattle between herds. But, contrary to the general expectation, in Africa, high prevalences can also be encountered in extensively managed cattle and other livestock because of the practice of housing them indoors at night in confined, roofed enclosures. Herd size, high animal densities, and the housing of cattle and other domesticated species are often proxies for the presence of and a high prevalence of the infection. Other herd-level risk factors in Africa include mixed rearing of domestic stock, such as goats, sheep, and camels, and intermingling with wildlife at the wildlife–livestock interface. Age is a risk factor and increasing numbers of infected animals are generally encountered with advancing age. This is particularly important in parts of Africa where cattle are kept for purposes other than the production of meat and milk; in those instances, the numbers of livestock owned are a sign of wealth, and they form part of a complex traditional social system. The role of gender as a predisposing factor varies according to management systems, and under specific circumstances, both cows and heifers, or oxen may present with higher prevalence rates. There were perceptions that breed may also play a role in susceptibility, and that African cattle breeds were more resistant to becoming infected with *M. bovis*, but ultimately it transpired that specific management practices were more important in determining the prevalence of the infection in the different breeds. Immune status may determine susceptibility to infection, and this too is an important risk factor in Africa, particularly in those animals owned by transhumant ethnic groups. Immunosuppression is a characteristic of animals kept under poor hygienic conditions, those subjected to climatic and feed stress, crowding and poor ventilation, and those with coinfections with pathogens that characteristically cause immunoincompetence.

Against the background of the very limited information about the extent and spread of BTB in Africa, it is clear that its distribution is not uniform. Its prevalence varies substantially between countries and also within countries where in certain regions the within-herd prevalence may be as high as 50%. This variation suggests the presence of hotspots of BTB within countries, and these may be the consequence of management practices, locality such as urban and periurban areas, and, in rural areas, factors such as communal water points and pastures and the aggregation of various herds during vaccination campaigns, at dip tanks, and at auction markets. Intermingling with wildlife maintenance hosts and contamination of the environment by feces or exudates following which mycobacteria may remain alive for varying periods of time that may be as long as 6 weeks during the winter, are additional sources of *M. bovis*.

The main epidemiological factors of importance in Africa determining persistence and spread of the infection include: the growing population and increasing demand for milk and meat, pastoral and transhumance husbandry practices, increasing herd sizes, confinement with high animal densities, and the practice of housing animals in homes during the night for security purposes. These practices are likely to persist indefinitely, and unless addressed, will serve to perpetuate the presence of BTB throughout Africa.

The use of molecular epidemiological techniques has become common practice in the developed countries to analyze outbreaks and the origin and spread of BTB. It is only recently that these techniques have been applied in some countries in Africa, but sufficient information has been generated to provide a broad understanding of the epidemiological dynamics of the disease on the continent. Using spoligotyping, deletion analysis, MIRU-VNTR, and RFLP as typing techniques, an evolving pattern of the distribution and movement of different *M. bovis* types could be determined.

A number of pivotal events that occurred over the last four millennia most likely determined the distribution of the various *M. bovis* types. The major events include, particularly, the occupation by the Romans and Arabs during earlier times, the migration of various ethnic groups within Africa, the more recent colonization by European countries, and the consequence of the devastating rinderpest epidemic that decimated cattle herds and caused a bottleneck in their numbers toward the end of the nineteenth century. Typing confirmed the presence of unique indigenous African strains of *M. bovis* that existed before colonization that has for long been assumed to introduce BTB into Africa. Based on archeological findings, it appears that *M. bovis* was already present in cattle and other livestock on the European continent about 4000 years ago and that it may thus also have been present in cattle in certain regions of Africa at that time. The distribution of these African strains was probably determined by the migration of ethnic groups in Africa, including the Bantu, from Western Africa, and the Luo, Mande, and Omoti people in Eastern Africa. Currently, there are two groups of *M. bovis* in Africa, those that are unique to the continent and those introduced by the settlers from the colonial countries. This latter group, having been introduced after the rinderpest epidemic, appears to be, with the exception of those in Western Africa, the major types of *M. bovis* in most of the African countries. In addition to the Af1, Af2, and the putative Af3 and five clonal complexes, a number of unique African spoligotypes also occur, for instance, in Chad, Tunisia, Ethiopia, Uganda, and Zambia to the south. Two complexes, Af1 and the putative AF5, occur in Western Africa particularly in Chad, Niger, and Cameroon, while most of the spoligotypes that occur in low frequency tend to be country-specific. The Af2 clonal complex is limited to the countries in eastern East Africa, as are some spoligotypes referred to as the "indigenous East African spoligotypes." The situation in Northern and South Africa is quite different, and in these regions, most of the spoligotypes belong to the Eur1 clonal complex, while in Madagascar all the isolates belong to the putative Af3 complex. Within this context, the dynamics and practices of livestock movements, the prevailing animal husbandry practices, and trade have had the biggest impact on the distribution of the various strains in the different African regions. Examples of these dynamics include the massive Sahel-Western African transhumant movement of humans and their livestock across international borders. These movements provide in-contact networks that promote the spread of the infection. In Eastern Africa, where the in-country profiles are more distinct, the effect of transhumance appears to be less, but it is, nonetheless, still apparent. Sudan appears to be the point of convergence of the two giant carousels of transhumant movement, and it may eventually play a pivotal role in the further dissemination of the various types of *M. bovis* in that part of Africa. In many of these countries, because of the changing dynamics in farming practices, and attempts to genetically improve the local breeds, there are also centrifugal and centripetal movements of cattle that, given the absence of any form of control of BTB, enhance the spread of the disease within these countries. Similar patterns of migration do not exist in Southern Africa, except for the localized migration in Zambia in the Kafue basin. In Southern Africa, as is the case in Zambia, and perhaps a developing trend in South Africa, the role of free-ranging wildlife in the epidemiology of the disease may become an important factor. Its role in the epidemiology of the disease in countries on the continent currently remains speculative because of the limited information about the disease dynamics in wildlife and their role in sustaining the infection in livestock. The extending wildlife–livestock interface, may yet, as in other countries, prove to complicate the control of BTB to the extent where it becomes impossible to control it. Application of molecular epidemiological techniques should equally address many of the questions about zoonotic TB, and the extent to which it occurs in those communities and groups of people at risk of contracting the infection.

A number of issues hamper progress in the management and control of zoonotic TB. In addition to the general assumption that it is not a problem in Africa, the lack of adequate disease control and eradication policies, the absence of continental reference databases and laboratory capacity for the isolation and genotyping of *M. bovis*, and the weak transdisciplinary and interlaboratory collaboration are major obstacles that will have to be overcome to get some degree of cooperation in regional and continental levels if the control of both BTB and zoonotic TB is to succeed in Africa.

Diagnosing BTB in live and dead animals is dealt with in Chap. 9. This is one of the most critical global impediments in the attempt to control BTB and zoonotic TB. The available tests and procedures all lack sensitivity and specificity, and are essentially herd, and not individual animal, tests. In spite of these deficiencies, some of the developed countries still managed to control and eliminate the disease from their cattle herds, albeit at a substantial cost, and over many years. In Africa, the lack of the technical capability and financial resources to use these tests commonly reduces the will of the politicians and regulatory authorities to control the disease.

As there is an international drive to eradicate human TB, by implication BTB and zoonotic TB must also be eradicated. Effectively eradicating the disease is dependent on the availability and use of detecting all the infected animals. In Africa, the lack of financial resources largely determines the diagnostic techniques that can be used, and meat inspection at slaughter remains the only affordable method by which the disease can be diagnosed and its spread and prevalence determined. Detecting BTB by meat inspection lacks sensitivity and specificity with the result that the information generated by its use is incomplete and unreliable. In many African countries, only a small percentage of carcasses for human consumption are processed, and most livestock are slaughtered informally and are not subjected to meat inspection. Using the data generated by the abattoirs largely becomes a process of counting numbers, without providing information that will augment the possibility of reducing the infection. There is a substantial variation in the epidemiology of BTB in individual countries and it is important that appropriate data are available to design effective control and eradication programs specifically for each individual country.

In addition to testing and applying the test-and-slaughter approach to deal with the disease, pasteurization of milk is the single most important intervention by which to reduce the prevalence of zoonotic TB. Although there is an increase in the provision of pasteurized milk, a large proportion of milk in Africa is still consumed raw, or after fermentation, particularly in rural areas where many ethnic groups are ill informed about the dangers associated with zoonotic TB and prefer raw instead of treated milk and meat. These practices remain a major risk, the extent of which is likely to increase with the increasing numbers of humans and cattle on the continent.

Tuberculosis is a disease without visible clinical signs, except for those in which it has reached an advanced stage. Even then, the signs are nonspecific and could be confused with those caused by a number of other diseases. This, too, is the situation in wildlife in which BTB may be asymptomatic, or cause nonspecific lesions that do not allow a definitive diagnosis. The intradermal skin tests (the single and comparative intradermal, and the caudal fold tests), and the INF-γ assay are the only ones approved by the OIE for use in cattle for diagnostic purposes. Although there is an increasing number of other tests, including serology, none of them has been validated and their interpretation remains uncertain, and many are less sensitive and specific than the skin tests. With the exception of African buffaloes for which the skin tests have been validated, the same situation prevails in wildlife.

In addition to being prohibitively expensive in Africa, the use of the skin tests in cattle and other livestock is also impaired by a number of other factors. One of the biggest problems in Africa is the lack of adequate road networks and other infrastructure to do the tests, and the reluctance of cattle owners to return their cattle to the testing site to evaluate the reaction. Further impediments include the limited availability of tuberculin, and the lack of trained human resources to do the tests. A number of other factors such as the impact of various types of stress and coinfection with diseases that cause immunosuppression and suppress the test reaction when doing the single intradermal test also play a role. Nonspecific reactions caused by infection with nontuberculous mycobacteria that cause nonspecific reactions in as many as 50% of animals tested in specific test cohorts are also major complicating factors in many parts of Africa.

Of all the available postmortal diagnostic techniques, meat inspection and necropsy remain the least expensive and, if correctly applied, one of the most specific and sensitive of the available diagnostic tests. For it to be used effectively, investigators must be adequately trained because of the extensive inter- and intraspecific variation in the macroscopical and histopathological appearance of the lesions and the difficulty of detecting them when only single lesions are present. In addition, the issue of no visible lesions and latent infections complicates the matter even further. Commonly when doing routine meat inspection in abattoirs about 50%, and in some reports, up to 90% of BTB cases are not detected. The meat inspection protocols in Africa are mostly insufficient for the purpose, and many of the meat inspectors are not well trained. If meat inspection is to be used as a diagnostic technique, the working conditions of meat inspectors will have to be improved and their workload reduced, the physical facilities in which they work will have to be improved, and quality control measures will have to be instituted.

The physical appearance of the lesions of BTB is not that typical. Hence, a final diagnosis cannot be made on visual inspection alone, and confirmation of the diagnosis should be made by histopathological examination, or preferably, by culture, which is considered the gold standard for diagnosing TB. The success of culturing is dependent on a number of factors including the way in which the specimens are collected, their storage and transport, and the media on which they are cultured. Culture in itself is only a means of isolating the bacteria, and they must be identified and typed using specific techniques. The commonly used molecular typing techniques in Africa include RFLP, spoligotyping, deletion analysis, and MVLA-VNTR, but their use is limited because of the cost, and the lack of trained human resources and adequate laboratory infrastructure. The interpretation of most of these tests also has to be adapted for specific countries because of the variation in the molecular biological characteristics of the strains circulating in each of them. Using combinations of tests is one way of solving this problem, but it is necessary to combine different tests for different countries, based on the variation, and the molecular characteristics of the *M. bovis* strains in each of them. Since most of the initial diagnoses are made during postmortal examination, confidently diagnosing BTB in African wildlife is even more challenging given the marked variation in the appearance of the lesions in the various species. This is an important issue because of the anticipated importance of wildlife in the epidemiology of the disease in many countries on the continent and the importance of determining their role in the epidemiology of the disease.

In Chap. 10, the available options for the control of BTB are dealt with. Because of the complexity of the disease, the lack of reliable diagnostic techniques, the absence of an effective vaccine, and the expected duration and cost of eradication programs, it is almost impossible for most of the African countries to attempt controlling the disease. The test-and-slaughter approach that has been applied by the developed countries is, however, currently the only available option, and poor countries will have no alternative when attempting to control BTB, but to apply the process in a selective way by focusing on hot spots of the infection, and those animals in areas that are of epidemiological significance. Testing alone is not the solution to the problem. To have any hope of success, surveillance to determine the extent and distribution of the infection, and to allow risk factor analysis, increasing the knowledge and attitude of farmers and policy makers, improving the competency of veterinary and abattoir officials, and improving movement restrictions and biosecurity measures should be addressed. Whether the African countries will be able to do this is a moot point, and without the support of international agencies and NGOs, it is unlikely that they will be close to attaining the international goal of eradicating both human tuberculosis, and by necessity, BTB because of its zoonotic potential.

Chapters 11–23 are reports of the situation in 13 of the 54 countries on the continent. In many of them, BTB is recognized as a major animal health problem. The variation in the status of the disease, its economic impact, the application (or lack) of control measures, and the zoonotic role of BTB is probably an adequate reflection of the situation in those countries that are not represented. These chapters provide a focused rendition of the situation that varies from a total lack of information, to the more sophisticated control programs applied by South Africa.

The information provided also emphasizes the marked differences in husbandry practices, ethnic traditions, and mostly the lack of control of the disease because of the lack of movement control within and across international borders. In addition, the lack of dealing with known BTB infections by not removing infected animals from the system and allowing free movement of them within countries and across international borders further complicate the process. These country reports also provide some insight into the ethnic practices that enhance the risk of contracting zoonotic TB and the general lack of knowledge about the disease, its dynamics, and the risk that it poses to human health.

If the content of this book manages to create awareness of an important neglected disease on the continent and to stimulate some form of activity to address the problems caused by *M. bovis* infections, it would contribute substantially to increasing the health and welfare of the humans and the health and productivity of the animal populations in Africa. We believe that this book provides useful information that will provide policy makers and regulatory authorities with a sound basis to enable them to make better-informed decisions about the importance of the disease and the need for its control and eventual eradication from the continent.

Chapter 2
The Current Status of Bovine Tuberculosis in Africa

Asseged B. Dibaba, C. J. Daborn, S. Cadmus, and A. Michel

Tuberculosis in humans and animals is an ancient contagious disease, with a worldwide distribution. The cause of the disease in cattle, *Mycobacterium bovis*, has a wide host range that include domesticated animals, wildlife, and humans in which it has been categorized as a neglected zoonosis. The World Organization for Animal Health (OIE) considers it to be a notifiable disease because of its socioeconomic impact and public health importance. It is one of the most challenging endemic diseases to control and eradicate because of its complex epidemiology, insidious nature, and multiple wildlife maintenance hosts that can sustain the infection within ecosystems.

Before the detection of the presence of indigenous Af1 and Af2 strains of *M. bovis* that existed in Western and Eastern Africa before colonization, it was for long assumed that BTB was introduced by the settlers into the various African countries during the time of colonialism. During that time, it became a well-known livestock disease in many parts of Africa. Currently, the disease is present in many African countries but, due to the lack of financial and human resources, and the political will, only a few of them attempt to apply adequate control measures. As

A. B. Dibaba (✉)
Department of Pathobiology, College of Veterinary Medicine, Tuskegee University, Tuskegee, AL, USA
e-mail: adibaba@tuskegee.edu

C. J. Daborn
Tigoni Veterinary Services, Nairobi, Kenya
e-mail: tvs@habari.co.tz

S. Cadmus
Tuberculosis and Brucellosis Research Laboratories, Department of Veterinary Public Health & Preventive Medicine, University of Ibadan, Ibadan, Nigeria

A. Michel
Department of Veterinary Tropical Diseases, Faculty of Veterinary Science, University of Pretoria, Onderstepoort, South Africa
e-mail: anita.michel@up.ac.za

A. B. Dibaba et al. (eds.), *Tuberculosis in Animals: An African Perspective*,
https://doi.org/10.1007/978-3-030-18690-6_2

a consequence, the extent and the epidemiology of BTB in Africa are largely unknown, and the disease, with few exceptions, is mostly diagnosed during meat inspection in abattoirs and by the very limited use of routine tuberculin skin testing.

According to the OIE's WAHID, 24 of the member countries failed to provide any information about the presence of BTB in their country during the course of time, and it remains unknown whether the infection is present in them. Of the countries that include cases or outbreaks of BTB in their reports to the OIE, only 28 consider it as important enough to deal with as a notifiable disease in livestock, and only 4 list it as a notifiable disease in wildlife that now appear to play a pivotal role in the epidemiology of the disease in certain countries and should always be considered when attempting to control the disease in cattle.

Bovine TB in Africa is a regional threat, particularly at the human-livestock-wildlife interface. Few countries in Africa are currently conducting BTB eradication schemes, and those that do, most do not adhere to the OIE-based norms and standards for the successful control of the disease. The limited attention given to BTB and zoonotic TB in Africa against the background of its projected, almost unbridled increase in the numbers of its inhabitants has the potential for them in the future to become overwhelming animal and public health problems. These important but neglected diseases are likely to have a detrimental effect on the general population far in excess of what is currently perceived to be the case; this is because, by choice, the extent and the importance of a disease in livestock and humans that has been present from antiquity remain largely unknown and uncontrolled in Africa.

2.1 Background

Bovine tuberculosis (BTB), caused by *Mycobacterium bovis*, is a corrosive, insidious, debilitating, and contagious infectious disease of domesticated and wild animals and of humans. It has a worldwide distribution and is one of the most important illnesses of cattle. It spreads extensively within animal populations before clinical signs and the effects of the disease become visible.

The World Organization for Animal Health (OIE) lists BTB as a notifiable disease as it considers it to be of socioeconomic or public health importance within countries, with a negative impact on the international trade of animals and animal products and hence local livelihoods (Benkirane 1998; Skuce and Neill 2004). Because of its spread to wildlife and the establishment of wildlife maintenance hosts, BTB has the potential to affect the well-being of wildlife populations and the sustainability of ecosystems (Renwick et al. 2006). Countries with a thriving cattle industry apply regulations that control the trade of animals and their products to protect the national herd from the introduction of BTB, its spread to wildlife, to limit the risk of zoonotic human infection, and to prevent reintroduction of the disease following its eradication (Cousins 2001).

The disease in the various species is characterized by the development of tuberculous granulomas in the lungs, lymph nodes, intestines, liver, kidneys, and other organs, thereby affecting their health that eventually has a detrimental effect on their productivity. Presentation of the disease is variable: clinical signs are absent in the early stages, whereas animals with advanced disease present with non-specific signs such as weakness, anorexia, emaciation, and coughing. The characteristics of the disease and its manifestation are species-dependent and are usually determined by the route by which the animals are infected. There is a marked variation in organ tropism between species, the distribution of the lesions throughout the body, and the physical appearance of lesions between and, often, within species.

Mycobacterium bovis, the etiological agent of BTB, is a member of the *Mycobacterium tuberculosis* complex (MTC) (Skuce et al. 2012). Contrary to *M. tuberculosis*, it has an exceptionally wide mammalian host range (domesticated animals, wildlife, and humans), and it also has the ability to persist in the environment, the time of its survival being dependent on local climatic and environmental conditions (O'Reilly and Daborn 1995). It is important too as a significant public health threat as the infection causes zoonotic TB in humans following their close contact with infected animals or after consuming *M. bovis*-containing raw milk and milk products (Cosivi et al. 1998). Based on experience gained to date in countries actively attempting to eradicate the disease, it is clear that BTB is one of the most challenging endemic diseases to deal with, because of its complex epidemiology, the insidious nature of the disease, and the multiple hosts that sustain the infection within ecosystems. It is generally considered to be one of the neglected zoonotic diseases, and it is increasingly attracting the attention of the international community because of its quest to eradicate human tuberculosis in the foreseeable future.

2.2 Historical Perspectives

Bovine TB is a worldwide animal health problem, and the infection was most likely disseminated with the movement of people as they migrated, taking their livestock with them, and settled on the various continents and islands across the globe (Smith 2012). In the 1800s and early1900s, BTB caused considerable losses in the cattle industry (Martin 1994), and with the development of an intensive livestock industry in Europe and America in the early part of the twentieth century, the disease was recognized as a significant problem when intensive cattle farming was practiced. At that time, BTB was considered characteristically to be a disease of housed animals, and it was virtually unknown in extensive, rural farming systems in which the opportunity for droplet transmission was remote.

Because BTB, associated with intensified cattle production in the post-industrial revolution era, was a threat to animal and human health, several countries (led by the USA) attempted its eradication by:

1. Implementing BTB control programs for cattle
2. Improving milk hygiene and implementing pasteurization of milk for human consumption

3. Increasing public awareness of the zoonotic and economic significance of an *M. bovis* infection in cattle

The campaign in the USA drastically reduced the number of cattle with BTB in the national herd from 5% in 1917 to less than 0.03% by 1985 (Martin 1994). Several countries in Europe had similar successes, but a few are still experiencing difficulties to keep the disease under control primarily because of the presence of wildlife reservoirs that act as sustainable sources of reinfection for livestock and make eradication of the disease almost impossible (Miller and Sweeney 2013). Although BTB internationally receives special attention in livestock because of its economic impact, it is mainly because of the zoonotic risk that it poses to humans that there is an international drive for its control and eradication.

2.3 The Situation in Africa

Before the detection of the presence of the indigenous Af1 and Af2 strains of *M. bovis* that, respectively, existed in Western and Eastern Africa before colonization, it was for long assumed that BTB was introduced by the settlers into the various African countries during the time of colonialism, particularly in the eighteenth and nineteenth centuries, by importing European cattle suffering from BTB (Michel 2008). During that time, it became a well-known livestock disease in many parts of Africa (Manley 1929).

The first recorded case of BTB in Africa was diagnosed in cattle in South Africa in 1880 (Hutcheon 1880), and it was subsequently diagnosed in 1888 in camels in Egypt (Mason 1912). Although there were differences of opinion, it was generally assumed that BTB was rare in indigenous livestock. Du Toit (1936) claimed that "judging by the reports of the veterinary authorities in Kenya, Anglo-Egyptian Sudan, the Gold Coast and Tanganyika," BTB is relatively unimportant in tropical Africa. As support for this opinion, only 18 cases of BTB were recorded in Sudan from 1933 to 1950 (Awad 1962). Carmichael (1938) though stated that "TB is not by any means a new disease to the tropics, and that neither man nor his domestic animals constitute that "virgin soil" in which the tubercle bacillus can seed itself." The general approach that BTB was not important was enhanced by the false assumption, based on anecdotal information, that the indigenous African zebu cattle were more resistant to contracting BTB compared to European cattle in which the infection progresses very rapidly. Known exceptions at the time were the straight-backed Ankole cattle in Uganda where during the early 1900s, more than 10% were found to be tuberculous (Carmichael 1938). However, as in Madagascar, and in India where a prevalence of BTB of up to 15% was recorded in zebu cattle during that time (Du Toit 1936), African zebus also proved not to be resistant to contracting the disease. This misconception about the zebu's perceived resistance to BTB contributed to the prevailing complacent attitude about the problem and the lack of its control in Africa (Awad 1962).

Through the course of time, BTB was detected in more and more African countries. The disease was diagnosed in the British Cameroon in 1929 following the importation from Germany of purebred stock cattle owned by a German company (Manley 1929). In a similar vein, BTB in cattle became widespread in the 1930s in Morocco soon after the importation of European cattle breeds by the French settlers (Berrada (1993). The same pattern was recorded in French West Africa, the British Cameroon, and Northern Rhodesia (now Zambia) where large numbers of cattle were affected (Du Toit 1936). From 1925 onward, BTB was recorded in cattle in an increasing number of African countries, such as in Cameroon, Mali, Senegal, Nigeria, Somalia, Sudan, Tanzania, Uganda, Zimbabwe (Carmichael 1939), Egypt, Uganda, and Sudan (Awad 1962) and in camels (Elmossalami et al. 1971) and cattle in Chad (Diguimbaye-Djaibé et al. 2006).

BTB became widely disseminated throughout Africa as this generally chronic disease, when left uncontrolled, spreads throughout a herd and between herds following the uncontrolled movement of cattle from infected to non-infected herds, often across international boundaries (Francis 1959). From the available data, it appears that its prevalence increased considerably on the continent over the years, primarily because of the lack of any form of control and the primitive methods of animal husbandry that were practiced in large parts of the continent (Carmichael 1939).

Not only did the disease spread throughout the continent, but its prevalence also increased, although in a patchy fashion. An example is the situation in Nigeria, where 42% of carcasses condemned between 1975 and 1977 were due to BTB (Opara 2005). Also in Nigeria in 1981 and 1990, 0.09% and 4.2% of carcasses examined, respectively, were fully or partially condemned due to BTB (Alaku and Moruppa 1993). Particularly high prevalences were detected in the Western Province of Zambia, where from 2000 to 2003, BTB was responsible for 83.1% ($n = 183$) of carcass condemnations (Munyeme et al. 2010). Losses because of the disease became extensive, and over 32 tons of beef and beef offal were condemned due to BTB in Togo from 1985 to 1987 (Domingo 2000) and this in countries that could ill afford the loss of protein of this dimension. Not dealing with the problem of BTB and the absence of animal identification to trace infection back to its source led to further dissemination of the disease and an increase in numbers of infected cattle. The mere counting of infected tissues, year after year, did not, and will not aid the control of the infection, and will not reduce the prevalence of BTB (Amanfu 2006).

In spite of the high prevalence in certain countries, such as Côte d'Ivoire, where nearly 50% of whole and 46% of partial bovine carcasses condemned in 1992 were attributed to BTB, the authorities appear not to be concerned as during that time, only one outbreak of BTB was officially reported (Benkirane 1998). This type of practice implies that because data on zoonotic diseases are collected in a passive manner and from a wide range of sources, including abattoir records (Table 2.1), they are unlikely to reflect the actual situation, and it is likely to result in a substantial underestimation of the magnitude of the problem. In addition, available data are not fully analyzed, and in most of the countries, little is done in terms of mitigation following the detection of tuberculous cattle.

Table 2.1 Status of bovine tuberculosis in 54 African countries (2016) (OIE 2017)

Country	Information available	Notifiable	Last occurrence	Type of control	Wildlife
Algeria	NI	–			
Angola	NI	–			
Benin	NI	–			
Botswana		+	Unknown	GTS	
Burkina Faso		+	Present		
Burundi	NI	–			
Cabo Verde		+	Unknown	GTS	
Cameroon	NI	–			
Central African Republic		+	Present		
Chad		+	Present		
Comoros	NI	–			
Democratic Republic of the Congo		+	Zonal occurrence		
Republic of the Congo		+	2015	GTS	
Côte d'Ivoire	NI	–			
Djibouti		+	Unknown	GS	
Egypt		+	Present		Notifiable
Equatorial Guinea	NI				
Eritrea		+	2012	GS	GS
Ethiopia		–	2015		
Gabon	NI	–			
Gambia	NI	–			
Ghana		+	Present		Absent since 2011
Guinea	NI	–			
Guinea-Bissau					
Kenya		+	Suspected		Suspected
Lesotho		–	2005	GTS	
Liberia	NI				
Libya	NI				
Madagascar		+	Present		
Malawi	NI	–			
Mali	NI	–			
Mauritania	NI	–			
Mauritius		–	2011		2000
Morocco		+	Present		
Mozambique		+	Present		Present
Namibia		+	1995	GTS	
Niger		+	2013		Suspected
Nigeria		+	Present		Present
Rwanda	NI	–			

(continued)

Table 2.1 (continued)

Country	Information available	Notifiable	Last occurrence	Type of control	Wildlife
Sao Tome and Principe	NI	–			
Senegal		–	Present		
Seychelles	NI	–			
Sierra Leone	NI	–			
Somalia	NI	–			
South Africa		+	Present	GTS; controlled	Present and notifiable
South Sudan	NI	–			Notifiable
Sudan		+	1992		
Swaziland		+	Present		
Tanzania		+	Present		Infection
Togo	NI	–			
Tunisia		+	Present		Absent; notifiable
Uganda		+	Present		Present
Zambia		+	Present		Present
Zimbabwe		–	1996		Present

NI, no information provided either over time for all diseases or specifically for BTB; +, notifiable disease; –, not notifiable; GS, general surveillance; GTS, general and targeted surveillance

The complacent attitude toward BTB is enhanced by the assumption that in extensive farming practices, close contact between animals is limited, and the spread of *M. bovis* in these herds is expected to be slow, and the prevalence low. Under such circumstances, it is argued that BTB will remain at low endemic levels, and there is thus scant motivation for regulatory authorities to apply the complex and expensive internationally accepted control measures required for its control and eradication (Benkirane 1998). In addition, the general apathy and lack of knowledge of African herdsmen about the importance of diseases that do not cause immediate losses complicate the processes of detection and reporting the occurrence of BTB in Africa (Carmichael 1939). This attitude still prevails, and in a Participatory Rural Appraisal (PRA) conducted in the Gambia, Senegal, Guinea-Bissau, and Guinea, for instance, participants ranked hemorrhagic septicemia and "diarrhea complex" (in cattle) and Peste des Petits Ruminants (PPR) (in small ruminants) high in the "general disease" survey. From the list of zoonotic diseases, only brucellosis (hygroma) made it to the "top 10" list in one country. Of the four zoonoses (rabies, anthrax, BTB, and brucellosis) incorporated in the analysis, only rabies was ranked high. Bovine TB was not known by any of the study subjects (Unger and Münstermann 2004). This is likely to be the situation in many, if not most, African countries.

Through the years, veterinary authorities in Africa were concerned with the more economically important, rapidly fatal, and usually transboundary animal diseases such as rinderpest, contagious bovine pleuropneumonia (CBPP), East Coast fever, and anthrax, which from time immemorial ravaged the herds in tropical countries

(Carmichael 1939; Alaku and Moruppa 1993). There was consensus that compared to the effect of the major endemic diseases prevalent in the developing countries, the overall economic impact of BTB on livestock production in Africa was assumed not to be important, and in most countries, this prompted regulatory authorities not to implement the costly control strategy at national and regional levels (Berrada 1993). The insidious nature of BTB enhanced the justification of this approach, but there seems little doubt that the disease has been quietly progressing for many years both in humans and animals (Carmichael 1938). Currently, the disease is present in many African countries, but due to the lack of financial and human resources, and the political will, only a few of them conduct any form of surveillance or attempt to apply adequate control measures. As a consequence, the extent and the epidemiology of BTB in Africa are largely unknown, and the disease, with few exceptions, is mostly diagnosed during meat inspection in abattoirs, which itself is a very rough measure of the extent of the infection, and by the very limited use of routine tuberculin skin testing.

According to the OIE's WAHID (OIE 2017), 24 of the member countries have not provided any information about the presence of BTB during the course of time, and it remains unknown whether the infection is present in these countries (Table 2.1). Of the countries that include cases or outbreaks of BTB in their reports to the OIE, only 28 consider it as important enough to deal with as a notifiable disease in livestock, and only four list it as a notifiable disease in wildlife.

2.4 A Paradigm Shift: Urbanization and Intensification

With the exception, perhaps, of Namibia, Botswana, and South Africa, there are few countries in Africa with well-developed sophisticated dairy and meat industries that participate in the international trade. In general, cattle farming on the continent utilizes the age-old traditional pastoral, extensive, or transhumant husbandry systems. These involve extensive seasonal movement, often over vast distances, within countries and across international borders, of people and animals in search of the limited sources of natural pasture and water.

In traditional African livestock husbandry, livestock is deeply embedded in the social system, and stock ownership is a sign of wealth; the more cattle you have, the wealthier you are. The role of livestock in the agropastoral populations of tropical Africa is much more complex than in temperate climates (Alaku and Moruppa 1993; Coulibaly and Yameogo 2000). Cattle are seldom culled or sold, but are often exchanged, in a complex social system of mutual obligation, within and between families and other social groups. Also as part of this close association between humans and livestock in many parts of Africa, the stockowners share their houses with shoats, fowls, and often cattle. These practices, which lack the application of the most basic of hygienic precautions and biosecurity measures, set the stage for the rapid transmission of *M. bovis* when the disease is present in a herd (Carmichael 1938).

In Africa, between 40 and 45% of the population live in abject poverty, while a further 30% is classified as extremely poor. There is a rapid population growth of 3% per annum, and urbanization is increasing at a rate of 6% per annum and is a driver of the increasing poverty and lack of food security across the continent. These are major causes of concern particularly in sub-Saharan Africa (SSA) (Unger and Münstermann 2004), and people in urban areas here attempt to improve their livelihoods by diversifying their income, and a significant number of them are entering the livestock sector and, in particular, small-scale dairy farming within and around the cities (Awah-Ndukum et al. 2010). As a result, in all parts of Africa, an increase in milk production is observed around major towns and cities to satisfy the needs of the increasing urban populations. These farmers are often organized into designated milk-shed areas or into smallholder cooperatives, also known as bulking groups (Tschopp et al. 2013). Authorities in Africa support these initiatives (Asseged et al. 2000; Tschopp et al. 2013), particularly in peri-urban centers where the demand for dairy products is very high, because some of their major challenges are to ensure a year-round supply of sufficient infrastructure for processing, preservation, and marketing of the products (meat and milk) and animal feed (Unger and Münstermann 2004). To improve productivity in the sector, there is, in addition, an increasing effort to introduce improved dairy breeds and to change the extensive rural farming practices to intensive management systems.

These changes are not without their challenges. Intensification of livestock production, particularly where regulations governing animal movement and importation are nonexistent or poorly enforced, often results in an increase in the incidence of BTB (Benkirane 1998; Boukary et al. 2012), and it has, in fact, become a serious problem in intensive dairy farms, impacting the productivity of the livestock industry in many of the developing countries. As an example, a recent study carried out on intensive dairy farms in and around Addis Ababa, Ethiopia, showed an overall prevalence of BTB of 10.3% and a prevalence of up to 90% in market-orientated farms that keep large numbers of improved Holstein/Friesians and their crosses (Asseged et al. 2000). This situation can destroy years of genetic improvement of desirable production traits in a herd (Skuce et al. 2012). Bovine TB also negatively impacts the welfare of affected farming communities due to its effect on profitability and trade and the inability to participate in the lucrative international trade in meat and dairy products.

An additional, and critical, problem is that veterinary services, including access to diagnostic laboratories, in most of the nations are inadequate. They also fail to regulate animal movement because of the lack of systematic and verifiable animal identification systems (Amanfu 2006). This situation also does not allow them to trace the sources of BTB-positive cattle identified by skin testing or by meat inspection at abattoirs, and they have limited opportunity to control the disease under these circumstances.

There is a strong link between poor human health and poverty and their likelihood of contracting zoonotic diseases (Unger and Münstermann 2004). Under these circumstances, farmers, abattoir workers, butchers, and veterinarians are at a high risk of acquiring the infection. Because of traditional ethnic practices in certain countries in Africa, humans and animals share the same microenvironment, and under these circumstances, cases of zoonotic *M. bovis* infection in humans are common (Cosivi et al. 1998). As would be expected, this occurs particularly in peri-urban, unsanitary, livestock production centers that increase the risk of contracting zoonotic diseases. Transhumant communities in rural areas in certain countries are, unexpectedly, also exposed and subject to contracting zoonoses because of their close association with their livestock.

In countries where BTB is common, *M. bovis* infection is estimated to be responsible for 10%–15% of human TB cases (Ashford et al. 2001). Large sectors of the African population drink unpasteurized milk that is a common source of the infection, and it is only in urban areas that there is an increasing supply of processed pasteurized milk. There is a perception that the souring of milk, a common practice throughout Africa because of lactose intolerance, kills mycobacteria and renders the milk safe to drink, but the microbicidal effect of souring is dependent on the stage of the souring process, since viable *M. bovis* organisms still occur in the milk during the early stages of the process (Michel et al. 2015).

2.5 BTB in Wildlife in Africa

There is a large and diverse, but dwindling, population of wildlife species in Africa, most of which are likely to be susceptible to *M. bovis* infections. However, with few exceptions, throughout the years, no attention was given to whether they were infected or to what extent they played a role in the epidemiology of the disease in the different countries on the continent. The fact that they could be infected and that they would then most likely play a role in the maintenance and dissemination of the disease was never in the past taken into consideration when designing BTB control programs. Most, if not all, African veterinary authorities approach the control of the disease in cattle as if they were the sole source and the only species that is infected.

Bovine TB has now been diagnosed in a range of African wild mammalian species (antelopes, omnivores, primates, and predators), which, depending on the circumstances, can be either maintenance or spillover hosts for *M. bovis*, and they constitute a source of infection for cattle and other susceptible species. The African continent, with its high densities of mammals, large cattle populations, and poorly resourced veterinary services, presents ideal circumstances for the bidirectional spread of *M. bovis* at the interface between cattle and wildlife populations as has been seen in a number of protected conservation areas in sub-Saharan Africa.

Reports dealing with BTB in wildlife over the years were few and far between. In African wildlife, BTB was first diagnosed during the 1920s in greater kudus (*Tragelaphus strepsiceros*) and a few other wildlife species during a destructive outbreak of the disease in South Africa (Paine and Martinaglia 1929). Further cases of BTB in wildlife were only reported in the 1960s when the disease was diagnosed in 13 free-ranging African buffaloes (*Syncerus caffer*) in a game park in Uganda (Guilbride et al. 1963). Later, in the 1970s the infection was reported in Kafue lechwe (*Kobus leche kafuensis*), an endemic aquatic swamp-dwelling antelope found on the Kafue Flats in Central Zambia. Here more than 33.3% of randomly shot, free-ranging lechwe had BTB (Gallagher et al. 1972). De Vos et al. (1977) described a single case of mycobacteriosis in an impala in South Africa. In the Masai Mara Game Reserve in Kenya, a troop of free-ranging olive baboons was infected and manifested clinical disease that they contracted from an unknown source (Tarara et al. 1985).

It was only after the disease was diagnosed in the Kruger National Park that more attention was given to the infection in wildlife, because of the role that some of them play as maintenance hosts and the impact on ecosystems in the long term. Following this event, the South African Veterinary Services developed a comprehensive protocol, the Veterinary Procedural Notice for Buffalo Disease Risk Management in South Africa (also referred to as the Buffalo Veterinary Procedural Notice), to regulate the movement of buffaloes in the country and to manage the risk posed by BTB-infected buffaloes to cattle that may come in contact with them (DAFF 2017). (For further information about BTB in wildlife see Chap. 5.)

It is clear that local BTB-infected cattle were the source of the infection for the lechwe in Zambia (Macadam et al. 1974) and the buffalo populations in the Kruger National Park (KNP) (Rodwell et al. 2001) and the Hluhluwe-iMfolozi game reserves in South Africa (Jolles et al. 2005), and it is likely that this situation prevails in other countries in which there is close contact and intermingling of cattle and infected wildlife. Following local investigations in South Africa, it appears that *M. bovis* can survive in the harsh African environment for up to 6 weeks in winter and up to 4 weeks under favorable conditions during the rest of the year (Tanner and Michel 1999). Infected wildlife species are thus another source of infection causing environmental contamination that is an important source of infection for a number of wildlife species, all scavengers, including chacma baboons (*Papio ursinus*), banded mongooses (*Mungos mungo*), and warthogs (*Phacocoerus africanus*) (Brüns et al. 2017).

Even a relatively low prevalence of the disease in any wildlife population may pose a substantial health risk to other wildlife species (particularly predators and scavengers) and domestic animals at the interface with infected wildlife. In this context, BTB caused by *M. bovis* is a regional threat, particularly at the human-livestock-wildlife interface in Africa, and it should always be kept in mind when attempting to control the disease in cattle.

2.6 Poor Application of Statutory Control Policies

The control of BTB in Africa is a tale of woe. From the early 1900s onward, TB, both in humans and animals, received little attention from the regulatory authorities and farming communities in Africa, mainly because they focused on the more visible and rapidly fatal epidemic and endemic diseases that periodically ravaged the livestock populations on the continent (Carmichael 1938). The situation has not changed, and almost a century later, the control of livestock diseases remains dismal because of the general inadequacy and incompetency of veterinary services in most of the African countries. This is the consequence of the provision of inadequate human, infrastructural, and financial resources and the lack of recognition by policymakers of the importance of the diseases in humans and animals and the effect that they have on the general well-being of the human population and productivity of its livestock.

In respect to BTB, the disease persists because most of the African countries do not conduct systematic national tuberculin testing of cattle. Due to socioeconomic limitations in many countries, culling of tuberculin test-positive reactors, as part of the national BTB control policy, is not routinely done, and the infected cattle remain within the system. Only 7 of the 28 African countries that consider BTB important enough to list it as a notifiable disease apply test-and-slaughter measure for disease control; the remaining countries control the disease inadequately or not at all (Cosivi et al. 1998). Only a few countries, such as South Africa (since 1969), Morocco (since 1982), Tunisia (since 1984), Nigeria (since 1988), and Tanzania (since 1997), have statutory BTB control programs though they are mostly poorly or inadequately resourced and implemented (Berrada 1993; Ben Kahla et al. 2011; Ejeh et al. 2014; Sahraoui et al. 2011; OIE 2017).

In many African countries, abattoir records are the only sources of information about the occurrence and prevalence of BTB. The situation in Ethiopia is a good example of the limited value of this approach. Here official slaughter provides only 28% of meat consumed; the remaining 72% comes from unsupervised, backyard slaughter (Etter et al. 2006), where the chances of detecting BTB are poor, and even if it should be recognized, the event will probably not be reported to the relevant veterinary officials. Even in the countries with a formal control program, the implementation is often limited, and the efforts to control the disease are inadequate as they often only focus on a small proportion of the national herd. In Tanzania, for instance, only intensive dairy farms supplying milk to the human populace in urban centers are subject to compulsory screening with the single comparative cervical test (CCT) for BTB (Jiwa et al. 1997). Although this approach makes sense from an epidemiological point of view if you have to deal with limited resources, such herds are relatively few, and a large portion of the cattle population is excluded from the control program. In some countries, tuberculin skin testing is required only when there is a need to relocate dairy cows (Berrada 1993), or to satisfy the requirements of the importing country, before being accepted for slaughter by export abattoirs (Quirin et al. 2001).

It is clear that few countries in Africa are currently applying BTB eradication schemes and those that are, generally do not meet the OIE-based standards as practiced in North America, Europe, Australia, and New Zealand. South Africa perhaps is the closest to meeting these standards, and it recently updated its control strategy (DAFF 2013), but there too, most of the emphasis is on commercial farms. Morocco [with technical support from the Food and Agricultural Organization (FAO)] may reach this goal as it completed a feasibility study for conducting a BTB control program (Berrada 2006).

The general lack of attention to controlling BTB also has an impact on humans because of their susceptibility to infection with *M. bovis*. Consequently diseases such as BTB, which are often associated with resource-poor communities, are now categorized in the neglected zoonoses group, and there is increasing international awareness about the risk of consuming raw dairy products (Drewe et al. 2014). The situation is compounded by the lack of knowledge about zoonotic TB. Several studies, such as in Ethiopia (Ameni et al. 2013), Tanzania (Swai and Schoonman 2012), Zambia (Munyeme et al. 2010), and Western Africa (Unger and Münstermann 2004), confirmed that farmers' awareness about the importance of zoonotic *M. bovis* is suboptimal.

Tschopp et al. (2013) summarized the imminent danger of zoonoses in the burgeoning intensive dairy farms in central Ethiopia, and this probably applies to many of the poorer African countries:

> Despite six decades of efforts by government and non-governmental organizations to promote dairying, general awareness of milk-borne diseases is poor. Even when farmers seem better informed of the disease situation, it does not seem to be linked with an informed control policy, and hence there is no guarantee that it will change. Currently, the main danger is the introduction of zoonotic diseases among upgraded Holstein and their crosses originating from high BTB prevalence areas such as Addis Ababa and its dairy belt.

The lack of understanding of the risk and the apathy of the regulatory authorities to control the disease are exemplified by the sale in central Ethiopia of stock from a dairy farm with a very high prevalence of BTB. The entire population of cattle, many suffering from BTB, was sold to unsuspecting local dairy farmers who scrambled to buy a few of the "high genetic value Holstein/Friesians." This situation allowed the uncontrolled movement of BTB-infected cows from the "depopulated farm" to multiple other farms and beyond. This is the consequence of a testing program that operates outside of any legal framework and not dealing adequately with BTB-positive reactors. One effective way to improve this situation is to inform the general public about the risk of *M. bovis* infections and the need to deal with the disease according to tested international norms and standards. Accordingly, public awareness campaigns and sensitization of farmers and the general public should be key components in the overall strategy for the control of BTB (Amanfu 2006).

Against the background of the rapid, and perhaps, overwhelming increase in the size of the African population, the current state of affairs as they relate to the threats and impact of BTB and zoonotic TB is a major cause of concern. The number of people in Africa nearly trebled from the estimated 478 million in 1980 to the current almost 1.2 billion. Urbanization increased rapidly from 27% in 1980 to the current 40%, and it is estimated to reach 56% by 2050 at which time, the continent's population is predicted to be 2.4 billion (UN 2016; World Population Review 2018). These figures are alarming given that the African countries as a group are the poorest in the world, and the current lack of infrastructure, financial resources, food, and space are crippling limitations when planning for the future. The little attention given to BTB and zoonotic TB has the potential for them to become overwhelming animal and public health problems, with a detrimental effect on the general population far in excess of what it is currently perceived to be; this because the extent and importance of important chronic diseases that have been present from antiquity are largely unknown and uncontrolled on the continent.

References

Alaku SO, Moruppa SM (1993) Tuberculosis condemnations in livestock slaughtered for meat in northeastern Nigeria. Prev Vet Med 15:67–72

Amanfu W (2006) The situation of tuberculosis and tuberculosis control in animals of economic interest. Tubercle 86:330–335

Ameni G, Tadesse K, Hailu E et al (2013) Transmission of *Mycobacterium tuberculosis* between farmers and cattle in central Ethiopia. PLoS One 8:e76891. https://doi.org/10.1371/journal.pone.0076891

Ashford DA, Whitney E, Raghunathan P et al (2001) Epidemiology of selected mycobacteria that infect humans and other animals. OIE Sci Tech Rev 20:105–112

Asseged B, Lubke-Becker A, Lemma E et al (2000) Bovine TB: a cross-sectional and epidemiological study in and around Addis Ababa. Bull Anim Health Prod Afr 67:71–80

Awad FI (1962) Studies on type-determination and epidemiology of tuberculosis among cattle in Sudan. Transbound Emerg Dis 9(9):890–898

Awah-Ndukum J, Kudi AC, Bradley G et al (2010) Prevalence of bovine tuberculosis in abattoirs of the littoral and western highland regions of Cameroon: a cause for public health concern. Vet Med Int. https://doi.org/10.4061/2010/495015

Ben Kahla I, Boschiroli ML, Souissi F (2011) Isolation and molecular characterization of *M. bovis* from raw milk in Tunisia. Afr Health Sci 11(S1):S2–S5

Benkirane A (1998) Bovine tuberculosis in Africa. Wild Anim Rev 90:54–56

Berrada J (1993) *Mycobacterium bovis* infection in cattle in Morocco: preparation and evaluation of chemical extracts for use in detection of immune responses. PhD Thesis, Iowa State University, USA

Berrada J (2006) Capacity building for surveillance and control of tuberculosis. In: Capacity building for surveillance and control of zoonotic diseases FAO/WHO/OIE Expert and Technical Consultation, Rome, 14–16 June 2005

Boukary AR, Thys E, Rigouts L et al (2012) Risk factors associated with bovine tuberculosis and molecular characterization of *M. bovis* strains in urban settings in Niger. Transbound Emerg Dis 59:490–502

Brüns AC, Tanner M, Williams MC et al (2017) Diagnosis and implications of *Mycobacterium bovis* infection in banded mongooses (*Mungos mungo*) in the Kruger National Park, South Africa. J Wildl Dis 53(1):19–29

Carmichael JL (1938) Tuberculosis of sheep in Uganda. Vet Rec 50:1138–1147

Carmichael JL (1939) Bovine tuberculosis in the tropics, with special reference to Uganda. J Comp Pathol Therap 52:322–335

Cosivi O, Grange JM, Daborn CJ, Raviglione MC et al (1998) Zoonotic tuberculosis due to *Mycobacterium bovis* in developing countries. Emerg Infect Dis 4:59–70

Coulibaly ND, Yameogo KR (2000) Prevalence and control of zoonotic diseases: collaboration between public health workers and veterinarians in Burkina Faso. Acta Trop 76:53–57

Cousins D (2001) *Mycobacterium bovis* infection and control in domestic livestock. Rev Sci Tech Off Int Epiz 20:71–85

DAFF (2013) Bovine tuberculosis scheme manual (Interim). Department of Agriculture, Forestry and Fisheries, Republic of South Africa

DAFF (2017) Department of Agriculture Forestry and Fisheries: veterinary procedural notice for buffalo disease risk management in South Africa. http://www.daff.gov.za/vetweb/pamphlets&Information/Policy/Buffalo%20Disease%20Risk%20Management%20VPN_Signed%202017-02-17.pdf

De Vos V, McCully RM, van Niekerk AWJ (1977) Mycobacteriosis in the Kruger National Park. Koedoe 20:1–9

Diguimbaye-Djaibé C, Hilty M, Ngandolo R et al (2006) *Mycobacterium bovis* isolates from tuberculous lesions in Chadian zebu carcasses. Emerg Infect Dis 12:769–771

Domingo M (2000) Current status of some zoonoses in Togo. Acta Tropica 76:65–69

Drewe JA, Pfeiffer DU, Kaneene JB (2014) Epidemiology of *M. bovis*. In: Thoen CO, Steele JH, Kaneene JB (eds) Zoonotic tuberculosis: *M. bovis* and other pathogenic mycobacteria, 3rd edn. Wiley, Chichester, pp 63–77

Du Toit PJ (1936) Bovine tuberculosis in South Africa. Tubercle 17:421–422

Ejeh EF, Raji MA, Bello M (2014) Prevalence and direct economic losses from bovine tuberculosis in Makurdi, Nigeria. Vet Med Int. https://doi.org/10.1155/2014/904861

Elmossalami E, Siam MA, Sergany ME (1971) Studies on tuberculous-like lesions in slaughtered camels. Zentralbl Veterinarmed B 18:253–261

Etter EMC, Ameni G, Roger FLM (2006) Tuberculosis risk assessment in Ethiopia: safety of meat from cattle slaughtered in abattoirs. In: Proceedings of the 11th international symposium on veterinary epidemiology and economics (ISVEE), Cairns, Australia

Francis J (1959) The work of the British Royal Commission on Tuberculosis, 1901–1911. Tubercle 40:124–132

Gallagher J, Macadam I, Sayer J et al (1972) Pulmonary tuberculosis in free-living lechwe antelope in Zambia. Trop Anim Health Prod 4:204–213

Guilbride PDL, Rollinson DHL, Mcanulty EG et al (1963) Tuberculosis in the free living African (Cape) buffalo *(Syncerus caffer caffer.* Sparrman*)*. J Comp Pathol 73:337–348

Hutcheon DT (1880) Consumption, Tables Mesenterica. Annual report, Colonial Veterinary Surgeon. Cape of Good Hope

Jiwa SFH, Kazwala RR, Aboud AA et al (1997) Bovine tuberculosis in the Lake Victoria Zone of Tanzania and its possible consequences for human health in the HIV/AIDS era. Vet Res Commun 21:533–539

Jolles AE, Cooper DV, Levin SA (2005) Hidden effects of chronic tuberculosis in African buffalo. Ecology 86:2258–2264

Macadam I, Gallagher J, Mckay J (1974) Experimental tuberculosis in lechwe antelope in Zambia. Trop Anim Health Prod 6:107–109

Manley FH (1929) A note on bovine tuberculosis in tropical Africa (British Cameroons). J Comp Pathol Ther 42:276–278

Martin SW (1994) Livestock disease eradication: evaluation of the cooperative state-federal bovine tuberculosis eradication program. National Academies

Mason FE (1912) Some observations on tuberculosis in camels in Egypt. J Comp Pathol Ther 25:109–111

Michel AL (2008) Tuberculosis in wild and domestic animals in South Africa. PhD thesis, Universiteit Utrecht, The Netherlands

Michel AL, Geoghegan C, Hlokwe T et al (2015) Longevity of *Mycobacterium bovis* in raw and traditional souring milk as a function of storage temperature and dose. PLoS One 10(6): e0129926. https://doi.org/10.1371/journal.pone.0129926

Miller RS, Sweeney SJ (2013) *Mycobacterium bovis* (bovine tuberculosis) infection in North American wildlife: current status and opportunities for mitigation of risks of further infection in wildlife populations. Epidemiol Infect 141:1357–1370

Munyeme M, Muma JB, Munang'andu HM et al (2010) Cattle owners' awareness of bovine tuberculosis in high and low prevalence settings of the wildlife-livestock interface areas in Zambia. BMC Vet Res 6:21. https://doi.org/10.1186/1746-6148-6-21

OIE (2017) World Animal Health Information Database (WAHIS Interface). https://www.oie.int/wahis_2/public/wahid.php/Countryinformation/Animalsituation

Opara M (2005) Pathological conditions of condemned bovine lungs from abattoirs in Akwa Ibom State, Nigeria. Anim Res Int 2:314–318

O'Reilly LM, Daborn CJ (1995) The epidemiology of *Mycobacterium bovis* infections in animals and man: a review. Int J Tuberc Lung Dis 76(Suppl 1):1–46

Paine R, Martinaglia G (1929) Tuberculosis in wild buck living under natural conditions. J Comp Pathol Ther 42:1–8

Quirin R, Rasolofo V, Andriambololona R et al (2001) Validity of intradermal tuberculin testing for screening of bovine tuberculosis in Madagascar. Onderstepoort J Vet Res 68:231–238

Renwick AR, White PCL, Bengis RG (2006) Bovine tuberculosis in southern African wildlife: a multi-species host-pathogen system. Epidemiol Infect 135:529–540

Rodwell TC, Kriek NP, Bengis RG et al (2001) Prevalence of bovine tuberculosis in African buffalo at Kruger National Park. J Wildl Dis 37:258–264

Sahraoui N, Muller B, Mamache B (2011) Tuberculosis in cattle and goats in the North of Algeria. Vet Res 4:100–103

Skuce RA, Neill SD (2004) Molecular epidemiology of *Mycobacterium bovis*, Chapter 6. In: Madkour et al (eds) Tuberculosis. Springer, Berlin pp 75–92

Skuce RA, Allen AR, McDowell SWJ (2012) Herd-level risk factors for bovine tuberculosis: a literature review. Vet Med Int. https://doi.org/10.1155/2012/621210

Smith NH (2012) The global distribution and phylogeography of *M. bovis* clonal complexes. Infect Genet Evol 12:857–865

Swai ES, Schoonman L (2012) A survey of zoonotic diseases in trade cattle slaughtered at Tanga city abattoir: a cause of public health concern. Asian Pac J Trop Biomed 2:55–60

Tanner M, Michel AL (1999) Investigation of the viability of *M. bovis* under different environmental conditions in the Kruger National Park. Onderstepoort J Vet Res 66:185–190

Tarara R, Suleman MA, Sapolsky R et al (1985) Tuberculosis in wild olive baboons, *Papio cynocephalus anubis* (Lesson), in Kenya. J Wildl Dis 21(2):137–140

Tschopp R, Abera B, Sourou SY et al (2013) Bovine tuberculosis and brucellosis prevalence in cattle from selected milk cooperatives in Arsi zone, Oromia region, Ethiopia. BMC Vet Res 9:163

UN (2016) The demographic profile of African countries. https://www.uneca.org/sites/default/files/PublicationFiles/demographic_profile_rev_april_25.pdf

Unger F, Münstermann S (2004) Assessment of the impact of zoonotic infections (BTB and brucellosis) in selected regions of the Gambia, Senegal, Guinea, and Guinea Bissau. - A scoping study - DFID Animal Health Program. Banjul, Gambia

World Population Review (2018) Africa Population. Retrieved 2018-04-02, from http://worldpopulationreview.com/continents/africa-population/

Chapter 3
Bovine TB Zoonosis in Africa

Paul D. van Helden and Anita Michel

3.1 Introduction

Tuberculosis (TB) is an infectious, chronic, and potentially fatal disease caused by members of the *Mycobacterium tuberculosis* complex (MTC) that are often able to cross species barriers between domestic and wild animals, and humans. *Mycobacterium tuberculosis* is the commonest cause of TB in humans, while *Mycobacterium bovis* is arguably the most common mycobacterial species causing TB in animals, and it probably has the most extensive host range of all the disease-causing mycobacteria. It has been known for many decades that *M. bovis* can also infect and cause zoonotic tuberculosis (zoonotic TB, zTB) in humans, and for this reason, and the socio-economic impact of the infection in livestock, efforts to control bovine tuberculosis (BTB) globally in wildlife and domestic animals are common and legislated for in many countries.

P. D. van Helden (✉)
Division of Molecular Biology and Human Genetics, Faculty of Medicine and Health Sciences, South African MRC Centre for TB Research, DST NRF Centre of Excellence for Biomedical Tuberculosis Research, Stellenbosch University, Tygerberg, South Africa
e-mail: pvh@sun.ac.za

A. Michel
Department Veterinary Tropical Diseases, Faculty of Veterinary Science, University of Pretoria, Onderstepoort, South Africa
e-mail: anita.michel@up.ac.za

A. B. Dibaba et al. (eds.), *Tuberculosis in Animals: An African Perspective*,
https://doi.org/10.1007/978-3-030-18690-6_3

3.2 Manifestation of the Disease in Humans

Pathways for contracting an *M. bovis* infection in humans include consumption of raw milk from diseased animals, inhalation of aerosol droplets during contact with infected animals, and consumption of undercooked contaminated meat. Although it has been possible for many decades to distinguish between *M. tuberculosis* and *M. bovis* using laboratory techniques, the process is time-consuming, laborious, and expensive, and therefore not routinely done, and the available data on the prevalence of *M. bovis* infections in humans are consequently patchy and incomplete. Furthermore, the disease caused by either *M. tuberculosis* or *M. bovis* is both clinically and pathologically indistinguishable, and neither the lesions nor their distribution can be used to determine the specific mycobacterial species causing the infection. Any study that focuses only on the pathology of TB to determine the specific cause will thus have some inherent bias.

As is the case with tuberculosis (TB) caused by *M. tuberculosis* in humans, not all *M. bovis* infections in humans progress from infection to active disease. When interpreting studies providing information about the disease in humans, one must thus distinguish carefully between those that measure infection and those that measure active disease. In human TB, for example, in the absence of immunosuppression, only 10% of infected people are thought to be likely to develop active disease. Studies that measure the rate of infection can thus not be used to extrapolate the prevalence of active disease.

Based on many early studies, it was initially widely accepted that zoonotic TB largely manifested as extra-pulmonary TB (ePTB), and that the occurrence of extra-pulmonary TB could therefore be used as an indicator of the prevalence of zoonotic TB. The death rate from abdominal TB in young children was one such surrogate measure. Evidence in records from various counties in the UK and elsewhere showed that after the introduction of the pasteurization of milk, the death rate due to abdominal TB declined dramatically. This decline was considered most likely to be due to the reduction in the zoonotic proportion of TB in children, and it was seen as further evidence for this assumption. Until the mid-twentieth century in Europe, 50% of tuberculous lymphadenitis was considered to be of bovine origin. This pattern changed over time: by 1975 only 8% of cases with tuberculous lymphadenitis in the UK were infected with *M. bovis* (Kleeberg 1975), and it further declined during the 1980s to 1.4% (Yates and Grange 1992). Recent studies confirmed that we cannot use ePTB as a surrogate for estimating bovine TB risk (Berg et al. 2015; Firdessa et al. 2013), and we cannot thus simply extrapolate from decades-old earlier findings and expect that ePTB today will have *M. bovis* as its major etiological component.

What is clear though is that zoonotic tuberculosis caused by *M. bovis* can be a major problem under certain circumstances, particularly when people consume unpasteurized milk. Studies done in Germany (Lange 1932; O'Reilly and Daborn 1995; Dürr et al. 2013) showed that many of the zTB cases were extra-pulmonary, with a strong bias toward alimentary TB that was most likely contracted by consuming *M. bovis*-infected unpasteurized milk or infected meat. In one study of the abdominal TB cases, 35% were due to *M. tuberculosis* and 65% due to *M. bovis* infection (Blacklock 1947). All ePTB are thus not caused by *M. bovis*, nor will pasteurization prevent all ePTB cases as people may become infected by other routes.

It is assumed that currently, although the number of cases may vary substantially across ethnic groups, a median of 63% of zTB cases globally are ePTB (Dürr et al. 2013). For Africa, the data are almost non-existent and totally inadequate, and an assessment of the situation there is largely based on small, local studies, and unreliable anecdotal information. No *M. bovis*, for instance, could be detected in humans in some studies (Ayles et al. 2013), but a recent study in Ethiopia found that 2 of 70 cases with tuberculous lymphadenitis (2.9%) were caused by *M. bovis* (Nuru et al. 2017). It is unknown what the current prevalence of zoonotic BTB is in African countries, or whether its prevalence will change over time, as it did in Europe. Given the high prevalence of HIV-AIDS in Africa, co-morbidity with HIV may alter the situation, but the actual situation is unknown, and the assumption that it may increase the number of zoonotic cases is highly speculative. Some light may be cast on the situation if the current prevalence of zoonotic TB in South Africa is considered. Based on the quoted figures, it can be expected that ePTB will occur in approximately 15–20% of immunocompetent adults and in 20–70% of HIV-infected, immune-incompetent individuals with TB caused by *M. tuberculosis* (Karstaedt 2014). However, currently most ePTB cases in humans in South Africa appear to be caused by *M. tuberculosis* or species of mycobacteria other than the TB complex (non-tuberculous mycobacteria, NTMs).

3.3 Estimating the Risk of Contracting Zoonotic TB

Many factors make it extremely difficult to determine the extent of zoonotic TB in populations. These include the lack of and the complexity of the required routine diagnostic procedures and the importance of obtaining appropriate samples from which to culture *M. bovis*.

Since there is no routine or even survey testing for *M. bovis* in the livestock populations of most of the African countries, and given the overwhelming burden of *M. tuberculosis*, attempts to quantify the burden of zoonotic TB are difficult, and, at best, a very rough estimate. Health care in all African countries is over-burdened and under-funded, and in terms of the priorities for allocating scare human and financial resources, introducing expensive testing for zoonotic TB on a routine basis for the assumed very few possible infections is not justified by policymakers. This approach resulted in the circular argument expressed to the authors by officials at a routine diagnostic laboratory in Africa that *we do not test for zoonotic TB, because we do not have any*. These problems are not likely to be resolved in the near future.

The risk of contracting bovine TB clearly depends on the level of exposure. The prevalence of the disease in cattle thus plays a critical role in this regard, and the marked variation in prevalence in various countries in Africa, and within countries will thus influence this risk, and the likely prevalence of zoonotic BTB. In Mali, for instance, 24% of cattle are infected (Diallo et al. 2016) while in Ethiopia the prevalence is 3.8% (Duguma et al. 2017). In Nigeria 11% of cattle had TB lesions in one area (Okeke et al. 2016), and in Zambia a herd prevalence of 50%, and up to 28% of individual cattle were reportedly infected (Malama et al. 2013). In Zambia despite these high prevalence rates, *M. bovis* could either not be recovered from human TB cases (Ayles et al. 2013), or only small numbers were infected (2 *M. bovis* isolates from 55 acid-fast-positive sputum smears, i.e., a prevalence of 3.6%) (Malama et al. 2014).

Close contact with infected and diseased animals and the consumption of raw milk are the main risk factors for the general population to contract zoonotic TB. For instance, the odds are eight times higher for persons in households where their cattle suffer from BTB to contract the disease, than those living with TB-negative livestock (Mengistu et al. 2015).

The process of pasteurization of food to kill pathogens was adopted for milk by 1886. The initial application of this process was limited, but by the early decades of the twentieth century it was used widely, although without proper quality controls or standards. The supply of pasteurized milk to urban areas in Africa improved during the latter half of the twentieth century, but the situation is far from optimal and varies from country to country. In rural African areas access to pasteurized milk is exceedingly limited, and the populations there consume large volumes of raw milk. Many rural African communities, however, do not consume fresh milk, as they prefer it to be naturally soured before consumption. The souring process limits mycobacterial growth, and even kills the bacilli, especially when souring takes place in hot climates (Macuamule et al. 2016; Michel et al. 2015). Many people of Bantu origin, such as in South Africa and elsewhere, suffer from lactose intolerance, and for this reason also prefer not to drink fresh milk.

The risk of contracting zoonotic TB from meat is extremely difficult to assess, and if meat is thoroughly inspected, rejected for consumption when containing tuberculous lesions, and cooked well enough, the risk is likely to be minimal, particularly if the internal organs and lymph nodes are not consumed. These precautionary measures, however, are generally not applied in many African countries. Large amounts of uninspected meat are consumed throughout the continent, and, even when meat inspection is done in abattoirs, the process is often flawed because in many instances the meat inspectors are poorly trained, and do not adequately inspect the carcasses for the presence of tuberculous lesions. Even under optimal conditions, routine meat inspection in abattoirs only detects about 50% of *M. bovis*-infected carcasses. There are also marked ethnic differences in the way in which meat is consumed in Africa. There is a large, often illegal, trade in bush meat that is not inspected for the presence of any pathogens or diseases. In Swaziland, 93% of livestock farmers consume high-risk tissues including lymph nodes from the head and internal organs (Dlamini 2013).

Great care must be taken when interpreting reports on the prevalence of TB in animals and humans. For example, the methodology used may vary, cut-offs for skin testing may not be the same, and there may be false-positive skin tests or INF-γ reactors. For example, in Ethiopia, cattle, goats, and wildlife in an intensive interface area were infected with many NTMs, but not by *M. bovis* (Tschopp et al. 2010). Sufficient studies were conducted in Tanzania, South Africa, Ethiopia, and Nigeria to know that there are considerable differences in the prevalence of the disease between regions within countries and that the results from a survey in one region do not necessarily represent the prevalence of the disease in the country as a whole.

Better information exists about the risk of transmission of *M. bovis* when humans were in contact with live, infected animals. Of humans with active TB, 16.4% of persons in contact with livestock had pulmonary BTB, while of those with no known contact with livestock, 1.6% had zoonotic pulmonary BTB (Cutbill and Lynn 1944). In the UK at that time (1944), between 0 and 5.7% of cases of pulmonary TB were infected with *M. bovis*. The more intensive the contact, the higher thus the risk. For example, in Denmark where in rural areas cattle were housed with their owners during winter, 41% of pulmonary TB cases were due to *M. bovis* infection (Sigurdsson and Andersen 1945). Similar conditions where livestock and humans live in such close association do not generally exist in Africa, and the risk of transmission is likely to be lower.

In general, the situation in Africa is unclear owing to the lack of published data and the lack of the ability in the past, and even today, to do large-scale, routine laboratory diagnosis of *M. bovis*. The likelihood of contracting the human form of TB (*M. tuberculosis*) remains extremely high in many areas, and it poses a far greater risk, and far out-strips the risk of becoming infected with *M. bovis* (Wood et al. 2011). Overall the zoonotic disease risk from reported cases for Africa appears to be surprisingly low, and many studies conducted in recent years reported either no or very few cases of zoonotic BTB (Ayles et al. 2013; de Garine-Wichatitsky et al. 2013; Dürr et al. 2013; Kazwala et al. 2001). A major weakness in these studies, however, is that Löwenstein-Jensen slants without pyruvate were commonly used for culture, and *M. bovis* does not grow well on this medium. In addition, laboratory-supported current human case diagnosis in South Africa and many other countries is largely smear- and GeneXpert-based, or MGIT culture-based. GeneXpert is an automated PCR test, unable to differentiate between *M. tuberculosis* and *M. bovis*. Likewise, smear and MGIT culture alone cannot differentiate *M. tuberculosis* from *M. bovis*. Thus, any sample sent for routine testing is likely to be diagnosed as *M. tuberculosis* unless special circumstances suggest that it must be processed differently or further. In the case of primary TB cases that give no indication of antibiotic resistance to isoniazid or rifampicin, further testing is highly unlikely. Furthermore, bovine TB frequently manifests as ePTB (often alimentary), and obtaining good samples for speciation is extremely difficult in this form of the disease. In addition, studies based on sputum sampling will considerably underestimate zoonotic BTB (Dürr et al. 2013), particularly in cases with HIV co-morbidity (Steingart et al. 2014). It is therefore likely that the diagnostic systems currently used in many African countries, are not geared to detect zoonotic *M. bovis*, and those cases will routinely be missed.

In 1994, the WHO estimated that *M. bovis* could be the causative organism globally of 3% of human TB cases (WHO 1994). This is a global estimate, and the number of cases may vary widely between countries and within countries. More recently it was estimated that zoonotic TB caused by *M. bovis* globally constituted <1.4% of all human TB cases but that in Africa there were seven zoonotic BTB cases/100,000 population per year (Müller et al. 2013). If this were true, for South Africa it would mean that there should be approximately 3850 zoonotic BTB cases, which is <1% of the total number of human TB cases. This expected rate, however, varies substantially from country to country. For example, in 10 African countries the median proportion of zoonotic cases was 3.5%, but in some countries, such as Tanzania, it was much higher, and the prevalence there ranged from 10 to 38%. The risk of infection was dramatically reduced in South Africa following the introduction of pasteurization more than half a century ago, and by the implementation in 1969 of a national bovine tuberculosis control and eradication scheme (Cousins et al. 2004).

3.4 Zoonotic TB in the Era of HIV

Concerns have been voiced that in the era of HIV, zoonotic TB poses a bigger risk. This seems likely, since there is ample evidence that even the avirulent form of *M. bovis*, *M. bovis* BCG, can cause BCGosis in HIV-positive children (Hesseling et al. 2003) and can occur at extremely high incidence rates of 992 per 100,000 vaccinated infants (Hesseling et al. 2009). It is possible that susceptibility to infection may vary with age, but this is by no means clear (Hesseling et al. 2015).

The epidemic of HIV infection in developing countries, particularly countries in which *M. bovis* infection is present in animals, and the conditions favoring zoonotic transmission could make zoonotic TB a serious threat to persons at risk. Even where pasteurization is effectively applied, and there is a strong attempt by the state to control or eradicate BTB, spillback from endemic wildlife reservoirs (Musoke et al. 2015) may create hotspots of infection at the interface in rural areas. However, so far, despite the ingress of HIV, zoonotic TB in Africa does not seem to be a major emerging disease. The caveat being that adequate, large-scale testing for bovine TB is not done, nor are the diagnostic practices for diagnosing TB sufficient to detect *M. bovis*.

Given the lack of adequate diagnostic techniques to detect *M. bovis* infections in humans, and the lack of BTB control in livestock in many, if not most, of the African countries, we may be living in a fool's paradise, a situation that may yet come to haunt us given the stated objective of globally eradicating zoonotic TB by 2030 (Anon 2017).

3.5 Antibiotic-Resistant *M. bovis*

All *M. bovis* strains are inherently resistant to pyrazinamide, one of the four first-line antibiotics used to treat TB in humans. In the absence of further diagnostic work, such as speciation or the determination of antibiotic resistance, primary TB cases will be given the first-line antibiotics, isoniazid, rifampicin, ethambutol, and pyrazinamide. This implies that only three and not four antibiotics will effectively treat zoonotic cases. The failure to properly identify a human TB case as BTB, and consequent treatment by three antibiotics instead of the recommended four increases the risk of treatment failure, and the acquisition of antibiotic resistance. There are very few reports of this happening, but this may be due to the failure of identifying *M. bovis* infections, and the lack of the ability to determine antibiotic resistance in the many economically poorly resourced countries in Africa. In one study 50% of unrelated *M. bovis* isolates were resistant to isoniazid and rifampicin (Vazquez-Chacon et al. 2015) with the result that these patients were treated with only one effective antibiotic. These isolates are well on their way to becoming pre-XDR and XDR *M. bovis* strains. Perhaps we should not over-react to this development, since evidence suggests that transmission of *M. bovis* between humans is relatively uncommon. However, there is a risk, and the potential exists that we may generate a new antibiotic-resistant epidemic unless the problem is managed well.

3.6 Zoonotic TB in a High TB Burden Setting: A Public Health Perspective

From a public health perspective, it is perhaps not surprising that there has been little or no concern about zoonotic TB. Questions that need to be asked are: What are the consequences for the individual who may be affected by zoonotic TB in such a high burden setting, and how much should be spent on making sure each TB case is caused by *M. tuberculosis* and not by *M. bovis*? In answering the first question one may argue that as long as such patients are diagnosed as a TB case and receive first line treatment, they are highly likely to cure as for a normal TB case with *M. tuberculosis*, because *M. bovis* is inherently susceptible to three of the four first-line drugs. Pyrazinamide is the exception, but it is given only for the first 2 months of the 6 months' treatment regimen. To address the second question: to change the diagnostic or treatment paradigm because of the possibility of the occasional misdiagnosis, and where the person is likely to be cured in any case, is arguably not financially justified in a high human TB disease setting. Failure to cure in a first-line case under conditions of adherence perhaps warrants further investigation, and the possibility of an *M. bovis* infection should not be overlooked.

It is likely that *M. bovis* can become a cause of morbidity and mortality in humans in certain target populations and settings. This possibility is supported by the extensive variation in the occurrence and prevalence of zoonotic TB across Africa.

We in Africa may face a situation similar to the unprecedented levels of zoonotic TB in people of Hispanic origin in the USA that is attributed to the consumption of unpasteurized, *M. bovis*-contaminated soft cheese imported from Mexico (Müller et al. 2013).

It is clear that the eradication of TB in humans (defined broadly and not just caused by *M. tuberculosis*) cannot be achieved unless zoonotic TB is also dealt with. When there is no control of BTB, also in wildlife populations, we can expect the disease to persist and spread, as has been experienced in the UK. Spread of the disease in wildlife populations in South Africa, albeit slow, where movement of *M. bovis* has been traced from wildlife to cattle (Musoke et al. 2015) creates the same risk for persistence of the disease in livestock because of spillback of the infection from wildlife to livestock, and humans. In a worst-case scenario in which bovine TB occurs in animals and particularly in cattle, we could expect zoonotic TB to rise to levels seen in an era prior to the use of antibiotics (Lange 1932). Under these circumstances children in whom the diagnosis is far more difficult than in adults will be the unfortunate indicators of the presence of the infection (Lange 1932; O'Reilly and Daborn 1995).

3.7 Conclusion

Although many challenges remain, attempts are being made to raise awareness of the problem of zoonotic TB, and to suggest some solutions (Carruth et al. 2016; Olea-Popelka et al. 2017). The necessary actions include control of bovine tuberculosis, and we propose the introduction of precision medicine, which will include specific diagnosis and appropriate treatment of zoonotic TB in humans. We have the tools to do it, but the health systems are not geared for it, and the logistics and cost implications are probably too daunting for African health authorities to implement it on a wide scale.

In summary, the extent of *M. bovis* infection in the global human population today is unknown, as is the significance of *M. bovis* in human tuberculosis in Africa. An integrated and inter-disciplinary One Health approach is therefore needed to investigate this problem in depth to ascertain whether it is a real public health issue or not, and to recommend how to deal with it in a cost-effective manner.

References

Anon (2017) Roadmap for zoonotic tuberculosis. World Health Organization, Geneva, p 20

Ayles H, Muyoyeta M, du Toit E et al (2013) Effect of household and community interventions on the burden of tuberculosis in Southern Africa: the ZAMSTAR community-randomised trial. Lancet 382:1183–1194

Berg S, Schelling E, Hailu E et al (2015) Investigation of the high rates of extrapulmonary tuberculosis in Ethiopia reveals no single driving factor and minimal evidence for zoonotic transmission of *Mycobacterium bovis* infection. BMC Infect Dis 15:1–10

Blacklock JWS (1947) The epidemiology of tuberculosis. Br Med J 1:707–712

Carruth L, Roess AA, Mekonnen YT et al (2016) Zoonotic tuberculosis in Africa: challenges and ways forward. Lancet 388:2460–2461

Cousins DV, Huchzermeyer HF, Griffin JF et al (2004) Tuberculosis. In: Coetzer JAW, Tustin RC (eds) Infectious diseases of livestock, vol 3. Oxford University Press, Cape Town, pp 1973–1993

Cutbill LJ, Lynn A (1944) Pulmonary tuberculosis of bovine origin. Br Med J 1(4338):283–285

de Garine-Wichatitsky M, Caron A, Kock R et al (2013) A review of bovine tuberculosis at the wildlife-livestock-human interface in sub-Saharan Africa. Epidemiol Infect 141(7):1342–1356

Diallo M, Diarra B, Sonogo M et al (2016) Molecular identification of *Mycobacterium bovis* from cattle and human host in Mali: expanded genetic diversity. BMC Vet Res 12(1):145

Dlamini M (2013) A study on bovine tuberculosis and associated risk factors for humans in Swaziland. Ph.D. thesis, University of Pretoria, Pretoria

Duguma A, Abera S, Zewdie W et al (2017) Status of bovine tuberculosis and its zoonotic implications in Borana zone, South Ethiopia. Trop Anim Health Prod 49(3):445–450

Dürr S, Müller B, Alonso S et al (2013) Differences in primary sites of infection between zoonotic and human tuberculosis: results from a worldwide systematic review. PLoS Negl Trop Dis 7(8): e2399. https://doi.org/10.1371/journal.pntd.0002399-e2399

Firdessa R, Berg S, Hailu E et al (2013) Mycobacterial lineages causing pulmonary and extrapulmonary tuberculosis, Ethiopia. Emerg Infect Dis 19(3):460–463

Hesseling AC, Schaaf HS, Hanekom WA et al (2003) Danish Bacille Calmette-Guérin vaccine-induced disease in human immunodeficiency virus-infected children. Clin Infect Dis 37 (9):1226–1233

Hesseling AC, Johnson LF, Jaspan H et al (2009) Disseminated Bacille Calmette-Guerin disease in HIV-infected South African infants. Bull World Health Organ 87(7):505–511

Hesseling AC, Jaspan HB, Black GF et al (2015) Immunogenicity of BCG in HIV-exposed and non-exposed infants following routine birth or delayed vaccination. Int J Tuberc Lung Dis 19 (4):454–462

Karstaedt AS (2014) Extrapulmonary tuberculosis among adults: experience at Chris Hani Baragwanath Academic Hospital, Johannesburg, South Africa. S Afr Med J 104(1):22–24

Kazwala RR, Daborn CJ, Sharp JM et al (2001) Isolation of *Mycobacterium bovis* from human cases of cervical adenitis in Tanzania: a cause for concern? Int J Tuberc Lung Dis 5(1):87–91

Kleeberg HH (1975) Tuberculosis and other mycobacteriosis. In: Hubbert WT, McCulloch WF, Schnurrenberger PR (eds) Diseases transmitted from animals to man, 6th edn. Charles C. Thomas, Springfield, IL, pp 303–360

Lange B (1932) The role played by bovine tubercle bacilli in human tuberculosis. BMJ 2 (3740):503–506

Macuamule CLS, Wiid IJ, van Helden PD et al (2016) Effect of milk fermentation by kefir grains and selected single strains of lactic acid bacteria on the survival of *Mycobacterium bovis* BCG. Int J Food Microbiol 217:170–176

Malama S, Bwalya H, Godfroid J (2013) A review of tuberculosis at the wildlife-livestock-human interface in Zambia. Infect Dis Poverty 2:13

Malama S, Johansen TB, Muma JB et al (2014) Characterization of *Mycobacterium bovis* from humans and cattle in Namwala District, Zambia. Vet Med Int 2014:187842

Mengistu A, Enquselassi F, Aseffa A et al (2015) Bovine tuberculosis (BTB) as a risk factor for developing tuberculosis in humans in the rural community of Ethiopia: a case-control study. Ethiop Med J 53(1):1–8

Michel AL, Geoghegan C, Hlokwe T et al (2015) Longevity of *Mycobacterium bovis* in raw and traditional souring milk as a function of storage temperature and dose. PLoS One 10(6): e0129926

Müller B, Dürr S, Alonso S et al (2013) Zoonotic *Mycobacterium bovis*-induced tuberculosis in humans. Emerg Infect Dis 19(6):899–908

Musoke J, Hlokwe T, Marcotty T et al (2015) Spillover of *Mycobacterium bovis* from wildlife to livestock, South Africa. Emerg Infect Dis 21(3):448–451

Nuru A, Mamo G, Zewude A et al (2017) Preliminary investigation of the transmission of tuberculosis between farmers and their cattle in smallholder farms in northwestern Ethiopia: a cross-sectional study. BMC Res Notes 10(1):31

O'Reilly LM, Daborn CJ (1995) The epidemiology of *Mycobacterium bovis* infections in animals and man: a review. Int J Tuberc Lung Dis 76(1):1–46

Okeke LA, Fawole O, Muhammad M et al (2016) Bovine tuberculosis: a retrospective study at Jos abattoir, Plateau State, Nigeria. Pan Afr Med J 25:202

Olea-Popelka F, Muwonge A, Perera A et al (2017) Zoonotic tuberculosis in human beings caused by *Mycobacterium bovis* – a call for action. Lancet 17:21–24

Sigurdsson JH, Andersen H (1945) Studies on the risk of infection with bovine tuberculosis to the rural population with special reference to pulmonary tuberculosis. Acta Tuberc Scand Suppl 15:26–40

Steingart KR, Schiller I, Horne DJ et al (2014) Xpert® MTB/RIF assay for pulmonary tuberculosis and rifampicin resistance in adults. Cochrane Database Syst Rev 1:CD009593. https://doi.org/10.1002/14651858.CD009593.pub3

Tschopp R, Asseffa A, Schelling E et al (2010) Bovine tuberculosis at the wildlife-livestock-human interface in Hamer Woreda, South Omo, Southern Ethiopia. PLoS One 5(8):1–7

Vazquez-Chacon CA, Martinez-Guarneros A, Couvin D et al (2015) Human multidrug-resistant *Mycobacterium bovis* infection in Mexico. Tuberculosis 95:802–809

Wood R, Lawn SD, Johnstone-Robertson S et al (2011) Tuberculosis control has failed in South Africa — time to reappraise strategy. S Afr Med J 101:111–114

World Health Organization (WHO) (1994) Report of WHO working group on zoonotic tuberculosis (*Mycobacterium bovis*) with the participation of FAO. World Health Organization, Geneva

Yates MD, Grange JM (1992) Bacteriological survey of tuberculous lymphadenitis in southeast England, 1981–1989. J Epidemiol Community Health 46(4):332–335

Chapter 4
The Control of *Mycobacterium bovis* Infections in Africa: A One Health Approach

S. I. B. Cadmus, P. I. Fujiwara, J. A. Shere, B. Kaplan, and C. O. Thoen

4.1 Introduction

For over 100 years, since Robert Koch discovered the causative agent in 1882, human tuberculosis (TB) caused by *Mycobacterium tuberculosis* remains a major global public health threat. In recent times, the problem posed by bovine TB (BTB) became of increasing concern to public health officials. *Mycobacterium bovis*, of which cattle is the primary maintenance host, is not easily differentiated from *M. tuberculosis*, as both species belong to the genetically related *M. tuberculosis* complex (MTC) (Brosh et al. 2002). What is more concerning is that *M. bovis* also infects humans, particularly those who consume unpasteurized dairy products and live in close contact with infected cattle (Cosivi et al. 1998; Thoen et al. 2009), hence the challenge of zoonotic TB (zTB). *Mycobacterium bovis* was hitherto believed to be transmitted only from cattle to humans (Ayele et al. 2004), but cases of human-to-human transmission via the pulmonary route have been reported recently (Gibson et al. 2004; LoBue et al. 2004; Evans et al. 2007; Thoen and LoBue 2007; Sunder et al. 2009; Etchechoury et al. 2010; Godreuil et al. 2010; Adesokan et al. 2012; Torres-Gonzalez et al. 2013).

Author "Thoen" was deceased at the time of publication.

S. I. B. Cadmus (✉)
University of Ibadan, Ibadan, Nigeria

P. I. Fujiwara
International Union Against Tuberculosis and Lung Disease (IUATLD), Paris, France
e-mail: Pfujiwara@theunion.org

J. A. Shere
United States Department of Agriculture, Washington, DC, USA
e-mail: Jack.A.Shere@aphis.usda.gov

B. Kaplan
One Health Initiative, Sarasota, FL, USA

A. B. Dibaba et al. (eds.), *Tuberculosis in Animals: An African Perspective*,
https://doi.org/10.1007/978-3-030-18690-6_4

The persistence of *M. bovis* at various prevalences in cattle in countries, such as in Mexico (13.8%) (Perez-Guerrero et al. 2008), Uganda (7%) (Oloya et al. 2008), Nigeria (5%) (Cadmus et al. 2006), the UK (0.17–0.5%) (Stone et al. 2012), France (0.5–2%) (Mignard et al. 2006), The Netherlands (1.4%) (Majoor et al. 2011), and most African countries where dairy products are generally not pasteurized or heat-treated, creates ongoing challenges to prevent zoonotic TB. Significantly, *M. bovis* infection has recently been categorized as an important neglected disease, affecting the livelihoods mostly of poor and marginalized communities, with increasing zoonotic implications (Hlavsa et al. 2008; Rodwell et al. 2008) and with serious ecological implications as the infection spreads between livestock, wildlife, and humans at the interface between them (Kriek 2014).

4.2 Bovine TB in Wildlife and Other Animals in Africa

Cattle are known to be the primary maintenance hosts of *M. bovis* (Brosh et al. 2002; Thoen et al. 2009), but certain species of wildlife are also maintenance hosts of this pathogen (Radunz 2006; Smith et al. 2006; Porphyre et al. 2008; Thoen et al. 2009). Essentially, the brushtail possum (*Trichosurus vulpecula*), European badger (*Meles meles*), American bison (*Bison bison*), African buffalo (*Syncerus caffer*), Kafue lechwe (*Kobus leche kafuensis*), and white-tailed deer (*Odocoileus virginianus*) are known maintenance hosts of *M. bovis* in the various countries of the world where they occur (De Lisle et al. 2002; Renwick et al. 2007). In some parts of Africa, particularly in Southern and Eastern Africa, several wildlife species have been identified as reservoirs of *M. bovis* at the livestock–wildlife interface (Michel et al. 2006). In Africa, African buffaloes and the Kafue lechwe are the primary wildlife maintenance hosts for *M. bovis*, but there are indications that a number of other species may also be able to sustain the infection (Michel et al. 2006; Renwick et al. 2007; Moiane et al. 2014). (For more information about BTB in wildlife, refer to Chap. 5.) Due to limited studies, and the lack of funding for BTB research and surveillance in wildlife, very scant information exists about the extent of the infection in African wildlife, the importance of wildlife reservoirs in Africa, and the role that they play in the epidemiology of the disease. Much of the available data deal with the situation in Southern and Eastern Africa because of the impact of BTB at the interface on wildlife–ecotourism and conservation in these areas. Bovine TB in wild animals has an impact on wildlife conservation, livestock production, public health, and the burgeoning, lucrative private game ranching enterprises in Southern Africa in particular (de Garine-Wichatitsky et al. 2013).

In Africa, other domesticated animals such as sheep (Houlihan et al. 2008; Mendoza et al. 2012), goats (Daniel et al. 2009; Hiko and Agga 2011; Naima et al. 2011), and camels (Kudi et al. 2012) have also been identified to play a significant role in certain countries in the transmission of *M. bovis* because of their very close interaction with cattle and joint cattle herding and management practices causing intermingling of these species.

Because of the largely unknown but increasing role that wildlife play in sustaining and transmitting BTB at the human–livestock–wildlife interface, it is important to adopt a holistic approach that includes wildlife populations when tackling the control and eventual eradication of BTB and zTB in Africa. This is critical given the close contact between wildlife and livestock populations in many areas and the known bi-directional transmission of BTB at the interface (Miller et al. 2007; Munyeme et al. 2008; Siembieda et al. 2011; Gortázar et al. 2012; Palmer et al. 2012; Kaneene et al. 2014a). In Africa the forced close interaction between wildlife and cattle, and sometimes humans, who enhance transmission of *M. bovis*, is the consequence of the scarce and limited water sources that they have to share.

4.3 Zoonotic BTB: Global Realities and Facts from the World Health Organization

The World Health Organization (WHO), in conjunction with the Food and Agriculture Organization (FAO) of the United Nations, previously recommended that the epidemiological data pertaining to human TB caused by *M. bovis* infections, particularly in populations at risk (e.g., livestock workers, veterinarians, hunters, and zoo workers) (WHO-FAO 1994), be collected. The African countries, unfortunately, did not respond to this request, and the information about the disease on the continent remains piecemeal and anecdotal (AU-IBAR 2013). Based on the available raw and incomplete data, Müller et al. (2013) estimated that in Africa there would be 7 zTB cases/100,000 population/year, caused by infection with *M. bovis*, which is approximately 1% of the total human TB burden. More recently, in 2015, the WHO estimated that 149,000 cases of *M. bovis* zTB with a mortality of 13,400 occurred globally (WHO 2015a) and that of these 76,300 cases and 10,000 deaths were expected to occur in Africa (WHO 2016a). They considered these figures to be an underestimation given the absence of routine reporting of the disease in most countries in Africa in which BTB is endemic.

The degree of under-investigation, and the likely underestimation of the number of zTB cases in Africa and in some of the other developing countries, implies that the actual number of human zoonotic BTB cases remains unknown but that it may be substantially more than the current estimate (Perez-Lago et al. 2014). With the persistence of *M. bovis* in cattle in most African countries (Cadmus et al. 2006; Perez-Guerrero et al. 2008; Oloya et al. 2008; Jenkins et al. 2011; Egbe et al. 2016) in association with the general poor standards of living and hygiene, there is a risk that zoonotic BTB will not remain an African problem but that it may become a global health threat (see also Chap. 3).

The increasing emphasis on addressing these issues is largely due to the attention that zTB received globally in recent times in a number of scientific meetings, workshops, symposia, and technical sessions on TB (WHO 2015b, c; IUATLD 2015). The WHO in 2016 (WHO 2016a) informed that:

> … accelerating the annual decline in TB incidence, reaching the 2020 milestone of a 35% reduction in TB deaths, requires reducing the global number of people with TB who will die from the disease (the case fatality ratio, or CFR) from 17% in 2015 to 10% by 2020.

The WHO's End TB Strategy (where every case of TB counts, no matter what its source) accepted by the World Health Assembly in 2014 (WHO 2015b) and the Global Plan to End TB in 2016–2020 (Stop TB Partnership 2015) further emphasize the need to address these deficiencies and to include the issue of zoonotic tuberculosis in planning the eventual eradication of TB. More importantly, the End TB objective was adopted as part of the global Sustainable Development Goals (UN 2015). Therefore, within this context, there were calls for better diagnosis and treatment of every human TB case, including those with zTB.

Apart from the aforementioned, the lack of global concern about zTB remains a problem, given that human TB is generally, and specifically in some African countries, considered to be caused only by *M. tuberculosis*, without taking into account the role played by other members of the MTC, particularly *M. bovis*. Consequently, the general consensus was that current global initiatives to control and eradicate TB must involve a more holistic approach. This is based on the knowledge that a critical mass of the world population lives in neglected communities where cohabitation with animals (particularly cattle) is common. Thus, as a way forward, a pragmatic approach to end TB must incorporate an all-inclusive strategy that will simultaneously focus on the disease in both humans and animals. Toward this end, a meeting of a committee of experts from academia, WHO, the World Organization for Animal Health (OIE), the FAO, International Union Against Tuberculosis and Lung Disease (IUATLD), and relevant research institutions was held at the WHO headquarters, Geneva, Switzerland, in April 2016 to deliberate on key priorities needed to reduce the burden of zTB. This culminated in the acceptance of zTB as a priority of the WHO at its Strategic Technical Advisory Group for TB (STAG-TB) meeting in June 2016 (Green 2006; WHO 2016b).

Because of the uncertainty about the actual number of zoonotic TB cases, the future prevention and control of *M. bovis* infections require the improvement of, and the use of rapid and accurate diagnostic tools, more comprehensive surveillance programs and greater collaboration between veterinary and medical health officials (Perez-Lago et al. 2014; WHO 2016b). These challenges provide an opportunity to reflect on the need for applying the principles of One Health to the control and eventual eradication of BTB and zTB.

4.4 The Burden of TB in Africa

As of 2016, Africa's 54 countries are home to approximately 1 billion people, constituting about 16% of the 7.4 billion people in the world. Although it has abundant natural resources and is showing signs of economic growth, Africa remains the world's poorest and most underdeveloped continent. Multiple factors have been advanced for its underdevelopment, which include corruption in government settings occasioned by serious human rights violations, civil wars, failing central planning, high levels of illiteracy, and poor healthcare services resulting in the spread of deadly diseases such as HIV/AIDS, malaria, and TB (UNDP/HDRO 2013). Worse still, Africa currently carries the highest burden of TB (28% of the world's cases in 2014) relative to its population (281 per 100,000 population), which is more than double the global average of 133 per 100,000 population (WHO 2015c). This situation is further worsened by the lack of surveillance and control measures to control BTB in the majority of the African countries (Cosivi et al. 1998; Thoen et al. 2009, 2010).

Overall, pertinent questions and key issues have yet to be addressed when BTB and zTB are considered in Africa (Thoen et al. 2010; El Idrissi and Parker 2012; Olea-Popelka et al. 2016). These include a lack of:

1. An estimate of the prevalence of BTB in most African countries
2. The burden of the disease in human populations at risk of infection
3. Optimal methods to document human-to-human transmission of the disease after possible zoonotic infection in both rural and urban settings
4. An understanding of the molecular epidemiology of BTB (cattle) and zTB (cattle and humans) for the purpose of developing adequate prevention and control strategies.

4.5 Concrete Steps Toward Setting an Agenda for the Control of Bovine and Zoonotic TB in Africa

A paradigm shift is required to address the challenge of BTB and its zoonotic burden on vulnerable human populations in Africa. To help achieve this, a concrete roadmap of implementable actions will be needed using multidisciplinary and interdisciplinary approaches that include governments, politics, education, health, and various interest groups within a practical national and international framework (Fig. 4.1). Borrowing from the expert submission made in June 2016 to the STAG-TB in Geneva by experts from WHO, FAO, OIE, and IUATLD, ten key areas for the roadmap were suggested (WHO 2016b):

1. Improved surveillance
2. Development of novel diagnostic tools
3. Coordinated research
4. Disease control in animals
5. Targeting key populations
6. Food safety
7. Raising awareness and engaging stakeholders

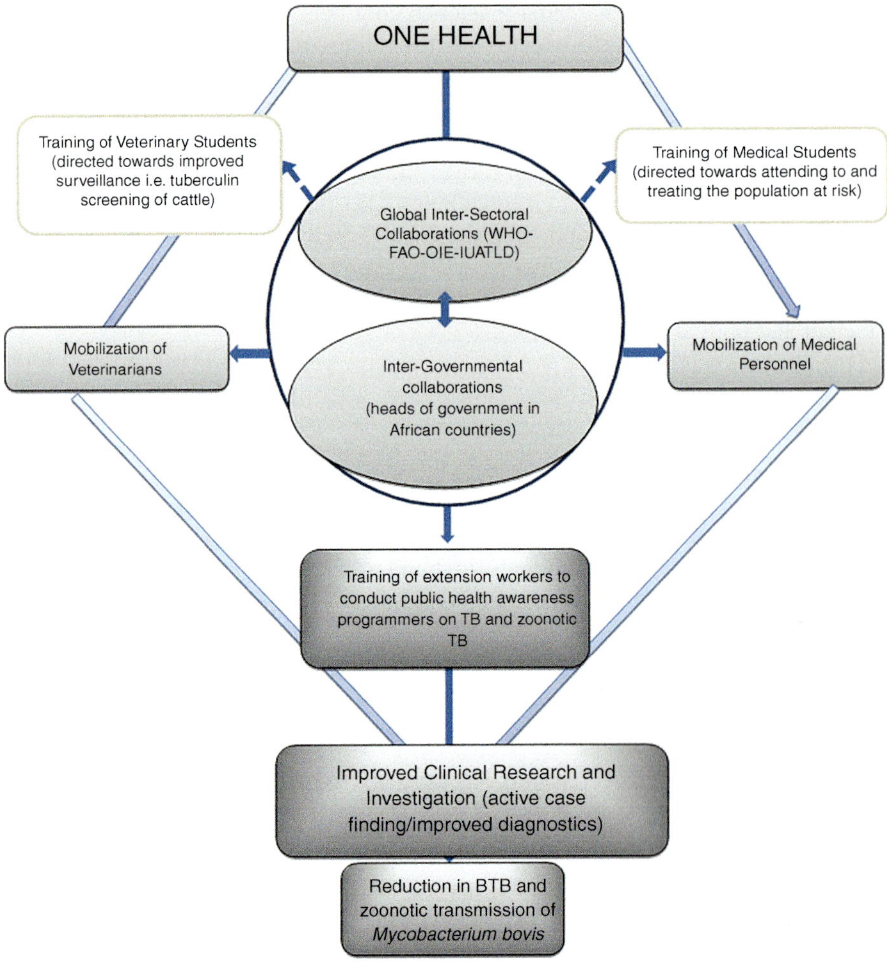

Fig. 4.1 Specific steps required for controlling bovine and zoonotic TB in Africa using a "One Health" approach

8. Developing policies and guidelines
9. Joint human/animal interventions
10. Commitment and funding by government and international organizations.

Veterinarians, physicians, sociologists, epidemiologists, geographers, public health experts, policy makers, and particularly livestock workers should be included. On the whole, the benefit accrued must be for all with the goal of controlling and reducing BTB and zTB to the barest minimum in Africa. In order to make progress, the One Health framework and principles must therefore be taken into consideration.

4.6 The Genesis and Principles of One Health

One Health evolved from the concept of One Medicine and focuses on health and ecosystems to achieve global public health for humans, healthy animals, and a stable and sustainable environment (Thoen et al. 2016). Health experts from around the world met in September 2004 for a symposium organized by the Wildlife Conservation Society of New York, hosted by The Rockefeller University, USA, that focused on the potential spread of diseases between humans, domestic animals, and wildlife populations, to address these issues. Using case studies on Ebola, avian influenza, and chronic wasting disease as examples, the assembled experts and panelists delineated priorities for an international, interdisciplinary approach for combating threats to global health. Thereafter, veterinarians, physicians, public health experts, sociologists, and epidemiologists globally supported the concept of One Health, which they believed would promote surveillance and enhance the prevention, control, and eradication of zoonotic diseases. The vision of One Health therefore is "dedicated to improving the lives of all species—human and animal—through the integration of human and veterinary medicine." Furthermore, "Recognizing that human and animal health are inextricably linked, One Health seeks to promote, improve, and defend the health and well-being of all species by enhancing cooperation and collaboration between physicians and veterinarians by promoting strengths in leadership and management to achieve these goals."

The importance of One Health is palpable from the benefits derived in public health through positive interventions on animal diseases (Roth et al. 2003; Zinsstag et al. 2009) by joint healthcare services (Thoen et al. 2016), as well as positive outcomes observed by joint disease surveillance (Mazet et al. 2009). Since most countries in Africa are burdened with a high prevalence of BTB (Gibson et al. 2004), a practical way forward, and to reduce the human burden of the disease, is to embrace and incorporate the One Health approach to help prevent and control human infections with *M. bovis*, for which cattle serve as its natural host.

4.7 Target Populations/Communities and Mitigations

An important strategy toward achieving community mobilization is the active engagement of the population/community of interest. These engagements will include periodical community outreaches involving sensitization and health awareness talks on zoonotic diseases.

In Africa, participation of three principal stakeholders/communities is important to stem the tide of BTB and its zoonotic implications, namely, the pastoralist community (the livestock producers), livestock marketers, and butchers (livestock processors) (Adesokan et al. 2012). These groups are the populations at risk and are most vulnerable to zoonotic infections due to *M. bovis*. Fundamentally, they are unlikely to have the education and awareness of important health and hygiene precautions, and they engage in habits and practices that expose them to infection with *M. bovis*. These risky practices and activities include:

1. Living closely with infected animals
2. Consumption of unpasteurized dairy products and of improperly cooked, contaminated meat products
3. Lack of the use of protective clothing/equipment during slaughtering
4. Other unhygienic habits during milk and meat processing

Consequently, to reduce the spread of zTB in Africa, the activities and actions to be implemented based on the One Health approach should be directed at these neglected and at-risk populations.

The implementable actions that will promote activities to reduce the problem currently being experienced by Africans as a result of BTB using the One Health approach are discussed below.

4.7.1 Intergovernmental Collaboration

To make meaningful progress in the fight against BTB in Africa from both the animal and human perspective, the heads of government of the entire region must first be made to appreciate the enormity of the problem (through local and international policy briefings as well as joint summits) and accept that BTB is a major challenge to animal productivity and human health. After this has been achieved, a central committee at the African Union (AU) secretariat, given its leadership and political clout to promote regional health initiatives, as has been done for HIV/AIDS, malaria, and TB, should be set up. This committee should, with the technical support of the OIE, focus on promoting and coordinating inter-sectorial collaboration between the Ministries of Health and Agriculture/Livestock Resources in the various African countries. Using this platform, the embracement of a holistic application of the One Health approach to prevent and control BTB in animal production and zoonotic transmission will be greatly enhanced. An initiative to deal with BTB similar to the current collaboration initiated between the medical, veterinary, and agriculture departments across Africa by the AU in tackling the epidemics of avian influenza and Ebola should also be established.

4.7.2 Global Inter-sectoral Support and Collaboration for Africa by the WHO–FAO–OIE Tripartite

Similar global efforts, coordination, and funding to stem the tide of *M. tuberculosis* in humans in Africa should be established to deal with zTB. In this regard, for effective prevention and control of BTB in the animal and human populations, the tripartite initiative involving the WHO, FAO, and the OIE must be strengthened to chart and coordinate a "practical and cost-effective" roadmap that can be implemented in all African countries. This tripartite collaboration must be seen to be working with individual African countries, taking into consideration key issues like prevalence and burden of the disease in both the animal and the human populations. This was reflected in the submission made by teams of experts at the June 2016 WHO-STAG-TB meeting (WHO 2016b). Other fundamental considerations that must be implemented should include the establishment of prevention and control policies and guidelines dealing with the movement of cattle and humans within countries and across their borders, the cattle trade, and clarifying the epidemiology of BTB. Consideration of, and understanding, these factors will be important when developing the roadmap since they will provide insights into the sociocultural and population dynamics characteristic of each local setting.

4.7.3 Joint Veterinary–Medical Education/Training

To promote One Health in relation to BTB and zTB prevention and control in Africa, there is a need for the development of joint One Health curricula in medical and veterinary faculties on the continent (Zinsstag et al. 2005; Monath et al. 2010). This should involve programs that can be conducted jointly focusing on pastoralist communities, livestock markets, and abattoirs for the purpose of training and community interventions. A practical example is the initiative of the MacArthur Foundation for Higher Education in Africa in establishing Centers of Excellence. Through this initiative, a Center of Excellence for the Control and Prevention of Zoonoses (CCPZ) was established at the University of Ibadan, Ibadan, Nigeria, with the aim of prevention and control of endemic, emerging, and reemerging zoonoses (including BTB and brucellosis) in Nigeria, Ghana, Sierra Leone, and Liberia. Through this platform, among other things, active surveillance and screening for BTB in cattle and livestock workers at the abattoirs and livestock markets were conducted by a team of veterinarians, community health physicians, and related disciplines. It is expected that similar initiatives will help to establish teams that can be mobilized to conduct tuberculin tests screening cattle for BTB. Following confirmation of BTB in a specific group of cattle, the medical team can then be invited to investigate possible zoonotic infections and to determine the burden of the disease in the affected human population. Such initiatives will further help to promote public awareness of the risk of consuming unpasteurized or non-heat-treated dairy products and contaminated meat and food products.

4.7.4 Public–Private Partnerships (PPPs)

In most African countries, the private sector plays a key role in TB control. The level of support given by the private sector, however, varies across the continent in terms of size, funding ability, level of organization, services rendered, and the communities supported. Using the One Health approach, African governments can partner with established nongovernmental organization (NGOs) that address TB, for example, the KNCV TB Foundation in the Netherlands, the German TB and Leprosy Foundation, the Belgian Damien TB and Leprosy Foundation, the IUATLD, and a host of well-established local groups to work actively with populations at risk. A successful model is the African Program for Onchocerciasis Control (APOC), which involved 28 African countries, a pharmaceutical company (Merck), and more than 30 development partners for the successful control of river blindness. Furthermore, Private Public Partnership (PPP) initiatives involving veterinary and medical teams working in different African countries can be established and mobilized, as was recently done during the Ebola outbreaks in some Western African countries (EU 2016). This becomes imperative in rural settings where information about the status of BTB has to be investigated to educate the inhabitants on practices related to pasteurization or heat treatment of dairy products before consumption. This is even more important given the need to mobilize the right personnel and experts to facilitate the necessary actions to control BTB in cattle and zTB in humans. Here, the international agencies listed above, including NGOs, and local groups will need to be mobilized both in rural and urban areas to increase public awareness about the risks of keeping and rearing cattle with BTB and, more importantly, the risk of consuming unpasteurized or non-heat-treated dairy products.

4.7.5 Clinical Research and Scientific Investigation

To achieve optimal care for patients with suspected zTB, clinical care is better achieved when there is cooperation and collaboration between the veterinary and medical investigators. A valid, rapid, point-of-care test that can differentiate between *M. bovis* and *M. tuberculosis* should be developed and deployed to screen patients (at hospitals and clinics) and "at-risk populations" selected from different livestock settings (Kaneene et al. 2014a, b). Based on the findings, appropriate and optimal care can then be given to the patient. Alternatively, routine samples at the hospitals and clinics should be inoculated on Lowenstein-Jensen media, one with glycerol and the other without glycerol but containing pyruvate to ensure growth for cases where *M. bovis* are suspected. In addition, active detection of zoonotic TB (i.e., through periodic visits by mobile clinics in at-risk populations in different livestock settings) should be encouraged and promoted to allow proper documentation of cases involving *M. bovis* as was the practice with the Health for Animal and Improved

Livelihood (HALI) project in Tanzania. With these measures in place, it would become easier for a joint team of veterinarians and physicians to conduct coordinated outreach clinical sessions and investigations among livestock workers in pastoralist communities, livestock markets, and abattoirs. Through such initiatives, there will be a better likelihood of detecting unreported cases of human infections with *M. bovis* and *M. tuberculosis*, which can subsequently be treated earlier, thus preventing further spread of the disease.

4.8 Advantages and Future Areas of Importance of the One Health Approach in the Control of Bovine and Zoonotic TB

The overall advantage of an integrated One Health approach to solving the problems of BTB and zTB will be the optimization of monitoring and surveillance systems to assess the overall burden of TB in animals and humans in Africa (Kaneene et al. 2014a, b). The impact of this will be demonstrated over time through the decline (measured through coordinated programs and targets) in the burden of the disease in both human and animal populations in local communities, individual countries, and Africa as a whole. Because of this, and equally important, the valuable money and time saved can be applied to developing the livestock sector on the continent, thereby creating wealth instead of enhancing poverty and death.

As we move along this new roadmap toward tackling BTB and zTB, further research areas that involve methods to control *M. bovis* must be proposed and pursued. This should include the disciplines of sociology (risk perception and hygiene), economics (the cost for the community and disability-adjusted life years—DALYs), and ecology (movement of animals and contact networks between species) (Roger 2012). As a follow-up to this, there is a need to form a coalition of experts within each country, and throughout the continent as a whole, in cooperation with other key stakeholders, to promote constant monitoring and surveillance, to gather comprehensive data at all epidemiological sites (pastoralist settings, livestock markets, abattoirs, hospitals/laboratories), and to work toward the reduction and control of the disease in animal and human populations. More importantly, this range of activities will ultimately inform the level of funding that will be required to support long-term goals at the community, state, and national levels in affected African countries.

4.9 Conclusion

After many years of inaction and poor coordination by the global community, it has now become imperative that the problems and challenges posed by BTB and zTB in Africa must be addressed. A similar need was vividly captured by the editorial published in the *American Journal of Public Health* almost a century ago (Anon 1932) where it was stated:

> Who can calculate the number of lives saved and the amount of crippling (tuberculosis ranks second as a crippling disease) avoided if we had followed the advice of Abraham Jacobi, great man and great physician, and had been as active in our efforts against bovine infection of children as we have been against the human? The facts have been before us for 30 years. They have been proved and re-proved. Is there any excuse for longer complaisance or inaction?

As we reflect on these words, and given the lack of progress to achieve the stated objectives, African governments, scientists, and key stakeholders must join hands with global agencies such as the WHO, the FAO, the OIE (to mention a few), and the IUATLD (that for many years housed a small but influential group of veterinarians and physicians dedicated to the issue of zoonotic TB) and map out strategies to reduce the prevalence and threat of BTB and the burden of zTB in Africa's human population (Olea-Popelka et al. 2016). More importantly, inter-sectorial collaboration particularly between veterinarians and the medical personnel across Africa must be strengthened to combat this disease. This collaboration should be directed at encouraging training programs at universities and related tertiary institutions and major stakeholders to engender advocacy (public health awareness), sustainability (continuous screening and monitoring), and progress (positive implementation of guidelines and policies by government agencies) in the fight against BTB and zTB particularly in neglected communities in the rural areas.

The way forward therefore is to develop a "Marshall Plan" of action (as highlighted earlier) that will help by employing a One Health approach to reduce the burden of BTB in livestock, wildlife, and zTB in humans. This roadmap is now contained in a recent document, Roadmap for Zoonotic Tuberculosis, which outlines the strategy to deal with the issue (WHO 2017). The overarching approach should focus on coordinated public education campaigns and interventions utilizing existing knowledge applied at a local level in a simple and practical way. Finally, cattle and certain wildlife species are maintenance hosts of *M. bovis* in the region, and unless BTB is controlled in Africa in all the infected species, the WHO's goal of ending all forms of TB will be impossible to accomplish.

Acknowledgment Partial funding support received by Simeon Cadmus from the John D. and Catherine T. MacArthur Foundation, USA, under the Higher Education Initiative in Africa (Grant No. 97944-0-800/406/99) for the establishment of the Center of Control and Prevention of Zoonoses (CCPZ) at the University of Ibadan is acknowledged.

References

Adesokan HK, Jenkins AO, van Soolingen D et al (2012) *Mycobacterium bovis* infection in livestock workers in Ibadan, Nigeria: evidence of occupational exposure. Int J Tuberc Lung Dis 16(10):1388–1392

Anon (1932) Bovine tuberculosis and human health. Am J Public Health 22(8):840–843

AU-IBAR (2013) Tuberculosis. http://www.au-ibar.org/tuberculosis. Accessed 28 July 2016
Ayele WY, Neill SD, Zinsstag J et al (2004) Bovine tuberculosis: an old disease but a new threat to Africa. Int J Tuberc Lung Dis 8:924–937
Brosh R, Gordon SV, Marmiesse VM et al (2002) A new evolutionary scenario for the *Mycobacterium tuberculosis* complex. Proc Natl Acad Sci U S A 99:3684–3689
Cadmus S, Palmer S, Okker M et al (2006) Molecular analysis of human and bovine tubercle bacilli from a local setting in Nigeria. J Clin Microbiol 44:29–34
Cosivi O, Grange JM, Daborn CJ et al (1998) Zoonotic tuberculosis due to *Mycobacterium bovis* in developing countries. Emerg Infect Dis 4:59–70
Daniel R, Evans H, Rolfe S et al (2009) Outbreak of tuberculosis caused by *Mycobacterium bovis* in golden Guernsey goats in Great Britain. Vet Rec 165:335–342
de Garine-Wichatitsky M, Caron A, Kock R et al (2013) A review of bovine tuberculosis at the wildlife-livestock-human interface in sub-Saharan Africa. Epidemiol Infect 141(7):1342–1356
de Lisle GW, Bengis RG, Schmitt SM et al (2002) Tuberculosis in free-ranging wildlife: detection, diagnosis and management. Rev Sci Tech 21(2):317–334
Egbe NF, Muwonge A, Ndip L et al (2016) Abattoir-based estimates of mycobacterial infections in Cameroon. Sci Rep 6:24320
El Idrissi A, Parker E (2012) Bovine tuberculosis at the animal-human-ecosystem interface. EMPRES Transb Anim Dis Bull 40:2–11
Etchechoury I, Valencia GE, Morcillo N et al (2010) Molecular typing of *Mycobacterium bovis* isolates in Argentina: first description of a person-to-person transmission case. Zoonoses Public Health 57:375–381
EU (2016) EU response to the Ebola outbreak in West Africa. European Commission-Fact Sheet. https://europa.eu/newsroom/highlights/special-coverage/ebola_en. Accessed 9 June 2016
Evans JT, Smith EG, Banerjee A et al (2007) Cluster of human tuberculosis caused by *Mycobacterium bovis*: evidence for person-to-person transmission in the UK. Lancet 369:1270–1276
Gibson AL, Hewinson G, Goodchild T et al (2004) Molecular epidemiology of disease due to *Mycobacterium bovis* in humans in the United Kingdom. J Clin Microbiol 42:431–434
Godreuil S, Jeziorski E, Banuls AL et al (2010) Intrafamilial cluster of pulmonary tuberculosis due to *Mycobacterium bovis* of the African 1 clonal complex. J Clin Microbiol 48:4680–4683
Gortázar C, Delahay RJ, McDonald RA et al (2012) The status of tuberculosis in European wild mammals. Mammal Rev 42:193–206
Green A (2006) Experts recognise zoonotic TB. Lancet Respir Med 4:433
Hiko A, Agga GE (2011) First-time detection of *Mycobacterium* species from goats in Ethiopia. Trop Anim Health Prod 43:133–139
Hlavsa MC, Moonan PK, Cowan LS et al (2008) Human tuberculosis due to *Mycobacterium bovis* in the United States, 1995–2005. Clin Infect Dis 47:168–175
Houlihan MG, Williams SJ, Poff JD (2008) *Mycobacterium bovis* isolated from a sheep during routine surveillance. Vet Rec 163:94–95
IUATLD (2015) International Union Against Tuberculosis and Lung Disease: world conference on lung health in Cape Town, South Africa, 2–6 December, 2015. http://edition.cnn.com/2015/12/23/health/tuberculosis-from-animals/index.html
Jenkins AO, Cadmus SI, Venter EH et al (2011) Molecular epidemiology of human and animal tuberculosis in Ibadan, Southwestern Nigeria. Vet Microbiol 151:139–147
Kaneene JB, Kaplan B, Steele JH et al (2014a) One Health approach for preventing and controlling tuberculosis in animals and humans. In: Thoen CO, Steele JH, Kaneene JB (eds) Zoonotic tuberculosis – *Mycobacterium bovis* and other pathogenic mycobacteria. Wiley-Blackwell, Ames, IA, pp 9–20
Kaneene JB, Miller RA, Kaplan B et al (2014b) Preventing and controlling zoonotic tuberculosis: a One Health approach. Vet Ital 50(1):7–22
Kriek N (2014) Tuberculosis in animals in South Africa. In: Thoen O, Steele JH, Kaneene JB (eds) Zoonotic tuberculosis: *Mycobacterium bovis* and other pathogenic mycobacteria, 3rd edn. Wiley-Blackwell, Ames, IA, pp 99–108
Kudi AC, Bello A, Ndukum JA (2012) Prevalence of bovine tuberculosis in camels in Northern Nigeria. J Camel Pract Res 19:81–86

LoBue PA, Betancourt W, Cowan L et al (2004) Identification of a familial cluster of pulmonary *Mycobacterium bovis* disease. Int J Tuberc Lung Dis 8:1142–1146

Majoor CJ, Magis-Escurra C, van Ingen J et al (2011) Epidemiology of *Mycobacterium bovis* disease in humans, The Netherlands, 1993–2007. Emerg Infect Dis 17:457–463

Mazet JAK, Clifford DL, Coppolillo PB et al (2009) A "One Health" approach to address emerging zoonoses: the HALI project in Tanzania. PLoS Med 6(12):e1000190

Mendoza MM, de Juan L, Menéndez S et al (2012) Tuberculosis due to *Mycobacterium bovis* and *Mycobacterium caprae* in sheep. Vet J 191:267–269

Michel AL, Bengis RG, Keet DF et al (2006) Wildlife tuberculosis in South African conservation areas: implications and challenges. Vet Microbiol 112(2–4):91–100

Mignard S, Pichat C, Carret G (2006) *Mycobacterium bovis* infection, Lyon, France. Emerg Infect Dis 12:1431–1433

Miller R, Kaneene JB, Schmitt SM et al (2007) Spatial analysis of *Mycobacterium bovis* infection in white-tailed deer (*Odocoileus virginianus*) in Michigan, USA. Prev Vet Med 82:111–122

Moiane I, Machado A, Santos N et al (2014) Prevalence of bovine tuberculosis and risk factor assessment in cattle in rural livestock areas of Govuro district in the Southeast of Mozambique. PLoS One 9(3):e91527

Monath TP, Kahn LH, Kaplan B (2010) Introduction: One Health perspective. ILAR J 51:193–198

Müller B, Dürr S, Alonso S et al (2013) Zoonotic *Mycobacterium bovis*-induced tuberculosis in humans. Emerg Infect Dis 19:899–908

Munyeme M, Muma JB, Skjerve E et al (2008) Risk factors associated with bovine tuberculosis in traditional cattle of the livestock/wildlife interface areas in the Kafue basin of Zambia. Prev Vet Med 85:317–328

Naima S, Borna M, Bakir M et al (2011) Tuberculosis in cattle and goats in the north of Algeria. Vet Res 4(4):100–103

Olea-Popelka F, Muwonge A, Perera A et al (2016) Zoonotic tuberculosis in human beings caused by *Mycobacterium bovis* – a call for action. Lancet Infect Dis 17(1):e21–e25

Oloya J, Opuda-Asibo J, Kazwala R et al (2008) Mycobacteria causing human cervical lymphadenitis in pastoral communities in the Karamoja region of Uganda. Epidemiol Infect 11 (136):636–643

Palmer MV, Thacker TC, Waters WR et al (2012) *Mycobacterium bovis*: a model pathogen at the interface of livestock, wildlife, and humans. Vet Med Int 2012:236205, 17. https://doi.org/10.1155/2012/236205

Perez-Guerrero L, Milian-Suazo F, Arriaga-Diaz C et al (2008) Molecular epidemiology of cattle and human tuberculosis in Mexico. Salud Publica Mex 50:286–291

Perez-Lago L, Navarro Y, Garciav de Viedma D (2014) Current knowledge and pending challenges in zoonosis caused by *Mycobacterium bovis*: a review. Res Vet Sci 97:S94–S100

Porphyre T, Stevenson MA, McKenzie J (2008) Risk factors for bovine tuberculosis in New Zealand cattle farms and their relationship with possum control strategies. Prev Vet Med 86(1–2):93–106

Radunz B (2006) Surveillance and risk management during the latter stages of eradication: experiences from Australia. Vet Microbiol 112:283–290

Renwick AR, White PC, Bengis RG (2007) Bovine tuberculosis in southern African wildlife: a multi-species host–pathogen system. Epidemiol Infect 135:529–540

Rodwell TC, Moore M, Moser KS et al (2008) Tuberculosis from *Mycobacterium bovis* in binational communities, United States. Emerg Infect Dis 14:909–916

Roger F (2012) Control of zoonotic diseases in Africa and Asia. The contribution of research to One Health. Perspect Policy Brief 18:4

Roth F, Zinsstag J, Orkhon D et al (2003) Human health benefits from livestock vaccination for brucellosis: case study. Bull World Health Organ 81:867–876

Siembieda JL, Kock RA, McCracken TA et al (2011) The role of wildlife in transboundary animal diseases. Anim Health Res Rev 12:95–111

Smith NH, Kremer K, Inwald J et al (2006) Ecotypes of the *Mycobacterium tuberculosis* complex. J Theor Biol 239(2):220–225

Stone MJ, Brown TJ, Drobniewski FA (2012) Human *Mycobacterium bovis* infections in London and Southeast England. J Clin Microbiol 50:164–168

Stop TB Partnership (2015) Global plan to end TB 2016–2020 – the paradigm shift. http://www.stoptb.org/assets/documents/global/plan/GlobalPlanToEndTB_TheParadigmShift_2016-2020_StopTBPartnership.pdf. Accessed 29 May 2016

Sunder S, Lanotte P, Godreuil S et al (2009) Human-to-human transmission of tuberculosis caused by *Mycobacterium bovis* in 11 immunocompetent patients. J Clin Microbiol 47:1249–1251

Thoen CO, LoBue PA (2007) *Mycobacterium bovis* tuberculosis: forgotten, but not gone. Lancet 369:1236–1238

Thoen CO, LoBue PA, Enarson DA et al (2009) Tuberculosis: a re-emerging disease in animals and humans. Vet Ital 45(1):135–181

Thoen CO, LoBue P, de Kantor I (2010) Why has zoonotic tuberculosis not received much attention? Int J Tuberc Lung Dis 14(9):1073–1074

Thoen CO, Kaplan B, Thoen TC et al (2016) Zoonotic tuberculosis: a comprehensive One Health approach. Medicina (B Aires) 76:159–165

Torres-Gonzalez P, Soberanis-Ramos O, Martinez-Gamboa A et al (2013) Prevalence of latent and active tuberculosis among dairy farm workers exposed to cattle infected by *Mycobacterium bovis*. PLoS Negl Trop Dis 7(4):e2177

UN (2015) Sustainable development goals. http://www.un.org/sustainabledevelopment/sustainable-development-goals/. Accessed 30 May 2016

UNDP/HDRO (2013) United Nations Development Programme, pp 144–147

WHO (2015a) WHO estimates of the global burden of foodborne diseases. Geneva: WHO, Foodborne diseases burden epidemiology reference group 2007–2015. http://www.who.int/foodsafety/publications/foodborne_disease/fergreport/en/. Accessed 24 Aug 2016

WHO (2015b) WHO end TB strategy. http://who.int/tb/post2015_TBstrategy.pdf?ua=1. Accessed 6 Nov 2015

WHO (2015c) End TB strategy (WHO/HTM/TB/2015.19). http://who.int/tb/post2015_TBstrategy.pdf?ua=1. Accessed 6 Nov 2015

WHO (2016a) World Health Organization global tuberculosis report. http://www.who.int/tb/; http://www.who.int/tb/publications/global_report/en/. Accessed 17 Oct 2016

WHO (2016b) Strategic and Technical Advisory Group for Tuberculosis (STAG-TB) Report of the 16th Meeting of the Strategic and Technical Advisory Group for Tuberculosis (STAG-TB), 13–15 June, 2016. WHO Headquarters, Geneva, pp 21–22. http://www.who.int/tb/advisory_bodies/stag_tb_report_2016.pdf?ua=1. Accessed 26 Jan 2017

WHO (2017) Roadmap for zoonotic tuberculosis. http://apps.who.int/iris/bitstream/10665/259229/1/9789241513043-eng.pdf. Accessed 20 Mar 2018

WHO-FAO (1994) Zoonotic tuberculosis (*Mycobacterium bovis*): memorandum from a WHO meeting (with the participation of FAO). Bull World Health Organ 72:851–857

Zinsstag J, Schelling E, Wyss K et al (2005) Potential of cooperation between human and animal health to strengthen health systems. Lancet 366:2142–2145

Zinsstag J, Dürr S, Penny MA et al (2009) Transmission dynamics and economics of rabies control in dogs and humans in an African city. Proc Natl Acad Sci U S A 106(35):14996–15001

Chapter 5
Tuberculosis in African Wildlife

Anita L. Michel and Paul D. van Helden

5.1 Introduction

Tuberculosis is an infectious, chronic, and usually fatal disease caused by members of the *Mycobacterium tuberculosis* complex (MTC) that are often able to cross species barriers between humans and domestic and wild animals. In wild animals, tuberculosis is primarily caused by spillover of *Mycobacterium bovis* infections from infected domestic cattle to free-ranging wildlife, while *M. tuberculosis* has long been known as, and still remains an important cause of death in captive wild animals in zoological collections worldwide, where close contact with humans facilitates its transmission to animals (Montali et al. 2001; de Lisle et al. 2001). *Mycobacterium tuberculosis* has been reported to cause generalized tuberculosis in a range of captive wildlife species. For more detailed information the reader is referred to recent literature focusing on tuberculosis in captive wildlife (Michel et al. 2013; Mikota et al. 2015; Miller and Lyashchenko 2015).

A growing number of MTC species, other than *M. bovis* and *M. tuberculosis*, able to cause lesions in infected animals has been identified in free-ranging wildlife in Southern and Western Africa (van Helden et al. 2009). The "Dassie bacillus" has been reported in rock hyraxes (*Procavia capensis*) as early as 1960, and appears to have established itself in this species (Parsons et al. 2008). A "member of the animal-adapted lineage of the MTC" was reported in free-ranging suricates (*Suricata*

A. L. Michel (✉)
Department Veterinary Tropical Diseases, Faculty of Veterinary Science, University of Pretoria, Onderstepoort, South Africa
e-mail: anita.michel@up.ac.za

P. D. van Helden
Division of Molecular Biology and Human Genetics, Faculty of Medicine and Health Sciences, South African MRC Centre for TB Research, DST NRF Centre of Excellence for Biomedical Tuberculosis Research, Stellenbosch University, Tygerberg, South Africa
e-mail: pvh@sun.ac.za

A. B. Dibaba et al. (eds.), *Tuberculosis in Animals: An African Perspective*,
https://doi.org/10.1007/978-3-030-18690-6_5

suricatta) in the Kalahari Desert of the Northern Cape (South Africa), was later characterized as a new species, and was named *M. suricattae* (Parsons et al. 2013). A rare and slow-growing member of the antelope clade of the MTC, the "oryx bacillus," was recently isolated from a free-ranging adult African buffalo (*Syncerus caffer*) in KwaZulu-Natal, South Africa (Gey van Pittius et al. 2012). Infection with *M. mungi* is considered to be an emerging disease, and it is the cause of a high death rate in free-ranging banded mongooses (*Mungos mungo*) living in close proximity to humans in northern Botswana (Alexander et al. 2010). In 2013, an MTC strain closely related to the human-associated lineage 6 (also known as *M. africanum* West Africa type 2) was isolated from a chimpanzee in Côte d'Ivoire (Coscolla et al. 2013). The gregarious social behavior common to all these species is the most likely reason for their ability to sustain tuberculosis (TB). The impact of these newly identified organisms on their hosts and their zoonotic potential is currently unknown. (See Chap. 6 for more detailed information about these mycobacterial species.)

In this chapter, we collate the current knowledge about bovine tuberculosis (BTB) caused by *M. bovis* in free-ranging wildlife species in Africa.

5.2 Bovine Tuberculosis in African Wildlife Species

Although BTB in wildlife in Africa has been diagnosed in South Africa as long ago as the mid-1920s, the extent of the infection, and the ability of some of these infected species to act as maintenance hosts of the disease, only transpired recently. The full extent of the infection in wildlife in Africa and the role that these infected species play in the epidemiology of the disease in cattle are unknown, but it is assumed that it is an increasingly expanding infection as is seen in wildlife in South Africa, where buffaloes, at least, appear not to have been infected before the mid-1950s in the Kruger National Park (KNP) from where BTB spread to other species in the Park and in the surrounding areas.

The slow, progressive nature of tuberculosis is characteristic of the development of the disease in individual animals, and of its relentless and progressive spread in free-ranging wild animal populations. The disease becomes established in an ecosystem only once a wildlife species with the ability to maintain the infection independently, and to transmit the infection to other susceptible animals within the specific ecosystem becomes infected. Many of the species that become infected are incidental (dead-end or spillover) hosts, and they are likely to play a very limited role in the epidemiology of BTB in these infected ecosystems.

In Africa, BTB both in cattle and in wildlife is underdiagnosed and underresearched, and its prevalence is largely unknown (de Garine-Wichatitsky et al. 2013; Ayele et al. 2004; FAO 2012). There is a substantial body of evidence suggesting that free-ranging wildlife species contracted BTB from cattle in many

Table 5.1 Notification and control status of BTB in countries in Africa that reported *M. bovis* in wildlife (2014)

Country	BTB status cattle	BTB status wildlife	Notifiable cattle	Control cattle	Notifiable wildlife	Surveillance wildlife
Cameroon	+	+	Yes	Yes	No	No
Ghana	+	+	Yes	Yes	Yes	No
Mauritius	+	+	No	Yes	No	No
Mozambique	+	+	Yes	N/A	No	No
Nigeria	+	+	Yes	S	No	No
South Africa	+	+	Yes	Yes	Yes	No
Sudan	+	+	Yes	N/A	Yes	Yes
Tanzania	+	+	Yes	S	Yes	Yes
Togo	+	+	Yes	N/A	No	No
Uganda	+	+	Yes	N/A	No	No
Zambia	+	+	Yes	No	No	No
Zimbabwe	Last reported 1996	+	Yes	S	Yes	No

+, BTB infection or disease present; *N/A* no information available, *S* surveillance

different countries including in Africa, and that this spillover infection often remains undetected for decades (Palmer et al. 2012). It is therefore not surprising that all 12 countries in Africa that reported *M. bovis* infection in wild animals between 2000 and 2014, also reported BTB in cattle, although not always concurrently (Table 5.1). The actual number of African countries in which BTB occurs both in cattle and wildlife is likely to be higher as no information is available for many of these countries (OIE 2017a). In addition, widespread intermingling of livestock and wildlife at the wildlife/livestock interface is common in large parts of Africa, and this is known to enhance the transmission of *M. bovis* especially during times when there is competition for limited pastures and water (Caron et al. 2013). A lack of diagnostic facilities to confirm the infection in remote wildlife areas is a further impediment to accurate reporting and the inadequacy of the available information.

Of the 54 countries in Africa, BTB in cattle was reported in 89% (34/38) of those reporting disease information to the World Organization for Animal Health (OIE) between 2000 and 2014. No information was available for 13 countries, while one country reported absence and three countries had their last outbreak before 2000. Only those that reported the disease listed BTB as a notifiable disease in cattle and only nine of them require notification of BTB in wildlife (Table 5.2). Only 11 African countries have national BTB disease control policies and/or culling strategies (OIE 2017b) but prevention and control are often poorly implemented due to political, economical, and sociological constraints (Abdalla and Nganwa 2014; Awah-Ndukum et al. 2012).

In many countries, there are no regulations requiring the reporting of BTB outbreaks in wildlife. This is a hazardous practice since once the disease has established itself in a wildlife host, disease notification, surveillance, and monitoring to

Table 5.2 African wildlife species (excluding maintenance hosts) infected with *Mycobacterium bovis*

Common name	Scientific name	Location	References
Common duiker	*Sylvicapra grimmia*	Agricultural farmland	Paine and Martinaglia (1929)
Lion	*Panthera leo*	GKNPC and other game parks, South Africa	Keet et al. (1996), Michel et al. (2006) and Hlokwe et al. (2011)
Cheetah	*Acinonyx jubatus*	GKNPC, South Africa	Keet et al. (1996)
Leopard	*Panthera pardus*	GKNPC, South Africa	De Vos et al. (2001) and Michel et al. (2006)
Lesser kudu	*Tragelaphus imberbis*	Northern Tanzania	Cleaveland et al. (2005)
Topi	*Damaliscus lunatus*	Northern Tanzania	Cleaveland et al. (2005)
Chacma baboon	*Papio ursinus*	GKNPC and other parks	Keet et al. (1996, 2000a)
Yellow baboon	*Papio cynocephalus*	Ruaha ecosystem, south-central Tanzania	Clifford et al. (2013)
Olive baboon	*Papio cynocephalus anubis*	Kenya	Tarara et al. (1985)
Kirk's dik-dik	*Madoqua kirkii*	Ruaha ecosystem, south-central Tanzania	Clifford et al. (2013)
Vervet monkey	*Chlorocebus pygerythrus*	Ruaha ecosystem, south-central Tanzania	Clifford et al. (2013)
Honey badger	*Mellivora capensis*	GKNPC	Michel (2002) and Michel et al. (2006)
Large spotted genet	*Genetta tigrina*	GKNPC Ruaha ecosystem, south-central Tanzania	De Vos et al. (2001) and Clifford et al. (2013)
African civet	*Civettictis civetta*	Serengeti ecosystem, Tanzania	Katale et al. (2017)
Warthog	*Phacochoerus africanus*	Uganda South Africa (multiple locations)	Woodford (1982a, b), Kalema-Zikusoka et al. (2005) and Michel et al. (2009)
Bush pig	*Potamochoerus larvatus*	Hhluhluwe-iMfolozi Park, South Africa	Michel et al. (2009)
Impala	*Aepyceros melampus*	GKNPC South Africa Ruaha ecosystem, south-central Tanzania	Michel et al. (2006) and Clifford et al. (2013)
Bushbuck	*Tragelaphus scriptus*	GKNPC, South Africa	Hlokwe et al. (2014)

(continued)

Table 5.2 (continued)

Common name	Scientific name	Location	References
Eland	*Taurotragus oryx*	South Africa	Michel et al. (2006)
Blue wildebeest	*Connochaetes taurinus*	GKNPC, South Africa Northern Tanzania	Hlokwe et al. (2014) and Clifford et al. (2013)
Banded mongoose	*Mungos mungo*	GKNPC, South Africa	Brüns et al. (2017)
Giraffe	*Giraffa camelopardalis*	GKNPC, South Africa	TM Hlokwe, unpublished data (2013)
African wild dog	*Lycaon pictus*	GKNPC, South Africa	A Michel, unpublished data (2016)
Nyala	*Tragelaphus angasii*	South Africa	Hlokwe et al. (2014)
Black rhinoceros	*Diceros bicornis*	GKNPC, South Africa	Miller et al. (2017)
White rhinoceros	*Ceratotherium simum*	GKNPC, South Africa	P Buss, personal communication (2016)

GKNPC Greater Kruger National Park Complex

determine spread of the disease in wildlife are needed to effectively manage the epidemic, and to protect livestock populations from spillback from the wildlife hosts. No information about the nature of the infection is available in most of the countries that reported BTB in wildlife to the OIE, as very few of the data have been published in scientific journals.

To date, BTB has been confirmed by culture in 29 free-ranging African wildlife species. Of those only African buffaloes (*Syncerus caffer*), greater kudus (*Tragelaphus strepsiceros*), and Kafue lechwe (*Kobus leche kafuensis*) are established maintenance hosts (Michel et al. 2006, 2015; Renwick et al. 2007; Munyeme et al. 2010; Clifford et al. 2013), but there are strong indications that warthogs (*Phacochoerus africanus*) and lions (*Panthera leo*) also have the potential to become maintenance hosts (Michel et al. 2015). The remainder of the species is considered to be spillover species (Table 5.2).

The characteristics of BTB in the maintenance hosts, and those that have the potential to become maintenance hosts are described in this chapter.

5.3 Maintenance Hosts of *Mycobacterium bovis*

*5.3.1 African Buffalo (*Syncerus caffer*)*

African buffaloes are probably the most important known wildlife maintenance host of BTB in Africa (Michel et al. 2006; Woodford 1982a, b). They are taxonomically closely related to domestic cattle, and exhibit the same gregarious social herd behavior (du Toit 2005). These characteristics are believed to create the ability for buffaloes to being infected with *M. bovis*, and to become a maintenance host.

Aerosol transmission of *M. bovis* because of the close contact between buffaloes within herds is the primary route of transmission of the infection in this species (Fitzgerald and Kaneene 2013). Herd sizes in the KNP vary substantially from as few as 50 to more than a 1000 buffaloes. These varying herd sizes are partly a function of seasonal fission and fusion events caused by environmental factors. These events ensure genetic exchange between herds, but they also facilitate the spread of BTB between herds (Cross et al. 2005). Normal dispersal behavior of single or small groups of buffalo heifers and bulls over considerable distances and fusion with new herds similarly cause gene flow and the spatial spread of BTB (de Vos et al. 2001; Michel et al. 2006; Caron 2014).

Once infected, the development of tubercles in buffaloes broadly follows the same stages of development and immunopathogenesis as has been described for domestic cattle. They appear to remain persistently infected, and are then, depending on the locality of the lesions, potential shedders of *M. bovis*.

In buffaloes, tuberculous lesions are more commonly found in the lymph nodes of the head and neck, tonsils, and in the lungs and associated lymph nodes. On many occasions, single, small lesions only are present in any one of the sites in which they usually occur. The appearance of the lesions in buffaloes varies substantially between animals, and when there is a florid reaction the lesions are poorly encapsulated and have a lardaceous appearance. Lesions often resemble abscesses caused by a number of pyogenic organisms, and these have often been dismissed as not tuberculous on the basis that the lesion is "merely an abscess." The lesions increase in size and number apparently as a function of the infectious dose and time. Generalization of infection following dissemination via the blood stream to multiple organ systems and serosal surfaces (miliary tuberculosis) is not uncommon and may occur in about 10% of diseased animals in a flock with a high BTB prevalence. The cumulative risk of exposure and infection over time results in a positively correlated association between age and the number of diseased animals (i.e., those with lesions; de Vos et al. 2001; Laisse et al. 2011). As the herd prevalence increases, the likelihood of recurrent exposure of individuals to invariably high doses of *M. bovis* increases, and more animals with advanced, generalized BTB can be expected to be present in the herd (de Vos et al. 2001). Additionally, with an increasing herd prevalence, BTB will be present more frequently in younger age groups, including calves. In very young calves with BTB, intrauterine transmission or transmission via milk is possible, but they appear to be rare events.

The long-term impact of BTB on the buffalo population is currently unknown as the disease is still spreading in a northerly direction following its entry in the south of the KNP. From the time of the estimated first infection of buffaloes in the south, it took about 50 years for the disease to reach the northern border of the Park, over a distance of about 500 km. Limited data about the effect of the disease on the buffalo population in the KNP reflect an increased vulnerability to drought and predation in herds with a high BTB prevalence (Jolles et al. 2005; Cross et al. 2009). These early indications could be a predictor of wider scale BTB-related ecological disruption that may be expected once the disease has developed to its full extent in that ecosystem (Caron et al. 2003). According to a recent mathematical model, it appears that BTB will have a more serious impact on cattle compared to buffaloes because of the difference in the transmission dynamics of the infection (Phepa et al. 2016).

5.3.2 *Greater Kudu (***Tragelaphus strepsiceros***)*

Greater kudus, one of the two subspecies of kudus, are large, gregarious antelopes that live in small groups in woodlands and bush lands where they browse on leaves of thorn trees of the genera *Acacia* and *Ziziphus*. Free-ranging greater kudus were among the first wild animals to be diagnosed with BTB in South Africa in the 1920s (Paine and Martinaglia 1929). Kudus are believed to contract the disease by different routes. One is the aerosol route in which case they develop multifocal tuberculous lesions in the lungs and associated lymph nodes with a distribution very similar to that seen in buffaloes and cattle. Generalization of the infection with hematogenous spread to organs, such as the spleen, liver, and kidneys, and to the serosal surfaces, has been observed. More frequently, and probably more importantly, *M. bovis* can be contracted percutaneously. In this instance, infection occurs at the site of lacerations in the oral and esophageal epithelium, and the thin skin of the inside of the ears by *M. bovis*-contaminated thorny twigs. Following infection of the parotid and retropharyngeal lymph nodes, typical pyogranulomas containing large numbers of mycobacteria develop. These lesions contain an abundant purulent exudate, and cause large irregular parotid and submandibular swellings. The pressure exerted by these lesions causes them often to rupture and to form draining sinuses from which bacilli-laden pus is shed into the environment including onto the browse.

The infection persists in localized kudu populations, at least for a considerable period of time (Michel et al. 2015), and for these reasons kudus are considered to be maintenance hosts, and they may become supershedders with the ability to effectively disseminate *M. bovis* within their home range and, perhaps more importantly, when they cross fences or are translocated to BTB-free ranches or conservation areas. Further support that they are maintenance hosts is provided by recent molecular epidemiological data that indicate that kudus maintain unique *M. bovis* strains in the absence of contact with buffaloes. In the KNP too, different populations of greater kudus are infected with either the common buffalo strain of *M. bovis* that occurs in the Park, or with a unique strain that only occurs in kudus in that region (Michel et al. 2009).

5.3.3 *Lion (*Panthera leo*)*

Bovine tuberculosis in free-ranging lions was first diagnosed in 1995 in the south of the KNP where, at that time, the prevalence of BTB in buffalo herds was the highest in the Park (Keet et al. 1996). Subsequent surveys showed that the prevalence of the disease in lions had the same south to north gradient as for the infection in buffalo herds, reflecting the spread of BTB in a northerly direction (Keet et al. 2000b; Rodwell et al. 2001). From 1993 to 2008, the prevalence of BTB in lions in the northern part of the KNP (low buffalo BTB prevalence zone) increased from 0 to 41% (Maas et al. 2012). In a study in 2017, the prevalence of BTB in the lion population was 33% in the central region, and 54% in the southern region where the infection appeared to have entered in the Park (Sylvester et al. 2017). *Mycobacterium bovis* infection in lions has also been reported in Tanzania where, compared to seronegative lions, serological data correlated with the presence of clinical signs and reduced survival times indicating the presence of the infection in lions, and its impact on *M. bovis*-infected prides (Cleaveland et al. 2005).

Lions are the only social felids, and they are the top predator in African ecosystems. Their status is listed as vulnerable by the International Union for Conservation of Nature (http://www.iucnredlist.org/apps/redlist/details/15951/0). Their sociality, communal cub rearing, communal hunting, their predilection to hunting buffaloes (a major prey species), and intraspecific aggression predispose them to becoming infected and to sustain the infection in prides once they become infected. Their preference for buffaloes as a prey species is greater during periods of drought when buffaloes are an easier prey due to their deteriorating body condition, weakness, and fission into smaller herds that make them more vulnerable to predation by lions (Mills 1995; Ferreira and Funston 2010; Tambling et al. 2013).

Tuberculosis is most frequently a subclinical but ultimately fatal disease in infected lions (Keet et al. 2000a), but there are currently insufficient data to calculate the mortality rate, the duration of the infection before death, or the average age at which they become infected, and how long after infection the onset of clinical signs is most likely to occur.

Lions may become infected by different routes, as reflected by the variable and inconsistent distribution of lesions in infected animals. During a kill and while feeding on infected buffalo carcasses, transmission of the infection may occur by the alimentary route, by aerogenous transmission of *M. bovis* while suffocating the prey, and during intraspecific aggressive behavior while feeding. Aggression while feeding also facilitates percutaneous transmission through bite wounds. The isolation of *M. bovis* from the mammary lymph nodes of some lionesses is not proof but an indication of a possible pseudo-vertical route of infection for suckling cubs. Similarly, the presence of a tuberculous endometritis in some females suggests that intrauterine transmission is also possible.

Dissemination of the infection followed by spread via the hematogenous and/or lymphatic route results in a generalized infection with the presence of miliary lesions mostly in the lungs but also in other organs (Keet et al. 2000a, b). Progression of an

M. bovis infection in lions, similar to other affected species, is generally slow, and body condition scoring during the early stages of the diseases is not useful to clinically detect lions with BTB (Maas et al. 2012). Exposure to consecutive high dose of *M. bovis* via different pathways may possibly contribute to a more rapid progression of the disease, with an earlier onset of the progressive emaciation and lethargy typically seen in advanced cases of the infection in lions. Bovine TB in lions is further characterized by the presence, in variable combinations in some animals, of swollen elbow joints caused by a tuberculous osteoarthritis, hygromas, lameness, corneal opacity, tuberculous panophthalmitis, a dull coat, and large, poorly healing skin lesions caused by the presence of subcutaneous tuberculous granulomas (Michel et al. 2015).

Visible lesions at necropsy in lions are limited to the lungs, skin, joints, and subcutaneous granulomas, and resemble those seen in domestic cats. The lesions are different from those seen in cattle and most other species. The lesions in the lungs are characterized by localized but poorly circumscribed areas of infection resembling an interstitial inflammatory reaction. Macroscopically the lesions in the lungs often contain cavities caused by bronchiectasis, and they contain a mucopurulent exudate containing large numbers of acid-fast bacilli. Histologically the lesions in the various tissues are characterized by a multifocal to confluent granulomatous inflammatory reaction seen as aggregates of predominantly macrophages lacking necrosis and the presence of Langhans giant cells typical of the tuberculous reaction in most other species (Keet et al. 1996). Tuberculous lesions in the lymph nodes of lions cannot be seen macroscopically, and the apparent absence of lesions there is thus not an indication of the absence of an *M. bovis* infection. When BTB is suspected in lions, it is mandatory to confirm the diagnosis by culture of a collection of specimens from the lungs, superficial and deep lymph nodes, the exudate from hygromas and affected joints, and from subcutaneous granulomatous lesions (Keet et al. 2000b).

The mounting evidence of the role of respiratory transmission of *M. bovis* infection in prides is currently the subject of an intense debate as to whether lions can be maintenance hosts under certain conditions. It is argued that once the infection has established itself in a critical mass of the population thus ensuring continuous pathogen circulation, that it could probably persist in lion prides in the absence of contact with buffaloes as the ongoing source of the infection. The likelihood that lions may become maintenance hosts is also dependent on their population size, and the long-term effects of BTB on the composition of the pride, and population dynamics.

Debilitating diseases, such as BTB, affect social species like lions in multiple ways. Apart from the direct effect on the health of individual animals, there is also an impact on the size of prides and their structure. Dominant male coalitions with BTB are probably evicted earlier because of the debilitating effects of the disease and the consequent increasing bouts of infanticide have an adverse effect on numbers and the composition of prides. The practice of communal hunting too provides support to lions with BTB allowing extended survival times and shedding of the pathogen over a longer period of time (Michel et al. 2006).

Lions are often translocated between conservation areas in Africa, and the risk of introducing lions subclinically infected with BTB into uninfected areas should always be considered.

5.3.4 *Warthog (***Phacochoerus africanus***)*

Bovine TB in warthogs was first diagnosed in the Queen Elizabeth National Park in Uganda by Woodford in 1982 who concluded that it was a spillover infection from African buffaloes (Woodford 1982a, b). It appears to have persisted both in buffaloes and warthogs in the Park since then, as BTB was still present in both populations when surveyed in 1997 (Kalema-Zikusoka et al. 2005). Warthogs with BTB were also found in several reserves and on private game farms in South Africa where they shared *M. bovis* strains carried by African buffaloes (Hlokwe et al. 2014). Lesions due to *M. bovis* are present both in the respiratory and gastrointestinal tracts. Warthogs are omnivores, and they opportunistically feed on carrion potentially containing *M. bovis* in addition to being exposed to environmental contamination with the pathogen. Once infected, close contact between members of family groups inside the confines of their burrows is thought to effectively facilitate aerosol transmission and the persistence of the infection within the family group. When warthog densities are low, BTB occurs as a sporadic spillover infection in them as their numbers are then too low to sustain the infection within the species. However, in the absence of predators, and when food is readily available and in abundance, the number of warthogs within an ecosystem can reach a threshold number at which they can sustain the infection within the population. Under these circumstances, the prevalence of the disease may vary from 5 to 30%, and they may then act as maintenance hosts (Michel et al. 2015).

5.3.5 *Kafue lechwe (***Kobus leche kafuensis***)*

Tuberculosis in Kafue lechwe was first diagnosed in 1946 (Munyeme et al. 2010). They are the dominant wildlife species in the Kafue Basin and live in very large herds. During the drier months of the year, cattle and lechwe intermingle on the dwindling pastures in the Basin, and congregate at watering points. The consistent high prevalence of BTB in lechwe (ranging from 20 to 36%) led to their classification as maintenance hosts of BTB (Gallagher et al. 1972; Clancey 1977). They have furthermore been identified as a risk factor for BTB in cattle (Munyeme et al. 2008), and they appear to be a source of spillback infection to cattle (Pandey 1998). The majority of tuberculous lesions is found in the lungs and associated lymph nodes, suggesting aerosol transmission. Generalization of the disease, following hematogenous spread, occurs in about 10% of diseased animals (Munyeme et al. 2010). The annual mortality rate ascribed to tuberculosis in a study conducted by Gallagher in

1972 was at least 20% (Gallagher et al. 1972), compared to the 10% in African buffaloes in the KNP (de Vos et al. 2001). (Refer to Chap. 23 for further information about tuberculosis in Kafue lechwe.)

5.4 Spillover Infection in African Wildlife Species

Transmission of BTB to spillover hosts is a "downstream" event in the epidemic curve of the disease in maintenance host(s), and its occurrence is dependent on a high prevalence of the disease in these hosts. In instances where an infected buffalo population has decreased markedly in size due to events such as drought or poaching, the prevalence of BTB in them, at best, was only slightly reduced or remained unchanged (de Vos et al. 2001; Kalema-Zikusoka et al. 2005), and they sustain the infection irrespective of the decrease in the number of animals.

Only a small number of the wildlife species diagnosed with BTB appears to play a role in the dissemination and maintenance of the disease. It is also difficult to predict whether infected spillover species may be responsible for transmission of the infection at the wildlife/livestock interface, but there is always a risk that they may transmit the disease to other species in one way or another.

To date, 26 spillover species have been reported (Table 5.1). It is likely that the disease may not have been reported in a number of species, while there may be those in which the disease has not been diagnosed. One can therefore expect that the number of African wildlife species infected with *M. bovis* will increase as time goes on.

Unless the conservation status of the infected spillover species is threatened or endangered, BTB is not likely to have a negative effect on their populations. Individual infected animals, however, will suffer the consequences of the infection.

5.5 Implications for Conservation and Trade

Infection of wildlife species with *M. bovis* has considerable implications for the national and international trade in livestock and wildlife because of the restrictions imposed by the international conventions governing trade with animals and their products from countries infected with *M. bovis*. These measures result in revenue losses for game farms and conservation areas particularly in Southern Africa where the different types of wildlife ranching are a profitable private enterprise. For the commercial wildlife industry that is rapidly evolving and expected to exceed the revenue generated from livestock farming in Southern Africa, quarantine regulations not only reduce the monetary value of breeding animals, but also prohibit the sale of live animals of the affected species (Munag'andu et al. 2006; de Garine-Wichatitsky et al. 2013; Michel et al. 2015; Hlokwe et al. 2016).

In some instances, BTB in ecosystems causes them to become conservation islands because of the limitations on the movement of BTB-infected wildlife from these areas, thus not only jeopardizing the conservation of endangered species but also preventing the free exchange of genetic resources between conservation areas.

5.6 Control of BTB in Free-Ranging Wildlife

While the long-term goal of eradicating BTB from domestic cattle has been achieved in many developed countries, it is unattainable in free-ranging wildlife populations, and very challenging in cattle, also in developed countries, in an environment where there is a coexistent wildlife maintenance host of the infection (Bengis et al. 2002; FAO 2012).

The choice of a suitable strategy for the control of BTB in wildlife depends on the primary conservation objectives for a particular ecosystem. African conservationists are faced with difficult choices and few options since they have an obligation to protect the species that host the pathogen, but they also have the responsibility of minimizing the risk of transmission of the disease to domestic cattle, other livestock, and humans at the interface. Additionally, they must also consider the potentially devastating impact on maintenance and spillover species, particularly when dealing with rare and endangered species.

Currently, *M. bovis* infection is not actively controlled in most affected African wildlife areas. In Southern Africa, fencing is used as a way of limiting the movement of BTB-infected animals (Jori et al. 2011), but flooding, elephant, and human activities disrupt them allowing BTB-infected animals to cross the interface at will. Game-deterrent fences are reasonably effective in restricting the movement of some animals, but a number of wildlife species including maintenance hosts of BTB, such as warthogs and kudus, cross intact fences at will either by jumping across them or by burrowing beneath them (Michel et al. 2006).

Control measures for BTB in free-ranging wildlife populations based on long-term interventions such as periodic mass capture followed by tuberculin testing and removal of positive animals can effectively reduce the prevalence of BTB and hence the infection pressure in the maintenance population in smaller conservation areas such as the Hluhluwe-iMfolozi Park (South Africa). Reduction in the prevalence rate reduces the risk of transmission to other wildlife species, and the possibility of spillback to cattle (Michel et al. 2015).

To effectively implement management strategies in wildlife populations, the BTB status of the infected population and neighboring domestic cattle should be monitored to gauge the impact of the control measures (Hlokwe et al. 2016).

5.7 Conclusion

In all likelihood vaccination remains the only control measure for BTB in wildlife populations. In the past decade, considerable progress has been made with the development of BTB infection models in various wild animal species including possums, badgers, African buffaloes, wild boars, and deer in which the efficacy of new vaccines can be tested. It remains to be seen whether a TB vaccine can be developed that will meet the requirements of efficacy, safety, affordability, and practicality to control BTB in African wildlife (Robinson et al. 2012; Buddle et al. 2013; Chambers et al. 2014; Gortazar et al. 2014; Palmer et al. 2014; Díez-Delgado et al. 2017).

Governments have an obligation to protect human and animal health at the interface of humans, domestic livestock, and wildlife. Given the significant economic impact of BTB on cattle farming and its zoonotic risk for human health, it is alarming that the situation in Africa over time has, at best, remained unchanged since the report by Cosivi et al. (1998) and despite experts calling for action to control the disease on the African continent (Olea-Popelka et al. 2017). This situation has now been exacerbated by the detection of BTB in a number of wildlife species, and their role as maintenance hosts of the disease.

African Governments must now take the expanding number of wildlife species infected with *M. bovis*, and the risk that they pose to the health and welfare of humans and livestock, into consideration if they want to act in the interest of their citizens and of conservation. Dealing with BTB, and attempting to control and eventually eradicate it, cannot be done without also taking the risks posed by the infection in wildlife, and the mostly unknown role that they play in the epidemiology of the disease in Africa, into consideration.

References

Abdalla E, Nganwa D (2014) Factors contributing to the transmission of bovine tuberculosis caused by *Mycobacterium bovis* and its control status in Sudan. In: Thoen CO, Steele JH, Kaneene JB (eds) Zoonotic tuberculosis: *Mycobacterium bovis* and other pathogenic mycobacteria, 3rd edn. Wiley, Hoboken, NJ, pp 159–174

Alexander KA, Laver PN, Michel AL et al (2010) Novel *Mycobacterium tuberculosis* complex pathogen, *M. mungi*. Emerg Infect Dis 16(8):1296–1299

Awah-Ndukum J, Kudi AC, Bah GS et al (2012) Bovine tuberculosis in cattle in the highlands of Cameroon: seroprevalence estimates and rates of tuberculin skin test reactors at modified cut-offs. Vet Med Int 2012:798502, p 13. https://doi.org/10.1155/2012/798502

Ayele WY, Neill SD, Zinsstag J et al (2004) Bovine tuberculosis: an old disease but a new threat to Africa. Int J Tuberc Lung Dis 8(8):924–937

Bengis RG, Kock RA, Fischer J (2002) Infectious animal diseases: the wildlife/livestock interface. Rev Sci Tech 21(1):53–65

Brüns AC, Tanner M, Williams MC et al (2017) Diagnosis and implications of *Mycobacterium bovis* infection in banded mongooses (*Mungos mungo*) in the Kruger National Park, South Africa. J Wildl Dis 53(1):19–29

Buddle BM, Parlane NA, Wedlock DN et al (2013) Overview of vaccination trials for control of tuberculosis in cattle, wildlife and humans. Transbound Emerg Dis 60(Suppl 1):136–146

Caron A (2014) Buffalo movements at the wildlife/livestock interface along the Limpopo River and consequences for pathogen transmission: preliminary results from the BUCATIN project 2010-2014, pp 12–15. RP-PCP and AHEAD GLTFCA Meeting, 12–15 May 2014, Dete, Hwange, Zimbabwe

Caron A, Cross PC, du Toit JT (2003) Ecological implications of bovine tuberculosis in African buffalo herds. Ecol Appl 13(5):1338–1345

Caron A, Miguel E, Gomo C et al (2013) Relationship between burden of infection in ungulate populations and wildlife/livestock interfaces. Epidemiol Infect 141(7):1522–1535

Chambers MA, Carter SP, Wilson GJ et al (2014) Vaccination against tuberculosis in badgers and cattle: an overview of the challenges, developments and current research priorities in Great Britain. Vet Rec 175(4):90–96

Clancey JK (1977) The incidence of tuberculosis in lechwe (marsh antelope). Tubercle 58 (3):151–156

Cleaveland S, Mlengeya R, Kazwala RR et al (2005) Tuberculosis in Tanzanian wildlife. J Wildl Dis 41:446–453

Clifford DL, Kazwala RR, Sadiki H et al (2013) Tuberculosis infection in wildlife from the Ruaha ecosystem Tanzania: implications for wildlife, domestic animals, and human health. Epidemiol Infect 141(7):1371–1381

Coscolla M, Lewin A, Metzger S et al (2013) Novel *Mycobacterium tuberculosis* complex isolated from a wild chimpanzee. Emerg Infect Dis 19(6):969–976

Cosivi O, Grange JM, Daborn CJ et al (1998) Zoonotic tuberculosis due to *Mycobacterium bo*vis in developing countries. Emerg Infect Dis 4(1):59–70

Cross PC, Lloyd-Smith JO, Getz WM (2005) Disentangling association patterns in fission-fusion societies using African buffalo as an example. Anim Behav 69:499–506

Cross PC, Heisey DM, Bowers JA et al (2009) Disease, predation and demography: assessing the impacts of bovine tuberculosis on African buffalo by monitoring at individual and population levels. J Appl Ecol 46(2):467–475

de Garine-Wichatitsky M, Caron A, Kock R et al (2013) A review of bovine tuberculosis at the wildlife-livestock-human interface in sub-Saharan Africa. Epidemiol Infect 141(7):1342–1356

de Lisle GW, Mackintosh CG, Bengis RG (2001) *Mycobacterium bovis* in free-living and captive wildlife, including farmed deer. Rev Sci Tech 20(1):86–111

de Vos V, Bengis RG, Kriek NP et al (2001) The epidemiology of tuberculosis in free-ranging African buffalo (*Syncerus caffer*) in the Kruger National Park, South Africa. Onderstepoort J Vet Res 68(2):119–130

Díez-Delgado I, Rodríguez O, Boadella M et al (2017) Parenteral vaccination with heat-inactivated *Mycobacterium bovis* reduces the prevalence of tuberculosis-compatible lesions in farmed wild boar. Transbound Emerg Dis 64(5):18–21

du Toit JG (2005) The African savanna buffalo, pp 78–105

FAO (2012) Tuberculosis. Animal Production and Health Division. http://www.fao.org/docrep/015/i2811e/i2811e.pdf

Ferreira SM, Funston PJ (2010) Estimating lion population variables: prey and disease effects in Kruger National Park, South Africa. Wildl Res 37(3):194–206

Fitzgerald SD, Kaneene JB (2013) Wildlife reservoirs of bovine tuberculosis worldwide: hosts, pathology, surveillance, and control. Vet Pathol 50(3):488–499

Gallagher J, Macadam I, Sayer J et al (1972) Pulmonary tuberculosis in free-living lechwe antelope in Zambia. Trop Anim Health Prod 4(4):204–213

Gey van Pittius NC, Perrett KD, Michel AL et al (2012) Infection of African buffalo (*Syncerus caffer*) by oryx bacillus, a rare member of the antelope clade of the *Mycobacterium tuberculosis* complex. J Wildl Dis 48(4):849–857

Gortazar C, Beltrán-Beck B, Garrido JM et al (2014) Oral re-vaccination of Eurasian wild boar with *Mycobacterium bovis* BCG yields a strong protective response against challenge with a field strain. BMC Vet Res 10(1):96

Hlokwe TM, Jenkins AO Streicher EM et al (2011) Molecular characterisation of *Mycobacterium bovis* isolated from African buffaloes (*Syncerus caffer*) in Hluhluwe-iMfolozi Park in KwaZulu-

Natal, South Africa. Onderstepoort J Vet Res 78(1):Art. #232, 6 pages. https://doi.org/10.4102/ojvr.v78i1.232

Hlokwe TM, van Helden P, Michel AL (2014) Evidence of increasing intra and inter-species transmission of *Mycobacterium bovis* in South Africa: are we losing the battle? Prev Vet Med 115(1–2):10–17

Hlokwe TM, de Klerk-Lorist LM, Michel AL (2016) Wildlife on the move: a hidden tuberculosis threat through introduction of untested species in an ecosystem. J Wildl Dis 52(4):837–843

Jolles AE, Cooper DV, Levin SA (2005) Hidden effects of chronic tuberculosis in African buffalo. Ecology 86(9):2358–2364

Jori F, Brahmbhatt D, Fosgate GT et al (2011) A questionnaire-based evaluation of the veterinary cordon fence separating wildlife and livestock along the boundary of the Kruger National Park, South Africa. Prev Vet Med 100(3–4):210–220

Kalema-Zikusoka G, Bengis RG, Michel AL et al (2005) A preliminary investigation of tuberculosis and other diseases in African buffalo (*Syncerus caffer*) in Queen Elizabeth National Park, Uganda. Onderstepoort J Vet Res 72(2):145–151

Katale BZ, Mbugi EV, Siame KK et al (2017) Isolation and potential for transmission of *Mycobacterium bovis* at human-livestock-wildlife interface of the Serengeti ecosystem, Northern Tanzania. Transbound Emerg Dis 64(3):815–825. https://doi.org/10.1111/tbed.12445

Keet DF, Kriek NPJ, Penrith ML et al (1996) Tuberculosis in buffaloes (*Syncerus caffer*) in the Kruger National Park: spread of the disease to other species. Onderstepoort J Vet Res 63 (3):239–244

Keet DF, Michel AL, Meltzer DGA (2000a) Tuberculosis in free-ranging lions (*Panthera leo*) in the Kruger National Park. In: Annual conference of the South African Veterinary Association, 26–28 September 2000, Pietermaritzburg, South Africa, pp 232

Keet DF, Kriek NPJ, Michel AL (2000b) Tuberculosis and its geographical distribution in free-ranging lions in the Kruger National Park. In: Third international conference on *Mycobacterium bovis*, 14–16 August 2000, Cambridge, UK

Laisse CJM, Gavier-Widén D, Ramis G et al (2011) Characterization of tuberculous lesions in naturally infected African buffalo (*Syncerus caffer*). J Vet Diag Invest 23(5):1022–1027

Maas M, Keet DF, Rutten VPMG et al (2012) Assessing the impact of feline immunodeficiency virus and bovine tuberculosis co-infection in African lions. Proc R Soc B Biol Sci 279 (1745):4206–4214

Michel AL (2002) Implications of tuberculosis in African wildlife and livestock. In: Gibbs EPJ, Bokma BH (eds) The domestic animal/wildlife interface. Issues for disease control, conservation, sustainable food production and emerging diseases, vol 969. Annals of the New York Academy of Sciences, New York, pp 251–255

Michel AL, Bengis RG, Keet DF et al (2006) Wildlife tuberculosis in South African conservation areas: implications and challenges. Vet Microbiol 112(2–4):91–100

Michel AL, Coetzee ML, Keet DF et al (2009) Molecular epidemiology of *Mycobacterium bovis* isolates from free-ranging wildlife in South African game reserves. Vet Microbiol 133 (4):335–343

Michel AL, Hlokwe TM, Espie IW et al (2013) *Mycobacterium tuberculosis* at the human/wildlife interface in a high TB burden country. Transbound Emerg Dis 60(s1):45–52. https://doi.org/10.1111/tbed.12099

Michel AL, de Klerk LM, Buss P et al (2015) Tuberculosis in South African wildlife: lions, African buffalo and other species. In: Mukundan H, Chambers M, Waters R et al (eds) Tuberculosis, leprosy and mycobacterial diseases of man and animals: the many hosts of mycobacteria. CABI, Oxfordshire, pp 365–385

Mikota SK, Lyashchenko KP, Lowenstine L et al (eds) (2015) Tuberculosis, leprosy and mycobacterial diseases of man and animals: the many hosts of mycobacteria. CABI, Wallingford, UK, pp 259–276

Miller MA, Lyashchenko KP (2015) Mycobacterial infections in other zoo animals. In: Mukundan H, Chambers M, Waters R et al (eds) Tuberculosis, leprosy and mycobacterial diseases of man and animals: the many hosts of mycobacteria. CABI, Wallingford, UK, pp 277–295

Miller MA, Buss PE, van Helden PD, Parsons SD (2017) Mycobacterium bovis in a free-ranging black rhinoceros, Kruger National Park, South Africa, 2016. Emerg Infect Dis 23(3):557–558. https://doi.org/10.3201/eid2303.161622

Mills MGL (1995) Notes on wild dog *Lycaon pictus* and lion *Panthera leo* population trends during a drought in the Kruger National Park. Koedoe 38(1):95–99

Montali RJ, Mikota SK, Cheng LI (2001) *Mycobacterium tuberculosis* in zoo and wildlife species. Rev Sci Tech Off Int Epiz 20(1):291–303

Munag'andu HM, Siamudaala VM, Nambota A et al (2006) Disease constraints for utilization of the African buffalo (*Syncerus caffer*) on game ranches in Zambia. Jpn J Vet Res 54(1):3–13

Munyeme M, Muma JB, Skjerve E et al (2008) Risk factors associated with bovine tuberculosis in traditional cattle of the livestock/wildlife interface areas in the Kafue basin of Zambia. Prev Vet Med 85(3–4):317–328

Munyeme M, Muma JB, Siamudaala VM et al (2010) Tuberculosis in Kafue lechwe antelopes (*Kobus leche kafuensis*) of the Kafue basin in Zambia. Prev Vet Med 95(3–4):305–308

OIE (2017a) http://www.oie.int/wahis_2/public/wahid.php/Diseaseinformation/statuslist

OIE (2017b) http://www.oie.int/wahis_2/public/wahid.php/Diseasecontrol/measures

Olea-Popelka F, Muwonge A, Perera A et al (2017) Zoonotic tuberculosis in human beings caused by *Mycobacterium bovis*—a call for action. Lancet Infect Dis 17(1):e21–e25

Paine R, Martinaglia G (1929) Tuberculosis in wild buck living under natural conditions. J Comp Pathol Ther 42:1–8

Palmer MV, Thacker TC, Waters WR (2012) *Mycobacterium bovis*: a model pathogen at the interface of livestock, wildlife, and humans. Vet Med Int 2012:236205

Palmer MV, Thacker TC, Waters WR et al (2014) Oral vaccination of white-tailed deer (*Odocoileus virginianus*) with *Mycobacterium bovis* Bacillus Calmette-Guerin (BCG). PLoS One 9(5): e97031

Pandey GS (1998) Studies of the infectious diseases of the Kafue lechwe (*Kobus leche kafuensis*) with particular reference to tuberculosis in Zambia, Dissertation, Azabu University, Tokyo

Parsons S, Smith SGD, Martins Q et al (2008) Pulmonary infection due to the dassie bacillus (*Mycobacterium tuberculosis* complex sp.) in a free-living dassie (rock hyrax—*Procavia capensis*) from South Africa. Tuberculosis 88(1):80–83

Parsons SD, Drewe JA, Gey van Pittius NC et al (2013) Novel cause of tuberculosis in meerkats, South Africa. Emerg Inf Dis 19(12):2004–2007

Phepa PB, Chirowe F, Govinder KS (2016) Modelling the role of multi-transmission routes in the epidemiology of bovine tuberculosis in cattle and buffalo populations. Math Biosci 277:47–58

Renwick AR, White PC, Bengis RG (2007) Bovine tuberculosis in southern African wildlife: a multi-species host-pathogen system. Epidemiol Infect 135(4):529–540

Robinson PA, Corner LAL, Courcier EA et al (2012) BCG vaccination against tuberculosis in European badgers (*Meles meles*): a review. Comp Immunol Microbiol Infect Dis 35(4):277–287

Rodwell TC, Kriek NP, Bengis RG et al (2001) Prevalence of bovine tuberculosis in African buffalo at Kruger National Park. J Wildl Dis 37(2):258–264

Sylvester TT, Martin LER, Buss P et al (2017) Prevalence and risk factors for *Mycobacterium bovis* infection in African lions (*Panthera leo*) in the Kruger National Park. J Wildl Dis 53 (2):372–376

Tambling CJ, Ferreira SM, Adendorff J et al (2013) Lessons from management interventions: consequences for lion-buffalo interactions. S Afr J Wildl Res 43(1):1–11

Tarara R, Suleman MA, Sapolsky R et al (1985) Tuberculosis in wild olive baboons, *Papio cynocephalus anubis* (Lesson), in Kenya. J Wildl Dis 21(2):137–140

Van Helden PD, Parsons SD, Gey van Pittius NC (2009) Emerging' mycobacteria in South Africa. J S Afr Vet Assoc 80(4):210–214

Woodford MH (1982a) Tuberculosis in wildlife in the Ruwenzori National Park Uganda (Part I). Trop Anim Health Prod 14(2):81–88

Woodford MH (1982b) Tuberculosis in wildlife in the Ruwenzori National Park, Uganda (Part II). Trop Anim Health Prod 14(3):155–160

Chapter 6
The *Mycobacterium tuberculosis* Complex in Africa

Sven D. C. Parsons, Michele A. Miller, and Paul D. van Helden

6.1 Introduction

The *Mycobacterium tuberculosis* complex (MCT) comprises a diverse collection of mycobacterial species that evolved from a common ancestral human pathogen of African origin (Wirth et al. 2008). While these organisms are over 99% genetically similar, their clonal evolution has, to a large extent, occurred through genomic deletions, and these genetic regions of difference (RDs) can be used to distinguish between lineages, strains, and species (Brosch et al. 2002; Huard et al. 2006; Warren et al. 2006). Notably, the lineage characterized by the deletion of RD9 is the only MTC clade to have given rise to strains that became permanently established in animal hosts (Huard et al. 2006; Smith et al. 2006). Many of these strains are strongly associated with specific host species and have been named after them. However, it is currently unclear if these pathogens have undergone evolutionary host adaptation or if their distinct genetic and phenotypic characteristics have arisen as a result of genetic drift following their ecological isolation within separate niches (Hershberg et al. 2008; Pepperell et al. 2013; Smith et al. 2006).

S. D. C. Parsons (✉) · M. A. Miller
DST-NRF Centre of Excellence for Biomedical Tuberculosis Research, Division of Molecular Biology and Human Genetics, Faculty of Medicine and Health Sciences, Stellenbosch University, Cape Town, South Africa

SAMRC Centre for TB Research, Division of Molecular Biology and Human Genetics, Faculty of Medicine and Health Sciences, Stellenbosch University, Cape Town, South Africa
e-mail: sparsons@sun.ac.za; miller@sun.ac.za

P. D. van Helden
Division of Molecular Biology and Human Genetics, Faculty of Medicine and Health Sciences, South African MRC Centre for TB Research, DST NRF Centre of Excellence for Biomedical Tuberculosis Research, Stellenbosch University, Tygerberg, South Africa
e-mail: pvh@sun.ac.za

A. B. Dibaba et al. (eds.), *Tuberculosis in Animals: An African Perspective*,
https://doi.org/10.1007/978-3-030-18690-6_6

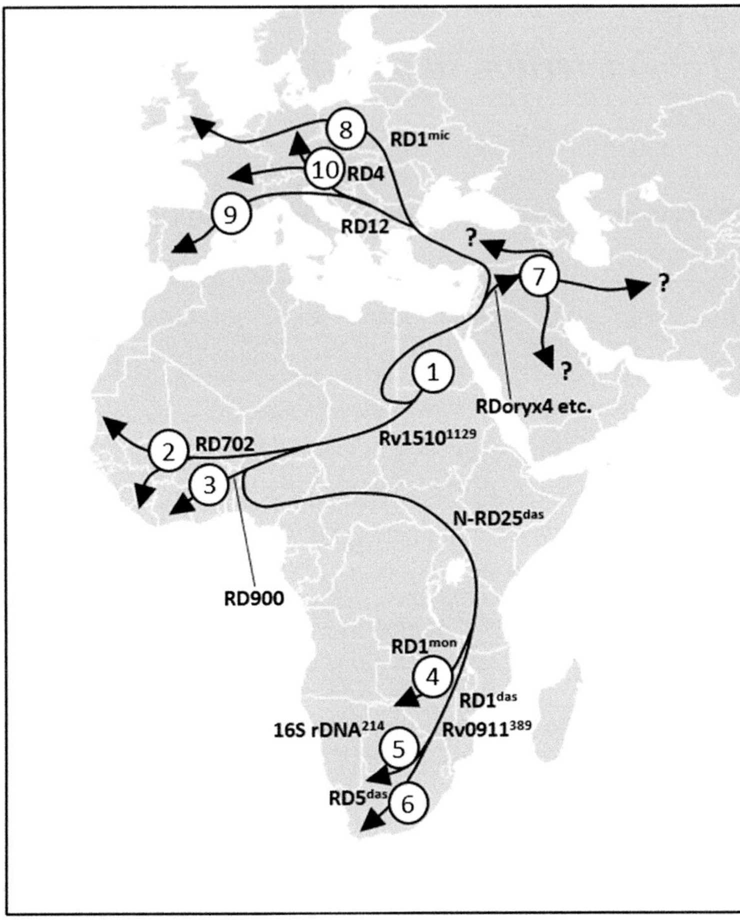

Fig. 6.1 Proposed geographic spread of the RD9-deleted clade of the *Mycobacterium tuberculosis* complex (MTC). Genetic regions of difference and single nucleotide polymorphisms, which characterize the molecular evolution of this lineage, are indicated. (1) Ancestral MTBC strain with deletions of RD7, RD8, RD9, and RD10, (2) *M. africanum*, (3) the chimpanzee bacillus, (4) *M. mungi*, (5) *M. suricattae*, (6) the dassie bacillus, (7) *M. orygis*, (8) *M. microti*, (9) *M. bovis*, (10) *M. caprae*

The RD9-deleted clade has given rise to two distinct sub-lineages (Fig. 6.1). The first includes species that are primarily found in Europe and Asia, and includes *M. orygis*, which has been isolated from antelopes, cattle, and humans (van Ingen et al. 2012); *M. caprae*, which principally infects goats (Aranaz 2003); and *M. microti*, which infects voles (*Microtus agrestis*) and other rodents, and llamas (*Lama glama*) (Cavanagh et al. 2002; Oevermann et al. 2004). The best-known member of this sub-lineage is *M. bovis* that is reviewed elsewhere in this book. The second sub-lineage of the RD9-deleted clade evolved in Africa and is the focus of this chapter.

6.2 Evolution of the MTC in Africa

The MTC is believed to have had its origins between 40,000 and 70,000 years ago as a pathogen of humans occupying parts of Northeast Africa (Comas et al. 2013; Wirth et al. 2008). Subsequently, the emergence of a number of distinct lineages of *M. tuberculosis* has resulted from the migration and isolation of human populations. Such migration, to West Africa, of people infected with an ancestral human-adapted pathogen appears to be the origin of the African RD9-deleted clade (Fig. 6.1). In modern times, this event is reflected by the occurrence in West Africa of a member of this lineage, *M. africanum*, which continues to be a significant cause of human tuberculosis (TB) in this region (Bentley et al. 2012; de Jong et al. 2010). The animal-associated members of this clade all share a common progenitor suggesting that the permanent establishment of this lineage in animal hosts has occurred on only a single occasion (Alexander et al. 2016a; Dippenaar et al. 2015; Coscolla et al. 2013). This event appears to have resulted in the ongoing transmission of this pathogen to novel host species, and thus far four distinct, animal-associated members of this clade have been described. These are the chimpanzee bacillus, the dassie bacillus, *M. mungi*, and *M. suricattae*.

6.2.1 Mycobacterium africanum

Mycobacterium africanum is the cause of up to 50% of all human TB cases in parts of West Africa but is uncommonly diagnosed elsewhere in the world despite human movement across the globe (Bentley et al. 2012; de Jong et al. 2010). For this reason, it has been postulated that a natural animal reservoir of this organism might exist in West Africa (Bentley et al. 2012; de Jong et al. 2010). However, to date, there is no evidence for this hypothesis, and *M. africanum* was not detected in an extensive survey of rodents and shrews in Benin (Durnez et al. 2010).

Moreover, despite widespread surveys of mycobacterial disease in livestock in West Africa, the isolation of *M. africanum* from domesticated animals has rarely been documented. In Ghana, the isolation of *M. africanum* from tissues of a skin-test-positive cow has been reported on a single occasion (Asante-Poku et al. 2014), while there are three reports of the isolation of this organism from tissues (Cadmus et al. 2006) and milk (Cadmus et al. 2010; Ofukwu et al. 2008) from cattle in Nigeria. It seems probable that these infections represent the transmission of this pathogen from human handlers that were in close contact with these animals (Asante-Poku et al. 2014). *Mycobacterium africanum* has also been isolated from three chimpanzees (*Pan troglodytes*) and a guenon (genus *Cercopithecus*) (Thorel 1980). However, as these animals were captive at the time of diagnosis, humans may again have been the source of these infections. Nonetheless, while *M. africanum* rarely appears to be associated with animal disease, such cases should probably be regarded as potential sources for reinfection of humans (Asante-Poku et al. 2014; Cadmus et al. 2010).

Mycobacterium africanum is a genetically diverse group of organisms comprising two major sub-lineages, i.e., *M. africanum* West African 1 (WAF1) and West African 2 (WAF2) (de Jong et al. 2010). The former sub-lineage is genetically characterized by the genomic deletion of RD711 but does not have deletions of RD7, RD8, or RD10 (de Jong et al. 2010). The WAF2 sub-lineage shares the deletions RD7, RD8, and RD10 with animal-associated members of the RD9-deleted clade and can be distinguished from these in harboring the unique deletion RD702 (Bentley et al. 2012). *Mycobacterium africanum* WAF2 displays a diversity of spoligotype patterns (de Jong et al. 2009) (Tables 6.1 and 6.2).

6.2.2 The Chimpanzee Bacillus

The chimpanzee bacillus has provisionally been named following its isolation, on a single occasion, from a chimpanzee with multi-organ TB from the Taï National Forest, Côte d'Ivoire (Coscolla et al. 2013). However, this infection may represent an incidental event in this species, as the organism was not detected in 28 other chimpanzee carcasses from the same geographic region (Coscolla et al. 2013). Nonetheless, this pathogen is genetically distinct from strains of *M. africanum* isolated from human TB patients, and it is more closely related to members of the RD9-deleted clade that have been isolated from animals (Dippenaar et al. 2015).

Table 6.1 Host species and geographic origin of selected African isolates of the MTC

Isolate	Host species	Geographic origin	Reference
M. orygis	African buffalo	South Africa	Gey van Pittius et al. (2012)
M. orygis	Oryx (unknown species)	South Africa	Mostowy et al. (2005)
M. orygis	Antelope (unknown species)	South Africa	van Ingen et al. (2012)
M. africanum SIT181	Human	The Gambia	de Jong et al. (2009)
M. africanum SIT326	Human	The Gambia	de Jong et al. (2009)
M. africanum	Cow	Nigeria	Cadmus et al. (2006)
M. africanum WAF1	Cow	Ghana	Asante-Poku et al. (2014)
Dassie bacillus	Rock hyrax	South Africa	Parsons et al. (2008)
Dassie bacillus 68/7171	Rock hyrax	South Africa	van Soolingen et al. (1998)
M. microti-like (dassie bacillus)	Rock hyrax	South Africa	Lutze-Wallace et al. (2006)
M. mungi	Banded mongoose	Botswana	Alexander et al. (2010)
M. suricattae	Meerkat	South Africa	Parsons et al. (2013)
Chimpanzee bacillus	Chimpanzee	Cote d'Ivoire	Coscolla et al. (2013)

Table 6.2 Spoligotype patterns of selected African isolates of the MTC

Isolate	Spoligotype pattern	Reference
M. orygis	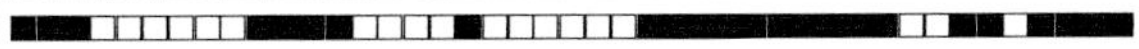	Gey van Pittius et al. (2012)
M. orygis	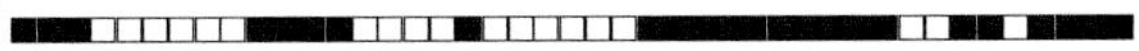	Mostowy et al. (2005)
M. orygis	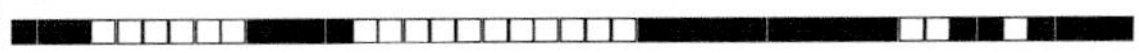	van Ingen et al. (2012)
M. africanum SIT181		de Jong et al. (2009)
M. africanum SIT326		de Jong et al. (2009)
M. africanum		Cadmus et al. (2006)
M. africanum WAF1		Asante-Poku et al. (2014)
Dassie bacillus	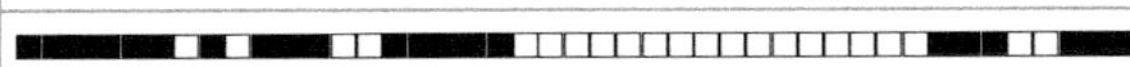	Parsons et al. (2008)
Dassie bacillus 68/7171	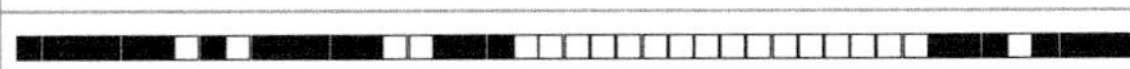	van Soolingen et al. (1998)
M. microti-like (dassie bacillus)	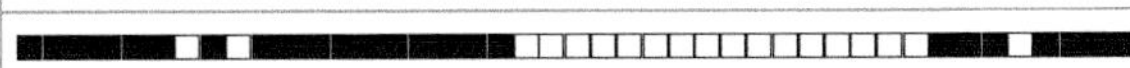	Lutze-Wallace et al. (2006)
M. mungi	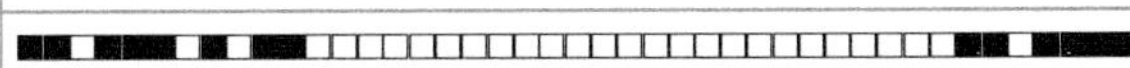	Alexander et al. (2010)
M. suricattae	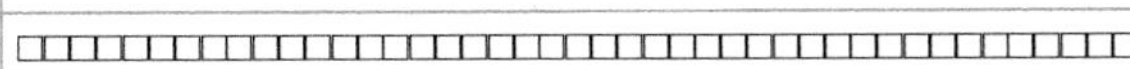	Parsons et al. (2013)
Chimpanzee bacillus	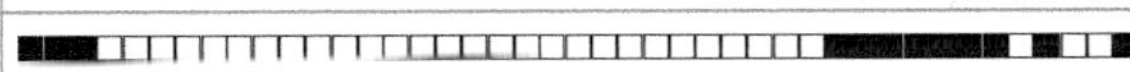	Coscolla et al. (2013)

This suggests that this species might have an as-yet unidentified preferred animal host. Chimpanzees are known to hunt and feed on a variety of animal species, and in the case described by Coscolla et al. (2013), transmission of the pathogen may have resulted from the ingestion of infected prey. Unlike other animal-adapted members of the African RD9-deleted clade, the RD1 locus has not been deleted in the chimpanzee bacillus, and this strain may represent an ancient lineage of the animal-adapted species. Its occurrence in Côte d'Ivoire provides further support for the assumed West African origin of the animal-associated lineage.

While the whole genome sequence of the chimpanzee bacillus is publicly available, only a limited number of defined genetic markers have been described for this organism. These include deletions of RD7, RD8, RD9, RD10, and RD900. Moreover, the organism shares a single nucleotide polymorphism (SNP) in the gene Rv1510 with other members of the African RD9-deleted clade (Coscolla et al. 2013).

6.2.3 The Dassie Bacillus

The dassie bacillus was first isolated in 1954 from a rock hyrax (*Procavia capensis*) in South Africa (Wagner et al. 1958). This hyrax is locally referred to as a *dassie*, hence the name given to this MTC member that has been isolated from rock hyraxes on subsequent occasions (Cousins et al. 1994; Parsons et al. 2008). The distribution of rock hyraxes includes much of sub-Saharan Africa and a number of North African countries, but the only confirmed isolation of this pathogen in Africa has been from animals that have originated from South Africa.

In rock hyraxes, infection with the dassie bacillus results in lesions typical of TB in many other species, and the respiratory tract appears to be the primary site of infection and disease (Parsons et al. 2008; Wagner et al. 1958). However, in severe cases, granulomas can occur in multiple organs (Cousins et al. 1994). In contrast, experimental infection of various laboratory animals with the dassie bacillus typically results in limited lesions suggesting that the organism is attenuated in these species (Cousins et al. 1994; Smith 1965). In part, this attenuated phenotype has been attributed to the deletion of $RD1^{das}$, which includes the genes Rv3874 and Rv3875 that encode the proteins culture filtrate protein 10 kDa (CFP-10) and early secretory antigenic target 6 kDa (ESAT-6), respectively (Mostowy et al. 2004).

There is only a single report of a natural infection with the dassie bacillus in a species other than the rock hyrax. This occurred in a suricate, or meerkat (*Suricata suricatta*), which was captured in South Africa and maintained in captivity with possible exposure to infected hyraxes (Mostowy et al. 2004; Parsons et al. 2008). Transmission of the pathogen to novel hosts is therefore possible, but it is currently unclear if the organism is confined to hyrax populations because of its attenuated phenotype or because of ecological isolation. While frank granulomatous disease has been detected in free-living hyraxes, there is currently no sound evidence that the pathogen might affect the population ecology of this species. Nonetheless, rock hyraxes can display significant fluctuations in population size (Barry et al. 2015), and TB in this species could conceivably contribute to this phenomenon.

The dassie bacillus can be genetically distinguished from other MTC members as having genomic deletions of $RD1^{das}$, $RD5^{das}$, $RDVirS^{das}$, and $N\text{-}RD25^{das}$ (Mostowy et al. 2004). Furthermore it displays a characteristic SNP in Rv1510 ($Rv1510^{1129}$), a single nucleotide deletion in Rv0911 ($Rv0911^{389}$) (Huard et al. 2006), and typical spoligotype patterns (Parsons et al. 2008; van Soolingen et al. 1994) (Tables 6.1 and 6.2).

6.2.4 Mycobacterium mungi

Tuberculosis in free-living banded mongooses (*Mungos mungo*) was first diagnosed in 1999 during an epidemic affecting populations along the Chobe River in northern Botswana (Alexander et al. 2002). Following the investigation of ongoing outbreaks of TB in mongooses between 2000 and 2010 in this region, the causative pathogen was described as a novel member of the MTC and named *Mycobacterium mungi* (Alexander et al. 2010). Surprisingly, despite the organism displaying a deletion within the RD1 region, it is highly pathogenic in mongooses, and causes severe morbidity and mortality in infected populations (Alexander et al. 2002, 2010). Uniquely, transmission of this infection appears to be via injuries and abrasions of the skin and the nasal planum (Alexander et al. 2010, 2015, 2016b; Flint et al. 2016). Clinical disease may first present as anorexia, cachexia, and distortion and erosion of the nasal planum (Alexander et al. 2002, 2010). Affected animals may show a lack of fear of humans (Alexander et al. 2002), and advanced stages of the disease may be associated with increased fecal glucocorticoid metabolite levels (Laver et al. 2012). Notably, once clinical signs are observed, progression to death usually occurs within 2–3 months (Alexander et al. 2010). At necropsy, granulomatous lesions may be found in all organs, but they are consistently present in the liver, spleen, lymph nodes, and lungs (Alexander et al. 2002).

It is unclear if the severity of the disease in mongooses is primarily a function of the pathogenicity of the organism or the susceptibility of the host. Whichever the case, TB in mongooses can lead to the collapse of infected populations (Alexander et al. 2010). There are no reports of *M. mungi* infection in species other than banded mongooses.

Mycobacterium mungi shares the deletion N-RD25^{das} with the dassie bacillus (Alexander et al. 2010). However, these organisms can be genetically distinguished from one another as the *M. mungi* genome contains a unique RD1 deletion (RD1^{mon}), which has to date been poorly characterized (Alexander et al. 2010). It also does not contain the SNP Rv0911^{389} and displays a characteristic spoligotype pattern (Alexander et al. 2010) (Tables 6.1 and 6.2).

6.2.5 Mycobacterium suricattae

Since the 1990s, TB has been known to occur in a population of well-studied free-living meerkats in the Kalahari Desert, South Africa (Alexander et al. 2002). Initially believed to have been caused by either *M. tuberculosis* or *M. bovis* (Alexander et al. 2002; Drewe et al. 2009a), the causative organism has since been genotyped as a distinct member of the MTC and named *M. suricattae* (Parsons et al. 2013). Transmission of the infection is primarily via the respiratory route (Drewe et al. 2009a); however, grooming and aggression are also associated with increased spread of TB in this species (Drewe 2010; Drewe et al. 2011). The infection in meerkats results in severe, generalized disease, but it is commonly first detected by the presence of distinctly enlarged submandibular lymph nodes (Alexander et al. 2002; Drewe et al. 2009a). Following the onset of lymph node involvement, progression of the disease is rapid, and clinical signs such as cachexia and lethargy precede death (Alexander et al. 2002). Consequently, the disease can have a devastating impact on meerkat colonies and can cause local extinctions of affected populations (Alexander et al. 2002). On postmortal examination, granulomas can be found in all organs, but they are especially prominent in the spleen, lungs, and lymph nodes of the head and neck (Drewe et al. 2009a). Immunodiagnostic tests of *M. suricattae* infection have been described for meerkats and include both serological assays (Drewe et al. 2009b) and an assay of cell-mediated immunity (Clarke et al. 2017).

Mycobacterium suricattae shares numerous genetic markers with the dassie bacillus, including the genomic deletions $RD1^{das}$, $N\text{-}RD25^{das}$, $RDVirS^{das}$, and the SNP $Rv1510^{1129}$ and SND $Rv0911^{389}$ (Parsons et al. 2013). However, because of a major deletion of the genomic region analyzed by spoligotyping, *M. suricattae* is exceptional in having no spoligotype pattern (Parsons et al. 2013; Dippenaar et al. 2015). Similarly, the 16S ribosomal DNA sequence of this pathogen differs from all other MTC members (Parsons et al. 2013).

6.2.6 Mycobacterium orygis

Mycobacterium orygis has been isolated from various captive antelopes in the Netherlands and an oryx in Saudi Arabia (van Soolingen et al. 1994), dromedaries (*Camelus dromedarius*) in the United Arab Emirates (Wernery et al. 2007), cattle and a rhesus macaque (*Macaca mulatta*) in Bangladesh (Rahim et al. 2007; van Ingen et al. 2012), and at least ten human patients originating from the South Asian subcontinent (Gey van Pittius et al. 2012; van Ingen et al. 2012). The pathogen is therefore similar to *M. bovis* in its ability to cause clinical TB in a wide variety of species. Isolation of this organism in Africa has been reported on three occasions, all from animals in South Africa. In two instances, the source of the isolate was described as a captive oryx (Mostowy et al. 2005) and an antelope from a zoo (van Ingen et al. 2012); however, it is unclear if these were indigenous animals (i.e., the gemsbok, *Oryx gazella*) or an imported species (e.g., an Arabian oryx, *Oryx leucoryx*). In the third case, isolation was from a semi-free-ranging African buffalo (*Syncerus caffer*) (Gey van Pittius et al. 2012). However, although this animal was born in South Africa, it originated from a herd that had been partly established with buffaloes imported from a Portuguese zoo (Gey van Pittius et al. 2012). As such, this finding may reflect the importation of this organism rather than its natural occurrence in Africa.

Given the small number of TB cases caused by *M. orygis*, and the diversity of species that it infects, it is currently unclear whether this organism should be regarded as primarily a pathogen of animals or of humans. Moreover, the geographical distribution of *M. orygis* may indicate the importance of global animal and human movement in its epidemiology. The organism shares the deletions RD7, RD8, and RD10 with the African RD9-deleted species; however, it is unique in having the deletions RD12oryx, RDoryx_wag22, RDoryx_1, and RDoryx4 (Mostowy et al. 2005) and a characteristic set of spoligotype patterns (Gey van Pittius et al. 2012; Rahim et al. 2007; van Ingen et al. 2012) (Tables 6.1 and 6.2).

6.3 Virulence of the MTC and Risk of Infection to Humans and Animals

The two distinct lineages of the RD9-deleted clade of the MTC appear to differ significantly in their virulence and the risk that they pose to humans and their domestic animals. *Mycobacterium bovis*, the dominant member of the Eurasian lineage, has become established in animal populations throughout the world and causes severe disease in numerous species. Other members of this lineage are also the cause of TB in multiple hosts, and *M. pinnipedii* and *M. orygis* have been documented to cause TB in a number of species, including humans and cattle (Cousins 2003; Loeffler et al. 2014; van Ingen et al. 2012). Moreover, *M. bovis* and *M. pinnipedii* are highly virulent in experimental animal TB models (Cousins 2003; Dannenberg Jr and Collins 2001).

In contrast, members of the African RD9-deleted lineage appear to display a less virulent phenotype, and the risk of transmission between host species may be lower. In humans, infection with *M. africanum* WAF2 is less likely to result in active TB than infection with *M. tuberculosis* (de Jong et al. 2008). Also, certain WAF2 strains show significant attenuation in laboratory animal models (de Jong et al. 2010), and despite the close association between humans and livestock, *M. africanum* is rarely isolated from domestic animals.

Similarly, the dassie bacillus, *M. mungi*, and *M. suricattae* are highly associated with specific host species. The genes encoding ESAT-6 and CFP-10 have been deleted from these pathogens, and such a loss has been associated with the attenuation of virulence of the vaccine strain *M. bovis* BCG and of *M. microti* (Mostowy et al. 2004). It is notable that the only known hosts of two members of this clade are mongoose species (the banded mongoose and the meerkat) and that the only known transmission of the dassie bacillus to a novel host has been to a meerkat (Mostowy et al. 2004). This phenomenon might merely reflect an epidemiological link between these hosts, but it might equally be indicative of a limited host range of these pathogens, possibly dependent on immunological susceptibility to mycobacterial disease (Clarke et al. 2016). While these organisms have not, to date, been associated with the infection of humans or livestock (Alexander et al. 2015), they are able to induce fatal disease in susceptible hosts, and the movement of potentially infected animals should be undertaken with care.

Since infection by members of the MTC typically presents as an insidious chronic disease, the risk for introduction into new populations is significant, for example, the risk presented by *M. pinnipedii*-infected sea lions (*Neophoca cinerea*) in zoos as a source of zoonotic disease (Cousins 2003). Similarly, the detection of the dassie bacillus in hyraxes in zoos outside of South Africa demonstrates the potential for introduction of these organisms to novel locations (Cousins et al. 1994; Mostowy et al. 2004). Zoos and other private facilities that import exotic wildlife may unknowingly import novel diseases that are not detected during the typical quarantine period. Since the susceptibility of native wildlife in these locations is undetermined, there is a potential threat of possible spillover to these species and to other zoo animals.

6.4 Culture of the African Animal-Associated Members of the MTC

The optimal in vitro growth conditions for most animal-associated members of the African lineage of the MTC are poorly defined. The chimpanzee bacillus has been grown on Lowenstein-Jensen (LJ) media and on Middlebrook 7H9 and 7H11 media supplemented with Oleic Albumin Dextrose Catalase (OADC) (Coscolla et al. 2013). The growth rate of this organism appears to be similar to that of *M. africanum* in it being somewhat slower than the growth of *M. tuberculosis* (Coscolla et al. 2013; Gehre et al. 2013).

The other members of this lineage display very slow in vitro growth. When first isolated in 1954, samples of the dassie bacillus were incubated for 4 months on Dorset's egg medium before growth was first detected (Wagner et al. 1958; Wagner and Bokkenheuser 1961). The difficulty in growing this organism was confirmed by Cousins et al. (1994) who reported that it grew poorly on egg and agar-based media but that growth was stimulated by pyruvate. Culture of the dassie bacillus from diseased tissues has also been achieved using the mycobacterial growth indicator tube (MGIT) system (Becton Dickinson, New York, NY, USA) (Parsons et al. 2008). However, several months of culture may be required to detect growth.

The MTC isolate originally recovered from banded mongooses, and subsequently named *M. mungi*, was grown on LJ slants with and without pyruvate (Alexander et al. 2002). Only a few acid-fast colonies of this organism were present after 6 weeks of incubation on this medium. Similarly, the culture of tissue homogenates from 52 meerkats with suspected TB for 10 weeks on LJ slants supplemented with either pyruvate or glycerol resulted in positive MTC growth in only 42% of cases (Drewe et al. 2009a,b). Cultures of *M. suricattae* have been established from tissue homogenates using the MGIT system; however, subsequent subculture on Middlebrook 7H10 agar for 8 weeks resulted in growth of only one of four of these isolates (Parsons et al. 2013). These results highlight the need for extended culture times of 2–3 months, or longer, for isolation of these species.

6.5 Conclusion

Advances in the genetic characterization of mycobacteria have provided ever-increasing insight into the evolution of the distinct lineages and species that comprise the MTC (Brosch et al. 2002; Comas et al. 2013; Huard et al. 2006). This has allowed for greater precision in defining the etiology of unusual cases of TB and has resulted in the recent naming of two novel members of the African RD9-deleted clade (Alexander et al. 2010; Parsons et al. 2013).

The precise genotyping of mycobacterial pathogens is particularly important for understanding the epidemiology of TB in humans and animals and for evaluating the risks associated with this disease. As such, it should be noted that a number of genetic markers that have historically been used to define lineages of the MTC may be ambiguous and may have resulted in the misclassification of species (Lutze-Wallace et al. 2006; Rahim et al. 2007). Knowledge of the diversity of the MTC in Africa and of the genetic characterization of these organisms will allow for greater insight into the consequences of TB in free-living and domestic animals.

References

Alexander KA, Pleydell E, Williams MC et al (2002) *Mycobacterium tuberculosis*: an emerging disease of free-ranging wildlife. Emerg Infect Dis 8:598–601. https://doi.org/10.3201/eid0806.010358

Alexander KA, Laver PN, Michel AL et al (2010) Novel *Mycobacterium tuberculosis* complex pathogen, *M. mungi*. Emerg Infect Dis 16:1296–1299. https://doi.org/10.3201/eid1608.100314

Alexander KA, Sanderson CE, Laver PN (2015) Novel *Mycobacterium tuberculosis* complex spp. in group-living African mammals. In: Tuberculosis, leprosy and mycobacterial diseases of man and animals: The many hosts of mycobacteria. CABI, pp 386–401

Alexander KA, Larsen MH, Robbe-Austerman S et al (2016a) Draft genome sequence of the *Mycobacterium tuberculosis* complex pathogen *M. mungi*, identified in a banded mongoose (*Mungos mungo*) in northern Botswana. Genome Announc 4:e00471–e00416. https://doi.org/10.1128/genomeA.00471-16

Alexander KA, Sanderson CE, Larsen MH et al (2016b) Emerging tuberculosis pathogen hijacks social communication behavior in the group-living banded mongoose (*Mungos mungo*). mBio 7:e00281-16. https://doi.org/10.1128/mBio.00281-16

Aranaz A (2003) Elevation of *Mycobacterium tuberculosis* subsp. *caprae* Aranaz et al. 1999 to species rank as *Mycobacterium caprae* comb. nov., sp. nov. Int J Syst Evol Microbiol 53:1785–1789. https://doi.org/10.1099/ijs.0.02532-0

Asante-Poku A, Aning KG, Boi-Kikimoto B et al (2014) Prevalence of bovine tuberculosis in a dairy cattle farm and a research farm in Ghana. Onderstepoort J Vet Res 81:E1–E6

Barry RE, Chiweshe N, Mundy PJ (2015) Fluctuations in bush and rock hyrax (Hyracoidea: Procaviidae) abundances over a 13-year period in the Matopos, Zimbabwe. Afr J Wildl Res 45:17–27. https://doi.org/10.3957/056.045.0102

Bentley SD, Comas I, Bryant JM et al (2012) The Genome of *Mycobacterium africanum* West African 2 reveals a lineage-specific locus and genome erosion common to the *M. tuberculosis* complex. PLoS Negl Trop Dis 6:e1552. https://doi.org/10.1371/journal.pntd.0001552

Brosch R, Gordon SV, Marmiesse M et al (2002) A new evolutionary scenario for the *Mycobacterium tuberculosis* complex. Proc Natl Acad Sci USA 99:3684–3689

Cadmus S, Palmer S, Okker M et al (2006) Molecular analysis of human and bovine tubercle bacilli from a local setting in Nigeria. J Clin Microbiol 44:29–34. https://doi.org/10.1128/JCM.44.1.29-34.2006

Cadmus SIB, Yakubu MK, Magaji AA et al (2010) *Mycobacterium bovis*, but also *M. africanum* present in raw milk of pastoral cattle in north-central Nigeria. Trop Anim Health Prod 42:1047–1048. https://doi.org/10.1007/s11250-010-9533-2

Cavanagh R, Begon M, Bennett M et al (2002) *Mycobacterium microti* infection (vole tuberculosis) in wild rodent populations. J Clin Microbiol 40:3281–3285. https://doi.org/10.1128/JCM.40.9.3281-3285.2002

Clarke C, van Helden PD, Miller MA et al (2016) Animal-adapted members of the *Mycobacterium tuberculosis* complex endemic to the Southern African sub region. J SA Vet Assoc 87(1):a1322. https://doi.org/10.4102/jsava.v87i1.1322

Clarke C, Patterson SJ, Drewe JA et al (2017) Development and evaluation of a diagnostic cytokine-release assay for *Mycobacterium suricattae* infection in meerkats (*Suricata suricatta*). BMC Vet Res 13:2. https://doi.org/10.1186/s12917-016-0927-x

Comas I, Coscolla M, Luo T et al (2013) Out-of-Africa migration and neolithic coexpansion of *Mycobacterium tuberculosis* with modern humans. Nat Genet 45:1176–1182. https://doi.org/10.1038/ng.2744

Coscolla M, Lewin A, Metzger S et al (2013) Novel *Mycobacterium tuberculosis* complex isolate from a wild chimpanzee. Emerg Infect Dis 19:969–976. https://doi.org/10.3201/eid1906.121012

Cousins DV (2003) Tuberculosis in seals caused by a novel member of the *Mycobacterium tuberculosis* complex: *Mycobacterium pinnipedii* sp. nov. Int J Syst Evol Microbiol 53:1305–1314. https://doi.org/10.1099/ijs.0.02401-0

Cousins DV, Peet RL, Gaynor WT et al (1994) Tuberculosis in imported hyrax (*Procavia capensis*) caused by an unusual variant belonging to the *Mycobacterium tuberculosis* complex. Vet Microbiol 42:135–145. https://doi.org/10.1016/0378-1135(94)90013-2

Dannenberg AM Jr, Collins FM (2001) Progressive pulmonary tuberculosis is not due to increasing numbers of viable bacilli in rabbits, mice and guinea pigs, but is due to a continuous host response to mycobacterial products. Tuberculosis 81:229–242. https://doi.org/10.1054/tube.2001.0287

de Jong BC, Hill PC, Aiken A et al (2008) Progression to active tuberculosis, but not transmission, varies by *M. tuberculosis* lineage in The Gambia. J Infect Dis 198:1037–1043. https://doi.org/10.1086/591504

de Jong BC, Antonio M, Awine T et al (2009) Use of spoligotyping and large sequence polymorphisms to study the population structure of the *Mycobacterium tuberculosis* complex in a cohort study of consecutive smear-positive tuberculosis cases in The Gambia. J Clin Microbiol 47:994–1001. https://doi.org/10.1128/JCM.01216-08

de Jong BC, Antonio M, Gagneux S (2010) *Mycobacterium africanum*—review of an important cause of human tuberculosis in West Africa. PLoS Negl Trop Dis 4:e744. https://doi.org/10.1371/journal.pntd.0000744

Dippenaar A, Parsons SDC, Sampson SL et al (2015) Whole genome sequence analysis of *Mycobacterium suricattae*. Tuberculosis 95(6):682–688

Drewe JA (2010) Who infects whom? Social networks and tuberculosis transmission in wild meerkats. Proc R Soc Lond B Biol Sci 277:633–642. https://doi.org/10.1098/rspb.2009.1775

Drewe JA, Foote AK, Sutcliffe RL et al (2009a) Pathology of *Mycobacterium bovis* infection in wild meerkats (*Suricata suricatta*). J Comp Pathol 140:12–24. https://doi.org/10.1016/j.jcpa.2008.09.004

Drewe JA, Dean GS, Michel AL et al (2009b) Accuracy of three diagnostic tests for determining *Mycobacterium bovis* infection status in live-sampled wild meerkats (*Suricata Suricatta*). J Vet Diagn Invest 21:31–39. https://doi.org/10.1177/104063870902100105

Drewe JA, Eames KTD, Madden JR et al (2011) Integrating contact network structure into tuberculosis epidemiology in meerkats in South Africa: Implications for control. Prev Vet Med 101:113–120. https://doi.org/10.1016/j.prevetmed.2011.05.006

Durnez L, Suykerbuyk P, Nicolas V et al (2010) Terrestrial small mammals as reservoirs of *Mycobacterium ulcerans* in Benin. Appl Environ Microbiol 76:4574–4577. https://doi.org/10.1128/AEM.00199-10

Flint BF, Hawley DM, Alexander KA (2016) Do not feed the wildlife: associations between garbage use, aggression, and disease in banded mongooses (*Mungos mungo*). Ecol Evol 6:5932–5939. https://doi.org/10.1002/ece3.2343

Gehre F, Otu J, DeRiemer K et al (2013) Deciphering the growth behaviour of *Mycobacterium africanum*. PLoS Neg Trop Dis 7(5):e2220. https://doi.org/10.1371/journal.pntd.0002220

Gey van Pittius NC, Perrett KD, Michel AL et al (2012) Infection of African buffalo (*Syncerus caffer*) by oryx bacillus, a rare member of the antelope clade of the *Mycobacterium tuberculosis* complex. J Wildl Dis 48:849–857. https://doi.org/10.7589/2010-07-178

Hershberg R, Lipatov M, Small PM et al (2008) High functional diversity in *Mycobacterium tuberculosis* driven by genetic drift and human demography. PLoS Biol 6:e311. https://doi.org/10.1371/journal.pbio.0060311

Huard RC, Fabre M, de Haas P et al (2006) Novel genetic polymorphisms that further delineate the phylogeny of the *Mycobacterium tuberculosis* complex. J Bacteriol 188:4271–4287. https://doi.org/10.1128/JB.01783-05

Laver PN, Ganswindt A, Ganswindt SB et al (2012) Non-invasive monitoring of glucocorticoid metabolites in banded mongooses (*Mungos mungo*) in response to physiological and biological challenges. Gen Comp Endocrinol 179:178–183. https://doi.org/10.1016/j.ygcen.2012.08.011

Loeffler SH, de Lisle GW, Neill MA et al (2014) The seal tuberculosis agent, *Mycobacterium pinnipedii*, infects domestic cattle in New Zealand: epidemiologic factors and DNA strain typing. J Wildl Dis 50:180–187. https://doi.org/10.7589/2013-09-237

Lutze-Wallace C, Turcotte C, Glover G et al (2006) Isolation of a *Mycobacterium microti*-like organism from a rock hyrax (*Procavia capensis*) in a Canadian zoo. Can Vet J 47:1011–1013

Mostowy S, Cousins D, Behr MA (2004) Genomic interrogation of the dassie bacillus reveals it as a unique RD1 mutant within the *Mycobacterium tuberculosis* complex. J Bacteriol 186:104–109

Mostowy S, Inwald J, Gordon S et al (2005) Revisiting the evolution of *Mycobacterium bovis*. J Bacteriol 187:6386–6395. https://doi.org/10.1128/JB.187.18.6386-6395.2005

Oevermann A, Pfyffer GE, Zanolari P et al (2004) Generalized tuberculosis in llamas (*Lama glama*) due to *Mycobacterium microti*. J Clin Microbiol 42:1818–1821. https://doi.org/10.1128/JCM.42.4.1818-1821.2004

Ofukwu RA, Oboegbulem SI, Akwuobu CA (2008) Zoonotic *Mycobacterium* species in fresh cow milk and fresh skimmed, unpasteurised market milk (nono) in Makurdi, Nigeria: implications for public health. J Anim Plant Sci 1:21–25

Parsons S, Smith SGD, Martins Q et al (2008) Pulmonary infection due to the dassie bacillus (*Mycobacterium tuberculosis* complex sp.) in a free-living dassie (rock hyrax—*Procavia capensis*) from South Africa. Tuberculosis 88:80–83. https://doi.org/10.1016/j.tube.2007.08.012

Parsons SDC, Drewe JA, Gey van Pittius NC et al (2013) Novel cause of tuberculosis in meerkats, South Africa. Emerg Infect Dis 19:2004–2007. https://doi.org/10.3201/eid1912.130268

Pepperell CS, Casto AM, Kitchen A et al (2013) The role of selection in shaping diversity of natural *M. tuberculosis* populations. PLoS Pathog 9:e1003543. https://doi.org/10.1371/journal.ppat.1003543

Rahim Z, Möllers M, te Koppele-Vije A et al (2007) Characterization of *Mycobacterium africanum* subtype I among cows in a dairy farm in Bangladesh using spoligotyping. Southeast Asian J Trop Med Public Health 38:706–713

Smith N (1965) Animal pathogenicity of the "dassie bacillus". Tubercle 46:58–64

Smith NH, Kremer K, Inwald J et al (2006) Ecotypes of the *Mycobacterium tuberculosis* complex. J Theor Biol 239:220–225. https://doi.org/10.1016/j.jtbi.2005.08.036

Thorel MF (1980) Isolation of *Mycobacterium africanum* from monkeys. Tubercle 61:101–104. https://doi.org/10.1016/0041-3879(80)90018-5

van Ingen J, Rahim Z, Mulder A et al (2012) Characterization of *Mycobacterium orygis* as *M. tuberculosis* complex subspecies. Emerg Infect Dis 18:653–655. https://doi.org/10.3201/eid1804.110888

van Soolingen D, de Haas PE, Haagsma J et al (1994) Use of various genetic markers in differentiation of *Mycobacterium bovis* strains from animals and humans and for studying epidemiology of bovine tuberculosis. J Clin Microbiol 32:2425–2433

van Soolingen D, van der Zanden AGM, de Haas PEW et al (1998) Diagnosis of *Mycobacterium microti* infections among humans by using novel genetic markers. J Clin Microbiol 36:1840–1845

Wagner JC, Bokkenheuser V (1961) The *Mycobacterium* isolated from the dassie *Procavia capensis* (Pallas). Tubercle 42:47–56

Wagner JC, Buchanan G, Bokkenheuser V et al (1958) An acid-fast bacillus isolated from the lungs of the cape hyrax, *Procavia capensis* (Pallas). Nature 181:284–285. https://doi.org/10.1038/181284b0

Warren RM, Gey van Pittius NC, Barnard M et al (2006) Differentiation of *Mycobacterium tuberculosis* complex by PCR amplification of genomic regions of difference. Int J Tuberc Lung Dis 10:818–822

Wernery U, Kinne J, Jahans KL et al (2007) Tuberculosis outbreak in a dromedary racing herd and rapid serological detection of infected camels. Vet Microbiol 122:108–115. https://doi.org/10.1016/j.vetmic.2007.01.012

Wirth T, Hildebrand F, Allix-Béguec C et al (2008) Origin, spread and demography of the *Mycobacterium tuberculosis* complex. PLoS Pathog 4:e1000160. https://doi.org/10.1371/journal.ppat.1000160

Part II
Epidemiology of Bovine Tuberculosis in Africa

Chapter 7
Epidemiology of Bovine Tuberculosis in Africa

Asseged B. Dibaba and C. J. Daborn

7.1 Background

In cattle, TB is generally a chronic wasting disease, and infected cattle survive for several months or years, often without manifesting clinical signs (Morris et al. 1994). The chronic, insidious nature of bovine tuberculosis (BTB) poses a significant challenge when attempting to contain the disease once it is introduced into a susceptible population. If, as is often the case, infection in a herd remains undetected, the disease spreads silently from individual to individual and ultimately also to in-contact, non-infected herds, particularly when infected animals are allowed to roam freely.

Chronic diseases such as BTB are difficult to investigate if one is dependent on the presence of clinical signs alone because diseased cattle often only manifest ill-thrift with depressed fecundity and productivity without regular mortalities (Jolles et al. 2005). Further challenges posed by BTB include its wide host range, the variety of interspecies transmission patterns (Skuce et al. 2012), and the large variation of different host species in their susceptibility to infection with *Mycobacterium bovis* (Drewe et al. 2014).

Africa is a vast continent, and the management practices vary substantially from the traditional nomadic and pastoral systems found in most of Africa to increasing small-scale urban farming and the sophisticated commercial farming systems found in countries such as South Africa, Namibia, and Botswana giving them access to international markets. Given the differences in disease dynamics in each of these systems, it should be expected that the epidemiology of BTB would accordingly

A. B. Dibaba (✉)
Department of Pathobiology, College of Veterinary Medicine, Tuskegee University, Tuskegee, AL, USA
e-mail: adibaba@tuskegee.edu

C. J. Daborn
Tigoni Veterinary Services (TVS), Nairobi, Kenya
e-mail: tvs@habari.co.tz

A. B. Dibaba et al. (eds.), *Tuberculosis in Animals: An African Perspective*,
https://doi.org/10.1007/978-3-030-18690-6_7

differ substantially, and it would be dangerous to extrapolate between the different systems.

There is an ample amount of epidemiological information available about BTB in cattle, and the other domestic and wild animals in Africa, but the information is often incomplete, fragmented, historical, contradictory, and difficult to access. The purpose of this chapter is to collate the available information, highlighting the complexities and contradictions, signposting the direction for further studies, and making the existing knowledge more readily available to those with an interest in the epidemiology and control of BTB in Africa.

In the developed countries, the interaction between the agent and hosts and prevailing farming systems are well documented, and the information is sufficient to predict the likelihood of BTB being contracted. Under these circumstances, risk factors determining the probability of infection and transmission of *M. bovis* are intimately linked to factors that affect the susceptibility of exposed animals (Skuce et al. 2012). Most livestock in Africa are farmed within regions where no formal control measures for BTB in the form of the conventional test-and-slaughter policy are applied (Cosivi et al. 1998). There are many gaps in the continent's datasets containing information about the distribution and prevalence of BTB. Likewise, there are only a few reports describing host and environmental factors that influence the transmission of *M. bovis* between various susceptible animal species and humans (Drewe et al. 2014).

Studies conducted in Africa elucidated some of the predominant risk factors, including a variety of farm management practices, the presence of wildlife reservoirs, and environmental factors that play a role in the transmission of the disease. Specific factors relevant to disease management, and those that are generally accepted to promote the introduction and spread of BTB, include the uncontrolled movement of animals (Oloya et al. 2007), intensification of husbandry systems (Elias et al. 2008), and contact with wildlife reservoirs (Michel 2008). Other locality-specific risk factors on the continent that will influence the application of practical preventative measures have been identified.

A recent analysis of wildlife-related risk factors dealt with the presence and spread of *M. bovis* in different wildlife populations (Gallagher et al. 1972; Bengis et al. 1996; Michel 2008) and the role that they may play in sustaining and disseminating the infection to other wildlife species and livestock. In Africa, this information is vital to consider when devising effective ways of controlling the disease and reducing interspecies transmission, should the approach be to eventually attempt to eradicate the disease from the continent.

It is clear that the risk of contracting BTB is multifactorial, and its investigation poses several, and often complex, challenges. To explore this complexity, the methodology of the Epidemiologic Problem Oriented Approach (EPOA) is applied to present the epidemiology of BTB in Africa.

7.2 Epidemiologic Problem Oriented Approach (EPOA) Methodology

The generic approach to problem-solving in any discipline consists of two steps: firstly to identify and characterize the problem and then to design a strategy to manage it (Habtemariam 1989). Problem identification and assessment of management options require a critical review of the biology, and preferably the epidemiology, of the disease in question. The EPOA methodology is based on epidemiological principles and comprises two interdependent components, the Problem Identification/Characterization triad and the Problem Management/Solution triad that are linked by the diagnostic procedure linkage (Fig. 7.1). The EPOA provides a scientific framework for systematic collection, organization, and analysis of epidemiological information, which is critical for developing a structured knowledge base of BTB and the difficulties surrounding its control. In this section, we deal with three component parts of the Problem Identification/Characterization triad, the Agent, Host, and Environmental pillars to develop a clearer understanding of the role of each component. The issues related to the diagnosis and control of BTB are dealt with in other chapters of this book (Chaps. 9 and 10).

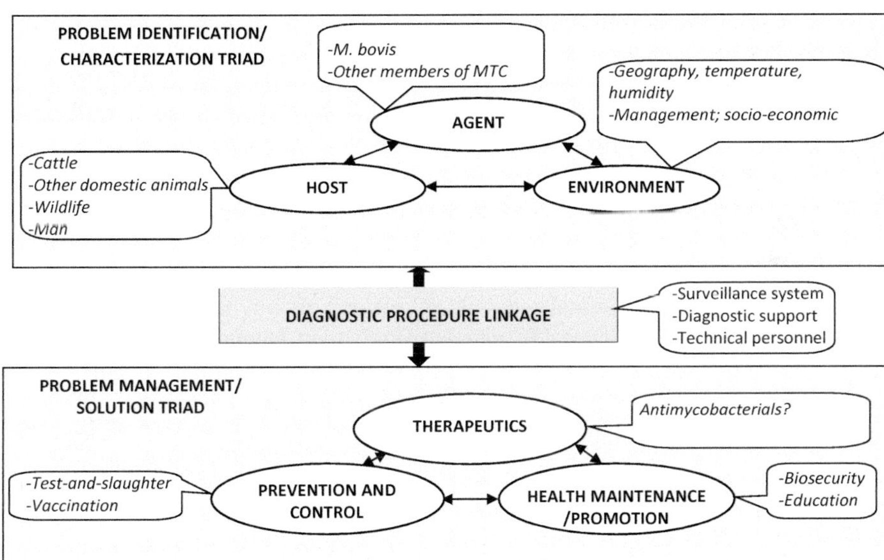

Fig. 7.1 The Epidemiologic Problem Oriented Approach (EPOA) for BTB

7.3 Problem Identification/Characterization Triad

7.3.1 Agent

Mycobacteria are grouped in the order *Actinomycetales*, along with other genera such as *Corynebacterium*, *Nocardia*, and *Rhodococcus*. These bacteria contain unique mycolic acids in their cell wall that confer distinct immune-stimulatory properties and resistance to drying, acidity/alkalinity, and many antibiotics (Sakamoto 2012). That tuberculosis remains such a challenging and enigmatic disease is principally a consequence of this unique cell wall construct. Although most of the identified mycobacterial species are facultative or opportunistic pathogens, *M. bovis* (the causative agent of BTB) and other members of the *Mycobacterium tuberculosis* complex (MTC) produce, in a variety of mammalian hosts, similar disease syndromes (TB) classically typified by the formation of an inflammatory granuloma referred to as a "tubercle". Other members of MTC include *M. tuberculosis*, the most important bacterial pathogen in humans, infecting more than one-third of the world's human population; *M. canettii* and *M. africanum*, usually isolated from African patients; *M. microti*, a rodent pathogen, usually isolated from voles but also from immune-compromised human patients; *M. caprae* and *M. orygis*, mainly isolated from goats and oryx, respectively; *M. pinnipedii*, isolated from seals and other marine animals; and *M. bovis* BCG, a laboratory-derived mutant of *M. bovis* (Müller et al. 2009) (for further details about these mycobacteria, refer to Chap. 6). We now know that members of the MTC have >99.9% similarity at the nucleotide level and that they have identical 16S rRNA sequences (Brosch et al. 2002).

Theobald Smith (1898) initially classified the mammalian tubercle bacilli into various types based primarily on the differences in virulence of cattle-derived (*M. bovis*) and human-derived (*M. tuberculosis*) strains. Although the differences in susceptibility are by no means absolute (Lewis and Sanderson 1927), calves and rabbits inoculated with minimal amounts of cultures of bovine-type bacilli developed progressive, usually fatal TB, whereas equivalent amounts of cultures of the human type only caused local, non-progressing lesions. Although *M. bovis* is the species most often isolated from tuberculous cattle, other members of the MTC may cause TB in animals and humans, and it is not uncommon to isolate *M. tuberculosis* from animals that have been in contact with humans. In a study conducted in Sudan, 7.4% ($n = 54$) of the mycobacteria isolated from tuberculous lesions of cattle at slaughter were *M. tuberculosis* (Sulieman and Hamid 2002).

7.3.1.1 Transmission

An understanding of the routes of transmission is critical to effectively control BTB. Determining the source of infection is often difficult, if not impossible (Brush 1898) because after becoming infected, it can take a year or even up to a decade for the animal to become visibly diseased. Although certain physiological characteristics are common to all of the susceptible hosts, the course of infection and the rate of the development of the disease and clinical signs can vary substantially in the various host species (Biet et al. 2005). The locality of the primary lesions observed in necropsied animals provides the best indication of the route of infection (Phillips et al. 2003). Tuberculous lesions are frequently found in more than one anatomical site in the same animal. However, both direct inoculation and in-contact infection in cattle resulted in lesions largely confined to the respiratory tract, confirming this location as the primary site of infection in cattle (Cassidy et al. 1998).

Respiratory Route (Aerosol Transmission) In the respiratory route of infection, mycobacteria, from an infected animal (or human) with open pulmonary lesions, are aerosolized in the respiratory tract and shed when exhaling. This creates a contaminated air space, putting any susceptible human or animal sharing that air space at risk (Renwick et al. 2007). That cattle serve as sources of *M. bovis* is supported by the bacteria consistently being isolated from the upper respiratory tract of 30–50% of tuberculin-test-positive reactors (Costello et al. 1998). Droplets (splashes) of contaminated water, eructation while grazing on infected pastures, or dust particles containing *M. bovis*-contaminated matter, such as desiccated feces or urine, can also contaminate the air space (Biet et al. 2005) and be a source of an infection. The risk of exposure to excreted *M. bovis* is enhanced by two important risk elements for animals sharing a confined air space: a crowded environment and a prolonged duration of contact with animals shedding the bacteria (Lurie 1930; Francis 1950; Michel 2008).

Tuberculous lesions in cattle are much more commonly found in the lungs and associated lymph nodes than in the abdominal organs or in the rest of the carcass (Table 7.1). That tuberculous lesions occur predominantly in the respiratory tract and associated lymph nodes of naturally (Langmuir 1961; Wilesmith et al. 1982; Collins and Grange 1983; Morris et al. 1994; O'Reilly and Daborn 1995; Whipple et al. 1996; Cleaveland et al. 2007) and experimentally infected cattle (Cassidy et al. 1998) have been accepted as sufficient evidence that the respiratory route is the primary portal of entry in this species. The minimum dose required to infect calves via the respiratory tract was up to 1000 times less than that required to infect animals via the oral route (Collins and Grange 1983). That few mycobacteria are required to establish an infection is also supported by mathematical models indicating that an *M. bovis* infection could be established in cattle by the inhalation of a single bacillus within an aerosol droplet (Neill et al. 1991).

The locality in the body of tuberculous lesions in livestock other than cattle and in wildlife species throughout Africa indicates that the respiratory route is also the primary route of infection in most of them (see Chap. 5 for a more detailed discussion about the situation in the various wildlife species).

***Respiratory Excretion of* M. bovis** The results of several epidemiological investigations suggest that naturally infected cattle in the early stages of the disease do not readily infect other cattle. The degree to which infected animals are sources of infection depends on factors such as the location of the lesions, the number of

Table 7.1 Distribution of visible lesions of BTB in naturally infected animals in Africa

Country	Species	No. of affected tissues	% positive			Reference
			Thorax	Abdomen	Head	
Algeria	Cattle	200	93.5	4.5	2.0	Sahraoui et al. (2011)
Burkina Faso		208	33.0	21.6	3.8	Tarnagda et al. (2014)
Cameroon		39	46.2[b]	10.3	12.8	Awah-Ndukum et al. (2010)
Chad		108	15.7[a]	NR	7.4	Ngandolo et al. (2009)
Egypt	Camel	20[c]	100.0	10.0	NR	Mason (1912)
Ethiopia	Cattle	1026	58.9	21.0	9.4	Asseged et al. (2014)
		108	61.1	3.0	36.1	Aylate et al. (2013)
	Camel	315	74.5	10.8	14.3	Mamo et al. (2011)
		24	54.2	37.5	8.3	Mamo et al. (2009)
		37	57.5	15.3	27.2	Beyi et al. (2014)
	Goats	83	95.2	1.2	3.6	Hiko and Agga (2011)
Kenya	Cattle	273	78.0	NR	NR	Gathogo et al. (2012)
Morocco		101	88.0	5.0	15.8	Berrada (1993)
Niger		207	61.4[a]	NR	NR	Boukary et al. (2012)
Nigeria		150	76.0[a]	NR	NR	Damina et al. (2011)
		46	41.3	17.4	23.9	Cadmus et al. (2004)
South Africa	African buffaloes	56	26.8	3.6	69.6	de Klerk-Lorist (2004)
Zambia	Kafue lechwe	65	67.7	12.3	NR	Gallagher et al. (1972)

NR not reported
[a]Lung lesions
[b]Lymph nodes
[c]Number of cases

organisms excreted, the duration and closeness of contact, and the production of aerosols of small particle size that contain viable mycobacteria (O'Reilly and Daborn 1995). Even when these criteria are satisfied, transmission of the disease is not rapid. This assumption is supported by the findings of an experiment where only 4 of 10 in-contact animals housed with 20 tuberculin-test-positive cattle for a year contracted the disease (Costello et al. 1998). A model simulating transmission within

New Zealand cattle herds estimated that each *M. bovis*-infected animal would infect an average of 2.6 cattle during the course of a year (Barlow et al. 1997).

That cattle-to-cattle transmission does not occur rapidly benefits the test-and-slaughter campaigns in countries in the absence of wildlife or other maintenance hosts (Anon 1994). The problem in most of Africa is aggravated by the practice of not immediately culling infected animals after they tested positive for BTB. Under these circumstances, those with advanced pulmonary lesions continue to shed large numbers of organisms, and they remain an ongoing source of infection.

Oral Route of Infection The distribution of lesions in tuberculous cattle, goats, and camels indicates that the oral route of infection may be significant in certain areas where traditional African livestock husbandry is practiced (Table 7.1). In Tanzania, for instance, the majority of visible lesions in cattle in one study were found in the gastrointestinal rather than in the respiratory tract (Cleaveland et al. 2007). Though oral transmission may occur in several ways, such as by the consumption of infected feed and water, during predator/prey interaction, pseudo-vertically, and percutaneously (Renwick et al. 2007), there are indications that, with some exceptions, this mode of transmission is uncommon and not important from an epidemiological perspective.

Feed and Water Environmental contamination by fecal and nasal excretion of *M. bovis* appears to be a less important mode of transmission as the bacteria are irregularly excreted (Neill et al. 1988). As an example, in an extensive study assessing the likelihood of transmission from infected pastures on BTB-positive farms, only mycobacteria other than *M. bovis* were cultured from the environmental substrates (Fine et al. 2011).

Mycobacteria can readily survive in water, and it is a major source of environmental mycobacteria that cause some of the non-tuberculous mycobacterioses in humans (Dailloux et al. 1999). Animals excreting *M. bovis* in fecal and nasal discharges contaminate running water (Humblet et al. 2009), and seasonal flooding in the environs of a village is a significant risk factor in the transmission of *M. bovis* in cattle (Cleaveland et al. 2007). Transmission by drinking *M. bovis*-contaminated water, however, is considered of little importance because the dilution of the numbers of mycobacteria in running water reduces the population density of *M. bovis* to a level where too few organisms occur to cause infection (Phillips et al. 2003). Stagnant water, however, can become heavily contaminated with animal waste containing *M. bovis*, and it constitutes a significant risk factor in smallholder/pastoral production settings (Oloya et al. 2007).

In terms of wildlife in South Africa, it appears that tuberculous buffaloes, some with advanced pulmonary lesions making them active shedders of *M. bovis*, do not commonly shed *M. bovis* in sufficient numbers in nasal or oral discharges to contaminate water and forage, and this potential source of infection appears also not to play an important role in the transmission of BTB in free-ranging ecosystems (Michel et al. 2007). Here too, non-tuberculous *Mycobacterium* species were cultured and appear to thrive under these conditions.

Contaminated feed and pastures may be a source of *M. bovis* but appear to play a negligible role in the transmission of BTB because *M. bovis* on fomites does not remain infective for long (Morris et al. 1994). Repeated environmental contamination does appear to increase the chances of oral infection (Maddock 1934) (refer to the section for further information). However, recognizing the normally rapid rate of dispersal of fecal matter and the resulting low bacterial numbers, oral transmission is unlikely to pose a major risk. There are examples, though not in Africa, of transmission through the sharing of feed and exposing cattle to pens that contained deer infected with *M. bovis* (Palmer et al. 2004) and the speculated role of latrines in the transmission of BTB from badgers to cattle (Humblet et al. 2009).

Other Sources of Oral Transmission Another type of oral transmission occurs during predator/prey encounters, which involves the consumption of infected tissues (lesions, blood, internal organs, etc.) by a susceptible predator. This type of transmission is a cause of concern in conservation areas (Michel et al. 2006), but dogs and cats can also fall victim to this type of transmission. Consumption of infected offal (and perhaps meat) poses a moderate risk because of the high oral dose that is required to establish an infection (Drewe et al. 2014).

Pseudo-vertical transmission can occur through consumption of infected milk or simply from close contact between lactating dams and their offspring. According to Michel (2008), social behavior, including regular and close physical contact such as licking, grooming, and suckling between members of the same herd over a protracted period of time, provides favorable conditions for pseudo-vertical transmission of *M. bovis*. Evangelista and De Anda (1996) examined the TB status of two groups of calves in *M. bovis*-infected dairy herds in Mexico. From 91 calves fed with pooled colostrum and raw milk from the bulk tank, 27% tested positive to the caudal fold test (CFT) at 6 months of age, whereas only 15% ($n = 279$) of calves fed with colostrum derived from tuberculin-negative cows and powdered milk were positive. Furthermore, the risk of skin-test-positive reactors increased from 11% in calves born from skin-test-negative cows to 33% in calves born from skin-test-positive cows. However, Francis (1950) noted that since only about 1% of tuberculous cows have TB of the udder, the risk of calves being infected by milk is considered low, unless they were fed pooled milk from the entire herd. Similarly, Gallagher et al. (1972) found no cases of TB in immature (under 18 months) lechwe (*Kobus leche kafuensis*) in Zambia, suggesting that neither congenital nor the consumption of infected milk was common. The offspring of infected cattle also do not appear to be at a significantly increased risk to become infected with *M. bovis* (Drewe et al. 2014).

Percutaneous Transmission A unique mode of transmission involving the percutaneous route has been documented in greater kudus (*Tragelaphus strepsiceros*) (Renwick et al. 2007), where transmission is caused by contact with thorns contaminated with purulent exudates discharged from fistulated parotid and submandibular lymph nodes containing tuberculous lesions.

7.3.2 *Host-Related Risk Factors*

Cowie et al. (2014) define risk factors as aspects of the system that influence the introduction of a disease and persistence of the disease in a population. Several studies have been conducted in Africa to provide information about the role of farm- and animal-level characteristics in the epidemiology of BTB. Most of the factors identified as consistently contributing to the occurrence of BTB were locality-specific (Awah-Ndukum et al. 2012) and related to the type of farm management (Berrada 1993). Specifically, the prevalence of BTB was high in peri-urban and urban areas where large numbers of dairy farms keep exotic cattle breeds (Tschopp et al. 2010). In contrast, low prevalence was recorded in pastoral settings, where indigenous cattle breeds are kept under a traditional management system (Gumi et al. 2012). At the livestock-wildlife interface (Munyeme et al. 2008), and in smallholder settings (Bernard et al. 2005), where animal movement between farms is uncontrolled (Firdessa et al. 2012), BTB was seen to spread very quickly. Many of these risk factors have been identified elsewhere (Marangon et al. 1998), and improving farm biosecurity is known to achieve effective control (Skuce et al. 2012; Cowie et al. 2014).

Animal-level risk factors also were extensively explored in Africa. It is, however, important to emphasize that BTB is a herd problem, and attempts to control the infection should be directed at herd/farm-level risk factors. In the following section, herd- and animal-level risk factors of BTB, with reference to African livestock production, are discussed.

7.3.2.1 Herd-Level Risk Factors

Herd Size A large herd size in association with a high cattle density and the housing of cattle, often exacerbated by between-herd movement, facilitates cattle-to-cattle transmission of BTB. In Africa (Cook et al. 1996; Asseged et al. 2000; Cleaveland et al. 2007) and on other continents (Griffin et al. 1996; Munroe et al. 1999; Karolemeas et al. 2010), herd size has been determined as one of the major risk factors for the transmission of BTB. However, in many regions of the African continent where animals constantly live in the open and under extensive conditions, BTB, irrespective of the herd size, is rare (Alhaji 1976; el Sanousi and Omer 1985). It is possible that herd size per se is not that important but that it is a proxy variable for other risk factors, such as a high animal density (Griffin et al. 1993). The detection of BTB reactors too is a problem in large herds. Since the tuberculin skin test is not perfect, the probability of a false-positive reactor would be greater in large herds (Humblet et al. 2009). Furthermore, the more animals that are skin-tested, the higher the probability to have a reactor. However, in a survey of cattle belonging to Somali and Guji pastoralists of southeastern Ethiopia (Gumi et al. 2012), herd size was not associated with the outcome of tuberculin tests. By analogy, the general practice of tuberculin skin testing of cattle and early removal of those infected with *M. bovis* has led to a progressively lower prevalence of BTB in all countries that have BTB control programs. In those countries, herd prevalence has effectively been reduced to less than the critical community size (CCS), and hence cattle are then no longer maintenance populations for BTB. The objective is to prevent spillover to cattle from other maintenance or spillover hosts. In Africa, where cattle are still the main source of *M. bovis* infection for all other species, including humans, this situation will prevail until such time that effective control programs have been implemented in those countries with infected livestock and wildlife populations.

Transmission and Host Dynamics Two categories of hosts are distinguished in BTB: maintenance and spillover hosts (Humblet et al. 2009). The maintenance of an *M. bovis* infection in a given species (a maintenance host) can be achieved only when sufficient intra-specific transmission occurs to sustain the disease at and beyond the threshold level. This level is determined by the basic reproductive number (R_0), which is defined as the expected number of new cases produced in a population by a single infected animal (Renwick et al. 2007). The threshold for disease maintenance by intra-specific transmission alone is met when $R_0 \geq 1$, meaning that each infected animal transmits the disease to one or more individuals (Palmer 2013); where $R_0 < 1.0$, the disease will disappear from the population. In addition to other factors, R_0 is determined by the abundance of the host species (Nugent 2011), referred to as the "critical community size" (CCS), which is the minimum size of a closed population within which a pathogen can persist indefinitely (Palmer 2013). In populations smaller than the CCS, the number and/or density of infected host levels are low enough for random extinction of the pathogen to occur.

Threshold values such as R_0 and CCS are not clear-cut, and they are also difficult to measure (Palmer 2013). The situation is also much more complex in multi-host ecosystems, such as those in some of the South African and Tanzanian national parks where these species-specific thresholds can be perturbed by the interspecies transmission of BTB from either maintenance or spillover hosts (Nugent 2011). In Africa, maintenance hosts include cattle (Cousins 2001) and wildlife species such as African buffaloes (Renwick et al. 2007), greater kudus (*Tragelaphus strepsiceros*), and Kafue lechwe (Gallagher et al. 1972), while several domesticated species, a number of wild ungulates and a few wild carnivores are spillover hosts (O'Reilly and Daborn 1995; Renwick et al. 2007). Since the common elements in maintenance hosts appear to be close family groupings and an aerogenous route of excretion of *M. bovis* (Fitzgerald and Kaneene 2012), it is possible that some of the other species may be able to become maintenance hosts of *M. bovis* under particular circumstances. Only a small proportion of these spillover species, however, have the required attributes to become maintenance hosts (Biet et al. 2005).

Infection persists by intra-specific transmission alone in maintenance hosts, and they are a source of infection for other species. The presence of a single maintenance host in an ecosystem is likely to cause the spread of TB to other susceptible species within that system. Conversely, in a spillover host, infection will not persist indefinitely unless there is regular reinfection from other species. It is important to keep in mind that transmission from spillover to maintenance hosts may occur and that both maintenance and spillover hosts may act as disease vectors in susceptible multi-species populations (Corner 2006).

The characteristics of an effective maintenance host include susceptibility to *M. bovis*, prolonged host survival after infection, the ability of infected females to reproduce, and the shedding of *M. bovis* via multiple routes (Drewe et al. 2014). Additionally, ethological factors such as gregarious behavior, and ecological characteristics such as eating behavior, population density, and interaction with other species are factors determining whether a particular species is likely to become a reservoir for *M. bovis* (Biet et al. 2005). *Mycobacterium bovis*-induced lesions such as extensive pulmonary lesions that drain into bronchi and bronchioles increase the likelihood of generating infectious respiratory aerosols that are required for the transmission and maintenance of the infection in a particular herd (Palmer 2013).

To cope with the complexity of BTB in a multi-host system, the threshold density theory is replaced by a threshold community configuration approach (Renwick et al. 2007). Within these complex systems, the rate of interspecies transmission is dependent on the contact rate between various host species (which is density-dependent) and on the way in which *M. bovis* is excreted. In a multiple host-species environment, the direction of flow of *M. bovis* varies according to the epidemiological role of the host species involved in the transmission of the infection.

Although cattle are the original and most common reservoir host (Cousins 2001), *M. bovis* also infects other domesticated mammalian species (O'Reilly and Daborn 1995). All these species, however, are not equally susceptible to *M. bovis*. In Africa, studies on animal TB focused mainly on cattle, and data about other domesticated species are scarce (Cosivi et al. 1998; Asseged et al. 2014).

Mixed Rearing of Domestic Stock Both pastoralists and smallholder crop-livestock farming communities in Africa employ a mixed livestock rearing system, in which small ruminants are important components and are herded together with cattle during the day (Boukary et al. 2012). At night, they are usually kept indoors in poorly ventilated farmers' houses for protection against theft and predators. Such multi-host husbandry practices and the sharing of air spaces are epidemiologically important for potential lateral spread of BTB within and between small and large ruminants and for potential zoonotic transmission (Tschopp et al. 2011).

Goats Tuberculosis in goats is widespread in certain regions of Africa where they co-graze with BTB-infected cattle (Table 7.2). Goats are susceptible to infection by *M. caprae*, *M. bovis* (Jenkins et al. 2011; Napp et al. 2013), and *M. tuberculosis* (Hiko and Agga 2011), and it appears that their resistance to become infected with *M. bovis* is not high (Anon. 2013; Napp et al. 2013). When exposed to a high infection pressure, large numbers of goats can be infected, and rapid spread of the disease within a flock can be expected. The involvement of the respiratory tract in tuberculous goats renders them suitable to act as a maintenance host for BTB and to disseminate the disease (Napp et al. 2013; Pesciaroli et al. 2014). TB caused by *M. caprae* is known to spread from infected flocks to in-contact cattle herds (Napp et al. 2013).

Table 7.2 Prevalence of BTB in different domestic animals across Africa

Country	Species	No. of sample	Diagnostic method	No. positive (%)	Reference
Algeria	Goats	995	PM	60 (6.0)	Sahraoui et al. (2011)
Egypt	Camels	1124	PM	6 (0.53)	Elmossalami et al. (1971)
		1786		– (2.8)	Mason (1917)[a]
		1579		51 (3.2)	Mason (1917)[b]
Ethiopia	Goats	1744	PM	76 (4.4)	Gumi et al. (2012)
		518	CIT	– (0.2)	
		630		48 (7.6)	Tafess et al. (2011)
		1536	PM	65 (4.2)	Hiko and Agga (2011)
	Camels	906		91 (10.0)	Mamo et al. (2011)
		694		81 (11.7)	Gumi et al. (2012)
		398		33 (8.3)	Beyi et al. (2014)
		480	CIT	29 (6.0)	
Niger	Sheep	124,759	PM	3 (0.0)	Boukary et al. (2012)
	Goats	19,731		– (0.01)	
	Camels	2604		7 (0.3)	
Nigeria	Goats	381,601	PM	986 (0.3)	Bala et al. (2011)
	Sheep	373,567		1423 (0.4)	
	Camels	219,308		732 (0.3)	
Uganda	Pigs	997	PM	93 (9.3)	Muwonge et al. (2010)

PM, necropsy; CIT, comparative intradermal TST; –, no info
[a]Data for 1910
[b]Data for 1916

Most BTB cases in goats in Africa occur as a result of contact with infected cattle (Cousins 2001). It appears, however, that mixing different species of domestic animals in the traditional, extensive farming practices does not play a major role in the transmission of BTB to goats. Exposure to *M. bovis* infection mostly occurs when sharing pastures, watering points, and night shelters with BTB-infected cattle (Deresa et al. 2013). Under those circumstances, the number of infected flocks is low, not exceeding 2.9%, and with an equally low herd prevalence of 0.2% (Tschopp et al. 2011; Gumi et al. 2012).

Sheep There is common belief that sheep are innately resistant to tuberculosis (Anon 2013). They do, on the contrary, appear to be quite susceptible to the infection and are at times infected with *M. bovis* or *M. caprae* with the development of lesions particularly in the respiratory tract, which are both macroscopically and histologically very similar to the tubercles seen in cattle with BTB (Cordes et al. 1981; Davidson et al. 1981; Malone et al. 2003). TB is infrequently reported in sheep in Africa (Tag el Din and el Nour Gamaan 1982; Fullerton 1902), and its prevalence in this species is usually very low (Fullerton 1902). Tuberculous sheep have been detected in a number of countries (that reported its presence) including Egypt

(Moustafa et al. 1964), Sudan (Tag el Din and el Nour Gamaan 1982), Niger (Boukary et al. 2012), (Kassa et al. 2012), and Nigeria (Bala et al. 2011) (Table 7.2). The reasons for the low prevalence of BTB in sheep probably are multiple (Allen 1988). Sheep appear not often exposed to high infection pressures of TB because of specific management and behavioral factors. They are usually extensively managed, and they mostly graze during daylight hours. These extensive conditions limit the spread of TB within flocks (McFadyean 1900). Furthermore, sheep tend to flock together both when grazing and when resting, thus evading contact with other species and limiting interaction with potentially tuberculous species to the minimum (Malone et al. 2003). They are more likely to acquire the infection when sharing pastures (Davidson et al. 1981), during communal housing with infected cattle (Malone et al. 2003), or during close contact with infected wildlife species (Deresa et al. 2013).

In sheep, lesions occur in the lungs, bronchial and mediastinal lymph nodes, and in the liver and spleen (Moustafa et al. 1964; Marianelli et al. 2010). In some instances, lesions are caseous, calcified, and well encapsulated, creating the impression that sheep have a degree of resistance to the infection causing it to remain localized.

The perception is that the low prevalence of BTB in sheep is likely to be the consequence of inadequate diagnostic facilities on most of the continent (Tag el Din and el Nour Gamaan 1982), and the prevalence of the disease in sheep may be substantially higher in those countries where it has been detected. Because tuberculous lesions develop in the respiratory tract, sheep are considered to be a potential maintenance host of *M. bovis*, particularly when they share grazing with a number of species (Malone et al. 2003).

Camels Tuberculosis in dromedary (Arabian) camels (*Camelus dromedarius*) was first reported in Egypt in 1888, and *M. bovis* was confirmed as the cause of the disease in 1911 (Mason 1912). A prevalence of BTB of 3.2% ($n = 1579$) was recorded in camels slaughtered at Cairo abattoir during the course of 1916 (Mason 1917).

While they are susceptible to both experimental and natural infection with *M. bovis* and *M. tuberculosis*, *M. bovis* is more commonly isolated from tuberculous camels. In two reports, 97.4% ($n = 76$) and 2.6% and 93.4% and 6.6%, respectively, of bovine and human bacilli were isolated from tuberculous lesions (Elmossalami et al. 1971; Kinne et al. 2006). The disease is more frequently seen in camels kept in close contact with cattle than in those living under extensive conditions (Mason 1917; Elmossalami et al. 1971). Accordingly, camels of the Bedouin in Egypt were seldom affected compared to those owned by farmers, the Fellaheen, who customarily at night keep their camels indoors in close contact with cattle. It is most likely that BTB-infected camels could be a source of infection when introduced into a herd of healthy camels (Mason 1917).

Camels are also found in sub-Saharan Africa where owing to their drought tolerance and multiple economic values, traditional cattle-rearing communities are increasingly replacing their cattle with camels. During the transitional period of phasing-in and phasing-out, herds of camels and cattle are reared together during the day and rounded in a confined space at night (Beyi et al. 2014), thus increasing the risk of camels contracting *M. bovis*. Currently, the prevalence of BTB in camels in these areas appears to remain low as at the Niamey abattoir in Niger, only 0.3% of 2604 slaughtered camels contained lesions consistent with BTB (Boukary et al. 2012). A similar low prevalence of 0.3% was recorded in Nigeria (Bala et al. 2011), but a prevalence of 8.3% was reported in Eastern Ethiopia (Beyi et al. 2014). The low prevalence of BTB in camels in many African countries is probably because most of the camels remain in the open and harsh environment where it is difficult for the pathogen to spread. The prevalence of BTB in camels is higher than that of pigs and small ruminants in all the African countries where they occur.

In camels with BTB, the lungs and the bronchial lymph nodes are frequently affected, and lesions are limited to these organs in 60% of cases (Cousins 2001). In a study conducted in Eastern Ethiopia (Mamo et al. 2011), the lungs were the most frequently affected organ (40.0%) followed by the mesenteric (38.0%), mediastinal (12.5%), and retropharyngeal lymph nodes (8.3%). Additionally, generalized BTB, with numerous lesions in the organs and lymph nodes throughout the carcass, was fairly common (Table 7.1). The predominant occurrence of lesions in the respiratory tract of camels suggests that airborne transmission is the principal route of infection in this species.

A unique spoligotype (SB 1433) was isolated from camels in Niger (Boukary et al. 2012), where they are customarily reared separately from other species, and this unique strain appears to be maintained in camels in this particular region. Because of their relatively longer life expectancy, camels are an ideal maintenance host, once the infection has established itself in a fairly large population.

Pigs Tuberculosis is highly prevalent in pigs in Africa owing to their indoor management, frequent contact with infected cattle, and access to the contaminated material of animal origin. A study conducted in Uganda (Muwonge et al. 2010) indicated that 9.3% ($n = 997$) of pigs had lesions compatible with those of TB, and 33.3% ($n = 93$) of the carcasses with lesions yielded mycobacteria on culture.

Pigs usually acquire *M. bovis* from shared grazing or by the intake of *M. bovis*-contaminated dairy products, especially milk. On rare occasions, *M. tuberculosis* has also been isolated from pigs in Nigeria (Jenkins et al., 2011). Pigs are considered to be spillover hosts (Cousins 2001), and the disease does not seem to spread in a herd, and it often disappears from a herd when the source of infection, such as contaminated milk, is removed (Anon 2013). Tuberculosis is now rare in domestic pigs in countries that successfully apply BTB control programs. In some countries, TB in pigs is re-emerging following the widespread adoption of outdoor pig farming systems (Pesciaroli et al. 2014).

Pigs appear to complicate the epidemiology of *M. bovis*, as they may be a source of infection for multiple species. Thus *M. bovis* strains isolated from the mesenteric lymph nodes of tuberculous pigs and from humans and cattle in the Northeastern part of Uganda had identical spoligotype patterns (SB 1469) (Muwonge et al. 2012).

The oral route appears to be the primary route of infection in pigs. Tuberculous lesions are thus common in the lymph nodes of the head, neck, and abdomen, but they may also be found in the lungs and abdominal viscera and in the mesenteric lymph nodes (Anon 1994). Tubercles often have an intense yellowish discoloration with a gritty texture. Although a rapidly progressive, disseminated disease with caseation and liquefaction of lesions has been reported in pigs, the lesions that usually develop tend to be localized and small and disappear or become inactive over time (Anon 2013).

After investigating TB in Sicilian black pigs, an endemic semi-free-ranging domestic pig breed, Di Marco et al. (2012) suggested that this species might act as a BTB reservoir in this ecosystem. The authors listed several facts to support their argument: that the pigs are infected, that the disease is active and self-maintaining in the pig population, and that conditions are favorable for black pigs to transmit *M. bovis* to cattle and humans. Furthermore, a spoligotype that was not previously described in the international spoligotyping database was isolated from pigs, suggesting that *M. bovis* with the potential of infecting cattle and humans circulates among pigs.

Warthogs (*Phacochoerus africanus*) may sustain the infection and similarly appear to be infected via the oral route and manifest lesions that are similar to those in domestic pigs (Woodford 1982).

7.3.2.2 Animal-Level Risk Factors

Age Although there appears to be some inconsistencies in the data, age is one of the most important animal-level risk factors identified in numerous studies in Africa (Berrada 1993; Katale et al. 2009; Ibrahim et al. 2012; Moiane et al. 2014). Animals of 6 years of age and older consistently had a higher prevalence (Vekemans et al. 1999; Tarnagda et al. 2014; Asseged et al. 2000; El-Olemy et al. 1985). When the average age of the herd was reduced and young stock was reared separately from older animals, the prevalence of BTB was much reduced (Awad 1962). It thus appears that the older the animal, the bigger the chances of being infected because of the prolonged and repetitive exposure to *M. bovis* that increases the likelihood of them becoming infected (de Vos et al. 2001; Cleaveland et al. 2007; Gumi et al. 2012).

A few reports exist where age was not significantly associated with BTB (Bedard et al. 1993). In a traditionally managed Moroccan herd, the number of tuberculin skin-test-positive animals did not differ between the age groups (Berrada 1993), and in Algeria, cattle aged between 2 and 5 years and goats <6 months of age were the most affected (Sahraoui et al. 2011). Despite these exceptions, in the majority of studies, the prevalence of BTB in older animals is significantly higher. These trends are consistent with those in Great Britain, where, on average, the incidence of BTB in cattle increases by 7.5% for every year of life, reaching 40% in the 5–6-year-old age group (O'Reilly and Daborn 1995). In intensive management systems, cows, heifers, and bullocks are, respectively, 10.3, 14.8, and 8.0 times more likely to fail a tuberculin test than calves (Griffin et al. 1996).

Studies involving other species revealed that age similarly is the most important risk factor. A logistical regression analysis (Rodwell et al. 2001) demonstrated that older African buffaloes were at greater risk for acquiring BTB than younger buffaloes but that the risk increased for all age groups as the prevalence in a herd increased. Older camels also have a higher prevalence of TB than young animals (Elmossalami et al. 1971).

Sex Sex is one of a few animal-level risk factors that has only been mentioned in studies conducted in Africa (Humblet et al. 2009), but other countries are also now assessing this issue (Phillips et al. 2002). In Africa, the findings of various studies about the role of sex as a risk factor for BTB are inconsistent. More females than males were positive in countries such as Ethiopia (Elias et al. 2008), Uganda (Inangolet et al. 2008), Ghana (Bonsu et al. 2000), Chad (Ngandolo et al. 2009), Morocco (Berrada 1993), and Nigeria (Damina et al. 2011). In other studies, the opposite was true for cattle in Algeria (Sahraoui et al. 2011), Ethiopia (Tschopp et al. 2010), Nigeria (Ibrahim et al. 2012), and Tanzania (Tanner et al. 2014) where males were significantly more affected than females, while in a number of studies there were no statistically significant association between the prevalence of BTB and sex (Cook et al. 1996; Oloya et al. 2006; Awah-Ndukum et al. 2012; Okeke et al. 2014).

In beef cattle, similar inconsistencies exist. The majority of beef cattle are slaughtered young, giving them a much reduced lifetime exposure compared to dairy cattle, and the prevalence of the disease in them is usually low (Driscoll et al. 2011). In smallholder production systems, higher prevalence in males is seen. Although farmers usually take good care of their oxen, in these systems oxen shoulder the burden of farm activities because they are essential for ploughing, threshing, and harvesting (Tschopp et al. 2010). They are therefore kept in the herd for longer, thereby increasing the chance of being exposed to infection compared to females (Ibrahim et al. 2012). As sharing of oxen is commonplace in smallholder, crop-livestock production systems (Laval and Ameni 2004), there is also a greater probability that they will be in contact with an *M. bovis*-infected herd.

The inconsistencies in the prevalence based on sex appear to be the consequence of the management system followed rather than sex and the number of animals tested in each of the systems. The higher prevalence of BTB in dairy cows, also in pastoral settings, may predominantly be due to production stress during lactation and a longer life expectancy than that of males thus increasing the time that they are exposed to cattle with BTB. Their usual close confinement generally plays a bigger role in causing the higher prevalence of the disease in dairy cows (Francis 1950; Humblet et al. 2009).

Breed Initially, during the 1930s, the perceived lower and higher prevalence of BTB in various indigenous and exotic European breeds, respectively, were ascribed to an innate resistance of the African zebu breeds to BTB. The effect of this perceived resistance was probably enhanced by the extensive farming practices employed in Africa that tended to reduce the prevailing prevalence and spread of BTB in the indigenous breeds (Du Toit 1936; Carmichael 1938).

There also appeared to be differences in susceptibility between some of the indigenous breeds, such as the short-horned zebu and the long-horned Ankole cattle sharing the same environmental conditions and management practices. However, no observable differences in susceptibility to BTB between sub-Saharan White Fulani (3.3%), Sokoto Gudali (3.7), N'dama (3.6%), and Bunaji (3.3%) breeds were detected. Similarly, indigenous Moroccan cattle breeds were also thought to be less susceptible to BTB than the exotic dairy breeds, but they too were found to be equally susceptible (Bedard et al. 1993). In Tanzania, a high prevalence (13.2%) of BTB has been reported in indigenous cattle kept under intensive husbandry practices (Katale et al. 2009).

The perception of resistance may have been strengthened by the impact of the limited exposure of indigenous cattle to the African strains of *M. bovis* prior to colonialism (Brosch et al. 2002), because of the prevailing management practices at the time that limited contact between cattle of different owners and tribes and the spread of the disease. It is known that autochthonous cattle breeds in Africa (zebus, N'dama, and Mtui) have several important positive production traits, including resistance to certain diseases, but according to current knowledge, it is clear that resistance to BTB is not one of them.

The results of more recent studies suggest that the effect of breed on BTB in Ethiopia (and elsewhere in Africa) on the difference in prevalence of BTB in imported, high-value, European breeds compared to local zebus was markedly confounded by differences in herd management (Asseged et al. 2014), and that when these are taken into consideration, breed itself played a minor role in determining the prevalence of BTB. There was, in addition, no significant difference in susceptibility between indigenous and exotic breeds after adjusting the data for age and sex (Berrada 1993; Asseged et al. 2000).

An example of the impact of management practices is seen in short-horned zebu and long-horned Ankole, which share the same environment and management conditions; however, Ankole cattle have a higher prevalence of tuberculin test positive reactors (54.9% compared to 4.6%) and gross lesions at slaughter (16% compared to 0.93%) (Carmichael 1937). Certain investigators reported no differences between the indigenous breeds (Oloya et al. 2007), while others reported lesion prevalence in abattoirs in the Ankole longhorn cattle of 12.5–51.1% compared to the 0.1% in autochthonous zebus (Opuda-Asibo 1995). Various factors appear to influence the prevalence of BTB in the indigenous breeds. The variation of prevalence in Ankole cattle exemplifies the concurrent influence of environmental and other risk factors. The high prevalence in Ankole cattle in Uganda was ascribed to a particular herd management style. In order to ward off biting insects, a smudge fire of dried cow dung is lit at the center of a kraal, and animals cluster around it, head-to-head, relishing the warmth and thus increasing the close contact between them. Under these circumstances, in infected herds, the coughing elicited by the smoke would facilitate the spread of BTB (Carmichael 1938). Elsewhere, Munroe et al. (1999) found no significant difference in risk based on breed between dairy and beef herds in BTB outbreaks in Canada. Francis (1950) concluded that "there is no reason to think that the various European breeds of cattle differ in their susceptibility to BTB." In the final analysis, it is clear that genetic variation in susceptibility to *M. bovis* infection exists between mammalian families and species, but not in breeds (Phillips et al. 2002).

There is also a lack of association with breed or strains and susceptibility to BTB in other species. There was similarly a perception that the susceptibility of various human races differed, but it became clear that the differences are primarily determined by the levels of disease in specific communities and levels of poverty prevailing in the different racial and ethnic groups (Opie 1931). No variation in susceptibility too was found in various mouse strains, and it was concluded that natural strain resistance simply does not exist (Gray et al. 1960).

A number of factors appear to influence the differences in prevalence that have been ascribed to the indigenous and exotic breeds, including the following:

1. The common practice, in Ethiopia, for example, to keep high-value Holstein cows in roofed cowsheds rather than in the open kraals (Elias et al. 2008).
2. The practice in the central highlands of Ethiopia for smallholders to keep dairy cows indoors, separately from other local zebus and crossbreds, which usually graze on communal pastures (Tschopp et al. 2013).
3. That genetically improved dairy animals are routinely kept under intensive conditions (Elias et al. 2008).
4. Different farms vary greatly in stocking density, extent of stock movement, the presence of *M. bovis*, and other risk factors (Driscoll et al. 2011).

Body Condition Score The body condition of cattle, assessed by using the body condition score (BCS), is another factor that has been used to subjectively assess the presence of BTB in cattle (Nicholson and Butterworth 1986). The scoring is based on visual inspection and palpation to assess the sharpness of the backbones and lumbar processes, and on occasion, the tail head, brisket, ribs, and hips, and the amount of muscle and fat covering these areas. The body condition score is typically categorized into poor (score 1–3), medium (score 4–6), and good (score 7–9) groups. The problem with using the body score as an assessment for the presence of BTB is that cattle in Africa often suffer from nutritional stress, evidenced by a low BCS, and that those with advanced clinical BTB characteristically also lose condition and manifest low BCSs.

It appears, based on available data, that BCS is not significantly associated with tuberculin test outcomes, and contradictory results have been recorded. In a study in Nigeria, the likelihood of detecting AFB in cattle increased as the BCS declined (Okeke et al. 2014). Similar results were obtained in Addis Ababa, Ethiopia (Elias et al. 2008), where, respectively, 36.1%, 30.9%, and 15.5% of 1869 of the cattle tested had poor, medium, and good BCSs. Similarly, in Zambia, a low BCS was associated with higher numbers of BTB-positive skin test cattle (Cook et al. 1996; Munyeme et al. 2008). The results using this parameter, however, are inconsistent as elsewhere a prevalence of 11.8%, 10.0%, and 13.1%, respectively, was recorded in cattle with poor, medium, and good scores (Asseged et al. 2000).

Immune Status Numerous factors impair the immune status of livestock, including poor hygienic conditions, nutritional deficiencies and imbalances, climatic stress, crowding, and poor ventilation, as well as those concomitant diseases that cause immune suppression (Berrada 1993). There appears to be no substantial evidence that the strain of milk production and calf bearing lower resistance to TB (Francis 1950; Ibrahim et al. 2012).

The rearing of cattle in Africa is mainly dependent on access to natural pastures, the availability of which varies seasonally thus causing a marked fluctuation in the nutritional status of animals during the dry periods of the year when they appear to become immunosuppressed (Ejeh et al. 2014). Malnutrition and nutritional stress are considered as important predisposing factors determining the susceptibility of cattle to BTB and are likely to play a significant role.

In certain African countries, co-infection with *M. bovis* and liver flukes appears to confound the interpretation of tuberculin test results, and it may also have an impact on the susceptibility of cattle to *M. bovis* (Munyeme et al. 2012) as the co-infection appears to influence the immune response to certain antigens. The extensive tissue damage caused by migrating metacercariae cause elevated immunoglobulin E levels, eosinophilia, and a T-lymphocyte-helper (Th) cell, type-2 immune response that has been shown to suppress the protective, *M. bovis*-specific, Th1 response (Phillips et al. 2002).

7.3.3 Environmental Factors

7.3.3.1 Geographic Distribution

It is generally accepted that the incidence of TB increased in industrialized countries in the seventeenth and eighteenth centuries, peaking at different times in different places. Looking at the trends, Lönnroth et al. (2009) suggested a temporal association between increased TB incidence and periods of rapid industrialization and urbanization. In a similar manner, many argue that BTB became widely disseminated in countries where cattle were housed for long periods and in large herds. For instance, the British Economic Advisory Council in 1934 reported that at least 40% of the cows in dairy herds in Great Britain were infected with *M. bovis* (Alhaji 1976; Collins and Grange 1983). A prevalence of 50% also was reported in Germany in the mid-twentieth century (Pittler and Steel 1995). The progress made toward the eradication of BTB in the developed countries has changed its global distribution. Countries that at the turn of the twentieth century had the highest prevalence of BTB (United States, Canada, Western Europe, Australia, and Japan) now have the lowest prevalence, while African countries, which probably acquired the bulk of the infection during European colonization (Alhaji 1976; El Sanousi and Omer 1985; Phillips et al. 2002), are today the problem areas. Accordingly, prevalences of up to 50% in some areas in Malawi (Bernard et al. 2005), 25% in parts of Northern Nigeria

(Alhaji 1976), 7.4% in Zambia (Cook et al. 1996), and 4.0% in Algeria (Sahraoui et al. 2009) have been reported.

Contrary to several opinions that BTB was absent in Africa before colonization (Du Toit 1936; Carmichael 1938), there is ample evidence, however, that BTB existed in the local cattle population before the introduction of European cattle breeds. Molecular typing of *M. bovis* isolates from Tanzania (Kazwala et al. 1998) showed two lineages of *M. bovis*: an aboriginal lineage with atypical properties and a lineage imported from Europe displaying a classical spoligotype profile. Müller et al. (2009) and Berg et al. (2011) later refined this observation. Accordingly, two clonal complexes Af1 and Af2 of *M. bovis* are geographically localized to Central-West and Eastern Africa. The East African clonal complex (Af2) is uniquely confined to Ethiopia, Kenya, Tanzania, and Uganda, although other non-Af2 strains were present at reasonably high frequencies (between 5% and 33%). The Central-West African clonal complex (Af1) is confined to Burkina Faso, Chad, Cameroon, and Nigeria. The introduction of improved European breeds of cattle and the advent of intensive dairying rapidly changed the distribution of BTB, and it is likely that some of the European cattle were infected with *M. bovis* European 1 clonal complex, which has a worldwide distribution, when imported, thereby amplifying the African burden of BTB (Smith 2012).

Mason (1912) paid particular attention to the origin of camels slaughtered in Cairo abattoir, Egypt. Although up to 40,000 camels were imported annually from Syria and Arabia, TB was found only in Egyptian camels and as far as it had been possible to ascertain, never in Syrian or Arabian camels.

7.3.3.2 Spatial Distribution

The pattern of BTB occurrence is not uniform: it can be classified as sporadic, persistent, or recurrent, while some herds may remain free of infection (Skuce et al. 2012). Biological and social mechanistic models have identified several risk factors, on a case-by-case basis, although some sporadic episodes offer few clues with respect to cause-and-effect relationships.

The prevalence of BTB varies widely in different areas within a country and in different herds within a given area. In South America (Cosivi et al. 1998), the highest prevalence of BTB was reported from areas surrounding major cities where intensive dairy production with large herd size was most common. In Spain, where about 97% of herds were officially free of BTB, bovine TB control programs that initially focused on dairy herds resulted in a dramatic reduction of the disease in that sector (Anon 2007a), and positive herds are now mainly located in the dry rural regions where there are more beef herds.

In the United States, 69% of the tuberculin reactors during 1917–1957 came from nine states: these nine states yielded 75% of the total reactors during 1955–1957 (Ranney 1958). In the United Kingdom, South West England and South West Wales have been consistently affected over the years (Anon 2007b). In New Zealand, which is divided into endemic and non-endemic areas based on difficulties in eradicating BTB, a total of 90% of tuberculous cattle comes from endemic areas (Collins et al. 1994). The common denominator in all of these reports is a localized, and possibly shared, external risk of the spread of BTB between neighboring herds, probably because of the presence of an additional maintenance host in the ecosystem (Johnston et al. 2011). In support of this argument, Costello et al. (1999) reported that isolates from cattle, badgers, and deer shared the same geographic range in Ireland, suggesting interspecies transmission. According to Skuce et al. (2012), historical incidence was a robust predictor of future breakdowns in UK and Irish herds, suggesting that the source was not totally removed. In intensive systems, this effect is expected to be less because of reduced exposure created by biological barriers to wildlife (Edwards et al. 1997).

The African Situation The relation between humans and their domestic animals in sub-Saharan Africa (SSA) is much closer than in the other regions. The human-animal interaction is growing in intensity due to the shift from extensive production system in the rural areas to intensified livestock husbandry in the peri-urban and urban centers, where even the most primitive of hygienic precautions are conspicuously absent (Carmichael 1938). Under these circumstances, it is likely that BTB spreads rapidly once an *M. bovis* infection is established on a farm. Furthermore, in rural and small country towns across Africa, the home is shared with goats, sheep, fowls, and often cattle, posing a significant risk for zoonotic TB transmission. Therefore, large variations in BTB occurrence between regions have been reported in Africa.

The prevalence of BTB in Nigeria ranges from 0.5% in Oyo (low livestock population) to 12.3% in Gombe (northeastern state), where the livestock concentration is high (Ejeh et al. 2014). In Ghana, BTB was seen to be highly prevalent (50%) in Ningo, a low-lying wetland (Bonsu et al. 2000), and in Tanzania, where the distribution of BTB is uneven, a significantly higher prevalence was seen in the southern highlands (14%) and southern regions (11.8%) compared to western (2.9%), northern (0.77%), lake (0.21%), and central (0.19%) regions (Daborn and Grange 1993). Consistent with the notion that the BTB prevalence is higher in humid and intensified conditions than in extensive, dryer areas, Bernard et al. (2005) reported a lower BTB prevalence in a pastoral area, which is dryer than the agropastoral zone where intensified conditions occurred. Much higher BTB prevalences have been reported from peri-urban areas where intensified dairy production is practiced (Bonsu et al. 2000). In South Africa, where the incidence of infection was about 6%, a stark difference in prevalence was noted: it was about 40% in the larger towns, 10% in the small country towns, and 2% in rural areas (du Toit 1936).

Geographic variation in the prevalence of BTB within a country may suggest the existence of foci of *M. bovis* (or hotspots), such as communal pastures, watering points, and auction markets (Katale et al. 2012; Shirima et al. 2003). Since high infection rates were recorded in rangelands traversed by major rivers (Mwakapuja et al. 2013), areas with good pasture and water sources can attract more stock resulting in overcrowding (Brahmbhatt et al. 2012) as seen in the Mediterranean habitats of southern Spain (Acevedo et al. 2007). Humblet et al. (2009) propounded the view that water points are a potential risk factor, because areas around ponds are generally moist, with greater amounts of shade, two favorable conditions for the survival of *M. bovis* in the environment.

Munyeme et al. (2008) suggested that area could be a proxy variable for contact with wildlife. To give some insight, Munyeme et al. (2011) assessed whether black lechwe (*Kobus leche smithemani*) on the Bangweulu swamps, an area occupied by wildlife, were infected with TB. For comparison, Kafue lechwe (*K. leche kafuensis*) dwelling in the Kafue basin also were included in the study. The difference between the Kafue basin and the Bangweulu swamp was the absence of a livestock/wildlife interface in the latter. As expected, *M. bovis* was cultured from 4 of 11 Kafue lechwe, whereas none of the 30 black lechwe showed gross lesions suggestive of TB. The grazing range of Kafue lechwe and cattle extensively overlapped, particularly during the dry season, thus increasing the frequency of intermingling at watering points and on ranges. According to the authors, this scenario, which is considered a day-to-day phenomenon in all livestock/wildlife interface areas in Africa (Guilbride et al. 1963; Munyeme et al. 2008; Brahmbhatt et al. 2012; Katale et al. 2012), increases the opportunity for interspecies transmission of TB. Another perspective provided by the authors was the fact that early settlers from South Africa (Gallagher et al. 1972) may have introduced TB into the Kafue basin with their cattle when they settled there. This scenario is similar to the situation in Kruger National Park (KNP), South Africa, where it is hypothesized that comingling with *M. bovis*-infected cattle herds along the southern border of the Park introduced BTB into the buffalo population (Bengis et al. 1996; de Vos et al. 2001). A subsequent follow-up study based on the presence of identical molecular fingerprints (IS6110 and PGRS) of *M. bovis* isolates clearly established an epidemiological link between BTB outbreaks in the buffaloes of KNP and a neighboring cattle herd (Michel et al. 2008). Based on the disease timeline and its geographical expansion northward (from its entry point in the south), de Vos et al. (2001) calculated that BTB would spread at about 6 km/year in a northerly direction in the KNP's buffalo population.

7.3.3.3 Climatic Factors

Following fecal contamination of pastures, *M. bovis* withstands nutrient deprivation and osmotic shock; however, the weather will influence its survival. According to Phillips et al. (2003), exposure to direct sunlight destroys the bacilli within about 12 h. The survival time of *M. bovis* was investigated in the KNP (Tanner and Michel 1999). Tuberculous lungs and/or lymph nodes of African buffaloes (*Syncerus caffer*) and spiked fecal specimens placed at seven different sites in the habitat of free-ranging wildlife were analyzed over a 1-year period. *Mycobacterium bovis* could only be isolated for a period of 6 (for tissues) and 4 (for feces) weeks although the assessment was done during winter and under moist conditions. The survival time of *M. bovis* was greatly reduced (to a maximum of 5 days) when specimens were buried underground.

In the United Kingdom, with its markedly different climatic conditions, Duffield and Young (1985) previously failed to re-isolate *M. bovis* from fecal substrates at 4 weeks following artificial inoculation into bovine feces held under various environmental conditions. Maddock (1934) cited studies conducted in 1931 and 1932 where virulent *M. bovis* could be recovered from infected materials exposed in the open for 178 and 152 days, respectively. However, a follow-up study conducted during the hot summer of 1933 failed to recover tubercle bacilli from artificially infected grass 21 days after infection. Nevertheless, tubercle bacilli were shown to survive for 63 days after infection of the grass during the autumn and winter. Adams et al. (2013) experimentally inoculated environmental substrates (hay, soil, corn, water) and then exposed them to natural weather conditions in Michigan, USA. In 128 samples tested monthly during the 12-month period, *M. bovis* was not detectable by culture after 2 months although its DNA was still detectable by PCR for at least 7 months.

7.3.3.4 Herd Management Factors

In Africa, both commercial dairy farming with specialized dairy breeds, often of European origin, and traditional milk and beef production with extensively managed herds are widely practiced. The pattern of BTB prevalence in recorded surveys also reflects the prevailing livestock husbandry systems. According to Humblet et al. (2009), the management system will define the extent of contact between cattle, and between cattle and environmental sources. It is therefore important to assess the linkage between the prevailing livestock husbandry system and the occurrence of BTB.

From the point of view of BTB risk, African livestock production can be grouped into three broad categories (Katale et al. 2012; Unger and Münstermann 2004):

1. Pastoral (also called transhumance or nomadic)
2. Peri-urban dairy
3. Agropastoral, involving the integration of crop and livestock subsistence farming

In the following section, a detailed review of each system, as it relates to BTB risk, is provided.

Pastoral Production system Pastoral (transhumance or nomadic) production is an extensive, traditional, and common husbandry practice in many parts of Africa. It is practiced in every corner of the continent, usually forced by unevenly distributed precipitation with marked spatial and temporal variation in scarce feed and water resources. Herders and their livestock in Arid and Semi-Arid Land (ASAL) areas are forced to migrate periodically in search of surface water and green pasture, which are often shared by entire communities. Bovine TB seldom reaches a high incidence in cattle kept in the open; it is only when they are housed for intensive milk production that the incidence increases (Francis 1950). Extensive husbandry, with no housing as in pastoral production, presents a low risk for within-herd BTB transmission (Morris et al. 1994; Costello et al. 1998; Menzies and Neill 2000).

Both transhumant and nomadic systems of livestock production involve annual movement of livestock over variable distances, in some cases many hundreds of kilometers. Herders practicing transhumance have a regular and pre-defined movement pattern. The Fulani herdsmen, for example, migrate with their cattle to the southern parts of Nigeria during the rainy season and return north when the rain starts there (Opara 2005). In Zambia, cattle owners take their herds to the Kafue plains during the dryer months (May to October) and return to their villages during the rainy season (November to April) (Cook et al. 1996). Not infrequently, several discrete herds are rounded up at a given site (for safety) before the annual migration starts, forming a pool of herds, referred to as a "super herd." Upward of eight large herds could be brought together for this purpose (Munyeme et al. 2008). This type of herd management involving "super herds" has also been described in Uganda (Oloya et al. 2007) and elsewhere (Ryan et al. 2006). A particular type of "super herd" was described in Botswana, where herds sharing the same range in the north congregate regularly (twice a month) at crush pens for health monitoring. In the process, cattle herds brought to the crush pens cohabit, share the same grazing spaces, and are managed under the same grazing strategy (Jori et al. 2013). Analogous to what has been described in African buffaloes (Michel and Bengis 2012), and dictated by seasonal availability of pasture and surface water, these alternating fission- and fusion-like events constitute a powerful driving force for pathogen dispersal among herds and across geographic niches as well as for bringing the source of infections and susceptible populations together.

In Northeastern Kenya, Southern Ethiopia, and parts of Somalia, two types of pastoral production are recognized: resident and migratory. The resident type incorporates a permanent base where a few milking cows are kept to provide milk for part of the family of herders, including children, women, and the elderly. The cows usually graze far away from home in a communal grazing land, whereas calves graze around the homestead and are rounded up in open bush kraals at night. In the migratory type of production, a few herds belonging to close relatives are organized into a mobile herding group (analogous to a super herd). Migration distances, which may be in any direction, are determined by tribal interfaces and often do not exceed a 150-mile radius (Sheik-Mohamed and Velema 1999). During the wet season, when surface water and pasture are abundant, pastoralists disperse over large areas, while in the dry season, they tend to remain within a day's reach of a water source. Individual herds are kept separate during the day (although several herds share ranges and water sources) and in separate bush kraals at night. Most of the cattle kept in this production setting are local zebus (Katale et al. 2012) or other indigenous breeds such as the N'dama (Unger and Münstermann 2004).

While it is true that animals provide milk for home consumption, draft power for cultivation, and are the main source of meat in many countries (Mwakapuja et al. 2013), livestock keeping in rural areas of SSA transcends purely economic reasons. In particular, cattle are an integral part of human social life, including the generation and accumulation of wealth and in maintaining a complex, deeply interwoven social system of mutual obligation (Michel et al. 2006). This is accomplished in part by the exchange of cattle within and between families and other social groups during traditional events such as marriage (dowries) and for other types of goods and services. As a result of this value system, animals are seldom sold or slaughtered (unless for funerals or other cultural ceremonies), and it is difficult to enforce culling, even for disease control purposes.

Bovine TB Risk Under extensive farming conditions, cattle live entirely in the open and are only rounded up at night in an open kraal made of thorny bushes, as a protection from theft and predators. In such areas, the prevalence of BTB is normally very low. For example, Paine and Martinaglia (1929) detected only 2 cases out of 17,263 cattle slaughtered at the Grahamstown, South Africa, abattoirs over 5.5 years. The authors listed several factors that might have led to the observed low BTB prevalence: (a) the cattle were mostly local grade stock, (b) they were not housed, and (c) the area was very lightly stocked. Carmichael (1938) similarly attributed the rarity of BTB in the tropics to the open-air existence of local livestock, whereas du Toit (1936) espoused the view that BTB is relatively unimportant in tropical Africa. However, a different perspective was brought to light when it was established that the long-horned Ankole breed were infected with BTB to an extent far in excess of any known in the worst infected regions of Europe (Carmichael 1938). According to the author, these animals constantly roamed over open range and were never housed; yet, 80% were tuberculin-test-positive, while abattoir inspection yielded a prevalence of more than 40%.

To better understand the high prevalence of BTB among southern cattle in Sudan, Awad (1962) assessed the method of husbandry and herd management. During the wet season, in accordance with transhumant practice, herders drove their cattle to the highlands where they had built huts for them. The animals were crowded into the humid, unhygienic, and unventilated huts full of wood smoke to repel insects. The situation is exacerbated by the general practice to keep animals indefinitely, in most cases, until they die. Since younger and older animals were not separated, older tuberculous cows would transmit *M. bovis* to the younger stock. Daniel et al. (2009) similarly noted the transmission of *M. bovis* associated with introduction of animals reared in the adult female group.

A number of factors might result in the unexpected high prevalence of *M. bovis* infection in cattle in the extensive production sector. Animal movement, either through trading or communal grazing, including exposure to a wildlife maintenance host, was the most frequently reported variable determining herd-to-herd spread of BTB (Anon 1994; Johnston et al. 2011). The unrestricted movement of animals in the extensive husbandry system involves mixing of herds on communally owned ranges and around watering points during the dry season, leading to overcrowding. Studies conducted in Zambia provided compelling evidence in this regard. Cattle herds from the Monze district in Zambia move to the Kafue plains during the dry season in search of surface water and pasture. In the process, they mix with large populations of the Kafue lechwe, an endemic antelope and a maintenance host of BTB (Gallagher et al. 1972). Furthermore, cattle from different locations around the basin congregate on the plains, thus increasing the risk of contact with other herds (Cook et al. 1996). de Vos et al. (2001) suggested that a short period of such comingling is enough for *M. bovis* transmission. As a result, a high prevalence of BTB was recorded in cattle originating from the Monze area. Ryan et al. (2006) stated that movement of animals in and out of super herds is a plausible cause of the spread of BTB between herds. Consistent with this assumption, in Zambia, the herd-level BTB prevalence in transhumant herds (TH) was comparably higher than in the village-resident herds (Munyeme et al. 2008). In a similar study conducted in Tanzania, significantly more farms in the extensive sector had tuberculin reactors compared to farms in the intensive sector (Durnez et al. 2009). The authors concluded that because of the higher turnover of animals (buying, selling, gift, and mixing) in the extensive sector, animals were at a greater risk of exposure to and contracting the disease. Added to this is the problem of social unrest due to political instability and skirmishes along tribal interface areas, resulting in displacement of human and animal populations, and stock rustling.

In a nutshell, transhumance/nomadism and the consequent intermingling of animals are a way of pastoral life in many parts of Africa, and it is not likely to change any time soon. Even in the developed nations, in modern livestock production, the concept of discrete herds does not seem to be practiced as extensively as in the past. A study conducted in Australia found that only 10% of cattle herds were "closed" and that 13% of animals were moved between farms over a year (Ryan et al. 2006). In such open herds, enhanced BTB surveillance was necessary to reduce the number of infected herds (Barlow et al. 1998).

Surveillance for BTB in Africa is generally inadequate. For example, despite the reported high prevalence of BTB in Chad, Cameroon, and Nigeria, there are no facilities to quarantine and/or test animals for BTB before they cross national borders. Similarly, many African veterinary authorities at state or local levels do not actively apply measures for the control of BTB (Aliyu et al. 2009). As a consequence of this, two conditions for the continued persistence of the disease are being met: uncontrolled animal movement and poor disease surveillance. Hence, the stage is set for the further rapid dissemination of BTB throughout the pastoral systems of SSA, with the risk of creating a large number of BTB-positive herds, albeit with relatively low within-herd prevalences (Bernard et al. 2005; Gumi et al. 2012).

Intensive (Urban/Peri-urban) Dairy Production As part of a poverty reduction strategy, several African governments have been encouraging increased milk production through the introduction and/or expansion of peri-urban and other dairy development schemes. The impetus for transitioning to market-oriented production has been the rising consumer demand due to increasing population growth, urbanization, and affluence (Unger and Münstermann 2004). Dairy production in Africa is largely characterized by unrestricted livestock keeping in towns and cities, and involves increasing intensification of large numbers of small production units. For instance, in Addis Ababa, Ethiopia, there were more than 5200 dairy farms in 1994, with an average herd size of 12 (Asseged et al. 2000). Importation of European breeds of cattle such as Holstein-Friesian was the centerpiece of these operations. Therefore, the majority of animals are exotic breeds or have a high proportion (>75%) of exotic blood. This was particularly evident in the peri-urban and urban areas (Shirima et al. 2003) owing to market access and availability of resources and services (Katale et al. 2012). Occasionally, peri-urban dairy cattle development projects in small towns may promote the use of milk and dairy products from indigenous cows (Bonsu et al. 2000). Commercial dairy farms tend to remain few in number (<10% of the total livestock population in many countries) where the trend is toward keeping animals indoors and feeding processed rations (Katale et al. 2012). Firdessa et al. (2012) studied 80 farms located in and around Addis Ababa, Ethiopia. They were all intensive dairy farms, with cattle herds kept indoors, separate from other herds ($n = 78$): three small herds shared the same house with their owners, and in one farm, animals were kept outdoors. Only one-third of the farmers reported purchasing replacement stock from within the area during 2008 and 2009, whereas about 50 herders sold stock locally and to clients from other regions.

Bovine TB Risk In areas where market-oriented livestock production is practiced, the tendency has been to manage more intensively. Thus, more and more animals are confined on smaller areas on cultivated pastures, and in feedlots, cattle sheds, milking parlors, and dairies around major cities (Anon 2013). Since most cases of BTB in cattle are due to inhalation of tubercle bacilli (Francis 1950), intensive management in covered and closed yards (Inangolet et al. 2008), with high animal densities, will lead to an increased risk of aerogenous *M. bovis* transmission. As a result, the prevalence of BTB can be extremely high in dairy herds in many parts of Africa. Based on several estimates, the overall prevalence of BTB, in urban and peri-urban production systems in Ethiopia, ranged from 21.1% to 46.8% in animals and from 48.2% to 53.6% at the herd level (Asseged et al. 2000). O'Reilly and Daborn (1995) and Bernard et al. (2005) similarly reported a prevalence as high as 50% in both Tanzania and Malawi. Manley (1929) previously suggested that BTB can quickly and intensely infect cattle in the tropics, and animals should not be imported into the tropics unless they have passed a rigorous tuberculin test.

Tuberculosis can enter a herd by two major routes: via purchase of infected animals and by contiguous spread across farm boundaries (Munroe et al. 1999). In one review (Menzies and Neill 2000), from 654 herd breakdowns with a known source, 42% involved purchased cattle. Introduction of new animals was also an important risk factor in Uganda (Oloya et al. 2007). Jiwa et al. (1997) reported that animals purchased from Dar es Salaam into Kagera had a significantly higher rate (2.12%) of BTB compared to local cattle (no reactors). Although several dairy farms appear discrete, animals commonly comingle with those from other herds. This comingling is enhanced by hiring and sharing bulls in peri-urban dairy production systems (Elias et al. 2008; Jiwa et al. 1997). These bulls may run with a number of different herds throughout the year, thus acting as a potential source of *M. bovis* transmission should they be infected. As the dairy sector itself is still developing, emergent dairy producers often purchase their foundation stock from established dairy farms, often without accompanying health certification (Shirima et al. 2003). In a report by Tschopp et al. (2013) where two cows reacted to the single intradermal test (SIT) from >500 tested in central Ethiopia, the reactors were purchased from areas around Addis Ababa in the previous 5 years. Since BTB persists on a farm unless some mitigation is put in place (Marangon et al. 1998), cattle movement, without preliminary BTB testing and certification, could spread the infection from the infected farms to the newly established ones. Two recent studies in the United States suggested that the sources of infection in BTB in California (McCluskey et al. 2014) and Minnesota (Ribeiro-Lima et al. 2015) dairies were the purchase of BTB-infected animals and their movement onto new premises. Similarly, a comparative study of open (in which mixing with animals from other herds is possible) and closed (isolated) herds conducted in Addis Ababa (Asseged et al. 2000) corroborates these facts. According to the study, open herds were significantly more affected (63.4% BTB prevalence) compared to closed herds (36.8% BTB prevalence).

Global trade agreements can contribute to significantly increasing the risk of the inter-regional spread of BTB. For example, BTB in Danish cattle first appeared after tuberculous cows were imported from Germany (Magnus 1966). A recent assessment of BTB infections in the United States, in areas without a wildlife reservoir, identified a variety of risk factors, including the importation and comingling of steers from Mexico (Ribeiro-Lima et al. 2015). Similarly, it was postulated that cattle imported from France (in 1935) and the United Kingdom (in 1945) were responsible for the introduction of European strains of *M. bovis* into Iran (Tadayon et al. 2013). A review by Karolemeas et al. (2011) underscored the problem associated with trading of infected, TST-negative animals: this happened in 2008 when infected calves were exported from the United Kingdom into the Netherlands.

In summary, there is every reason to consider that the main cause of the high prevalence of BTB in dairy cattle is not their greater natural susceptibility to infection, but rather their crowded indoor existence, which increases their chances of exposure to infection. Francis (1950) emphasized the significance of inadequate ventilation and confining of livestock to the same house: "calves, which were housed with cows were exposed to a constant risk of infection by the aerogenous route, but when calves were not housed with cows, about 90% of them reached maturity without being infected, even if they were pastured with infected cows." Opie (1931) reached a similar conclusion with regard to the transmission *M. tuberculosis* in humans: "the prevalence of TB in Jamaica was in large part referable to the crowded dwellings of the poorer people, where whole families, consisting of as many as six children, lived in small huts or yard rooms." Therefore, although BTB occurs in both intensive and pastoral farming systems, the expanding dairy sector, along with intensive livestock management, will alter BTB dynamics in Africa in the future (Shirima et al. 2003).

To meet the rising demand for animal products in an increasingly growing population, there seems no better alternative than keeping improved stocks and changing the husbandry practices. There is, however, sufficient evidence to suggest that the application of broad principles of biosecurity will reduce the risk of cattle becoming infected by other animals, including wildlife (Skuce et al. 2012). Most of these emerging dairy producers are organized into associations (Asseged et al. 2000) or discrete bulk milk collection centers (Tschopp et al. 2013). These associations and centers coordinate requests for supplies and other services and also facilitate the acquisition of biotechnological inputs provided by livestock development agents. The silver lining is that the collected bulk milk is delivered to processing plants where it is pasteurized or processed at temperatures known to destroy mycobacteria. The risk of zoonotic TB through consumption of infected milk is therefore greatly reduced.

Smallholder Crop-Livestock Production System

Smallholder, crop-livestock production is a low input/low output production system that operates at the subsistence or semi-subsistence level, relying on family labor for growing crops. Livestock are kept largely as a secondary activity to support work on crop fields, with multiple other objectives including sales, inputs to agriculture (traction and manure), and subsistence (milk and milk products). Herd sizes are characteristically small (up to 15 head of cattle), and animals consist mainly of autochthonous breeds, such as the N'dama and short-horned zebu, which are typically adapted to inferior feeding and health regimens and to tropical climatic stressors that place serious limitations on their productivity levels. The animals are reared around the homestead and fed on natural pasture, cut grass, farm leftovers, and other by-products; only selected animals are nutritionally supplemented during the dry season (Unger and Münstermann 2004). Since infrastructures, such as road networks, are poorly developed, severely restricting access to market and other farm inputs, smallholder farmers have no incentives to invest in innovative production technologies. As of late, smallholders have been increasingly responding to new challenges such as an increasing demand for milk and other livestock products, by moving to more intensified and integrated mixed farming systems (Bonsu et al. 2000). As the availability of land is the main constraint, comingling of individual herds on communally owned grazing land is a common practice. Therefore, livestock owned by several owners living in the same village or neighborhood constitutes an epidemiological unit (Moiane et al. 2014).

Moiane et al. (2014) suggested that a high prevalence of BTB was observed in almost all livestock areas where small-scale farming was practiced. Bonsu et al. (2000) examined 400 heifers and 747 cows in the Dagme District of Ghana. The prevalence of BTB in heifers was 9.8% and 16.7% in cows. Similarly, Bernard et al. (2005) conducted a BTB survey in the dairy-producing areas of Mbarara District, Uganda. In total, 252 of 340 herds had at least 1 BTB-positive cow. Furthermore, the BTB prevalence was higher in herds (79.3%) and animals (8.3%) from the agropastoral zone, compared to the pastoral areas. In Tanzania, 5.7% (n = 105) of smallholder dairy farms contained tuberculin reactors compared to none from traditionally managed herds (Swai and Schoonman 2012). According to Moiane et al. (2014), trading of animals among smallholders is frequently performed without regard to BTB status. The management system in the smallholder sector, which involves sharing water source and grazing areas, and mixing of animals from different herds, increases the likelihood of the spread of BTB. A study conducted in the central Ethiopian highlands indicated that 94% of owners kept their animals, predominantly zebu and low-grade crosses, on communal grazing land (Tschopp et al. 2013). Additionally, during vaccination campaigns and other animal health service activities such as deworming and tick control, animals from different herds use the same dip tanks, thereby comingling in holding areas. Cattle regularly attending a dip tank had a higher prevalence of BTB than cattle in bulking groups who more commonly practiced on-farm disease control measures (Bedard et al. 1993).

7.4 Conclusion

A substantial amount of epidemiological information pertaining to BTB, both in domestic and in wild animals, exists in Africa. Four aspects stand out clearly:

1. Because of the growing population and the rising demand for animal products, there is an increasing trend to keep improved stock and to change the husbandry practice. Although BTB had been present in the local cattle population before improved cattle breeds were introduced into Africa, these activities have amplified its burden in Africa.
2. Pastoralism, in which cattle (and camels in certain areas) are an integral part of human social life, will remain a way of life for the foreseeable future in many parts of Africa. Because of its value system, the desire to maximize the headcount of stock, rather than the productivity of animals, remains a norm in this way of life.
3. Two conditions, i.e., increased herd size and confinement with high animal densities, have been created in the virtual absence of the application of any biosecurity measures and control of animal movement. Under such circumstances, BTB disseminates rapidly into new hosts (including wildlife) and new territories, once *M. bovis* infection has been established on the farm.
4. In small towns and in the countryside, the above conditions occur where homes are shared with shoats, calves, and fowls, all posing a significant risk for zoonotic TB transmission in rural Africa.

We will conclude by borrowing a segment from Mr. Ken Follett (World without End): "as it was in the beginning, it is now, and ever shall be; world without end."

There has been bovine and zoonotic TB in the past. However, all existing reports concur that its prevalence and potential implications have worsened over time in Africa. Unless we change course, the situation will only get worse by the day.

There is light at the end of the tunnel though, as several African experts have accumulated an encyclopedic knowledge about the biology, epidemiology, and diagnostic and immunologic aspects of BTB. All the necessary ingredients to launch effective BTB control programs are available, except the political will to allocate the necessary resources to implement the required control and eradication programs. The experts will have to make a concerted effort to rock the boat to create public alarm about the situation and to bring the politicians on board.

References

Acevedo P, Vicente J, Höfle U et al (2007) Estimation of European wild boar relative abundance and aggregation: a novel method in epidemiological risk assessment. Epidemiol Infect 135 (3):519–527

Adams AP, Bolin SR, Fine AE et al (2013) Comparison of PCR versus culture for detection of *M. bovis* after experimental inoculation of various matrices held under environmental conditions for extended periods. Appl Environ Microbiol 79(20):6501–6506

Alhaji I (1976) Bovine tuberculosis: a general review with special reference to Nigeria. Vet Bull 46 (11):829–839

Aliyu MM, Adamu JY, Bilyaminu YA (2009) Current prevalence of tuberculous lesions among slaughtered cattle in northeastern states of Nigeria. Revue Élev Méd vét Pays Trop 62(1):13–16

Allen GM (1988) Tuberculosis in sheep—a very rare disease. Surveillance 15(5):8–9

Anon (1994) Livestock disease eradication evaluation of the cooperative state–federal bovine tuberculosis eradication program. Committee on Bovine Tuberculosis. National Academy Press, Washington

Anon (2007a) Report of the meeting of bovine tuberculosis sub-group of the task force for monitoring animal disease eradication held in Seville, Spain, 14–15 November 2007. European Commission Health & Consumer Protection Directorate-General

Anon (2007b) Bovine TB: the scientific evidence. Final report of the independent scientific group on cattle TB. Department of Environment, Food and Rural Affairs, London

Anon (2013) Bovine tuberculosis scheme manual (Interim). Department of Agriculture, Forestry and Fisheries, Republic of South Africa. pp 80

Asseged B, Lübke-Becker A, Lemma E et al (2000) Bovine tuberculosis: a cross-sectional and epidemiological study in and around Addis Ababa. Bull Anim Health Prod Afr 48:71–80

Asseged B, Tamiru B, Habtemariam T (2014) Status and control of bovine tuberculosis in Ethiopia. In: Thoen CO, Steel JH, Kaneene JB (eds) Zoonotic tuberculosis: *Mycobacterium bovis* and other pathogenic mycobacteria, 3rd edn. Wiley-Blackwell, Ames, IA, pp 109–132

Awad FI (1962) Studies on type-determination and epidemiology of tuberculosis among cattle in Sudan. Transbound Emerg Dis 9(9):890–898

Awah-Ndukum J, Kudi AC, Bradley G et al (2010) Prevalence of bovine tuberculosis in abattoirs of the littoral and western highland regions of Cameroon: a cause for public health concern. Vet Med Int 2010:495015. p8. https://doi.org/10.4061/2010/495015

Awah-Ndukum J, Kudi AC, Bradley GI et al (2012) Prevalence of bovine tuberculosis in cattle in the highlands of Cameroon based on the detection of lesions in slaughtered cattle and tuberculin skin tests of live cattle. Vet Med 57(2):59–76

Aylate A, Shah SN, Aleme H et al (2013) Bovine tuberculosis: prevalence and diagnostic efficacy of routine meat inspection procedure in Woldiya municipality abattoir north Wollo zone, Ethiopia. Trop Anim Health Prod (3):855–864

Bala AN, Garba AE, Yazah AJ (2011) Bacterial and parasitic zoonoses encountered at slaughter in Maiduguri abattoir, Northeastern Nigeria. Vet World 4(10):437–443

Barlow ND, Kean JM, Hickling G et al (1997) A simulation model for the spread of bovine tuberculosis within New Zealand cattle herds. Prev Vet Med 32:57–75

Barlow ND, Kean JM, Caldwell NP et al (1998) Modelling the regional dynamics and management of bovine tuberculosis in New Zealand cattle herds. Prev Vet Med 36:25–38

Bedard BG, Martin SW, Chinombo D (1993) A prevalence study of bovine tuberculosis and brucellosis in Malawi. Prev Vet Med 16(3):193–205

Bengis RG, Kriek NPJ, Keet DF et al (1996) An outbreak of bovine tuberculosis in a free-living African buffalo (*Syncerus caffer* Sparrman) population in the Kruger National Park: a preliminary report. Onderstepoort J Vet Res 63:15–18

Berg S, Garcia-Pelayo MC, Muller B et al (2011) African 2, a clonal complex of *M. bovis* epidemiologically important in East Africa. J Bacteriol 193(3):670–678

Bernard F, Vincent C, Matthieu L et al (2005) Tuberculosis and brucellosis prevalence survey on dairy cattle in Mbarara milk basin (Uganda). Prev Vet Med 67:267–281

Berrada J (1993) *Mycobacterium bovis* infection in cattle in Morocco: preparation and evaluation of chemical extracts for use in detection of immune responses. PhD Thesis, Iowa State University

Beyi AF, Gezahegne KZ, Mussa A et al (2014) Prevalence of bovine tuberculosis in dromedary camels and awareness of pastoralists about its zoonotic importance in Eastern Ethiopia. J Vet Med Anim Health 6(4):109–115

Biet F, Boschiroli ML, Thorel MF et al (2005) Zoonotic aspects of *M. bovis* and *Mycobacterium avium-intracellulare* complex (MAC). Vet Res 36:411–436

Bonsu OA, Laing E, Akanmori BD (2000) Prevalence of tuberculosis in cattle in the Dangme-West district of Ghana, public health implications. Acta Tropica 76:9–14

Boukary AR, Thys E, Rigouts L et al (2012) Risk factors associated with bovine tuberculosis and molecular characterization of *M. bovis* strains in urban settings in Niger. Trans Emerg Dis 59 (6):490–502

Brahmbhatt DP, Fosgate GT, Dyason E et al (2012) Contacts between domestic livestock and wildlife at the Kruger National Park interface of the Republic of South Africa. Prev Vet Med 103:16–21

Brosch R, Gordon SV, Marmiesse M et al (2002) A new evolutionary scenario for the *Mycobacterium tuberculosis* Complex. Proc Natl Acad Sci USA 99(6):3684–3689

Brush OF (1898) The association of human and bovine tuberculosis. Wynkoop Hallenbeck Crawford Co. Printers, New York, p 140

Cadmus SIB, Atsanda NN, Oni SO et al (2004) Bovine tuberculosis in one cattle herd in Ibadan in Nigeria. Vet Med Czech 49(11):406–412

Carmichael J (1937) A brief note on tuberculosis in Tropical Africa with special reference to Uganda. JCPT 50:383–385

Carmichael JL (1938) Tuberculosis of sheep in Uganda. Vet Rec 50:1138–1147

Cassidy JP, Bryson DG, Pollock JM et al (1998) Early lesion formation in cattle experimentally infected with *M. bovis*. J Comp Pathol 119:2744

Cleaveland S, Shaw DJ, Mfinanga SG et al (2007) *Mycobacterium bovis* in rural Tanzania: risk factors for infection in human and cattle populations. Tubercle 87:30–43

Collins CH, Grange JM (1983) The bovine tubercle bacillus: a review. J Appl Bacteriol 55:13–29

Collins D, Radford AJ, de Lisle GW et al (1994) Diagnosis and epidemiology of bovine tuberculosis using molecular biological approaches. Vet Microbiol 40:83–94

Cook AJC, Tuchili LM, Buvet A et al (1996) Human and bovine tuberculosis in the Monze District of Zambia—a cross-sectional study. Br Vet J 152:37

Cordes DO, Bullians JA, Lake DE et al (1981) Observations on tuberculosis caused by *Mycobacterium bovis* in sheep. N Z Vet J 29:60–62

Corner LAL (2006) The role of wild animal populations in the epidemiology of tuberculosis in domestic animals: how to assess the risk. Vet Microbiol 112:303–312

Cosivi O, Grange JM, Daborn CJ et al (1998) Zoonotic tuberculosis due to *Mycobacterium bovis* in developing countries. Emerg Infect Dis 4(1):59–70

Costello E, Doherty ML, Monaghan TML et al (1998) A study of cattle-to-cattle transmission of *M. bovis* infection. Vet J 155:245–250

Costello E, O'Grady D, Flynn O et al (1999) Study of restriction fragment length polymorphism analysis and spoligotyping for epidemiological investigation of *M. bovis* infection. J Clin Microbiol 37(10):3217–3222

Cousins D (2001) *Mycobacterium bovis* infection and control in domestic livestock. Rev Sci Tech Off Int Epiz 20(1):71–85

Cowie CE, Marreos N, Gortázar C et al (2014) Shared risk factors for multiple livestock diseases: a case study of bovine tuberculosis and brucellosis. Res Vet Sci 97:491–449

Daborn CJ, Grange JM (1993) HIV/AIDS and its implications for the control of animal tuberculosis. Br Vet J 149:405–417

Dailloux MC, Laurain C, Weber M et al (1999) Water and nontuberculous mycobacteria. Water Res 33(10):2219–2228

Damina MS, Owoludun OA, Chukwukere S et al (2011) The use of deletion analysis in the detection of *Mycobacterium bovis*, *Mycobacterium tuberculosis* and *Mycobacterium africanum* among slaughtered cattle in Plateau State, north central Nigeria. Niger Vet J 32(1):9–15

Daniel ZRH, Evans S, Rolfe R et al (2009) Outbreak of tuberculosis caused by *M. bovis* in golden Guernsey goats in Great Britain. Vet Rec 165:335–342

Davidson RM, Alley MR, Beatson NS (1981) Tuberculosis in a flock of sheep. N Z Vet J 29:1–2

De Klerk-Lorist L-M (2004) The evaluation of a BCG vaccine against bovine tuberculosis in African buffalo (*Syncerus caffer*). MSc dissertation. University of Pretoria

Deresa B, Conraths FJ, Ameni G (2013) Abattoir-based study on the epidemiology of caprine tuberculosis in Ethiopia using conventional and molecular tools. Acta Vet Scand 55(1):15

de Vos V, Bengis RG, Kriek NPJ et al (2001) The epidemiology of tuberculosis in free-ranging African buffalo (*Syncerus caffer*) in the Kruger National Park, South Africa. Onderstepoort J Vet Res 68:119–130

Di Marco V, Mazzone P, Capucchio MT et al (2012) Epidemiological significance of the domestic black pig (*Sus scrofa*) in maintenance of bovine tuberculosis in Sicily. J Clin Microbiol 50 (4):1209–1218

Drewe JA, Pfeiffer DU, Kaneene JB (2014) Epidemiology of *M. bovis*. In: Thoen CO, Steele JH, Kaneene JB (eds) Zoonotic tuberculosis: *M. bovis* and other pathogenic mycobacteria, 3rd edn. Wiley, Oxford, pp 63–77

Driscoll EE, Hoffman JI, Green LE et al (2011) A preliminary study of genetic factors that influence susceptibility to bovine tuberculosis in the British cattle herd. PLoS One 6(4):e18806

Duffield BJ, Young DA (1985) Survival of *Mycobacterium bovis* in defined environmental conditions. Vet Microbiol 10:193–197

Durnez L, Sadiki H, Katakweba A et al (2009) The prevalence of *Mycobacterium bovis* infection and atypical mycobacterioses in cattle in and around Morogoro, Tanzania. Trop Anim Health Prod 41:1653–1659

Du Toit PJ (1936) Bovine tuberculosis in South Africa. Tuberculosis 5:421–422

Edwards DS, Johnston AM, Mead GC (1997) Meat inspection: an overview of present practices and future trends. Vet J 154:135–147

Ejeh EF, Adeshokan HK, Raji MA et al (2014) Current status of bovine tuberculosis in Otukpo, Nigeria. J Anim Prod Adv 4(8):501–507

Elias K, Hussein D, Asseged B et al (2008) Status of bovine tuberculosis in Addis Ababa dairy farms. Rev Sci Tech Off Int Epiz 27(3):915–923

Elmossalami E, Siam MA, Sergany ME (1971) Studies on tuberculous-like lesions in slaughtered camels. Zoonoses Public Health 18:253–261

El-Olemy GM, El-Bassiouni AAA, Negm S (1985) Tuberculosis in Toukh-Tanbisha, Menufia, Egypt. In: Proceedings of the 4th international symposium on veterinary epidemiology and economics. www.sciquest.org.nz

El Sanousi SM, Omer EE (1985) Bovine tuberculosis in Benghazi Cow Project (Libya). Int J Zoonoses 12(3):203–206

Evangelista TBR, De Anda JH (1996) Tuberculosis in dairy calves: risk of *Mycobacterium* spp. exposure associated with management of colostrum and milk. Prev Vet Med 27:23–27

Fine AE, O'Brien DJ, Winterstein SR et al (2011) An effort to isolate *M. bovis* from environmental substrates during investigations of bovine tuberculosis transmission sites (cattle farms and wildlife areas) in Michigan, USA. ISRN Vet Sci. p 11. https://doi.org/10.5402/2011/787181

Firdessa R, Tschopp R, Wubete A et al (2012) High prevalence of bovine tuberculosis in dairy cattle in central Ethiopia: implications for the dairy industry and public health. PLoS One 7(12): e52851

Fitzgerald SD, Kaneene JB (2012) Wildlife reservoirs of bovine tuberculosis worldwide: hosts, pathology, surveillance, and control. Vet Pathol 50(3):488–499

Francis J (1950) Control of infection with the bovine tubercle bacillus. Lancet 255(6593):34–39

Fullerton AGR (1902) A case of tuberculosis in a sheep. J Comp Pathol 15:102–104
Gallagher J, Macadam I, Sayer J et al (1972) Pulmonary tuberculosis in free-living lechwe antelope in Zambia. Trop Anim Health Prod 4:204–213
Gathogo SM, Kuria JKN, Ombui JN (2012) Prevalence of bovine tuberculosis in slaughter cattle in Kenya: a postmortem, microbiological and DNA molecular study. Trop Anim Health Prod 44:1739–1744
Gray DF, Graham-Smith H, Noble JL (1960) Variations in natural resistance to tuberculosis. J Hyg Camb 58:215–227
Griffin JM, Hahesy T, Lynch K et al (1993) The association of cattle husbandry practices, environmental factors and farmer characteristics with the occurrence of chronic BTB in dairy herds in the Republic of Ireland. Prev Vet Med 17:145–160
Griffin JM, Martin SW, Thorburn MA et al (1996) A case-control study on the association of selected risk factors with the occurrence of bovine tuberculosis in the Republic of Ireland. Prev Vet Med 27:75–87
Guilbride PDL, Rollinson DHL, Mcanulty EG et al (1963) Tuberculosis in the free living African (Cape) Buffalo (*Syncerus caffer caffer*. Sparrman). J Comp Pathol 73:337–348
Gumi B, Schelling E, Firdessa R et al (2012) Low prevalence of bovine tuberculosis in Somali pastoral livestock, southeast Ethiopia. Trop Anim Health Prod 44(7):1445–1450
Habtemariam T (1989) Utility of epidemiologic simulation models in the planning of trypanosomiasis control programs. Ann de la Societe Belge de. Med Trop 69:109–124
Hiko A, Agga GE (2011) First-time detection of *Mycobacterium* species from goats in Ethiopia. Trop Anim Health Prod 43:133–139
Humblet MF, Boschiroli ML, Saegerman C (2009) Classification of worldwide bovine tuberculosis risk factors in cattle: a stratified approach. Vet Res 40:1–24
Ibrahim S, Cadmus SIB, Umoh JU et al (2012) Tuberculosis in humans and cattle in Jigawa State, Nigeria: risk factors analysis. Vet Med Int. https://doi.org/10.1155/2012/865924
Inangolet FO, Demelash B, Oloya J et al (2008) A cross-sectional study of BTB in the transhumant and agro-pastoral cattle herds in the border areas of Katakwi and Moroto districts, Uganda. Trop Anim Health Prod 40:501–508
Jenkins AO, Cadmus SIB, Venter EH (2011) Molecular epidemiology of human and animal tuberculosis in Ibadan, Southwestern Nigeria. Vet Microbiol 151:139–147
Jiwa SFH, Kazwala RR, Aboud AAO et al (1997) Bovine tuberculosis in the Lake Victoria Zone of Tanzania and its possible consequences for human health in the HIV/AIDS era. Vet Res Commun 21:533–539
Johnston AJK, Conlan CA, Donnelly JLN et al (2011) Recurrence of bovine tuberculosis breakdowns in Great Britain: risk factors and prediction. Prev Vet Med 102:22–29
Jolles AE, Cooper DV, Levin SA (2005) Hidden effects of chronic tuberculosis in African buffalo. Ecology 86(9):2258–2264
Jori F, Mokospasetso M, Etter E et al (2013) Preliminary assessment of bovine tuberculosis at the livestock/wildlife interface in two protected areas of northern Botswana. Trans Emerg Dis 60 (Suppl. 1):28–36
Kassa GM, Abebe F, Worku Y et al (2012) Tuberculosis in goats and sheep in Afar pastoral region of Ethiopia and isolation of *Mycobacterium tuberculosis* from goat. Vet Med Int 2012: 8. https://doi.org/10.1155/2012/869146
Karolemeas K, McKinle TJ, Clifton-Hadley RS et al (2010) Predicting prolonged bovine tuberculosis breakdowns in Great Britain as an aid to control. Prev Vet Med 97:183–190
Karolemeas K, McKinle TJ, Clifton-Hadley RS et al (2011) Recurrence of bovine tuberculosis breakdowns in Great Britain: risk factors and prediction. Prev Vet Med 102:22–29
Katale BZ, Mbugi EV, Karimuribo ED et al (2009) The prevalence of *M. bovis* infection and atypical mycobacterioses in cattle in and around Morogoro. BMC Vet Res 9:267
Katale BZ, Mbugi EV, Kendal S et al (2012) Bovine tuberculosis at the human-livestock-wildlife interface: is it a public health problem in Tanzania? A review. Onderstepoort J Vet Res 79 (2):463. https://doi.org/10.4102/ojvr.v79i2.463

Kazwala RR, Daborn CJ, Kusiluka LJ et al (1998) Isolation of *Mycobacterium* species from raw milk of pastoral cattle of the Southern Highlands of Tanzania. Trop Anim Health Prod 30 (4):233–239

Kinne J, Johnson B, Jahans KL et al (2006) Camel tuberculosis—a case report. Trop Anim Health Prod 38:207–213

Langmuir AD (1961) Epidemiology of airborne infection. Bacteriol Rev 25(3):173–181

Laval G, Ameni G (2004) Prevalence of bovine tuberculosis in Zebu cattle under traditional animal husbandry in Boji district of western Ethiopia. Rev Med Vet (Toulouse) 155:494–499

Lewis PA, Sanderson ES (1927) The histological expression of the natural resistance of rabbits to infection with human and bovine type tubercle bacilli. J Exp Med 45(2):291–304

Lönnroth K, Jaramillo E, Williams BG et al (2009) Drivers of tuberculosis epidemics: the role of risk factors and social determinants. Soc Sci Med 68:2240–2246

Lurie MB (1930) Experimental epidemiology of tuberculosis: the effect of crowding upon tuberculosis in guinea pigs, acquired by contact and by inoculation. J Exp Med 51(5):729–741

Maddock ECG (1934) Further studies on the survival time of the bovine tubercle bacillus in soil, soil and dung, in dung and on grass, with experiments on feeding guinea pigs and calves on grass artificially infected with bovine tubercle bacilli. J Hygiene 34(3):372–379

Magnus K (1966) Epidemiological basis of tuberculosis eradication: 3. Risk of pulmonary tuberculosis after human and bovine infection. Bull World Health Organ 35:483–508

Malone E, Wilson EC, Pollock JM et al (2003) Investigations into an outbreak of tuberculosis in a flock of sheep in contact with tuberculous cattle. J Vet Med B 50:500–504

Mamo G, Kassaye A, Sanni M et al (2009) A cross sectional study of camel tuberculosis in Ethiopia. Bull Anim Health Prod Afr 57:13–20

Mamo G, Bayleyegn G, Sisay TT (2011) Pathology of camel tuberculosis and molecular characterization of its causative agents in pastoral regions of Ethiopia. PLoS One 6(1):15862

Manley FH (1929) A Note on bovine tuberculosis in tropical Africa (British Cameroons). J Comp Pathol Therap 42:276–277

Marangon S, Martini M, Dalla PM et al (1998) A case-control study on bovine tuberculosis in the Veneto Region (Italy). Prev Vet Med 34:87–95

Marianelli C, Cifani N, Capucchio MT et al (2010) A case of generalized bovine tuberculosis in a sheep. J Vet Diagn Invest 22:445–448

Mason FE (1912) Some observations on tuberculosis in camels in Egypt. J Comp Pathol Ther 25:109–111

Mason FE (1917) Tuberculosis in camels. J Comp Pathol Ther 30:80–84

McCluskey B, Lombard J, Strunk S et al (2014) *Mycobacterium bovis* in California dairies: a case series of 2002–2013 outbreaks. Prev Vet Med 115:205–216

McFadyean J (1900) Tuberculosis of the sheep. J Comp Pathol 13:59–60

Menzies FD, Neill SD (2000) Cattle-to-cattle transmission of bovine tuberculosis. Vet J 160:92–106

Michel AL (2008) Tuberculosis in wild and domestic animals in South Africa. PhD thesis, Universiteit Utrecht, The Netherlands

Michel AL, Bengis RG (2012) The African buffalo: a villain for inter-species spread of infectious diseases in southern Africa. Onderstepoort J Vet Res 79(2):453. https://doi.org/10.4102/ojvr.v79i2.453

Michel AL, Bengis RG, Keet DF et al (2006) Wildlife tuberculosis in South African conservation areas: implications and challenges. Vet Microbiol 112:91–100

Michel AL, de Klerk L-M, van Pittius NCG et al (2007) Bovine tuberculosis in African buffaloes: observations regarding *M. bovis* shedding into water and exposure to environmental mycobacteria. BMC Vet Res 3(1):23

Michel AL, Hlokwe TM, Coetzee ML et al (2008) High *M. bovis* genetic diversity in a low prevalence setting. Vet Microbiol 126:151–159

Moiane I, Machado A, Santos N et al (2014) Prevalence of bovine tuberculosis and risk factor assessment in cattle in rural livestock areas of Govuro District in the southeast of Mozambique. PLoS One 9(3):e91527

Morris RS, Pfeiffer DU, Jackson R (1994) The epidemiology of *Mycobacterium bovis* infections. Vet Microbiol 40:153–177

Moustafa H, Mostafa AMB, Zayed I (1964) Some observations on the incidence and histopathology of pneumonia in Egyptian sheep with special reference to two cases with tuberculosis. Zentralbl Veterinarmed B 11(3):231–239

Müller BM, Hilty S, Berg MC et al (2009) African 1, an epidemiologically important clonal complex of *M. bovis* dominant in Mali, Nigeria, Cameroon, and Chad. J Bacteriol 191 (6):1951–1960

Munroe FA, Dohoo IR, McNab WB et al (1999) Risk factors for the between-herd spread of *Mycobacterium bovis* in Canadian cattle and cervids between 1985 and 1994. Prev Vet Med 41:119–133

Munyeme M, Muma JB, Skjerve E et al (2008) Risk factors associated with bovine tuberculosis in traditional cattle of the livestock/wildlife interface areas in the Kafue basin of Zambia. Prev Vet Med 85:317–328

Munyeme M, Muma JB, Munang'andu HM (2011) Failure to detect tuberculosis in Black lechwe antelopes (*Kobus leche smithemani*) in Zambia. BMC Res Notes 4:233. http://www.biomedcentral.com/1756-0500/4/233

Munyeme M, Munang'andu HM, Nambota A et al (2012) The nexus between bovine TB and fasciolosis infections in cattle of the Kafue Basin ecosystem in Zambia: implications on abattoir surveillance. Vet Med Int. https://doi.org/10.1155/2012/921869

Muwonge A, Kankya C, Godfroid J et al (2010) Prevalence and associated risk factors of mycobacterial infections in slaughter pigs from Mubende district in Uganda. Trop Anim Health Prod 42:905–913

Muwonge A, Johansen TB, Vigdis E et al (2012) *Mycobacterium bovis* infections in slaughter pigs in Mubende district, Uganda: a public health concern. BMC Vet Res 8:168

Mwakapuja RS, Makondo ZE, Malakalinga J et al (2013) Molecular characterization of *M. bovis* isolates from pastoral livestock at Mikumi-Selous ecosystem in the eastern Tanzania. Tuberculosis 93:668–674

Napp S, Allepuz A, Mercader I et al (2013) Evidence of goats acting as domestic reservoirs of bovine tuberculosis. Vet Rec 172:663

Neill SD, Hanna J, O'Brien JJ et al (1988) Excretion of *Mycobacterium bovis* by experimentally infected cattle. Vet Rec 123:340–343

Neill SD, O'Brien JJ, Hanna J (1991) A mathematical model for *M. bovis* excretion from tuberculous cattle. Vet Microbiol 28:103–109

Ngandolo BNR, Müller B, Diguimbaye-Djaïbe C et al (2009) Comparative assessment of fluorescence polarization and tuberculin skin testing for the diagnosis of bovine tuberculosis in Chadian cattle. Prev Vet Med 89:81–89

Nicholson MJ, Butterworth MH (1986) A guide to condition scoring of zebu cattle. ILRI, Addis Ababa, Ethiopia

Nugent G (2011) Maintenance, spillover and spillback transmission of bovine tuberculosis in multi-host wildlife complexes: a New Zealand case study. Vet Microbiol 151:34–42

Okeke LA, Cadmus S, Okeke IO et al (2014) Prevalence and risk factors of *Mycobacterium tuberculosis* Complex infection in slaughtered cattle at Jos South Abattoir, Plateau State, Nigeria. Pan Afr Med J 18(S1):5. https://doi.org/10.11694/pamj.supp.2014.18.1.3841

Oloya J, Opuda-Asibo J, Djønne B et al (2006) Responses to tuberculin among Zebu cattle in the transhumance regions of Karamoja and Nakasongola district of Uganda. Trop Anim Health Prod 38:275–283

Oloya J, Muma JB, Opuda-Asibo J et al (2007) Risk factors for herd-level BTB seropositivity in transhumant cattle in Uganda. Prev Vet Med 80:318–329

Opara M (2005) Pathological conditions of condemned bovine lungs from abattoirs in Akwa Ibom State, Nigeria. Anim Res Int 2(2):314–318

Opie EL (1931) The epidemiology of tuberculosis of Negroes. Tubercle 12(5):207–214

Opuda-Asibo J (1995) Regional and Country Status Reports, Uganda. In: Thoen CO, Steel JH (eds) *Mycobacterium bovis* infection in animals and humans, 1st edn. Iowa State University Press, Ames, IA, pp 299–303

O'Reilly L, Daborn CJ (1995) The epidemiology of *Mycobacterium bovis* infections in animals and man: a review. Tuber Lung Dis 7:1–46

Paine R, Martinaglia G (1929) Tuberculosis in wild buck living under natural conditions. J Comp Pathol Ther 42:1):1–1):8

Palmer MV (2013) *Mycobacterium bovis*: characteristics of wildlife reservoir hosts. Trans Emerg Dis 60:1–13

Palmer MV, Waters WR, Whipple DL (2004) Investigation of the transmission of *M. bovis* from deer to cattle through indirect contact. Am J Vet Res 65:1483–1489

Pesciaroli M, Alvarez J, Boniotti MB et al (2014) Tuberculosis in domestic animal species. Res Vet Sci 97:S78–S85

Phillips CJ, Foster CR, Morris PA et al (2002) Genetic and management factors that influence the susceptibility of cattle to *Mycobacterium bovis* infection. Anim Health Res Rev 3(1):3–13

Phillips CJC, Foster CRW, Morris P et al (2003) The transmission of *Mycobacterium bovis* infection to cattle. Rev Res Vet Sci 74:1–15

Pittler DR, Steel JH (1995) Germany: regional and country status reports. In: Thoen CO, Steel JH (eds) *M. bovis* infection in animals and humans. Iowa State University Press, Ames, IA

Ranney F (1958) Status of federal-state cooperative tuberculosis eradication. Dis Chest 34 (6):577–585

Renwick R, White PCL, Bengis RG (2007) Bovine tuberculosis in southern African wildlife: a multi-species host-pathogen system. Epidemiol Infect 135:529–540

Ribeiro-Lima J, Enns EA, Thompson B et al (2015) From network analysis to risk analysis—An approach to risk-based surveillance for bovine tuberculosis in Minnesota, US. Prev Vet Med 118(4):328–340

Rodwell TC, Whyte IJ, Boyce WM (2001) Evaluation of population effects of bovine tuberculosis in free-ranging African buffalo (*Syncerus caffer*). J Mammal 82(1):231–238

Ryan TJ, Livingstone PG, Ramsey DSL et al (2006) Advances in understanding disease epidemiology and implications for control and eradication of tuberculosis in livestock: the experience from New Zealand. Vet Microbiol 112:211–219

Sahraoui N, Müller B, Guetarni D et al (2009) Molecular characterization of *M. bovis* strains isolated from cattle slaughtered at two abattoirs in Algeria. BMC Vet Res 5:4. https://doi.org/10.1186/1746-6148-5-4

Sahraoui N, Müller B, Mamache B et al (2011) Tuberculosis in cattle and goats in the north of Algeria. Vet Res 4(4):100–103

Sakamoto K (2012) The pathology of *M. tuberculosis* infection. Vet Pathol 49(3):423–439

Sheik-Mohamed A, Velema JP (1999) Where health care has no access: the nomadic populations of sub-Saharan Africa. Trop Med Int Health 4(10):695–707

Shirima GM, Kazwala RR, Kambarage DM (2003) Prevalence of bovine tuberculosis in cattle in different farming systems in the eastern zone of Tanzania. Prev Vet Med 57:167–172

Skuce RA, Allen AR, McDowell SWJ (2012) Herd-level risk factors for bovine tuberculosis: a literature review. Vet Med Int 2012:10. https://doi.org/10.1155/2012/621210

Smith T (1898) A comparative study of bovine tubercle bacilli and of human bacilli from sputum. J Exp Med 3:451–511

Smith NH (2012) The global distribution and phylogeography of *M. bovis* clonal complexes. Infect Gen Evol 12:857–865

Sulieman M, Hamid ME (2002) Identification of acid-fast bacteria from caseous lesions in cattle in Sudan. J Vet Med B 49:415–418

Swai SE, Schoonman L (2012) Differences in prevalence of tuberculosis in indigenous and crossbred cattle under extensive and intensive management systems in Tanga Region of Tanzania. Trop Anim Health Prod 44:459–465

Tadayon K, Mosavari N, Feizabadi MM (2013) An epidemiological perspective on bovine tuberculosis spotlighting facts and dilemmas in Iran, a historically Zebu-dominant farming country. Iranian J Microbiol 5(1):1–13

Tafess K, Dawo F, Sori T et al (2011) Prevalence of caprine tuberculosis in Mid-Rift Valley area of Oromia, Ethiopia. Afr J Microbiol Res 5(12):1473–1478

Tag el Din MH, el Nour Gamaan I (1982) Tuberculosis in sheep in the Sudan. Trop Anim Health Prod 14(1):26

Tanner M, Michel AL (1999) Investigation of the viability of *M. bovis* under different environmental conditions in the Kruger National Park. Onderstepoort J Vet Res 66:185–190

Tanner M, Inlameia O, Michel A et al (2014) Bovine tuberculosis and brucellosis in cattle and African buffalo in the Limpopo National Park, Mozambique. Transbound Emerg Dis 62 (6):632–638

Tarnagda Z, Kanyala E, Zingué D et al (2014) Prevalence of tuberculosis in bovine carcasses in two slaughterhouses of Burkina Faso. Int J Microbiol Immunol Res 2(6):92–100

Tschopp R, Schelling E, Hattendorf J et al (2010) Repeated cross-sectional skin testing for bovine tuberculosis in cattle kept in a traditional husbandry system in Ethiopia. Vet Rec 167:250–256

Tschopp R, Bobosha K, Aseffa A et al (2011) Bovine tuberculosis at a cattle-small ruminant-human interface in Meskan, Gurage region, Central Ethiopia. BMC Infect Dis 11(1):318

Tschopp R, Abera B, Sourou SY et al (2013) Bovine tuberculosis and brucellosis prevalence in cattle from selected milk cooperatives in Arsi zone, Oromia region, Ethiopia. BMC Vet Res 9 (1):163

Unger F, Münstermann S (2004) Assessment of the impact of zoonotic infections (bovine tuberculosis and brucellosis) in selected regions of the Gambia, Senegal, Guinea, and Guinea Bissau. - A scoping study - DFID Animal Health Programme. Banjul, the Gambia

Vekemans M, Cartoux M, Diagbouga S et al (1999) Potential source of human exposure to *Mycobacterium bovis* in Burkina Faso, in the context of the HIV epidemic. Clin Microbiol Infect 5:617–621

Whipple DL, Bolin CA, Mille JM (1996) Distribution of lesions in cattle infected with *Mycobacterium bovis*. J Vet Diagn Invest 8:351–354

Wilesmith JW, Little TWA, Thompson HV et al (1982) Bovine tuberculosis in domestic and wild mammals in an area of Dorset. I. Tuberculosis in cattle. J Hyg 89:195–210

Woodford MH (1982) Tuberculosis in wildlife in the Ruwenzori National Park, Uganda (Part II). Trop Anim Health Prod 14(3):155–160

Chapter 8
Molecular Epidemiology of *Mycobacterium bovis* in Africa

Adrian Muwonge, Franklyn Egbe, Mark Bronsvoort, Demelash B. Areda, Tiny Hlokwe, and Anita Michel

8.1 Introduction

Bovine tuberculosis (BTB) is a chronic, systemic disease caused by *Mycobacterium bovis*. It is characterized by gradual physical deterioration, lymphadenopathy, and development of tuberculous lesions that may be extensive, mainly in the lungs and associated lymph nodes. The presence of lesions in other body systems is possible and varies according to the route of infection and the extent of dissemination of the

A. Muwonge (✉)
The Roslin Institute, College of Medicine and Veterinary Medicine, University of Edinburgh, Edinburgh, Midlothian, UK
e-mail: Adrian.muwonge@roslin.ed.ac.uk

F. Egbe
Microbiology and Parasitology Unit, Faculty of Allied Medical Sciences, University of Calabar, Calabar, Nigeria

M. Bronsvoort
Genetics and Genomics, Roslin Institute, Edinburgh, Midlothian, UK
e-mail: mark.bronsvoort@roslin.ed.ac.uk

D. B. Areda
College of Science Engineering and Technology (CSET), Grand Canyon University, Phoenix, AZ, USA

Department of Biological Sciences, Minnesota State University, Mankato, Mankato, MN, USA
e-mail: demelash.areda@mnsu.edu

T. Hlokwe
Zoonotic Diseases Section, ARC-Onderstepoort Veterinary Institute, Onderstepoort, South Africa
e-mail: hlokwet@arc.agric.za

A. Michel
Department Veterinary Tropical Diseases, Faculty of Veterinary Science, University of Pretoria, Onderstepoort, South Africa
e-mail: anita.michel@up.ac.za

A. B. Dibaba et al. (eds.), *Tuberculosis in Animals: An African Perspective*,
https://doi.org/10.1007/978-3-030-18690-6_8

pathogen via the bloodstream (Neill et al. 2005). Bovine TB is endemic in most African countries, and 80% (33/43) of the OIE member countries on the African continent reported its presence (Daborn et al. 1996). These infections result in an unknown but probably substantial economic loss by the livestock industry, mostly by failing to participate in the international trade in livestock and livestock products because of the presence of BTB in the herds. Apart from the domestic animal species, BTB also occurs in a wide range of wild animal species, some of which have become maintenance hosts that pose a threat of maintenance and transmission of the diseases at the wildlife-livestock interface. From a public health perspective, zoonotic tuberculosis is transmitted through the consumption of raw, infected milk and aerosol exposure during close physical contact with *Mycobacterium bovis*-infected live or dead animals (Cosivi et al. 1998).

Despite its known endemic status, data accurately describing the distribution and prevalence of BTB and its causal agent, *M. bovis*, in Africa are lacking or inadequate since most countries do not maintain functional and timely disease recording and reporting systems. The limited available data have, however, been reviewed to document its occurrence and distribution in animals and humans on the continent (Ayele et al. 2004).

8.2 A Brief History of Cattle Movements in Africa

Animal population movements are known to be important in the spread of infectious diseases (Fèvre et al. 2006). Disease dynamics are a function of host-population characteristics and dynamics, and within this context, pathogens are "hitchhikers" on these waves of activities and events that occur in animal populations (Muwonge et al. 2016). Animal population movements are known to be important in the spread of infectious diseases (Fèvre et al. 2006). It is therefore critical to analyze the drivers, numbers, direction of movement, and associated disease risks over time in relation to changing environmental factors.

8.2.1 Ancient Cattle Movements

Archaeological findings from the neolithic time indicate that hunter-gatherers in the Nilo-Saharan communities that occupied the dry grasslands of Northern Africa about 6000 years ago also started practicing agriculture. The Mandé and Omotic people, a century later, also domesticated a number of species such as cattle, chickens, dogs, sheep, goats, and camels (Clark and Brandt 1984; Diamond 1999; Christopher 2002).

The earliest eastern and western ethnic migration of the Bantu in Africa started about 4000 years ago. This migration saw them leave their homeland in Western Africa, from where they moved through Central Africa to Eastern Africa and then

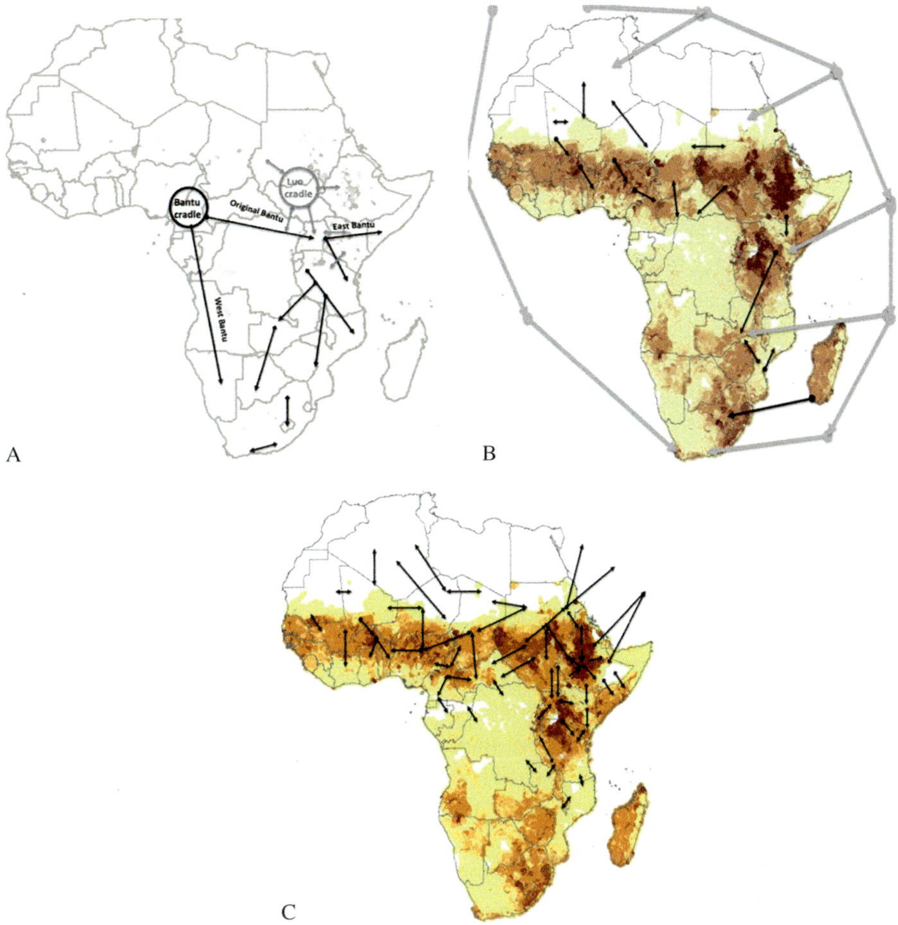

Fig. 8.1 Precolonial and ancient animal movements. Map **a** shows the great Bantu and Luo migration with their livestock from their homeland in Southern Cameroon and Sudan (~2000–1000 BC), respectively. Maps **b** and **c** represent the livestock movement in the colonial and postcolonial era. The thick gray arrows in **b** show the routes of exotic cattle breed introduction. The color gradient represents the cattle population density reflected in Fig. 8.3

into Southern Africa (Adler et al. 2007). Since this great migration occurred after the domestication of animals, it is reasonable to assume that they took their livestock with them. This notion was supported by language cognates associated with livestock in the eastern and southern Bantu, which suggest that they had livestock in their settlements following migration (Fig. 8.1) (Barbieri et al. 2014; Castrì et al. 2009).

The Bantu was not the only ethnic group in Africa that migrated as, a century later, the cattle-keeping Luo ethnic group started their migration from Southern Sudan into Uganda in the south, westward to the Congo, and eastward into Ethiopia and Kenya (Fig. 8.1) (Clark and Brandt 1984). This phase in the history of humanity (human settlement, livestock domestication, and agricultural development) is reported to have led to a surge in zoonotic and crowd-related diseases (Perry et al. 2011).

The Greeks and Romans, during the classical antiquity era, exerted the earliest foreign influence on the continent when they entered the Mediterranean North African region. The Arabs later followed them during the seventh century and extended their occupation to Madagascar (Boardman 1965). At that time, the European presence was limited to inhabited islands like the Cape Verde Islands (Khapoya 2012), and they only ventured into the interior of the continent during the nineteenth century. Prior to the arrival of the Europeans, most African societies were organized into Kingdoms or Chiefdoms in which agriculture and animal husbandry provided an integral part of household income, and cattle were used as payment for dowry (Goody 1976; Olson 1996).

8.2.2 Introduction of Exotic Cattle Breeds into Africa

The arrival of the Europeans in Africa initiated the second wave of animal movement across the continent (Fig. 8.1). This event linked the livestock of inland Africa to those of the European cattle-breeding societies that at the time were developing new breeds of cattle (Mäki-Tanila et al. 2010). Some of the African countries, such as South Africa, established breeding societies by 1910 that served as "conduits" through which exotic genetic material was introduced into Africa. In addition to developing local breeds such as the Afrikaner and Bonsmara, these societies also imported exotic breeding bulls of breeds such as the Limousin, Charolais, Angus, Guernsey, Hereford, and Brahman from France, the United Kingdom, and the United States (Afrikaner 2009; Stackyard 2009). The importation of bulls and heifers continues to date with the most prominent being the recent importation of Romagnola bulls into South Africa from Italy (Armando 1995). This influx of European cattle breeds was not limited to South Africa, but they were also imported into Kenya, Zambia, Uganda, and Zimbabwe, the animals mostly originating from the United Kingdom and France (Taneja 1999; Hansard and Harrison 1958; Bothar 2008). African breeds such as the Boran from East Africa were also brought to Southern Africa for breeding purposes (Fratkin 2001; Esther 2012).

8.2.3 Commercial Cattle Markets

Colonial times also saw the growth and establishment of commercial markets for cattle and their products. In South Africa, thousands of animals were herded from Pondoland to be sold in Pietermaritzburg, driven by the emerging hides industry in the 1860s (Bienart 1989). This was also the time when rinderpest was decimating cattle populations on the continent. In East Africa, up to 95% of the cattle population was lost and led to the great Ethiopian famine at the end of the century (African Union 2010; Tambi et al. 1999). This population bottleneck not only changed the structure of the cattle population on a continental level but also opened up new cattle trade routes in a bid to continue satisfying the needs of the hides industry (Anon 1858; McKenna 2011). The outbreak of rinderpest in Southern Africa later led to the importation of animals into the then Natal Province in South Africa from Madagascar. These new introductions likely also occurred in the rest of Africa where the need for cattle products both at subsistence and commercial levels caused an influx of local cattle breeds into new territories in a bid to replace the lost cattle. Epidemics of other diseases in livestock, whether caused by introduced conditions such as CBPP (contagious bovine pleuropneumonia), or indigenous pathogens such as anthrax caused by *Bacillus anthracis*, and the tsetse fly-transmitted protozoa further contributed to the bottleneck in the growing African cattle population (Zeleza 1993).

8.2.4 Transhumant Cattle Movements

Natural seasonal changes in Africa cause the cyclical transhumant movement of people and their animals in search of pastures and water (Fig. 8.1). These activities predated the colonial era and still exist to date. For example, the Sahel-West African transhumant movement constitutes a giant carousel in which cattle move from as far west as Mali to Cameroon and the Central African Republic (Cour 2001; SWAC 2007). Although the existence of this carousel predates the colonial era, it is reasonable to assume that the colonial boundaries that were established had an influence on the volume, direction, and duration of these human and animal movements (Huillery 2009). Similarly, in Eastern Africa, cattle movements are linked to the movement of pastoral ethnic groups in the Turkana, Karamoja, and Borana regions, respectively, of Kenya, Uganda, and Ethiopia (Fig. 8.1) (Fratkin 2001; Homewood et al. 2012). These, too, predate the colonial era, but their movement has been greatly reduced by the current human population explosion and urbanization.

8.3 Bovine Tuberculosis in African Cattle Populations

Bovine tuberculosis is endemic in the African cattle population, and it is critical to examine the demographics of the most affected host. Africa is home today to approximately 300 million head of cattle, about 24% of the world's cattle (Figs. 8.2 and 8.3). According to the Food and Agricultural Organization (FAO) of the United Nations, this number was approximately 200 million in 1999 (FAO 2005), and, by then, this was estimated to be almost double the cattle population on the continent in the 1950s (Lunde and Lindtjørn 2013).

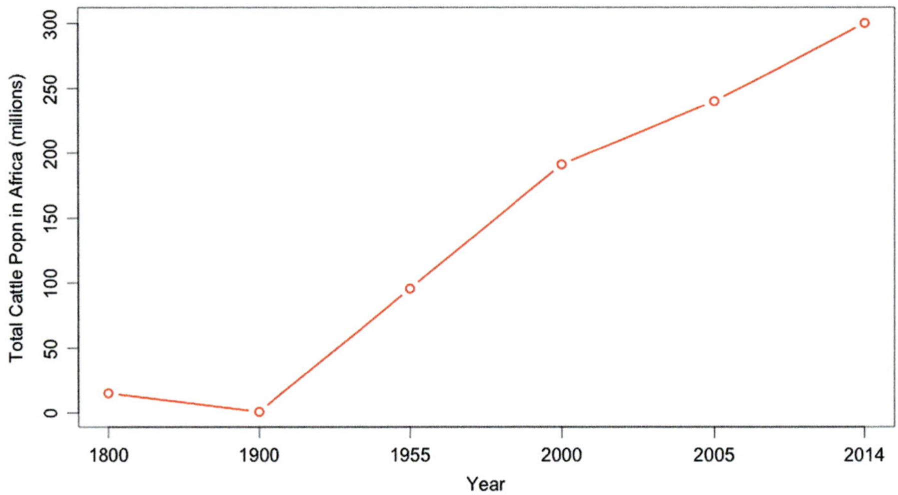

Fig. 8.2 Estimated cattle population in the African continent between 1800 and 2014. The graph was generated based on data from the ILRI and FAO (2005) and reports on rinderpest in Africa (AU 2010) population hind-casting from 1955 to 1800

East Africa has by far the largest cattle population on the continent, of which 50 million are found in Ethiopia (HarvestChoice 2015). The West and East African zebus are the most common cattle breeds in Africa and are mostly farmed with in the arid and semiarid regions, away from the humid, tsetse-infested, forested areas where only the tsetse-tolerant N'Damas survive (Deshler 1963). The Sanga and Afrikaner are present in Eastern and Southern Africa, respectively, while the Kuri, whose numbers are steadily declining, is the only known heat-tolerant breed and is farmed with around Lake Chad (Tawah et al. 1997). The variation in cattle density (Fig. 8.3) on the continent is influenced by many factors that shall not be explored in this chapter. It is important to note that the distribution of the different breeds in Africa is a reflection of the migration of its human population with their cattle and the introduction of new cattle breeds onto the continent (Deshler 1963).

8.4 Molecular Typing for Epidemiologic Studies of *Mycobacterium bovis*

Molecular epidemiology is distinct from the other subspecialties as it uses molecular biological tools to characterize nucleic acid- or amino acid-based content of microbial pathogens to study their distribution and the determinants of disease occurrence in host populations (Foxman 2001). Typing is a process by which different organisms within a species are identified (Sabat et al. 2013). Originally this process was based on different phenotypes such as serotypes, biotypes, or antibiograms, but

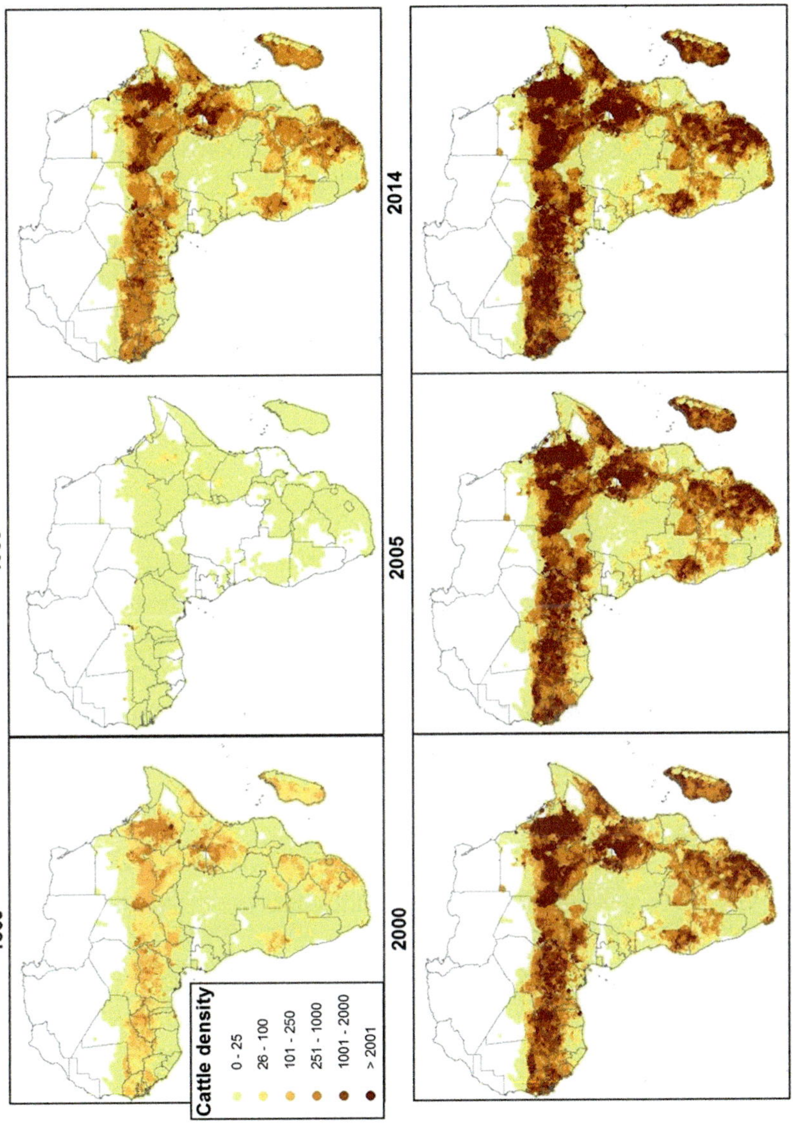

Fig. 8.3 Cattle population density in sub-Saharan Africa. The map was generated, using R and ArcGIS, based on population hind- and forecasting (FAO 2005), with 2005 as the point of reference

recent technological advances allow the use of molecular markers to determine the relatedness of isolates by using differences in conserved polymorphic genomic segments (Allix et al. 2004; Razanamparany et al. 2006; Sabat et al. 2013). Generally, the observed rate of the development of polymorphism (stability of the marker) and the genetic diversity of strains in the population are the key aspects when choosing an adequate molecular tool for studying the epidemiology of TB (Sabat et al. 2013). This implies that the rate of change of a marker must be sufficiently rapid to be able to distinguish between epidemiologically unrelated strains but slow enough to reliably link the related strains. These factors, together with the general background information of the prevalence of bovine tuberculosis, are normally taken into consideration when choosing molecular typing tools (Mathema et al. 2006).

The information in this chapter is based on the results of spoligotyping, MIRU-VNTR, and deletion analysis because they are the most commonly used techniques in Africa (Allix et al. 2004; Razanamparany et al. 2006; Sabat et al. 2013). In addition, spoligotypes are reported to mutate relatively slowly compared to the attributes detected by MIRU-VNTR that mutate relatively more rapidly (Reyes and Tanaka 2010). When used in combination with deletion analysis, the two techniques are suitable for examining the molecular epidemiology of BTB within historical and contemporary perspectives.

8.4.1 Deletion Analysis

Screening for the deletion or presence of defined segments of DNA by Polymerase chain reaction (PCR) methods is widely used as a typing tool to identify different strains of the *Mycobacterium tuberculosis* complex (MTC). The presence or absence of region of difference (RD) has been used to infer the evolutionary process of members of the MTC of which *M. bovis* is a member. These host-specific strains arose from a common ancestor by the successive loss of segments (RD) of DNA (Fig. 8.4) (Brosch et al. 2002).

Mycobacterium bovis has further been classified into different clonal complexes based on deletion of specific, defined segments of its genome. Deletion of RDAf1, RDAf2, and RDEu1, respectively, characterizes the African 1 (Af1), African 2 (Af2), and European 1 (Eur1) clonal complexes, respectively, found in Western Africa, Eastern Africa, and the British Isles and their former colonies (Müller et al. 2009; Berg et al. 2011; Smith et al. 2011). These deletions are also highly correlated with the absence of spacers 30, 3–7, and 11 in Af1, Af2, and Eur1 spoligotypes, respectively.

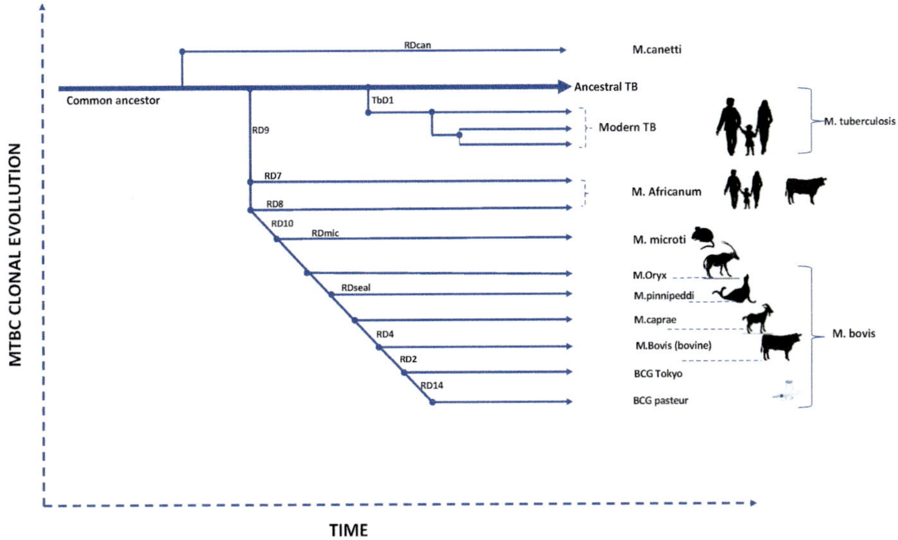

Fig. 8.4 Successive loss of DNA material. The scheme is based on the presence or absence of conserved deleted regions and on sequence polymorphism in five selected genes. The blue arrows indicate that strains are characterized by katG463 CTG (Leu) and gyrA95 ACC (Thr), typical for Group 1 organisms. The green arrows indicate that strains belong to Group 2 and are characterized by katG463 CGG (Arg) and gyrA95 ACC (Thr). The red arrow indicates that strains belong to Group 3, characterized by katG463 CGG (Arg) and gyrA95 AGC (Ser) (Brosch et al. 2002)

8.4.2 *Spoligotyping*

Spacer oligotyping (spoligotyping) is a rapid, PCR-based method for identifying and genotyping members of the MTC. It is a robust technique, and the data are repeatable and easily exchangeable between laboratories making it the most widely used of the typing techniques. Its usage has been adopted globally because the data are presented in a binary digital form, and it is the largest current collection of the molecular characteristic of *M. bovis* (Driscoll 2009). This method exploits 31–41-bp-long, nonrepetitive spacers interspersed within short repetitive units of clustered, regulatory, short, palindromic repeats (CRISPR), referred to as the direct repeat (DR) region (Van der Zanden et al. 2002). Although there are more than 90 spacers in the DR unit of the MTC, the universally accepted panel contains only 43 spacers (Kamerbeek et al. 1997; He et al. 2012) that are immobilized by hybridization onto a membrane by southern blotting that produces a characteristic binary pattern depending on the presence or absence of each of the 43 spacers. Identification of members of the MTC (*M. tuberculosis*, *M. bovis* BCG, *M. bovis*, *M. canettii*, *M. microti*, *M. africanum*, and *M. bovis* subsp. *caprae*) is based on whether a sample contains the DR unit, in which case, there is a positive result on amplification. Genotyping is based on the different patterns produced by the presence or absence of the spacers. For example, the presence of all the spacers, except spacers 3, 9, 16, and 39 through 43, characterizes *M. bovis*. The resulting patterns are compared to those stored in the global spoligotype database (www.mbovis.org), and if unique, identity numbers are awarded to the new spoligotype patterns.

8.4.3 *Interstitial Repetitive Unit-Variable Number of Tandem Repeat (MIRU-VNTR) Analysis*

MIRU-VNTR is also a PCR-based typing tool that detects the number of repeats or alleles (Fig. 8.5) at each of the micro-repetitive satellites (loci) found on the mycobacterial genome (Allix et al. 2004). The number of repeats at each locus is determined by the weight of the PCR product for that particular locus and is designated by a number digit. The number of repeats at each of the loci chosen in a panel is a unique signature (expressed as integers) that is used for typing purposes (Fig. 8.6). The use of two panels (number of loci), the 15 and 24 loci panels, is recommended (van Soolingen et al. 1994; Frothingham and Meeker-O'Connell 1998; Roring et al. 2002; Supply et al. 2006). The choice of panel is dependent on cost, the required discriminatory power, and the polymorphism of the samples, which is usually a function of the geographical areas in question. Because of the likelihood of convergent evolution (homoplasy), it is advisable not to use this technique as a stand-alone molecular tool.

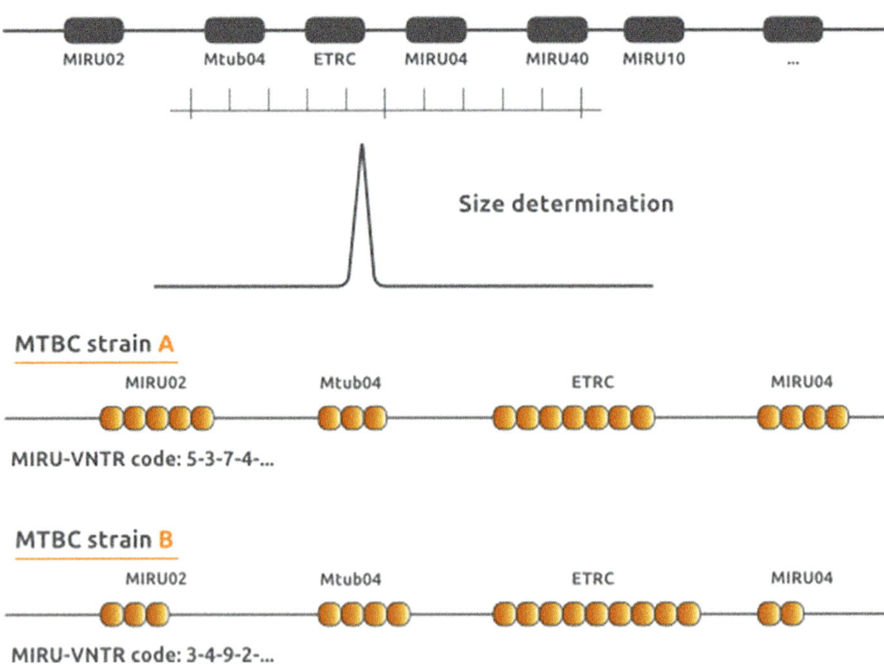

Fig. 8.5 MIRU-VNTR: The schematic representation of how repeats at micro-satellites (MIRU) are deduced

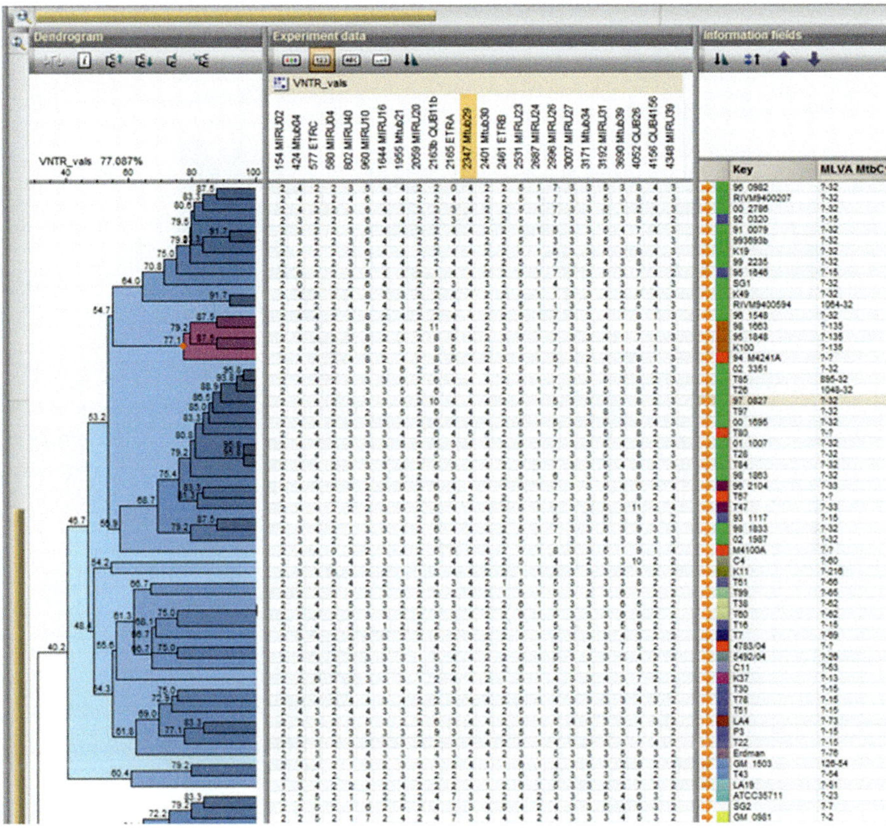

Fig. 8.6 An example of a typical MIRU-VNTR database showing how this tool is used for typing and phylogenetic purposes

8.4.4 Restriction Fragment Length Polymorphism (RFLP)

Restriction fragment length polymorphism (RFLP) is a partial genomic analysis technique that identifies DNA sequence polymorphism characterized by specific, nontandem, repeat units within specific regions of the mycobacterial genome (Durr et al. 2000). The detection of RFLPs is based on restriction enzyme digestion, and fragmentation of mycobacterial DNA in a genomic region where a specific short sequence is known to occur, followed by physical fragment separation by length using agarose gel electrophoresis (Fig. 8.7). A fragment length profile is considered an allele, and each one varies with an individual isolate thus forming the base for identification (Lin et al. 2014).

Insertion sequence 6110 (IS6110) is the most extensively used genomic region (molecular marker) for epidemiological studies of *M. tuberculosis* including outbreak investigation, nosocomial infection, and multi-drug resistance studies (van Embden et al. 1992). Because of the presence of a relatively higher copy numbers

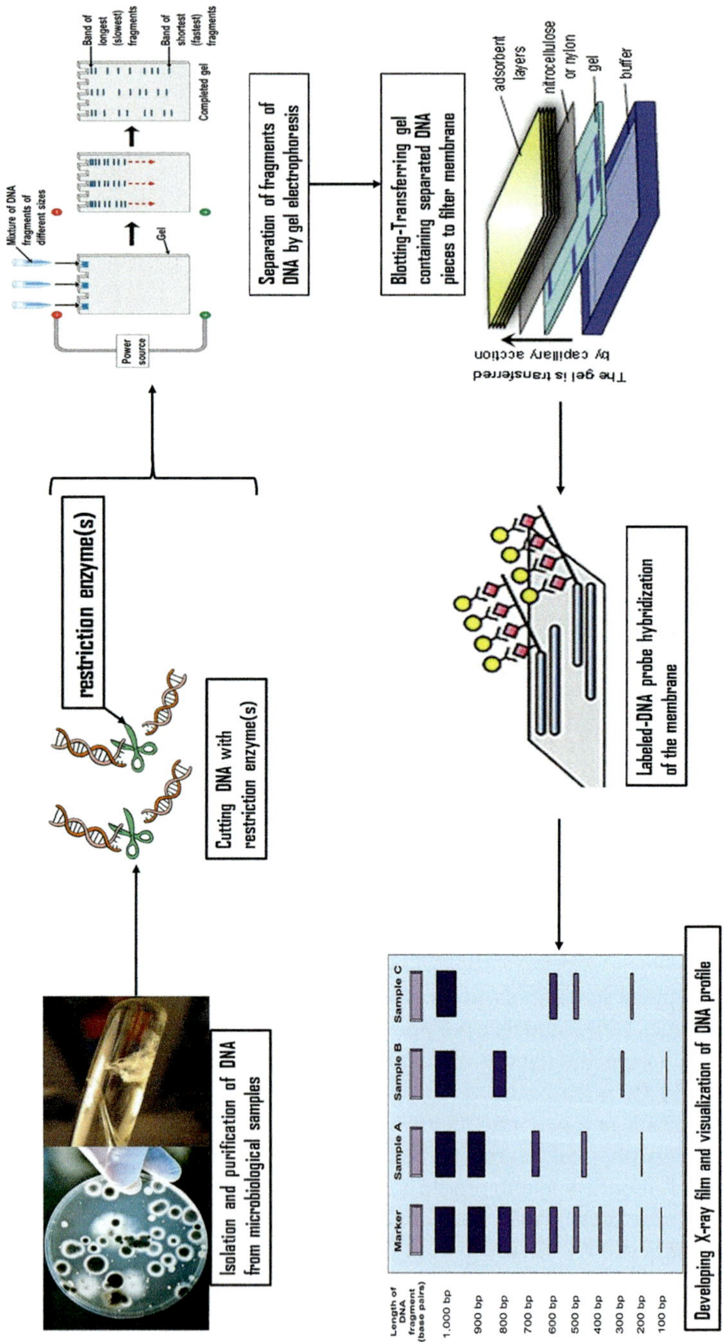

Fig. 8.7 Flow chart showing the steps involved in RFLP typing. This figure is an adaptation based on Komal (2014)

(8–15) of the insertion sequence in the *M. tuberculosis* genome, RFLP analysis has a higher discriminatory power. By comparison, the majority of *M. bovis* isolates have fewer copies of IS6110 in their genomic structure (Kanduma et al. 2003). To compensate for the low discriminatory power of IS6110 because of the low copy number of insertion sequences, researchers proposed the use of the polymorphic GC-rich repeat sequence (PGRS) technique, as it is useful to differentiate between strains with fewer than six copies of IS6110 (van Embden et al. 1993; Yang et al. 2000). This PGRS-based RFLP probe is the most discriminatory of the probes currently available for *M. bovis* strain typing (Seva et al. 2014). Another novel RFLP probe, the pUCD probe that produces a highly polymorphic yet simple-to-analyze banding pattern, is also an effective tool for molecular typing of *M. bovis strains* (O'Brien et al. 2000; Cameron et al. 2001).

8.5 The Use of Molecular Epidemiology in Understanding the Dynamics of *M. bovis*

Molecular epidemiology is a field that emerged from the integration of molecular biology, clinical veterinary medicine, statistics, and epidemiology. It focuses on the role of genetic changes in organisms by linking their distribution within a population to events and niche-specific dynamics (Mathema et al. 2006). The data synthesis that underpins this multidisciplinary field is pivotal to the development and implementation of effective risk management and control strategies. This is especially so for a disease like BTB where a multidisciplinary approach to identifying disease determinants in human and animal populations is critical for resource-limited human and animal health systems.

8.5.1 Understanding Genetic Diversity

Epidemiological studies in most countries are conducted to assess the genetic diversity and population structure of *M. bovis*. This information can then be used to explain the spread of a disease in different animal species. In Africa, these studies show varying levels of *M. bovis* diversity with the same typing tools (Oloya et al. 2007; Müller et al. 2008; Biffa et al. 2010; Hlokwe et al. 2011; Sanou et al. 2014). This reflects not only niche-specific genetic changes in pathogens but also host-population dynamics given that pathogens are "hitchhikers" on the back of these events (Muwonge et al. 2016). This population-based diversity is a functional aggregate of molecular marker site diversity for the typing tool used and is assessed using the allelic diversity index. For example, the allelic diversity of MIRU-VNTR provides a measure of the discriminatory power of each locus and how it ultimately contributes to the total discriminatory power of MIRU-VNTR as a typing tool. The

Table 8.1 The discriminatory power of five exact tandem repeat (ETR) loci measured as the allelic diversity using 246 samples (Muwonge et al. 2016)

VNTR loci	Number of samples	Allelic diversity
ETRA	246	0.780
ETRB	246	0.729
ETRC	225	0.543
ETRD	240	0.595
ETRE	246	0.254

Table 8.2 The discriminatory power of 18 VNTR loci measured as the allelic diversity for 121 *Mycobacterium bovis* isolates (Muwonge et al. 2016)

VNTR loci	Number of samples	Allelic diversity
ETRA	121	0.710
ETRB	121	0.686
ETRC	121	0.581
ETRD	121	0.410
ETRE	121	0.041
MIRU10	121	0.102
MIRU16	121	0.554
MIRU20	121	0.350
MIRU23	121	0.204
MIRU24	121	0.072
MIRU26	121	0.729
MIRU27	121	0.341
MIRU39	121	0.257
MIRU40	121	0.326
Mtub04	121	0.104
Mtub21	121	0.643
Mtub30	121	0.391
Mtub39	121	0.322

allelic diversity (h) for each locus is calculated using the formula (Selander et al. 1986):

$$h_i = \sum x_i^2 \left(\frac{n}{n-1}\right); \tag{8.1}$$

where x is the frequency of the ith allele at the locus, n is the number of isolates in the analysis and $\left(\frac{n}{n-1}\right)$ being the correction factor for bias in small samples.

Based on two databases (Muwonge et al. 2016), ETRA, ETRB, MIRU26, and Mtub21 are the most diverse allelic loci in Africa, which, by this measure, are the most desirable loci to include in a panel when MIRU-VNTR is used as a typing tool to investigate the diversity of *M. bovis* in Africa. ETRE, MIRU 24, Mtub04, and MIRU10 are the least allelic diverse loci, and the value of their inclusion in a panel to determine the diversity in *M. bovis* in Africa would be minimal (Tables 8.1 and 8.2).

There are a number of methods available to determine spoligotype diversity (Biffa et al. 2010), but we evaluated it by assessing the number of unique spoligotypes detected in a specific country. Based on this approach, Nigeria, Ethiopia, South Africa, and Algeria have the highest spoligotype diversity (Fig. 8.8). At a

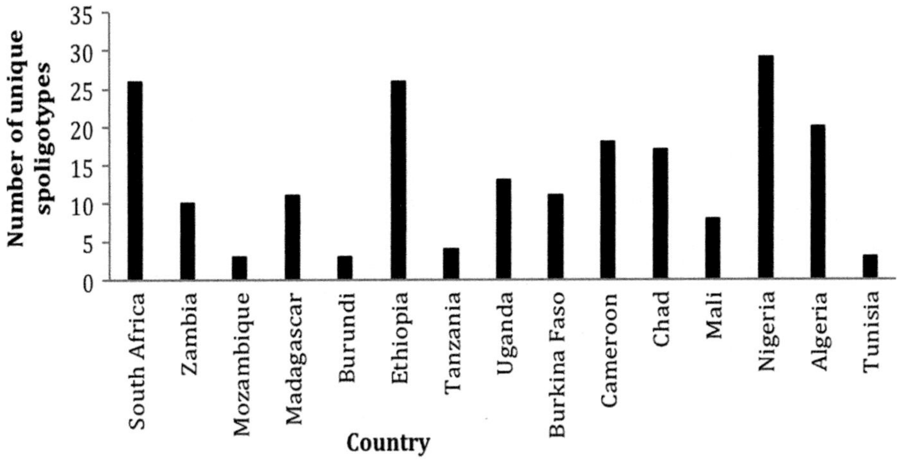

Fig. 8.8 Spoligotype diversity based on the number of unique spoligotypes recovered from each country in the last 20 years

regional level though, Western African isolates are almost twice as diverse as those present in Eastern and Southern Africa. The significance of this diversity will be discussed in Sect. 8.6.3.

8.5.2 Tracking the Sources and Routes of Transmission

In molecular epidemiology, transmission is inferred when homologous genotypes are recovered from two or more individuals linked spatially and temporally. These reflect genotypic clustering on phylogenetic visualization of isolates, and numerous molecular epidemiological studies in Africa reflect this attribute. For example, using clustering, in South Africa the increasing inter- and intraspecies transmission of BTB have been deduced based on this homology (Hlokwe et al. 2014). In Zambia, where transmission of *M. bovis* between cattle and lechwe (*Kobus leche kafuensis*) occurs, currently, lechwe are believed to be the largest BTB wildlife reservoir (Hang'Ombe et al. 2012). Similarly cross-species transmission has been inferred in Burkina Faso (Sanou et al. 2014), Ethiopia (Biffa et al. 2010), Madagascar (Razanamparany et al. 2006), and Uganda (Muwonge et al. 2012; Oloya et al. 2008). Additionally, in Ibadan, Nigeria, there was homology between *M. bovis* genotypes isolated from slaughter cattle and from abattoir workers (Jenkins et al. 2011; Lawson et al. 2012).

In a wider geographical context, identical genotypes have been reported in Europe and Africa, especially in Northern and Southern Africa (Smith et al. 2011). The presence of these strains in these two African regions is not linked as their occurrence violates the space requirement for transmission. This matter will be dealt with in Sect. 8.6.3.

Source tracing is one of the aspects of disease investigation that benefits from the use of molecular epidemiological tools. This information can be beneficial in detecting on-farm sources of infection or contamination and tracing multiple-strain outbreaks (Biffa et al. 2010). The use of molecular markers to trace the sources of infections has been used in a number of other African countries including Madagascar (Razanamparany et al. 2006), Ethiopia (Biffa et al. 2010), Burkina Faso (Sanou et al. 2014), and Tanzania (Kazwala et al. 2001). In South Africa, for example, recent studies identified on-farm infections caused by two strains that were traced back to animals purchased and introduced from another farm (Michel et al. 2008; Hlokwe et al. 2014). Similarly, it has recently been shown that free-ranging African buffaloes play a vital role as a source of infection for livestock of neighboring communities in South Africa (Michel et al. 2009) and that wildlife to livestock spillover of BTB does occur (Musoke et al. 2015). In Zambia, infections in indigenous Zambian cattle in five of the six studied districts were traced to a common source (Munyeme et al. 2009a), while the zoonotic transmission of *M. bovis* in Namwala District has been traced back to cattle. Additionally, reverse zoonotic transmission of *M. tuberculosis* from humans to cattle has been detected in the same area (Malama et al. 2014). In Uganda, most zoonotic tuberculosis has been traced back to the local cattle populations (Oloya et al. 2007). Recently a potential alternative source of infection for humans has been identified in the form of spillover of the infection from cattle to pigs and then to humans (Muwonge et al. 2012).

8.5.3 Elucidation of the Effect of Infection Pressure on the Severity of the BTB

Numerous studies have been carried out to determine the association between the virulence of genotypes of *M. tuberculosis* and the severity of the infection in humans. To date, there have been only a few studies conducted to characterize the severity of *M. bovis* infections in animals. The increasing availability of molecular tools in recent years allowed researchers not only to record the prevalence but also to describe the severity and extent of the lesions in infected herds. In this respect, molecular epidemiologic tools are used to demonstrate superseding genotype(s) in the area, the presence of polyclonality (simultaneous infection with multiple genotypes), and how these phenomena affect the severity of infection. Linking these data to spatial and temporal epidemiologic data allows researchers to understand the propagation of the infection and the persistence and evolution of the causal agent. For example, the combined use of MLVA and spoligotyping techniques provided an explanation of the increased severity of BTB lesions in Ethiopian cattle linked to a multiple-genotype infection and the spatial clustering of those genotypes. Among factors contributing to these phenomena were the rapid growth in livestock numbers, the expansion of intensive dairy farming, and the stock co-mingling, favoring the persistence and spread of infection in the area (Biffa et al. 2014). Similar studies in human populations in Uganda revealed a high prevalence of multiple-genotype/strain/clonal infections in human TB patients (Muwonge et al. 2013), reaffirming the need for viewing tuberculosis lesions and the severity of clinical manifestations as highly dynamic processes that are more likely caused by coinfection with a number of strains in a single host.

8.6 Inferences on the Origin of *M. bovis*

This riddle cannot be solved without revisiting the turbulent history of humans on the African continent. The data used are not exhaustive (not all countries in Africa are represented), but it is the most comprehensive collated data set to date using three genotyping tools as discussed earlier (Sect. 8.4).

In this section, we circumstantially link observed molecular marker patterns of *M. bovis* to events during three periods in the history of the African continent: the precolonial, colonial, and postcolonial eras.

8.6.1 Indigenous African Strains

Archaeological and osteo-forensic studies show that the Shurmuk and Uyuk people who occupied the southern Siberian region during the second century BC (~4000 years ago) were plagued by tuberculosis caused by *M. bovis* (Taylor et al. 2007). Based on this report, if we assume that cattle at that time were infected with *M. bovis*, and that they were the most probable source of this infection in the Shurmuk and Uyuk communities, then it is logical to assume that the disease was not limited to that region but that it could have been present in any settled community that was rearing cattle at that time. Donoghue (2009), using ancient biomolecules to study the MTC, concluded that this human infection could have been the consequence of prolonged close interaction between these semi-nomadic pastoralists and their cattle. If this were indeed the case, then the transhumant communities of Northern Africa, the Bantu, Luo, Mandé, and the Omotic people who lived at the same and more recent times (Clark and Brandt 1984) could have farmed with cattle that were infected with *M. bovis*.

There is a small, but unique, collection of spoligotypes that are only found in Africa (Muwonge et al. 2016). This suggests that they are most likely genotypes that were native to Africa and that they were present before European settlement and the importation of *M. bovis*-infected cattle. These spoligotypes include SB1100 and SB1102 from Chad (Müller et al. 2008); SB1003 and SB1200 from Tunisia (Ben Kahla et al. 2011); SB1265, SB1476, SB1517, and 1520 from Ethiopia (Biffa et al. 2010); SB1405 and SB1406 from Uganda (Oloya et al. 2007); and SB1536, SB1572, and SB1765 from Zambia (Munyeme et al. 2009b).

8.6.2 Introduced Strains

It has been assumed for some time that *M. bovis* was introduced onto the African continent during the European colonization (Müller et al. 2009; Berg et al. 2011; Smith et al. 2011). Given the detection of the unique African strains, it is likely that *M. bovis* infections predate the European presence on the Africa continent. It seems to be more likely that the colonial era saw an introduction of new (additional) strains that influenced the evolutionary course of *M. bovis* as reflected by the variety and distribution of the strains that are currently present. These introductions during the colonial times are likely to have occurred in four ways:

1. The introduction of large numbers of cattle by European settlers following the colonization of South Africa (and other African countries)
2. By the importation of bulls for breed improvement purposes during the colonial and postcolonial times (Armando 1995; Taneja 1999; Stackyard 2009)
3. The introduction of exotic female cattle for commercial dairy farming (colonial and postcolonial era) (Stackyard 2009)
4. For restocking the decimated African cattle populations following the rinderpest epidemic during the course of and until the end of the nineteenth century (Bienart 1989; Mariner et al. 2012)

The introduced genotypes (spoligotypes) are likely to have the following characteristics: They are present in both colonial and colonized territories, and they are the dominant spoligotypes given that they are from regions with intensive husbandry systems (maximum number of animals per unit area) and intensive disease control (test-and-slaughter) strategies. These genotypes originated from a system in which there was increasing transmission pressure and a population bottleneck and had to be better adapted (Smith et al. 2006). For example, France and Britain started their attempts to eliminate BTB in the 1930s at the height of their colonial power (More and Good 2006; Smith et al. 2006; Berdah 2010). These interventions were increased during the 1950s and up to the 1980s. The time could not have been better for these new *M. bovis* introductions as the African cattle herd was in the process of recovering from the population bottleneck caused by the rinderpest epidemic (Tambi et al. 1999; African Union 2010). This aggregate of factors likely created a conducive environment for the introduced new *M. bovis* strains to stamp their dominance wherever they were introduced. The current distribution of spoligotypes seemingly reflects this situation in that SB0944 is localized in Western Africa; SB0133 is prevalent in Eastern Africa; and SB0120, SB0134, SB0140, SB0121, SB0130, and SB0961 are found in Southern and Northern Africa (Muwonge et al. 2016). The most recent common ancestor (MRCA) of the clonal complexes documented in Africa (Müller et al. 2009; Berg et al. 2011; Smith et al. 2011) is also in this category, which corroborates this generalization in terms of origin.

Clonal evolution is discussed in the next section (Sect. 8.6.3), but it may be argued that the African clones may have been introduced into Europe as some of the "spoils of the colonial conquest," as has been suggested for other diseases (Jackson

2003). This notion, however, is at variance with the history of animal and human movements in Africa. In addition, SB0944 has only been reported in French cattle (Haddad et al. 2001) and livestock and humans in Francophone West Africa. It was reported outside this range only once in a man in Great Britain, whose origin was traced back to an Anglophone Western African country (Müller et al. 2009).

8.6.3 Understanding the Evolution and Phylogeny of M. bovis

In the past few decades, several robust molecular biological tools have been developed and applied in combination with conventional epidemiologic approaches. These have enabled researchers to carry out extensive, in-depth investigations to generate valuable information that enhanced the understanding of the molecular genetic characteristics of *M. bovis*. Such approaches revealed previously unknown genetic and microbiological characteristics of *M. bovis*, the genetic relatedness of the different types, and how the various genotypes with different virulence, and with geographic clustering, host preference, and adaptability evolved.

Subtractive genomic hybridization (Mahairas et al. 1996) revealed the genetic relatedness of the BCG vaccine strain with known virulent strains of *M. bovis*. The BCG strain is avirulent due to deletion of three genomic regions of difference (RD1, RD2, and RD3), suggesting a possible association between the evolutionary genomic conservation of genes existing in those regions and the extent of strain virulence. The loss of virulence by the BCG strain may be linked to the evolutionary loss of a gene regulatory function as a result of deleted RDs, particularly RD1. This finding paves the way for the possibility of developing new diagnostic tools to differentiate the immune response to BCG vaccination from infection with virulent strains (Mostowy et al. 2002).

The complete genomic sequence of *M. bovis* provided information about the evolution, molecular genetics, and phenotypic characteristics of the pathogen. The previously held dogma that *M. bovis* was the progenitor of the human pathogen, *M. tuberculosis*, was challenged by these data (Garnier et al. 2003). While *M. bovis* was initially assumed to have crossed the species barrier to the human host after the domestication of cattle, the deletion of certain genes from the genomic structure of *M. bovis*, which are present in *M. tuberculosis*, led to the current status where *M. tuberculosis* is believed to be closer to the ancestral progenitor of MTC than *M. bovis* is (Brosch et al. 2002; Garnier et al. 2003). Further clarification of the variation in genomic characteristics of various *M. bovis* strains should assist researchers to understand how these phenomena influence the variation in virulence, host and geographic selection, and the adaptation of mycobacterial strains to host immune responses, and environmental pressures.

8.6.3.1 Clonal Complexes

A clonal complex is defined as a group of strains belonging to a bacterial species deemed to have descended from a single bacterial species cell (the MRCA) because they all bear characteristics derived from this single cell (Berg et al. 2011; Müller et al. 2009; Smith et al. 2011). In this section, we adapted this grouping, based on the findings of Berg et al. (2011), Müller et al. (2009), and Smith et al. (2011) who characterized *M. bovis* isolates from the African and European continents using spoligotyping and deletion analysis. There are currently three confirmed and two putative clonal complexes:

1. African 1 (Af1) is characterized by the genomic deletion of RDAf1, which is the absence of spacer 30 in the spoligotype pattern; SB0944 is the MRCA, and it is predominantly present in Western Africa.
2. African 2 (Af2) is characterized by the genomic deletion of RDAf2, which is the absence of spacers 3–7 in the spoligotype pattern; SB0133 is the MRCA, and it is predominantly present in Eastern Africa.
3. European 1 (Eur1) is characterized by the genomic deletion of RDEur1, which is the absence of spacer 11 in the spoligotype pattern, and it is predominant on the British Isles, but it is present worldwide.
4. Putative Af3 and Af5 clonal complexes are characterized, respectively, by the loss of spacers 3–5 and 3–5 and 8–10 in their spoligotype patterns and have been, respectively, reported in Mali and Burkina Faso, and Madagascar. The status of these two clonal complexes has not been fully confirmed, pending deletion analysis.

Given the strong correlation between spoligotype pattern, the region-of-difference-based molecular markers (RDAf1, RDAf2, and RDEur1) (Müller et al. 2009; Berg et al. 2011; Smith et al. 2011), and geographical homology, spoligotypes in this section are discussed as pseudomolecular markers of evolution. This is based on the assumption that homoplasy is rare within a confined geographical area.

8.6.3.2 West African Clonal Complexes

There are two clonal complexes in Western Africa: Af1 and the putative Af5 clonal complexes (Müller et al. 2009; Sanou et al. 2014). The former is the most dominant in this region, and spoligotype SB0944 is prevalent in all the countries in Western Africa. This spoligotype is central in the minimum spanning tree to which all the other spoligotypes in the region are rooted, and there is consensus that this pattern is the same as Af1's MRCA (Figs. 8.9 and 8.10). Although many countries have very similar spoligotype profiles, there are subtle country-specific differences, most of which will occupy less central positions (Fig. 8.9). To put it in numbers, about 90%–95% of *M. bovis* isolates from Chad, Nigeria, and Cameroon belong to the Af1 clonal complex. In these three countries, SB0944 is the predominant spoligotype. However, other spoligotypes including SB1027 and SB1025 are frequently also

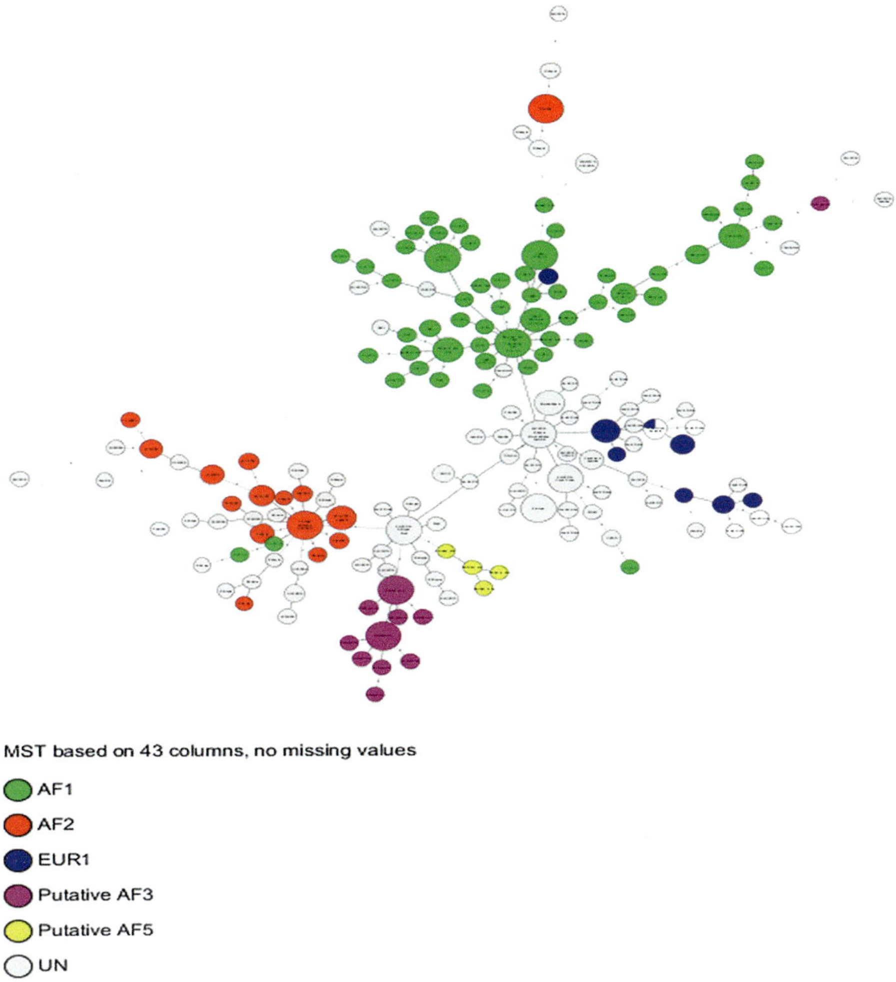

Fig. 8.9 African clonal complexes of *M. bovis*. (A) The minimum spanning tree (MST) was generated based on 43 spacer spoligotypes; each node of the MST represents a unique spoligotype, and the edge length and number represent the number of spacers deleted between any two connected nodes. UN denotes isolates that have not yet been assigned to a clonal complex

observed in Nigeria, but they are less common in Chad (Müller et al. 2009). Spoligotype SB0953, however, has been isolated exclusively in the Northwestern Region of Cameroon (Awah-Ndukum et al. 2013). Most of the spoligotypes that occur at low frequencies also tend to be country-specific (Figs. 8.11 and 8.12).

Another interesting feature of *M. bovis* isolates from West-Central Africa is the observed westwardly decreasing dominance of the Af1 clonal complex. Thus, only 76% and 60% of *M. bovis* isolates, respectively, from Burkina Faso and Mali belong to the Af1 clonal complex (Müller et al. 2008; Sanou et al. 2014). Unlike in the rest of the Western African countries, SB0300 that has only been reported in Burkina

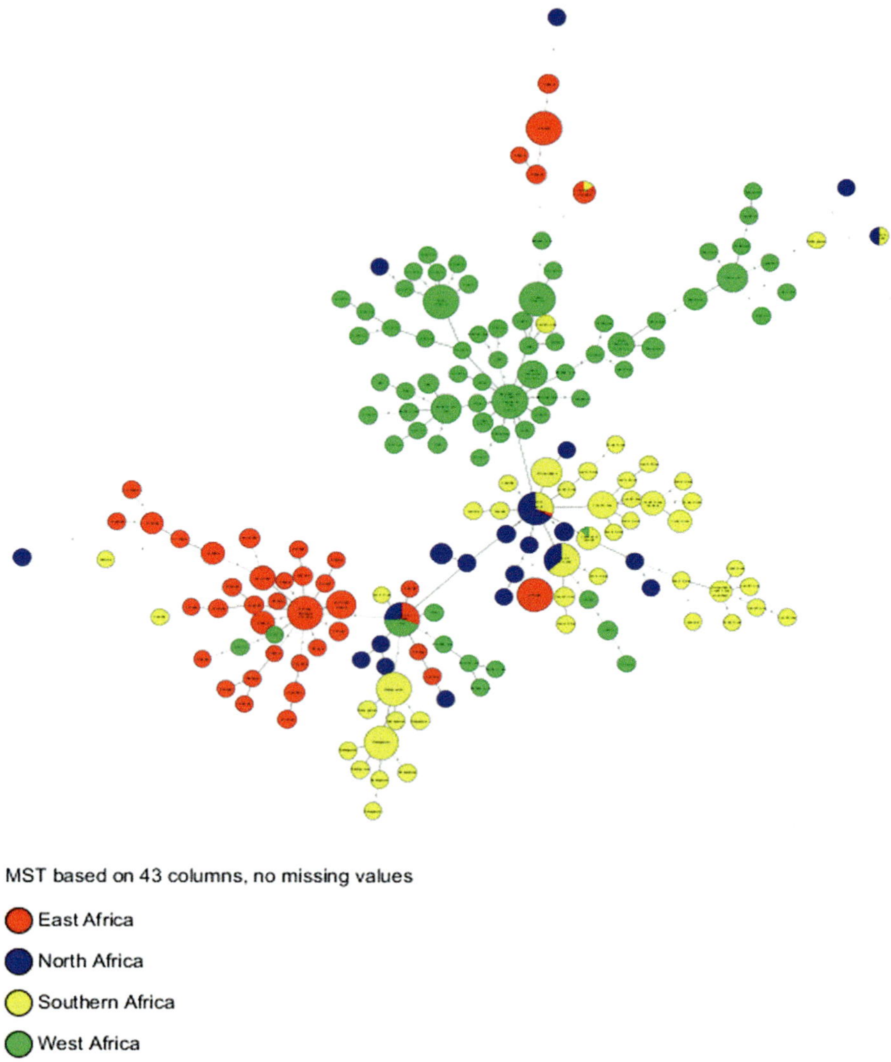

Fig. 8.10 Geo-distribution of clonal complexes in Africa. The same parameters and calibration as in Fig. 8.9 were used, but they were color-coded by region

Faso and France (Haddad et al. 2001) is the most prevalent spoligotype in Mali. It is interesting to note that all spoligotypes, except for SB0944 from Mali, lack spacer 6 in their spoligotype pattern (Müller et al. 2008). The non-Af1 spoligotypes in these two countries belong to the putative Af5 clonal complex of which SB0134 is the most ubiquitous in Mali. The latter spoligotype has also been reported in Algeria (Sahraoui et al. 2009), Ethiopia (Biffa et al. 2010), France (Haddad et al. 2001), and Spain (Mateos et al. 1996).

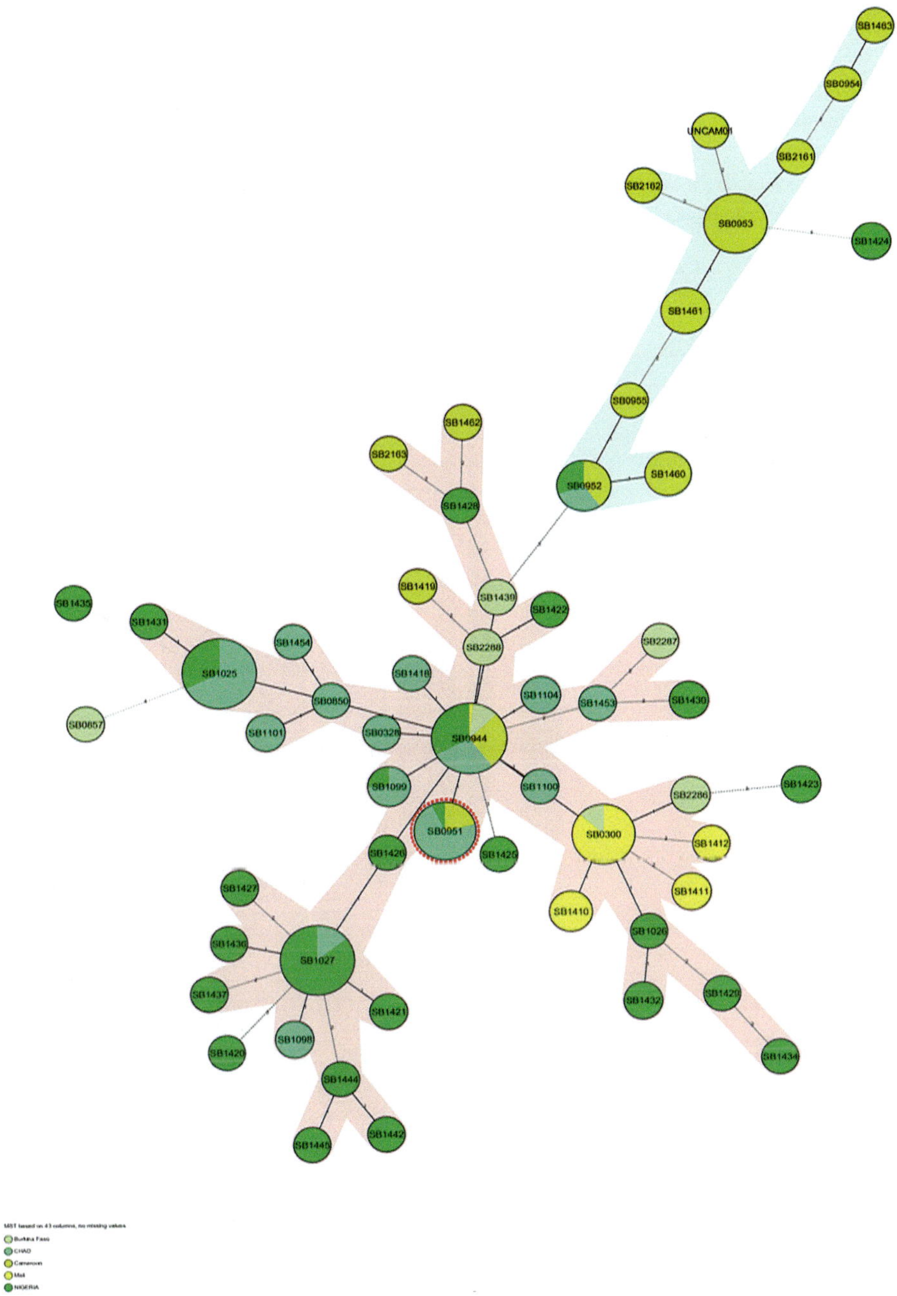

Fig. 8.11 The minimum spanning tree showing the distribution of clones (spoligotypes) belonging to the Af1 clonal complex

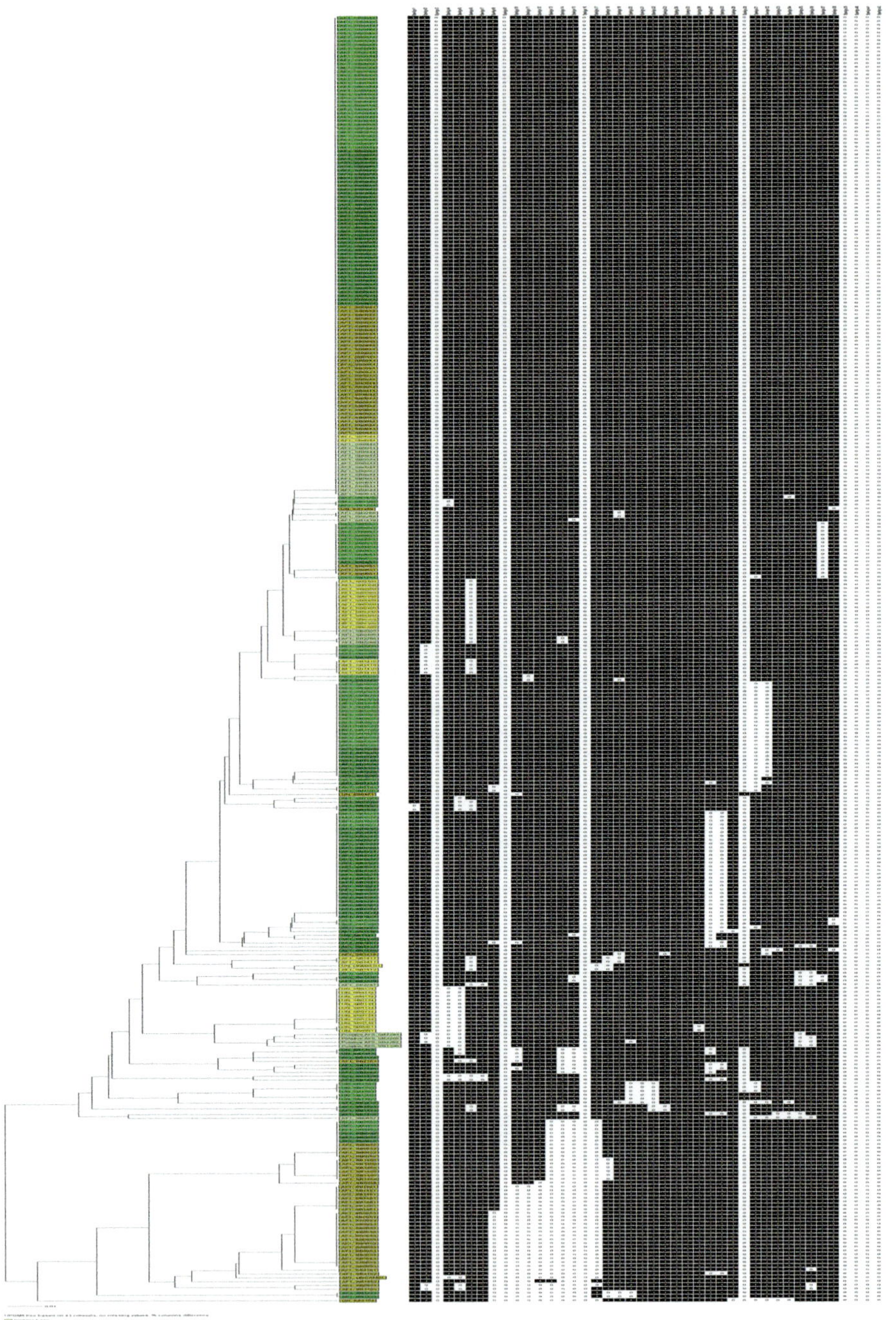

Fig. 8.12 Dendrogram showing the putative Af5 and unknown clonal complexes in five West African countries

8.6.3.3 East African Clonal Complexes

Af2 is the dominant clonal complex in East Africa. Unlike its Western African counterpart (Af1), where the MRCA is also the dominant spoligotype (SB0944), SB0133, Eastern Africa's MRCA, is not the dominant spoligotype (Fig. 8.13) (Berg et al. 2011). This spoligotype has been recovered in samples collected from all Eastern African countries except from Burundi, although the results in Burundi are limited and are based on a single study that is not representative of the entire country (Rigouts et al. 1996; Berg et al. 2011). There is a score of non-Af2 isolates unique to Eastern African countries, such as SB1405 and SB1406 in Uganda, SB0425 in Tanzania, and SB1476 in Ethiopia. These have putatively been named the "indigenous east African spoligotypes." Unlike in Western Africa, a BCG-like spoligotype has also been reported from Ethiopia (Biffa et al. 2010) and Zambia (Munyeme et al. 2009b).

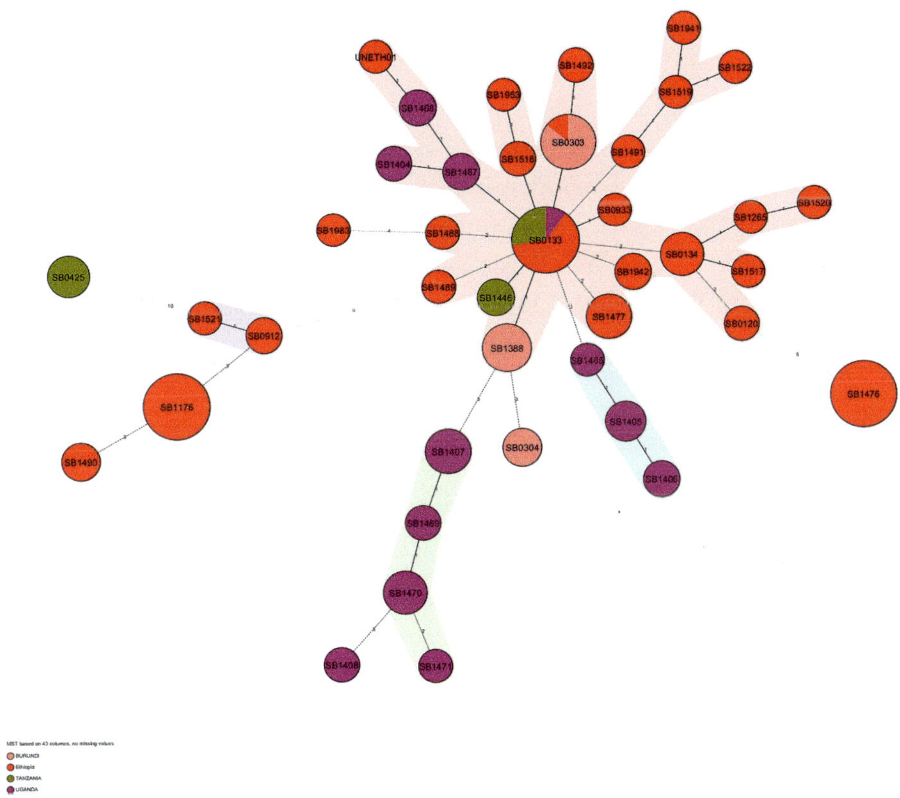

Fig. 8.13 The minimum spanning tree showing the distribution of clones (spoligotypes) belonging to the Af2 clonal complex in four Eastern African countries

8.6.4 Southern and Northern African Clonal Complexes

The spoligotype profiles of Southern and Northern Africa are very similar, although the regions are thousands of miles apart (refer to Sect. 8.6.2) (Figs. 8.14, 8.15, and 8.16). A significant proportion of the spoligotypes observed in these two regions belong to the Eur1 clonal complex, which is globally ubiquitous (Hlokwe et al. 2011; Sahraoui et al. 2009). This is specifically true for Algeria and South Africa, while a BCG-like spoligotype occurs in Zambia, and all the isolates from

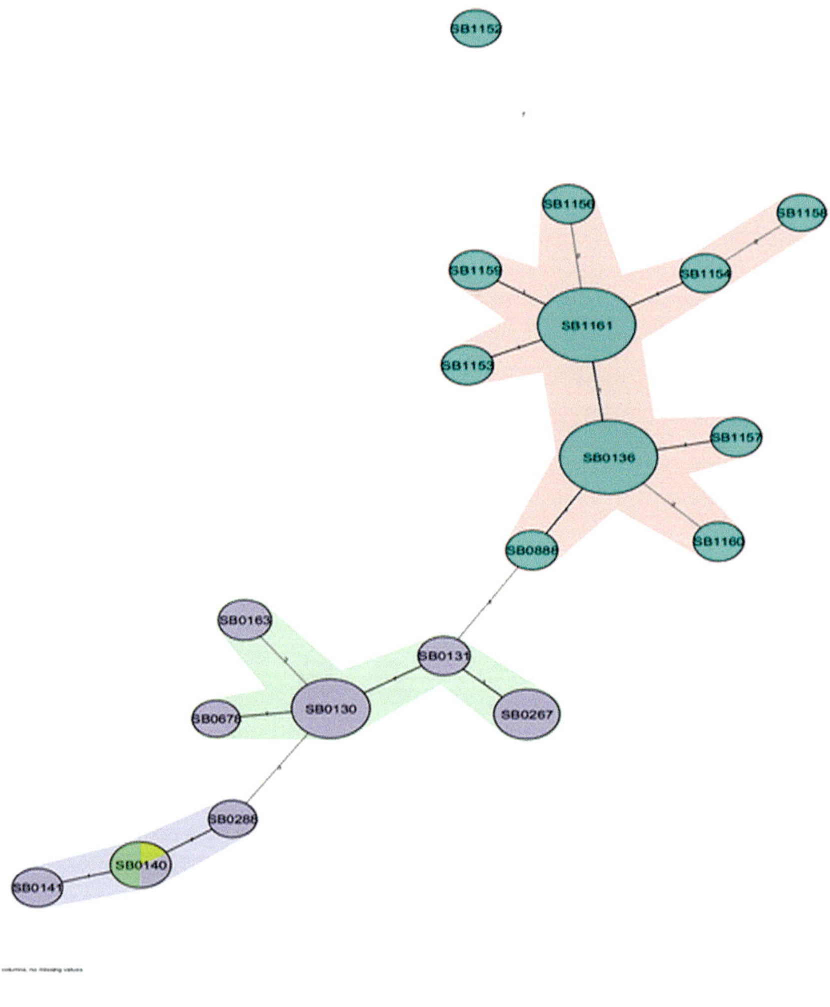

Fig. 8.14 The minimum spanning tree showing the distribution of clones (spoligotypes) belonging to the Eur1 and putative Af3 clonal complexes in four countries in Southern Africa

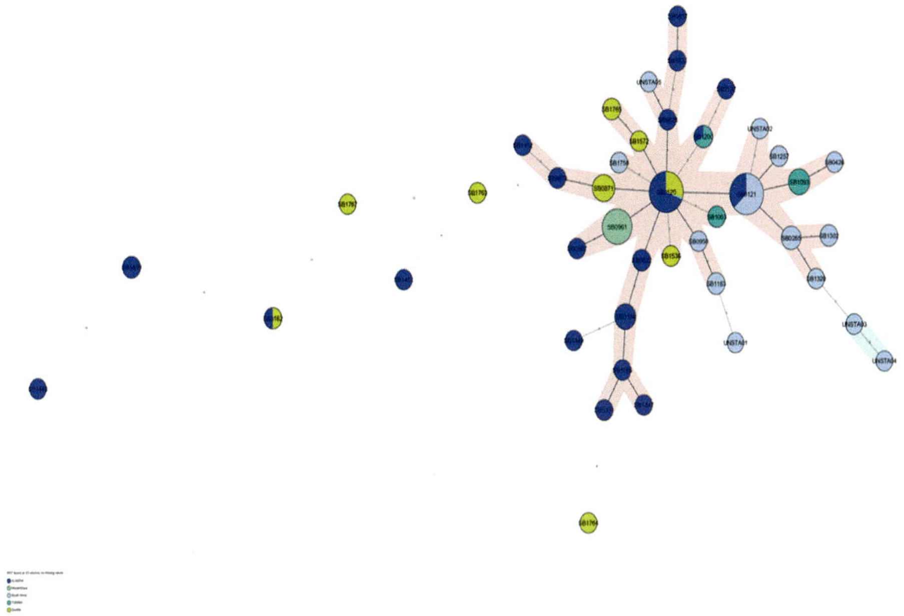

Fig. 8.15 The minimum spanning tree showing the distribution of clones (spoligotypes) that have not yet been assigned to a clonal complex in two Northern and three Southern African countries

Madagascar belong to the unique, putative Af3 clonal complex (Razanamparany et al. 2006; Müller et al. 2009; Munyeme et al. 2009b).

8.6.5 *The Clonal Expansion and Spread of* M. bovis

The various clonal complexes in Africa appear to be limited to specific geographical regions. The within-region distribution of the strains of specific clonal complexes, however, seems to be more complex. Although indigenous and introduced (exotic) genotypes determined the molecular landscape of *M. bovis* in Africa, it is more likely that the dynamics and practices of livestock movements, management, and trade had the biggest impact on the distribution of the various strains.

BTB was first documented in Cameroon in West Africa in 1913 when it was still a German-occupied territory. However, it was not uncommon to encounter BTB-infected cattle from Cameroon in Nigerian abattoirs in the 1940s, well into the time of French and British colonial rule (Alhaji 1976). These cross-boundary animal movements, driven by Sahel-West African transhumance, extended well beyond these two countries. The Sahel-West African transhumant movement is a giant carousel, involving the movement of between 70% and 90% of the regional cattle population, which provides 65% and 70%, respectively, of the meat and milk in the region (SWAC 2007). This cyclical animal movement occurs in the arid belt of

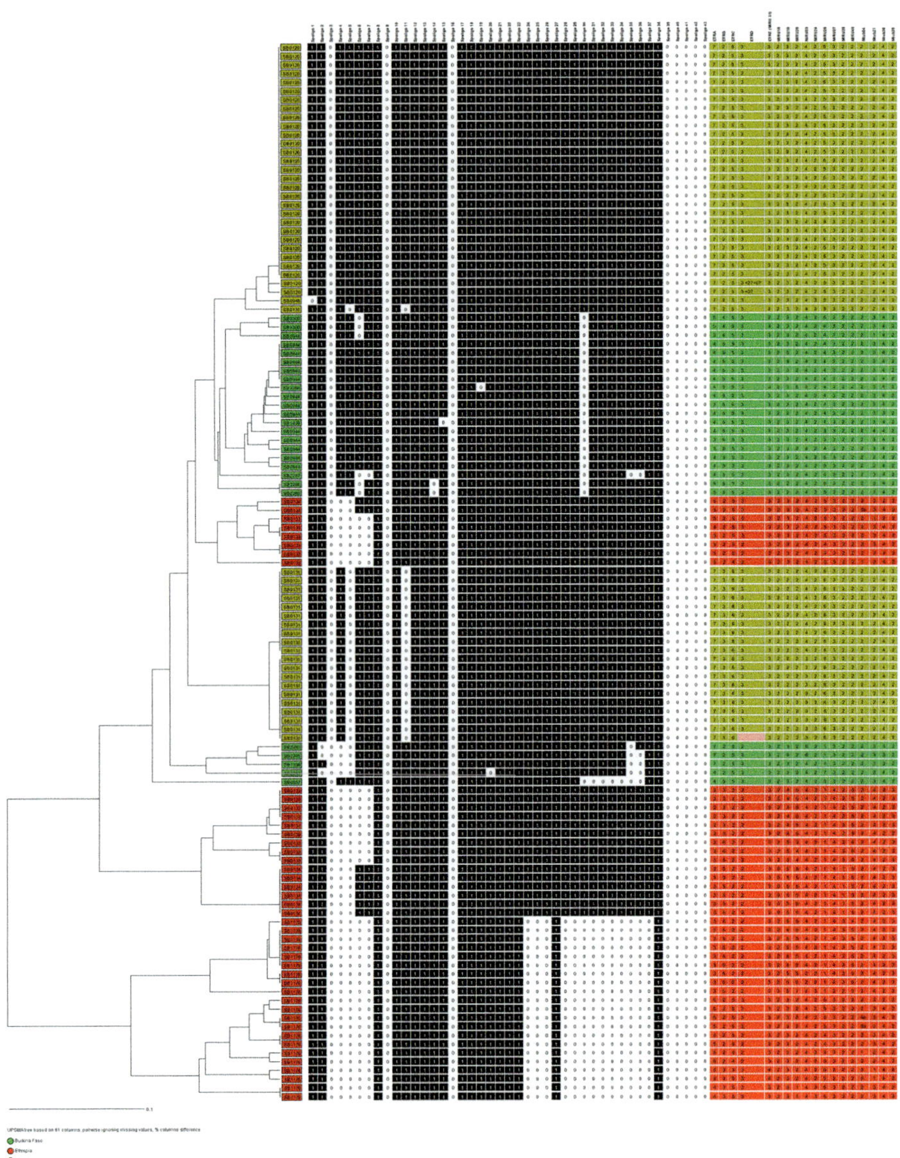

Fig. 8.16 A dendrogram showing diversity within selected spoligotypes in the east (Ethiopia in red), south (Zambia in yellow), and west (Burkina Faso in green) based on a MIRU-VNTR 18 loci panel using the Ridon version (Munyeme et al. 2009b; Biffa et al. 2014; Sanou et al. 2014)

Western Africa, involving half of Chad, Northern Cameroon, Nigeria, Burkina Faso, and most of Mali. Livestock system experts only recently acknowledged that this way of life was indeed profitable and competitive and that it played a pivotal role in maintaining the culture of the Peulh communities, who have existed in this way for thousands of years (Cour 2001; SWAC 2007). It is plausible that transhumance, the size of the cattle population, and marketing channels of beef and dairy cattle and their products were the likely mechanisms by which the Af1 clonal complex established itself in Western Africa. The granular movement of cattle in this giant carousel is, however, likely to be diffuse, and it is unlikely that cattle are herded over a distance of about 2800 km, from Kidal in Mali to Ngaoundere in Northern Cameroon, without encountering common sources of water and pastures and markets. These encounters from in-contact networks and herds from Ngaoundere are thus indirectly connected to Malian herds through degrees of separation while the cattle are herded through this great, arid expanse. The opposite is also true, and it is the likely explanation for the gradual decrease in the number of clones belonging to the Af1 clonal complex, when moving west from Chad to Mali (Müller et al. 2008, 2009; Sanou et al. 2014). At a more granular level, there are no identical genotypes (combined spoligotypes and MIRU-VNTR) in Chad and Mali as opposed to the numerous shared clones in Mali and Burkina Faso and in Nigeria and Chad (Müller et al. 2008, 2009; Jenkins et al. 2011; Sanou et al. 2014).

In Eastern Africa, although the countries share a common MRCA (Berg et al. 2011), there is a preponderance of country-specific spoligotype profiles, suggesting that the Eastern African countries are not as connected as those in Western Africa. Ethiopia has the most elaborate animal movement network in Eastern Africa, and the country's cattle movement network is connected to Eritrea, Kenya, and Sudan (see Chap. 14). No data exist for the *M. bovis* genotypes in Eritrea and Sudan, but it appears that Sudan is also connected to the Sahel-West Africa transhumance network (Fig. 8.1c), and although the Western and Eastern African clonal complexes appear to be limited to their respective geographical regions, Sudan may be the point of convergence of their distribution (Müller et al. 2009; Berg et al. 2011).

At a local level, Ethiopia has a centripetal animal movement network in which cattle filter through rural markets into the urban centers, especially Addis Ababa (Firdessa et al. 2012). Since most studies were conducted at abattoirs, the data reflect cross-sectional observations of the entire region rather than the molecular composition in the urban areas. The distribution of molecular markers in Ethiopia (Biffa et al. 2010, 2014; Gumi et al. 2012) and in Uganda (see Chap. 22) supports the existence of this centripetal movement. Recently, a centrifugal movement of cattle developed in Ethiopia where improved cattle breeds are moved from the urban dairies into rural areas as part of the dairy development program (Staal et al. 2008). This is likely to have a significant impact on the profile and distribution of genotypes of *M. bovis* and other disease agents in Ethiopia and other countries linked to its livestock movement network.

In Southern Africa, there is not an elaborate transhumance movement network as exists in Western Africa, but in recent years wildlife appear to be playing an increasing role in the epidemiology of *M. bovis* in the region (Michel et al. 2008; Munyeme et al. 2008; Hlokwe et al. 2011). Historically in Southern Africa, wildlife migration, prompted by seasonal changes that influence the abundance of water and pasture, occurred uninterrupted (de Garine-Wichatitsky et al. 2013). The pattern and extent of this migration, however, changed and decreased over the past century mostly due to increasing livestock farming activities. Veterinary cordon fences erected to contain wildlife and the spread of transboundary animal diseases blocked these ancient migratory routes in countries such as Botswana, South Africa, and Mozambique (Gadd 2012). Contact between *M. bovis*-infected cattle and wildlife, especially lechwe (*Kobus leche kafuensis*) in Zambia and African buffaloes (*Syncerus caffer*) in South Africa led to spillover of BTB to wildlife in the absence, or well before the erection, of fences (Michel et al. 2008; de Garine-Wichatitsky et al. 2013).

During the last two decades, about 20 transfrontier conservation areas were established in sub-Saharan Africa. In addition, in South Africa the wildlife industry developed into the fastest growing agricultural sector, resulting in a rapidly expanding human-livestock-wildlife interface, and the associated risk of multidirectional transmission of diseases (Bengis et al. 2002; Cloete et al. 2007; De Garine-Wichatitsky et al. 2013). In South Africa, *M. bovis* has a broad host spectrum, and the disease is endemic in several of its ecosystems. Currently 21 different wildlife species are known to be infected. Although the majority of these are spillover hosts, which do not play a substantial role in the epidemiology of BTB, there is a risk of spillback to cattle, as was recently demonstrated (Musoke et al. 2015). There is an increasing risk to uninfected wildlife populations because of the translocation of *M. bovis*-infected, but undiagnosed wildlife at this expanding wildlife-wildlife interface in South Africa (Hlokwe et al. 2014).

These changes are likely to have an influence on the composition of molecular strains of *M. bovis* isolates in the following ways:

1. Cattle-specific genotypes of *M. bovis* transmitted to wildlife undergo clonal expansion in the wildlife populations in geographically restricted areas, with periodic spillover to other wildlife species. These genotypes characteristically have minimal diversity because of their restriction to these areas (Michel et al. 2008; Hlokwe et al. 2011).
2. Genotypes established in wildlife following spillover from cattle increasingly spread between geographically linked wildlife populations, the rate of spread being a function of the extent of contact between these wildlife populations (De Garine-Wichatitsky et al. 2013; Hlokwe et al. 2014).

In Zambia's Kafue flood plains, cattle herders still practice transhumance, and animals are moved to the lower plains at the commencement of the dry season in search of adequate pastures. These animals, tended by herdsmen, can remain in these areas for months-on-end depending on the length of the dry season. The majority of the people practicing this type of cattle grazing system are traditional pastoralists who lived in these areas for decades. Particularly in the Liuwa plains, a huge migration of wildebeest, the second largest migration of land mammals (the largest being in the Serengeti), takes place between Zambia and Angola (De Garine-Wichatitsky et al. 2013; Muma et al. 2013). In these events, the risk of spillback from infected wildlife to cattle and to the human populations enhances the risk of contracting the disease both for humans and animals and emphasizes the need to embrace the "One Health" approach in the control and management of BTB.

As evidence of the historical connectedness in Southern Africa, shared spoligotypes (SB0140 and or SB0131) occur in South Africa, Zambia, and Mozambique (Fig. 8.17). This may infer that they shared a common source of infection (Müller et al. 2009; Smith et al. 2011), that independent introductions into each of the countries occurred, or that a single introduction into one country then spread to the others due to the migration of *M. bovis*-infected wildlife. The role of wildlife in

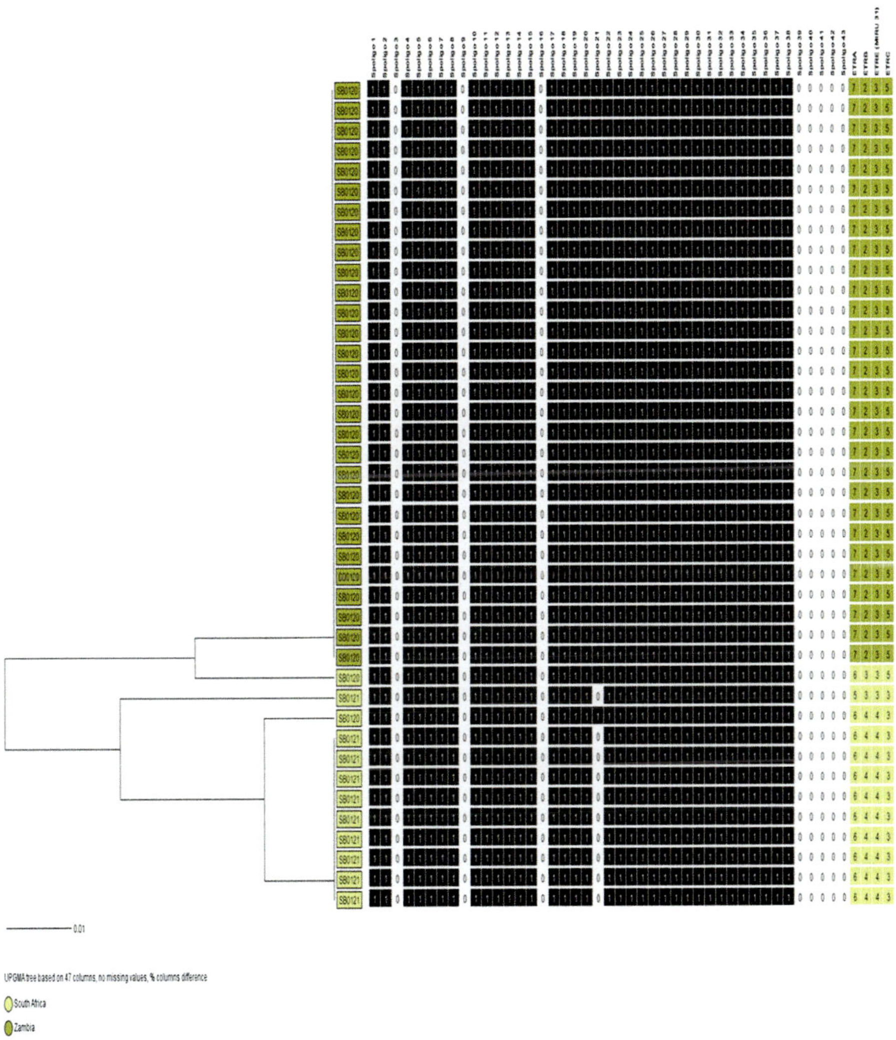

Fig. 8.17 Divergent evolution of SB0120 and SB0121 in South Africa (yellow) and Zambia (brown). The dendrogram was generated using spoligotypes and four exact tandem repeat loci (A–C and E)

this dissemination is currently speculative as it is not known whether they were infected with *M. bovis* at that time. However, when spoligotyping is combined with four loci of MIRU-VNTR for South Africa and Zambia (Figs. 8.18 and 8.19), it is evident that these apparently shared spoligotypes evolved divergently. If this situation also applies to countries such as Mozambique, Swaziland, and Botswana, it is possible that the restricted contact because of fencing, and international import control requirements that were more strictly applied on those countries, contributed to the divergent evolution of the strains in the respective countries. There are approximately 45, 42, and 73 unique *M. bovis* spoligotypes, respectively, in the Southern, Eastern, and Western regions of Africa, and the significant differences in

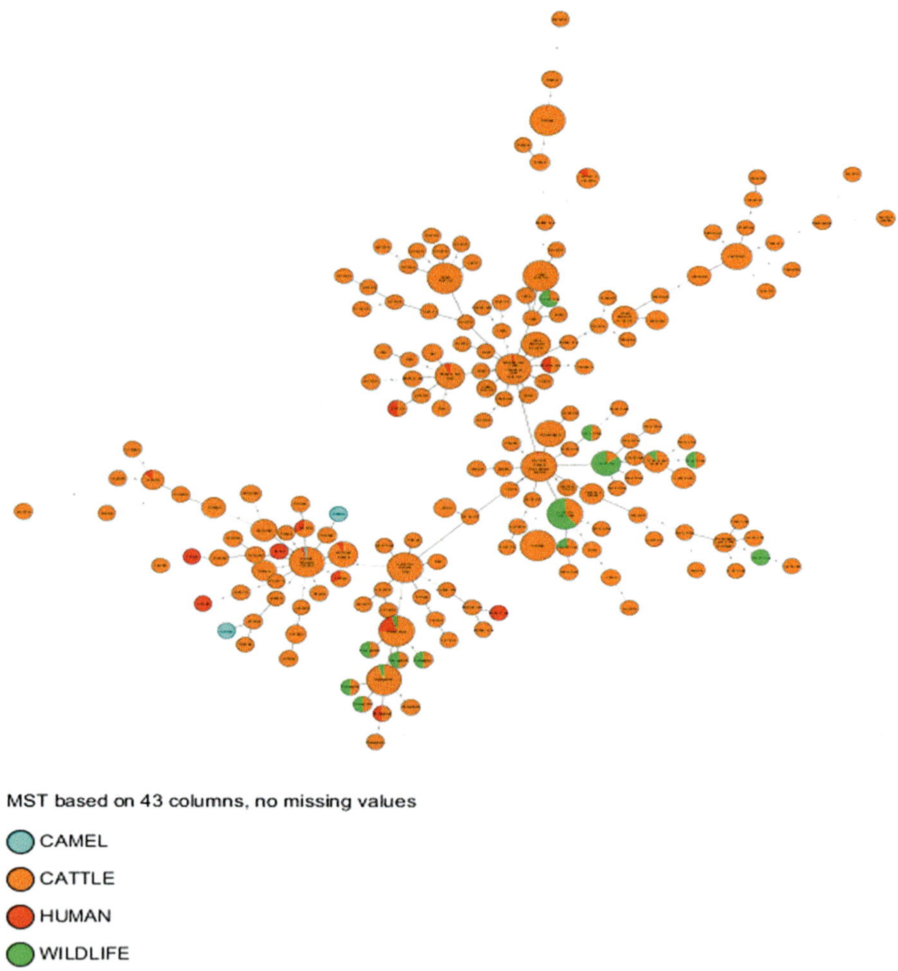

Fig. 8.18 The epidemiology of bovine and zoonotic TB in Africa: distribution according to host. The MST was generated using the same parameters and calibration as in Figs. 8.9 and 8.10 but color-coded by host

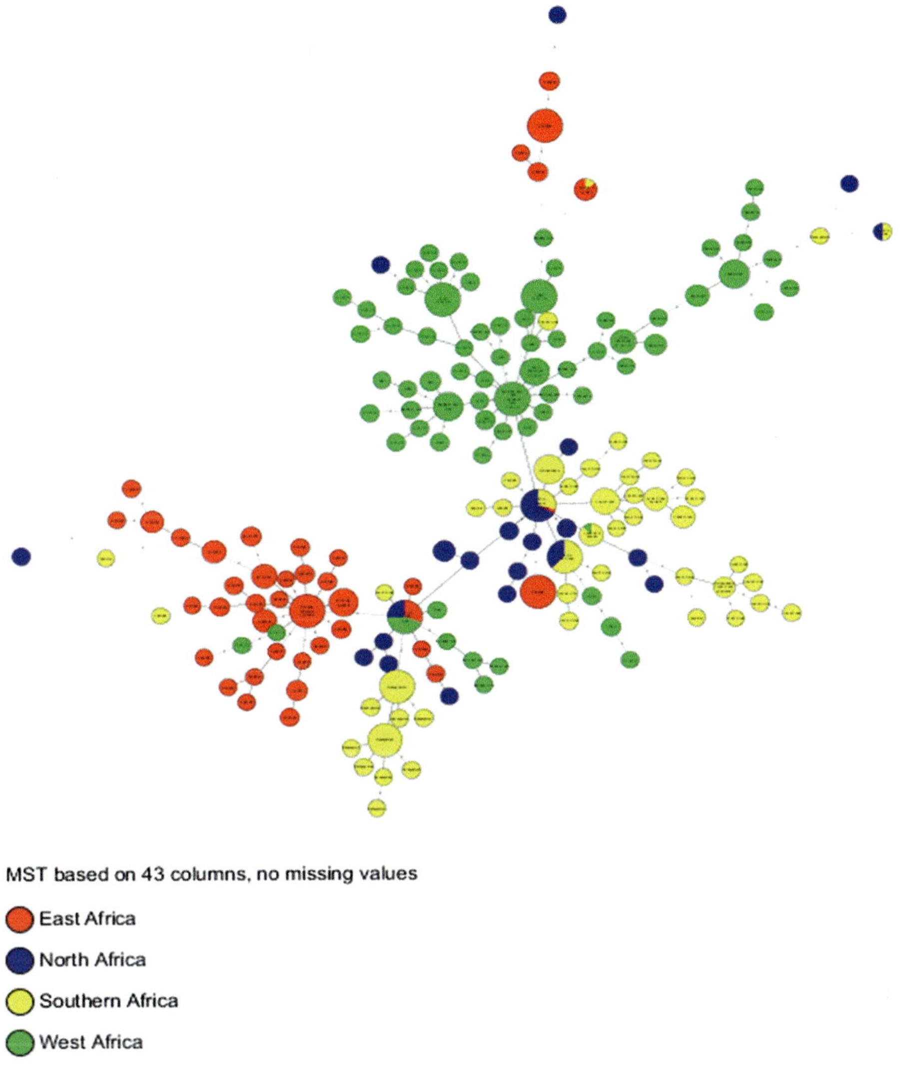

Fig. 8.19 The epidemiology of bovine and zoonotic TB in Africa: distribution according to country. The MST was generated using the same parameters and calibration as in Figs. 8.9 and 8.10 but color-coded by country

the spoligotype diversity between Western and Southern Africa support the bottleneck theory (Sect. 8.2.3). It is apparent from the molecular markers (Fig. 8.17) that countries such as Zambia and South Africa have had independent epidemiological dynamics for some time.

8.7 Molecular Epidemiology of *M. bovis* in Humans and Other Hosts in Africa

Although *M. bovis* mainly infects ungulates, it is known to infect other homoeothermic animals (Cosivi et al. 1995, 1998). In humans, it causes zoonotic tuberculosis that affects an estimated 70,000 people every year in Africa (Müller et al. 2013). The real contribution of zoonotic TB to the African tuberculosis burden is, however, yet to be determined given that the current diagnostics are not geared to detecting this form of tuberculosis (Cosivi et al. 1995; Müller et al. 2013). Molecular markers show that zoonotic TB occurs in all the regions in Africa (Figs. 8.18 and 8.19). Although most of these cases were caused by genotypes prevalent in the animal populations, there were a few genotypes that have only been recovered from humans, most of whom had pulmonary TB (Sunder et al. 2009; Sanou et al. 2014). This raises the question as to whether human-to-human transmission of zoonotic TB also occurs in Africa as has been documented elsewhere (Müller et al. 2013).

In Southern Africa, wildlife plays an increasing role in the epidemiology of BTB (Figs. 8.18 and 8.19). Lions are the most important nonungulate in South Africa infected by *M. bovis*. They contract the infection primarily from infected prey species such as African buffaloes, as evidenced by the shared *M. bovis* genotypes with buffaloes (Michel et al. 2008). The role of wild suids such as warthogs (*Phacochoerus africanus*) and bush pigs (*Potamochoerus larvatus*) in the epidemiology of BTB is still unknown, but it is believed to be potentially important, as they move more freely between farmed land and game reserves than most ungulates (Hlavsa et al. 2008). In Eastern Africa, especially in Ethiopia, camels may play a role in their BTB epidemiology. However, the molecular markers isolated from camels, there have not yet been detected in the cattle populations, and camels could thus have an independent disease dynamic.

8.8 Limitations and Opportunities for the Use of Molecular Epidemiological Tools in Africa

8.8.1 Limitations

8.8.1.1 Lack of or Inadequate Disease Control and Eradication Policies

Having a well-thought-out and achievable livestock disease control policy is critical not only for achieving maximum benefits from the livestock industry but also for safeguarding consumers' health and the national economy. Disease control and prevention policies pertinent to infectious diseases need to embrace, among other mitigations:

1. Mandatory isolation and quarantine measures
2. Strict import and export regulations

3. Restriction of uncontrolled movement of animals between farms and locations
4. Establishing surveillance and monitoring activities including keeping records of animal health status
5. Identification of individual animals
6. Adequate biosecurity and maintenance of sanitation

Distinctive features of BTB that include its high economic impact, zoonotic transmission, involvement of multiple hosts, complex ecology, chronic course of infection, and difficulty to diagnose unquestionably demand special consideration in designing and implementing a sustainable national BTB control policy.

However, to date, only few countries in Africa have adopted and implemented a functional national BTB control policy. According to an assessment by Cosivi et al. (1995), BTB control is almost nonexistent in the majority of dairy and other cattle on the African continent. The BTB control policy gap in Africa extends across all activities, including the absence of regular tuberculin testing, slaughtering of test-positive animals, extensive uncontrolled movement of livestock, lack of contact tracing, and absence of abattoir surveillance. It is perplexing to conduct an epidemiological investigation of BTB under these circumstances since the findings are ultimately shelved and not implemented, due to the absence of an appropriate disease control policy.

8.8.1.2 Absence of a Continental Reference Database and Laboratory Capacity for Mycobacterial Genotype Characterization

Because of the lack of control policies, there is an inherent disparity and weakness in surveillance standards among countries, which compromise the source and quality of specimens from which *M. bovis* could be isolated for molecular epidemiological discourse. There has been some work carried out in laboratories in Africa on the presence and distribution of *M. bovis*, funded by the European Research Council, but it is critical to note that there is not an established central collaborative laboratory in Africa as is the case for other continents. The current situation is probably also responsible for the lack of a critical mass of professionals required to sustain the discipline of molecular epidemiology in Africa.

8.8.1.3 Weak Transdisciplinary and Interlaboratory Collaboration

The discipline of molecular genetics gendered enormous support and collaboration by a wide range of disciplines to unravel the intricate interaction between host, pathogen, and environment and the resultant diseases and complications that threaten the existence of humans and animals. Unlike with conventional epidemiology, with molecular epidemiology there is extensive collaboration with various professionals such as statisticians, environmental experts, industrial hygienists, clinicians, biochemists, molecular biologists, geneticists, pathologists, and immunologists (Schulte and Perera 1993). This interdisciplinary network in most African countries is in its nascent stage, and it is not robust enough to generate the critical mass required to sustain this discipline.

The distinctive pathobiological features of BTB, including its chronic course, an almost unlimited host range, various routes of transmission, lack of an effective vaccine, difficulty to accurately diagnose the infection, and its impacts on both animal and human health, support the case for transdisciplinary collaboration between veterinarians, physicians, economists, microbiologists, and public health and environmental science experts.

8.8.2 Opportunities

Most African countries do not have state-of-the-art veterinary diagnostic facilities to routinely make definitive diagnoses, but they rely on submitting samples or isolates to laboratories in Europe or the United States to confirm certain diseases or conditions. Although this has some drawbacks, such collaboration also has benefits that allow technology transfer, producing skilled professionals and establishing laboratory facilities to do molecular typing of mycobacteria, in addition to providing invaluable insight into the molecular genetics and epidemiology of mycobacterial diseases, especially of the MTC organisms. A good example of the outcome of such a joint venture is the recent identification of the two major clonal complexes of *M. bovis* strains, Africa 1 (Af1) and Africa 2 (Af2). The typing process involved collecting *M. bovis* isolates from various countries in Africa and transporting them to a number of laboratories abroad, including the Animal Health and Veterinary Laboratories Agency (VLA), Weybridge, United Kingdom, for molecular typing (spoligotyping, VNTR typing, deletion analysis, and IS6110 RFLP typing). In addition to creating further opportunities for collaboration, the findings of the study revealed important insights into the phylogeography of *M. bovis* including the detection of distinct clonal complexes in Western (Af1) and Eastern Africa (Af2).

There is a need to emphasize the weak capacity in pathobiological diagnostics of disease, as this is likely to attract the much-needed funding to complement clinical- and public health-based research at the animal-human interface. Investment in pathobiological capacity would inherently lead to the development of biobanks of pathogens from animals and humans at the human-animal interface. This would give most African countries better opportunities to collaborate and share their experiences allowing them to build and strengthen technical capacity that include skilled manpower and facilities to conduct molecular typing and epidemiologic analyses of the causes of the mycobacterial diseases. Furthermore, such a resource will serve many purposes including innovation in diagnostic and vaccination technologies, standardization of genotyping methods to support the long-term goal of controlling, and eradicating TB both from animals and humans on the African continent.

8.8.3 Future Challenges

The contribution of zoonotic TB to human TB is poorly investigated particularly in sub-Saharan Africa, where a combination of endemic BTB, an expanding human-animal interface, an expanding dairy industry, the lack of appropriate diagnostic tools, and cultural norms that discourage health-seeking behavior present considerable challenges to poor communities. This complex of the drivers of the disease is likely to hamper progress toward achieving the World Health Organization's ambitious goal of eliminating TB by 2035. This, in effect, means that now more than ever, it is imperative to be able to make a definitive diagnosis in each case when dealing with TB, regardless of the causative agent, as each individual case matters if eradication is to be successful. Without the much-needed investment in developing faster and cheaper molecular assays, it will be difficult to properly understand this ever-changing landscape of TB in general, but specifically, BTB and zoonotic TB in Africa.

8.9 Conclusions

The use of molecular epidemiological tools is increasing in Africa. Its application is still hampered by the limited investment by governmental authorities in laboratory facilities and the lack of adequately trained human resources. Synthesis of the available historical and contemporary livestock-related *M. bovis* molecular data indicates that the popular belief that this pathogen was first introduced onto the African continent during the European colonial period is wrong. It is more likely that a small but unique set of genotypes representing an *M. bovis* population native to Africa was present on the continent before that time. The geographical clustering of molecular markers in former European colonial territories supports the notion that the dominant genotypes were introduced onto the continent following the cattle population bottleneck caused by rinderpest during and toward the end of the nineteenth century.

We acknowledge that this synthesis only represents a "coarse" molecular epidemiological analysis of the epidemiology of *M. bovis* in Africa. It is hoped that a more "granular" picture will be brought to light in the near future by being able to incorporate next-generation, high-throughput data.

Acknowledgments The authors are grateful to all contributors to this chapter, especially Dr. Paul Bessel, at the Roslin Institute, who did the Global Information Systems (GIS) and mapping of population data used in this chapter.

References

Adler P, Pouwels J, Randall L (2007) World civilizations: since 1500, 5th edn. Cengage Learning, Vienna, pp 1–25
African Union (2010) History of rinderpest eradication from Africa: impact, lessons learnt and way forward. African Union (AU), Addis Ababa
Afrikaner (2009) The Afrikaner cattle breed. http://www.afrikanerbees.com/Society-History.htm. Accessed 29 July 2015.
Alhaji I (1976) Bovine tuberculosis: A general review with special reference to Nigeria. Vet Bull 46:829–841
Allix C, Supply P, Fauville-Dufaux M (2004) Utility of fast mycobacterial interspersed repetitive unit-variable number tandem repeat genotyping in clinical mycobacteriological analysis. Clin Infect Dis 39:783–789
Anon (1858) Examination of Nonqause before the Chief Commissioner: British Kaffraria Government Gazette
Armando B (1995) Romagnola breeds. Hekpoort, South African Breeding Society, pp 1–2
Awah-Ndukum J, Kudi AC, Bradley G et al (2013) Molecular genotyping of *Mycobacterium bovis* isolated from cattle tissues in the North West Region of Cameroon. Trop Anim Health Prod 45:829–836
Ayele W, Niel S, Zinsstag J et al (2004) Bovine tuberculosis: an old disease but new threat to Africa. Int J Tuberc Lung 8(8):924–937
Barbieri C, Vicente M, Oliveira S (2014) Migration and interaction in a contact zone: mtDNA variation among Bantu-speakers in Southern Africa. PLoS ONE 9(6):e99117
Ben Kahla I, Boschiroli ML, Souissi F et al (2011) Isolation and molecular characterisation of *Mycobacterium bovis* from raw milk in Tunisia. Afr Health Sci 11(3):2–5
Bengis R, Kock R, Fisher J (2002) Infectious animal diseases: The wildlife/livestock interface. Sci Tech Rev OIE 21:53–65
Berdah D (2010) Vaccinating cattle against bovine tuberculosis in France, 1921–1963: Between the epistemic value of the animal model and an alternative to sanitary policies. Rev d'Etudes en Agric Environ 91:393–415
Berg S, Garcia-Pelayo MC, Müller B et al (2011) African 2, a clonal complex of *M. bovis* epidemiologically important in East Africa. J Bacteriol 193(3):670–678
Bienart W (1989) Introduction: The politics of colonial conservation. J South Afr Stud 15:143–163
Biffa D, Asseged B, Skjerve E (2010) Diagnostic efficiency of abattoir meat inspection service in Ethiopia to detect carcasses infected with *Mycobacterium bovis*: Implications for public health. BMC Public Health 10(1):46210(462)
Biffa D, Johansen TB, Godfroid J et al (2014) Multi-locus variable-number tandem repeat analysis (MLVA) reveals heterogeneity of *Mycobacterium bovis* strains and multiple genotype infections of cattle in Ethiopia. Infect Genet Evol 23:13–19
Boardman J (1965) The Greeks overseas, 4th edn. Classical Association of Canada, pp 11–25
Bothar (2008) Bothar—Our history. http://www.bothar.ie/index.jsp?p=101&n=115. Accessed 29 July 2015
Brosch R, Gordon SV, Marmiesse M et al (2002) A new evolutionary scenario for the *Mycobacterium tuberculosis* complex. Proc Natl Acad Sci USA 99(6):3684–3689
Cameron H, O'Brien R, Murray A et al (2001) Evaluation of the *Mycobacterium bovis* restriction fragment length polymorphism probe pUCD, in combination with the direct repeat probe, for molecular typing of *Mycobacterium tuberculosis* strains in Ireland. J Clin Microbiol 39:4404–4406
Castrì L, Tofanelli S, Garagnani P et al (2009) mtDNA variability in two Bantu-speaking populations (Shona and Hutu) from Eastern Africa: Implications for peopling and migration patterns in sub-Saharan Africa. Am J Phys Anthropol 140:302–311
Christopher E (2002) The civilizations of Africa. University of Virginia, pp 12–59

Clark JD, Brandt SA (1984) From hunters to farmers: the causes and consequences of food production in Africa, 1st edn. University of California Press, Berkeley, CA, p 33
Cloete PC, Taljaard PR, Grové B (2007) A comparative economic case study of switching from cattle farming to game ranching in the Northern Cape Province. South Afr J Wildl Res 37:71–78
Cosivi O, Meslin FX, Daborn CJ et al (1995) Epidemiology of *Mycobacterium bovis* infection in animals and humans, with particular reference to Africa. Rev Sci Tech OIE 14:733–746
Cosivi O, Grange JM, Daborn CJ et al (1998) Zoonotic tuberculosis due to *Mycobacterium bovis* in developing countries. Emerg Infect Dis 4:59–70
Cour JM (2001) The Sahel in West Africa: Countries in transition to a full market economy. Glob Environ Change 11:31–47
Daborn CJ, Grange JM, Kazwala RR (1996) The bovine tuberculosis cycle—An African perspective. J Appl Bacteriol 81:S27–S32
De Garine-Wichatitsky M, Caron A, Kock R et al (2013) A review of bovine tuberculosis at the wildlife-livestock-human interface in sub-Saharan Africa. Epidemiol Infect 141:1342–1356
Deshler W (1963) Cattle in Africa: distribution, types, and problems. Geogr Rev 53:52–58
Diamond J (1999) Guns, germs, and steel, 1st edn. Norton, New York, pp 15–39
Donoghue HD (2009) Human tuberculosis - an ancient disease, as elucidated by ancient microbial biomolecules. Microbiol Infect 11:1156–1162
Driscoll J (2009) Spoligotyping for molecular epidemiology of the *Mycobacterium tuberculosis* complex. In: Caugant D (ed) Methods in molecular biology, 1st edn. Humana, New York, pp 117–140
Durr PA, Hewinson RG, Clifton-Hadley RS (2000) Molecular epidemiology of bovine tuberculosis. I. *Mycobacterium bovis* genotyping. Rev Sci Tech 19:675–688
Esther H (2012) Boran Cattle Breeders' Society of South Africa. http://www.boran.org.za. Accessed 10 Sept 2015
FAO (2005) Cattle population. Food and Agricultural Organization of the United Nations (FAO), Rome
Fèvre EM, Bronsvoort B, Hamilton K et al (2006) Animal movements and the spread of infectious diseases. Trends Microbiol 14:125–131
Firdessa R, Tschopp R, Wubete A et al (2012) High prevalence of bovine tuberculosis in dairy cattle in central Ethiopia: Implications for the dairy industry and public health. PLoS One 7(12): e52851
Foxman B (2001) Molecular epidemiology: Focus on infection. Am J Epidemiol 153:1135–1141
Fratkin E (2001) East African pastoralism in transition: Maasai, Boran, and Rendille cases. Afr Stud Rev 44:1–25
Frothingham R, Meeker-O'Connell WA (1998) Genetic diversity in the *Mycobacterium tuberculosis* complex based on variable numbers of tandem DNA repeats. Microbiology 144:1189–1196
Gadd ME (2012) Barriers, the beef industry and unnatural selection: A review of the impact of veterinary fencing on mammals in southern Africa. In: Somers MJ, Hayward M (eds) Fencing for conservation: Restriction of evolutionary potential or a riposte to threatening processes. Springer, New York, pp 153–186
Garnier T, Eiglmeier K, Casmus JC et al (2003) The complete genome sequence of *Mycobacterium bovis*. Proc Natl Acad Sci USA 100(13):7877–7882
Goody J (1976) Production and reproduction: A comparative study of the domestic domain, 1st edn. Cambridge University Press, Cambridge, pp 6–12
Gumi B, Schelling E, Berg S et al (2012) Zoonotic transmission of tuberculosis between pastoralists and their livestock in South-East Ethiopia. Ecohealth 9:139–149
Haddad N, Ostyn A, Karoui C et al (2001) Spoligotype diversity of *Mycobacterium bovis* strains isolated in France from 1979 to 2000. J Clin Microbiol 39:3623–3363
Hang'Ombe MB, Munyeme M, Nakajima C et al (2012) *Mycobacterium bovis* infection at the interface between domestic and wild animals in Zambia. BMC Vet Res 8:221
Hansard A, Harrison JH (1958) Kenya white highlands. House of commons records 598:463–474

HarvestChoice (2015) Cattle population. International Food Policy Research Institute, Washington, DC. http://harvestchoice.org/data/an05_catt. Accessed 29 July 2015
He L, Fan X, Xie J (2012) Comparative genomic structures of *Mycobacterium*. J Cell Biochem 113:2464–2473
Hlavsa MC, Moonan PK, Cowan LS et al (2008) Human tuberculosis due to *Mycobacterium bovis* in the United States, 1995-2005. Clin Infect Dis 47(2):168–175
Hlokwe TM, Jenkins AO, Streicher EM et al (2011) Molecular characterisation of *Mycobacterium bovis* isolated from African buffaloes (*Syncerus caffer*) in Hluhluwe-iMfolozi Park in KwaZulu-Natal, South Africa. Onderstepoort J Vet Res 78(1):39–44
Hlokwe TM, van Helden P, Michel AL (2014) Evidence of increasing intra- and inter-species transmission of *Mycobacterium bovis* in South Africa: Are we losing the battle? Prev Vet Med 115:10–17
Homewood KM, Trench P, Brockington D (2012) Pastoralist livelihoods and wildlife revenues in East Africa: a case for coexistence? Pastoralism 2:19
Huillery E (2009) History matters: The long-term impact of colonial public investments in French West Africa. Am Econ J Appl Econ 1:176–215
Jackson S-A (2003) Disease and biomedicine: Colonial strategies in Southern Africa. University of California, Riverside, pp 303–308
Jenkins O, Cadmus SIB, Venter EH et al (2011) Molecular epidemiology of human and animal tuberculosis in Ibadan, Southwestern Nigeria. Vet Microbiol 151:139–147
Kamerbeek J, Schouls L, Kolk A et al (1997) Simultaneous detection and strain differentiation of *Mycobacterium tuberculosis* for diagnosis and epidemiology. J Clin Microbiol 35:907–914
Kanduma E, McHugh TD, Gillespie SH (2003) Molecular methods for *Mycobacterium tuberculosis* strain typing: A users guide. J Appl Microbiol 94:781–791
Kazwala RR, Kambarage DM, Daborn CJ et al (2001) Risk factors associated with the occurrence of bovine tuberculosis in cattle in the Southern Highlands of Tanzania. Vet Res Commun 25:609–614
Khapoya VB (2012) The African experience, 2nd edn. Taylor and Francis, New York
Komal BP (2014) DNA fingerprinting process. http://www.buzzle.com/articles/dna-fingerprinting-process.html. Accessed 19 Oct 2015
Lawson L, Zhang J, Gomgnimbou MK et al (2012) A molecular epidemiological and genetic diversity study of tuberculosis in Ibadan, Nnewi and Abuja, Nigeria. PLoS One 7(6):e38409
Lin T, Lin L, Zhang F (2014) Review on molecular typing methods of pathogens. J Med Microbiol 4:147–152
Lunde TM, Lindtjørn B (2013) Cattle and climate in Africa: How climate variability has influenced national cattle holdings from 1961-2008. PeerJ 1:e55. https://doi.org/10.7717/peerj.55
Mahairas CG, Sabo PJ, Hickey MJ et al (1996) Molecular analysis of genetic differences between *Mycobacterium bovis* BCG and virulent *M. bovis*. J Bacteriol 178(5):1274–1282
Mäki-Tanila A, Fernandez J, Toro M et al (2010) Local cattle breeds in Europe. Development of policies and strategies for self-sustaining breeds. Wageningen Academic, Wageningen, p 151
Malama S, Muma J, Munyeme M et al (2014) Isolation and molecular characterization of *Mycobacterium tuberculosis* from humans and cattle in Namwala District, Zambia. Ecohealth 11 (4):564–570
Mariner JC, House JA, Mebus CA et al (2012) Rinderpest eradication: Appropriate technology and social innovations. Science 337(6100):1309–1312
Mateos ANA, Dominguez L, Vidal D et al (1996) Spacer oligonucleotide typing of *Mycobacterium bovis* strains from cattle and other animals : A tool for studying epidemiology of tuberculosis. J Clin Microbiol 34:2734–2740
Mathema B, Kurepina NE, Bifani PJ et al (2006) Molecular epidemiology of tuberculosis: Current insights. Clin Microbiol Rev 19:658–685
McKenna A (2011) The history of Southern Africa Britannica guide to Africa. Rosen, New York, p 220

Michel AL, Hlokwe TM, Coetzee ML et al (2008) High *Mycobacterium bovis* genetic diversity in a low prevalence setting. Vet Microbiol 126:151–159

Michel AL, Coetzee ML, Keet DF et al (2009) Molecular epidemiology of *Mycobacterium bovis* isolates from free-ranging wildlife in South African game reserves. Vet Microbiol 133:335–343

More S, Good M (2006) The tuberculosis eradication programme in Ireland: A review of scientific and policy advances since 1988. Vet Microbiol 112(2–4):239–251

Mostowy S, Cousins D, Brinkman J et al (2002) Genomic deletions suggests phylogeny for the *Mycobacterium tuberculosis* complex. J Infect Dis 186(1):74–80

Müller B, Steiner B, Bonfoh B et al (2008) Molecular characterisation of *Mycobacterium bovis* isolated from cattle slaughtered at the Bamako Abattoir in Mali. BMC Vet Res 4:26

Müller B, Hilty M, Berg S et al (2009) African 1, an epidemiologically important clonal complex of *Mycobacterium bovis* dominant in Mali, Nigeria, Cameroon, and Chad. J Bacteriol 191:1951–1960

Müller B, Dürr S, Alonso S et al (2013) Zoonotic *Mycobacterium bovis*-induced tuberculosis in humans. Emerg Infect Dis 19:899–908

Muma JB, Syakalima M, Munyeme M et al (2013) Bovine tuberculosis and brucellosis in traditionally managed livestock in selected districts of Southern Province of Zambia. Vet Med Int. https://doi.org/10.1155/2013/730367

Munyeme M, Muma JB, Skjerve E et al (2008) Risk factors associated with bovine tuberculosis in traditional cattle of the livestock/wildlife interface areas in the Kafue basin of Zambia. Prev Vet Med 85:317–328

Munyeme M, Muma J, Samui K et al (2009a) Prevalence of bovine tuberculosis and animal level risk factors for indigenous cattle under different grazing strategies in the livestock/wildlife interface areas of Zambia. Trop Anim Health Prod 41(3):345–352

Munyeme M, Rigouts L, Shamputa IC et al (2009b) Isolation and characterization of *Mycobacterium bovis* strains from indigenous Zambian cattle using spacer oligonucleotide typing technique. BMC Microbiol 9:144

Musoke J, Hlokwe T, Marcotty T et al (2015) Spillover of *Mycobacterium bovis* from wildlife to livestock, South Africa. Emerg Infect Dis 21:448–451

Muwonge A, Johansen TB, Vigdis E et al (2012) *Mycobacterium bovis* infections in slaughter pigs in Mubende district, Uganda: a public health concern. BMC Vet Res 8:168

Muwonge A, Kankya C, Olea-Popelka F et al (2013) Molecular investigation of multiple strain infections in patients with tuberculosis in Mubende District, Uganda. Infect Genet Evol 17:16–22

Muwonge A, Motto P, Nkongho EF et al (2016) Predicting cattle movement networks: *Mycobacterium bovis* spatial-genotyping versus gravity modelling. Paper presented at The Colorado Mycobacteria Conference 2016, Fort Collins, United States, 7 June 16, Accessible here https://www.research.ed.ac.uk/portal/en/publications/predicting-cattle-movement-networksmycobacterium-bovis-spatialgenotyping-versus-gravity-modelling(a62cafb7-8e5c-4b15-9e25-c1dbabc16a1f)/export.html

Neill SD, Skuce R, Pollock JM (2005) Tuberculosis—new light from an old window. J Appl Microbiol 98:1261–1269

O'Brien R, Danilowicz BS, Bailey L et al (2000) Characterization of the *Mycobacterium bovis* restriction fragment length polymorphism DNA probe pUCD and performance comparison with standard methods. J Clin Microbiol 38:3362–3369

Oloya J, Kazwala R, Lund A et al (2007) Characterisation of mycobacteria isolated from slaughter cattle in pastoral regions of Uganda. BMC Microbiol 7:95

Oloya J, Opuda-Asibo J, Kazwala R et al (2008) Mycobacteria causing human cervical lymphadenitis in pastoral communities in the Karamoja region of Uganda. Epidemiol Infect 136:636–643

Olson J (1996) The peoples of Africa: An ethnohistorical dictionary, 1st edn. Greenwood, Santa Barbara, CA, pp 2–45

Perry BD, Grace D, Sones K (2011) Livestock and global change special feature: Current drivers and future directions of global livestock disease dynamics. Proc Natl Acad Sci USA 110:20871–20877

Razanamparany VR, Quirin R, Rapaoliarijaona A et al (2006) Usefulness of restriction fragment length polymorphism and spoligotyping for epidemiological studies of *Mycobacterium bovis* in Madagascar: Description of new genotypes. Vet Microbiol 114:115–122

Reyes JF, Tanaka MM (2010) Mutation rates of spoligotypes and variable numbers of tandem repeat loci in *Mycobacterium tuberculosis*. Infect Genet Evol 10:1046–1051

Rigouts L, Maregeya B, Traore H et al (1996) Use of DNA restriction fragment typing in the differentiation of *Mycobacterium tuberculosis* complex isolates from animals and humans in Burundi. Tuber Lung Dis 77:264–268

Roring S, Scott A, Brittain D et al (2002) Development of variable-number tandem repeat typing of *Mycobacterium bovis*: Comparison of results with those obtained by using existing exact tandem repeats and spoligotyping. J Clin Microbiol 40:2126–2133

Sabat AJ, Budimir A, Nashev D et al (2013) Overview of molecular typing methods for outbreak detection and epidemiological surveillance. Euro Surveill 18(4):20380. http://www.eurosurveillance.org

Sahraoui N, Müller B, Guetarni D et al (2009) Molecular characterization of *Mycobacterium bovis* strains isolated from cattle slaughtered at two abattoirs in Algeria. BMC Vet Res 5:4

Sanou A, Tarnagda Z, Kanyala E et al (2014) *Mycobacterium bovis* in Burkina Faso: Epidemiologic and genetic links between human and cattle isolates. PLoS Negl Trop Dis 8:e3142

Schulte P, Perera F (1993) Molecular epidemiology: Principals and practices. In: Schulte P (ed) Concepts and historical frameworks for molecular epidemiology, 1st edn. Academic, San Diego, CA, p 608

Selander RK, Caugant DA, Ochman H (1986) Methods of multilocus enzyme electrophoresis for bacterial population genetics and systematics. Appl Environ Microbiol 51:873–884

Seva J, Sanes JM, Ramis G et al (2014) Evaluation of the single cervical skin test and interferon gamma responses to detect *Mycobacterium bovis* infected cattle in a herd co-infected with *Mycobacterium avium* subsp. *paratuberculosis*. Vet Microbiol 171:139–146

Smith NH, Gordon SV, de la Rua-Domenech R et al (2006) Bottlenecks and broomsticks: The molecular evolution of *Mycobacterium bovis*. Nat Rev Microbiol 4:670–681

Smith NH, Berg S, Dale J, Allen A et al (2011) European 1: A globally important clonal complex of *Mycobacterium bovis*. Infect Genet Evol 11:1340–1345

Staal SJ, Nin Pratt A, Jabbar MA (2008) Dairy development for the resource poor. Part 2: Kenya and Ethiopia. Dairy development case studies. http://192.156.137.110/Link/Publications/Theme3/DairyDevForResourcePoor-2.pdf.

Stackyard (2009) South African Breeds. http://www.stackyard.com/pedigree/html/sacattle.html. Accessed 29 July 2015

Sunder S, Lanotte P, Godreuil S et al (2009) Human-to-human transmission of tuberculosis caused by *Mycobacterium bovis* in immunocompetent patients. J Clin Microbiol 47:1249–1251

Supply P, Allix C, Lesjean S et al (2006) Proposal for standardization of optimized mycobacterial interspersed repetitive unit-variable-number tandem repeat typing of *Mycobacterium tuberculosis*. J Clin Microbiol 44:4498–4510

SWAC (2007) Promoting and supporting change in transhumant pastoralism in the Sahel and West Africa. OECD, Paris, pp 2–4. https://www.oecd.org/swac/publications/38402714.pdf

Tambi EN, Maina OW, Mukhebi AW et al (1999) Economic impact assessment of rinderpest control in Africa. Rev Sci Tech 18:458–477

Taneja VK (1999) Chapter 5: Dairy breeds and selection. In: Falvey L, Chantalakhana C (eds) Smallholder dairying in the tropics,. pp 71. ILRI (International Livestock Research Institute), Nairobi

Tawah C, Rege J, Aboagye S (1997) A close look at a rare African breed—The Kuri cattle of Lake Chad Basin: Origin, distribution, production and adaptive characteristics. Afr J Anim Sci 27:31–40

Taylor GM, Murphy E, Hopkins R et al (2007) First report of *Mycobacterium bovis* DNA in human remains from the Iron Age. Microbiology 153:1243–1249

Van der Zanden AGM, Kremer K, Schouls LM et al (2002) Improvement of differentiation and interpretability of spoligotyping for *Mycobacterium tuberculosis* complex isolates by introduction of new spacer oligonucleotides. J Clin Microbiol 40(12):4628–4639

Van Embden JDA, van Soolingen D, Small PM et al (1992) Genetic markers for the epidemiology of tuberculosis. Res Microbiol 143:385–391

Van Embden JDA, Cave MD, Crawford JT et al (1993) Strain identification of *Mycobacterium tuberculosis* by DNA fingerprinting: Recommendations for a standardized methodology. J Clin Microbiol 31:406–409

Van Soolingen D, de Haas P, Hermans P et al (1994) DNA fingerprinting of *Mycobacterium tuberculosis*. Methods Enzymol 235:196–205

Yang ZH, Ijaz K, Bates JH et al (2000) Spoligotyping and polymorphic GC-rich repetitive sequence fingerprinting of *Mycobacterium tuberculosis* strains having few copies of IS6110. J Clin Microbiol 38:3572–3576

Zeleza T (1993) A modern economic history of Africa: The nineteenth century, 1st edn. East African Educational Publisher, Nairobi, p 51

Chapter 9
The Diagnosis of Bovine Tuberculosis

Nicolaas P. J. Kriek, Demelash B. Areda, and Asseged B. Dibaba

9.1 Introduction

Bovine TB (BTB) caused by *Mycobacterium bovis* is a complex, transmissible, infectious disease. Cattle are the main hosts of the infection, but BTB also occurs in a wide range of domesticated and wild animals, and in humans. The clinical and pathological manifestations, and immunological responses in this chronic disease differ substantially within and between species, the stage of the disease, the pathogenicity of the specific *M. bovis* strain, and environmental factors. In cattle, whether it is a recently infected herd, a herd with long-standing established infection with advanced and generalized, chronic cases, or a herd that has been subjected to eradication of BTB and elimination of most of the infected cattle for some time, the detection of all BTB-positive cattle is notoriously challenging.

Because of the lack of sensitivity (Se) and specificity (Sp) of all the currently available diagnostic methods, there is not a single technique for individual animals to confirm a diagnosis of BTB with 100% certainty; all the tests are herd tests. Herein lies the challenge of controlling the disease, since the inability to reliably make a diagnosis in individual animals adds to the risk of infected animals remaining in herds in which attempts are being made to control and eradicate the disease, or to not detect its presence when the herd prevalence is low and/or unknown.

N. P. J. Kriek
Department of Paraclinical Sciences, Faculty of Veterinary Science, University of Pretoria, Onderstepoort, South Africa

D. B. Areda
Department of Biological Sciences, Minnesota State University, Mankato, Mankato, MN, USA
e-mail: demelash.areda@mnsu.edu

A. B. Dibaba (✉)
Department of Pathobiology, College of Veterinary Medicine, Tuskegee University, Tuskegee, AL, USA
e-mail: adibaba@tuskegee.edu

A. B. Dibaba et al. (eds.), *Tuberculosis in Animals: An African Perspective*,
https://doi.org/10.1007/978-3-030-18690-6_9

To successfully treat, control, and eradicate a disease, it is critical to make a correct diagnosis to allow the collection of information about the origin, presence, extent, distribution, species involved, and the sources of the infection on which to base the development and execution of successful control and eradication programs (van Soolingen et al. 1994; Ryan et al. 2006). These are the challenges faced by most of the African countries in which the disease remains uncontrolled, and its status unknown. The recent development of molecular techniques to address these issues is a great step forward, but the costs involved, the laboratory infrastructure required, and the availability of human resources with the necessary expertise to do the tests, remain scarce commodities in Africa.

To increase the likelihood of diagnosing BTB in infected animals and herds, it is prudent to use a combination of diagnostic methods. These include clinical signs and tests based on the delayed hypersensitivity reaction such as the tuberculin skin and gamma interferon (IFN-γ) tests; serology, postmortal, and histopathological examination, and microscopic examination of various clinical specimens likely to contain the organism, culture, and DNA-based molecular techniques. The Se and Sp of each one of these tests vary according to circumstances and species involved, and the limitations of each one should be taken into consideration when using them.

It is also important to keep in mind that only results of validated tests for specific species should be used for diagnostic purposes because of the marked differences in serological and immunological responses between species. When using unvalidated tests, culturing the organism is the only reliable method to confirm a diagnosis of BTB in species other than cattle and particularly in wildlife in which the disease is suspected. Because of the complexity of many of the diagnostic methods, the variation in techniques used in laboratories and the observations of field staff when assessing tuberculin test reactions, it is critical that a system of quality control be instituted to monitor the various diagnostic procedures, be they field or laboratory based.

Although BTB is considered to be one of the most important diseases of cattle and because of its zoonotic potential, the specificity and sensitivity of surprisingly few ante- and postmortal tests have been adequately assessed for the various species in which they are commonly used. These include the single intradermal skin test (SIT), single intradermal comparative cervical test (CCT), caudal fold skin test (CFT), the IFN-γ assay, enzyme-linked immunosorbent assay (ELISA), latex bead agglutination assay, multiplex immunoassay, fluorescence polarization assay, necropsy (including abattoir meat inspection), microscopic examination (including histopathology), culture, and the polymerase chain reaction (PCR) (Downs et al. 2017). The available tests that can be used to confirm the diagnosis of BTB in the various animal species are discussed in the following sections. It is not intended to deal with the detailed methodology of the techniques but only to outline the applicability, practicality, and reliability of each of the tests for surveillance and diagnostic purposes, with emphasis on the needs of the African countries.

9.2 Diagnosis Based on Clinical Signs

Cattle Contrary to some diseases, bovine tuberculosis cannot be diagnosed with any degree of confidence based on its clinical signs. *Mycobacterium bovis* infection in animals causes a predominantly slowly progressive, often-asymptomatic disease in which clinical signs are usually not apparent until during the advanced stages of the disease (Fig. 9.1). The infection can remain subclinical for months (Thom et al. 2004), and even years, until such time that infected organs are sufficiently affected and/or when the disease becomes generalized. The disease mostly affects the lungs and their draining lymph nodes, eventually causing ill-defined respiratory signs, emaciation, lethargy, lymphadenitis, fluctuating fever, anorexia, and ultimately death (Hines et al. 2006). This complex of clinical signs is well-known in parts of Africa, as in West Africa, BTB in Fulani cattle with coughing and loss of condition has for long been known as *sondarou* (Martrenchar et al. 1993). Advanced cases only occur in herds and in countries in which the disease has not been or is not controlled. Not all infected cattle develop the advanced or generalized disease, and more than 90% of cattle in a herd (Bonsu et al. 2000; Asiak et al. 2007; Menin et al. 2013) may remain asymptomatic, some of them for the duration of their life, with the infection only being detected at slaughter (Brush 1898). There is often no difference in the appearance and condition of cattle in herds with and without tuberculosis (Laval and Ameni 2004), and in Africa this may be the consequence of the presence of other chronic debilitating diseases such as contagious bovine pleuropneumonia, trypanosomosis, endoparasites, and malnutrition that often cause similar,

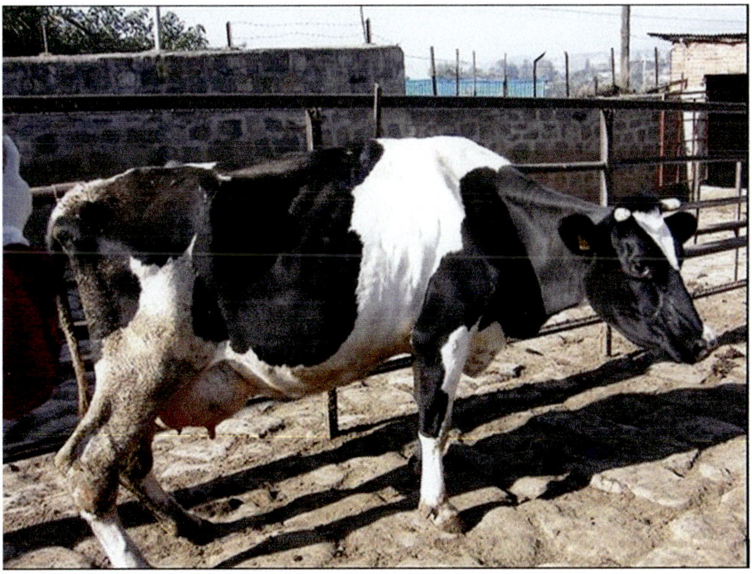

Fig. 9.1 A typically emaciated cow with advanced BTB in Addis Ababa, Ethiopia

non-specific clinical signs (Tschopp et al. 2010). When the infection localizes in organs other than the lungs, the clinical signs will vary accordingly. The inguinal lymph nodes in cows with mammary TB are usually enlarged, and the udder, depending on the nature of the reaction, may be visibly affected (Brush 1898).

Small Stock, Pigs, and Camels Sheep, goats, and pigs have a similar distribution of lesions to those described for cattle, and they similarly develop clinical signs of marked, unexplained, chronic ill thrift, and occasionally diarrhea, with high mortality rates, and do not respond to symptomatic treatment (Hines et al. 2006; Bezos et al. 2012; Di Marco et al. 2012). In camels, a persistent high body temperature, cough, a profuse glairy discharge from the nostrils, pleurisy and/or peritonitis, cold and hard swelling of the superficial lymph nodes (Mason 1917), and gradual weight loss with inappetence occur (Kinne et al. 2006).

Wildlife Clinical signs in African wildlife vary substantially. Severe emaciation, lethargy, coughing, and death are characteristics in animals with advanced disease. These clinical signs are seen in greater kudus (*Tragelaphus strepsiceros*) (Keet et al. 2001), Kafue lechwe (*Kobus leche kafuensis*) (Gallagher et al. 1972), African buffaloes (*Syncerus caffer*), chacma baboons (*Papio ursinus*) (Keet et al. 2000), and lions (*Panthera leo*) (Keet et al. 1996). In African buffaloes, 75% or more of tuberculous buffaloes do not manifest clinical signs, and only 5% of BTB-positive animals usually have generalized lesions that would cause them to die within the course of a year (Jolles et al. 2005). In infected buffalo herds and particularly those with a high prevalence of BTB, coughing, that can be heard from a distance, is a distinct and regular feature (de Vos et al. 2001).

In greater kudus and bushbuck (*Tragelaphus scriptus*) with BTB, some have enlarged parotid and submandibular lymph nodes creating a large irregular swelling below the ear and at the angle of the jaw (Fig. 9.2) that justifies a presumptive diagnosis of BTB (Keet et al. 2001). Kudus are sometimes referred to as "roarers," when lymph nodes containing tuberculous lesions press on parts of the respiratory tract and cause a roaring sound that can be heard at a distance when they breathe.

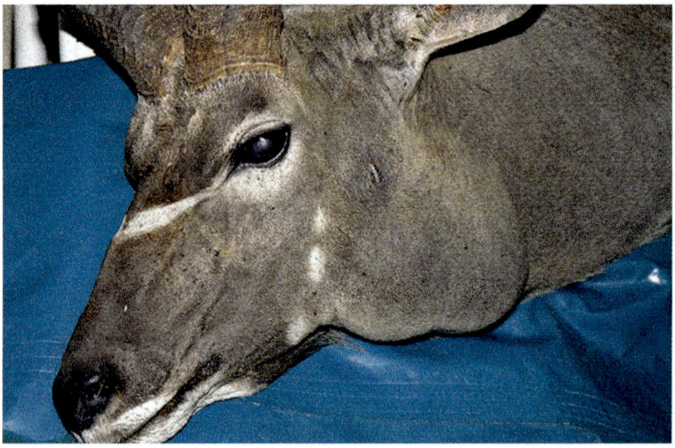

Fig. 9.2 Irregular swelling in the parotid and submandibular region of a kudu with BTB

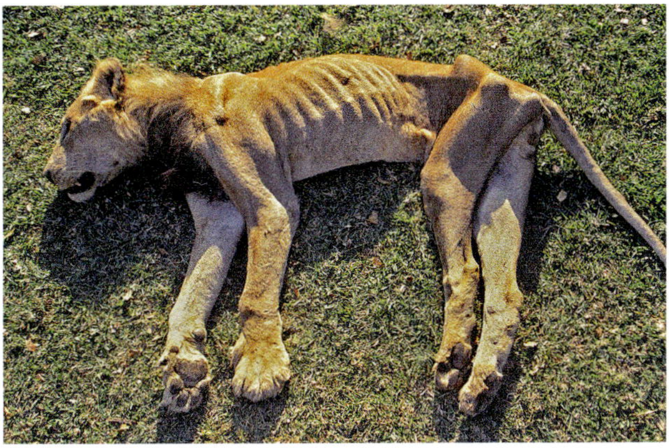

Fig. 9.3 A severely emaciated lion with BTB in the terminal stage of the disease

Lions, in addition to becoming severely emaciated (Fig. 9.3), also develop chronic arthritis with swelling of the elbow joints and large, ulcerated subcutaneous granulomas.

9.3 Diagnosis Based on the Use of Ante-mortal Tests

9.3.1 The Intradermal Tuberculin Tests

The intradermal tuberculin test is one of two, the other being the IFN-γ assay, of the standard ante-mortal diagnostic procedures for detecting *M. bovis* infections in cattle accepted by the Office International des Epizooties (OIE) for BTB surveillance and eradication purposes. The most common tuberculin tests used currently are the caudal fold test (CFT), the single intradermal test (SIT) and the single intradermal comparative cervical tuberculin test (CCT) done in the skin of the neck. Should an animal be suffering from BTB, the resulting reaction is the consequence of a cell-mediated hypersensitivity response causing a local infiltration of a succession of different types of T lymphocytes. It is characterized by the development of an inflammatory edema limited by a meshwork of fibrin deposited around the collagen bundles of the reticular dermis, to the area where the tuberculin was administered (Doherty et al. 1996). The test result of the skin tests is based on the measurement of the increase in the skin thickness caused by this localized delayed-type hypersensitivity response 72 h after intradermal injection of bovine tuberculin. The reaction is characterized by the presence of local erythema, swelling, heat, pain, and edema. The CFT is assessed by palpation of the injection site. The comparative test (CCT), when tuberculin derived from *Mycobacterium avium* is simultaneously injected at a different site on the neck to the bovine tuberculin, is assessed on the extent of the difference between the two reactions. Depending on the nature of the reaction, and the increase in the thickness of the skin at the injection site, the reaction will be classified as positive, negative, or inconclusive (suspicious), thus giving an indication of the BTB status of the specific animal and, by implication, of the herd to which it belongs. Because of the lack of Se and Sp of the tuberculin tests, it is essentially a herd test, and individual animals cannot be tested with confidence.

The tuberculin skin test has only been validated for use in cattle and perhaps, for diagnostic use, in South Africa in African buffaloes (*Syncerus caffer*). Using these tests in other species of wildlife, in particular, is risky since their reactions and responses have not been validated, and the results may be erroneous leading to incorrect and unreliable diagnoses (OIE 2017).

Superficially, using the intradermal tests appears to be a simple matter, but because of the complex nature of the disease, it is a major challenge making a diagnosis in all instances. The procedure is influenced by the relative and variable lack of Se and Sp of the intradermal tests and a host of other factors including the time since infection, how the test was executed, host factors, environmental variables, and the characteristics and potency of the tuberculin used in each case.

It is not possible to detect all infected animals in a BTB-infected herd by only using the intradermal tuberculin test, and this creates major problems. When attempting to achieve disease-free status, the test is combined with other tests and abattoir surveillance to detect those cases that do not respond to tuberculin. Even with all the combined diagnostic procedures, some infected cattle may still not be detected and remain in the herd. Not identifying and removing all positive cattle make it most likely that the disease will eventually re-emerge, a situation that requires ongoing surveillance also in disease-free herds, if the disease is to be adequately controlled. In countries certified BTB-free, abattoir surveillance is used on an ongoing basis to monitor the situation.

9.3.1.1 Tuberculin

Tuberculin, initially known as Koch's old tuberculin (OT), and then as the heat-concentrated synthetic medium (HCSM) tuberculin, is now known as purified protein derivative or PPD, and is prepared from heat-killed cultures of *M. bovis* (bovine PPD) and *M. avium* (avian PPD). Tests utilizing tuberculin remain the most acceptable diagnostic techniques used internationally for the diagnosis of BTB and their use has been sanctioned by the OIE, for use in the intradermal tests, and for the INF-γ assay.

Tuberculin is a complex mixture of poorly defined proteins (both secreted and somatic), lipids, sugars, and nucleic acids that include a mixture of antigens that are common to several mycobacterial species (Monaghan et al. 1994; Van Pinxteren et al. 2000). Tuberculin currently is a mixture of refined, heat-treated products of the growth and lysis of specific strains of *M. bovis* (strain AN5) and *M. avium* (strain D4ER and TB 56) internationally produced commercially by different companies and institutions. The culture broths of the two mycobacteria contain more than 800 individual proteins that, dependent on several factors, vary in concentration. Commercially available tuberculin consists of a mixture of small, water-soluble proteins, following removal of the high-molecular-weight proteins. Bovine PPD is usually available at a concentration of 2000 IU/ml for cattle for routine diagnostic purposes, while that of avian PPD is 2500 IU/ml. The use of higher concentrations of up to 5000 IU/ml of bovine PPD is required for cattle with a decreased sensitization and in national eradication programs when there is a need to increase the Se of the test. Though the protein content in different batches of PPD may be similar, the correct protein concentration does not predict adequate and a similar biological activity. The variable concentration of specific proteins determines the potency of specific batches of tuberculin that may vary between manufacturers and between batches from the same manufacturer (Downs et al. 2017). The potency of each of the different batches must be assessed, usually in rabbits and periodically in cattle under conditions in which the tuberculin will be used (Good and Duignan 2011). There have been numerous attempts to improve the performance of PPD and its use in tuberculin skin tests (TSTs), but a formulation that would satisfy all the requirements of a good test has yet to be developed (Schiller et al. 2010a).

Tuberculin should be produced according to the standards determined by the OIE and the European Union (EU) and should be of a specified potency. However, tuberculins of varying and lower concentrations are available in the market. Consistency of the potency of the tuberculins used in a specific country is critical as it has an impact on the consistency of test performance and interpretation of the test results. For this reason, attention should be given to the quality of the product in addition to ensuring that there is evidence that the potency of each batch was evaluated and certified (Good and Duignan 2011). For best practice, regulatory authorities in specific countries should ensure that tuberculin from the same manufacturer is used for the duration of control and eradication campaigns.

9.3.1.2 Tuberculin Skin Tests (TSTs)

An ideal screening test must be able to identify all tuberculous animals tested in a group, and it must not identify non-diseased animals as tuberculous. Thus an ideal diagnostic test would detect a signal that is uniquely and specifically associated with an *M. bovis* infection, that is present in the animal throughout the course of the infection, and that can be detected in a reproducible manner (Anon 1994). An additional feature of the skin test, is also the requirement that it should differentiate between animals that are naturally infected, and those vaccinated with a Bacillus Calmette–Guérin (BCG) vaccine (Buddle et al. 1995; Whelan et al. 2010).

While the tuberculin skin tests have been the mainstay of the immunological diagnosis of TB in humans and many animal species (particularly cattle), there is a marked variation in the nature and extent of the response to tuberculin between the various species. The induced reaction is also not fully specific since various other mycobacteria and other closely related bacteria share many of the epitopes of *M. bovis* and thus also those contained in PPD. However, in spite of these limitations, using these tests resulted in the eradication of BTB from a number of countries and for them to sustain a BTB-free status.

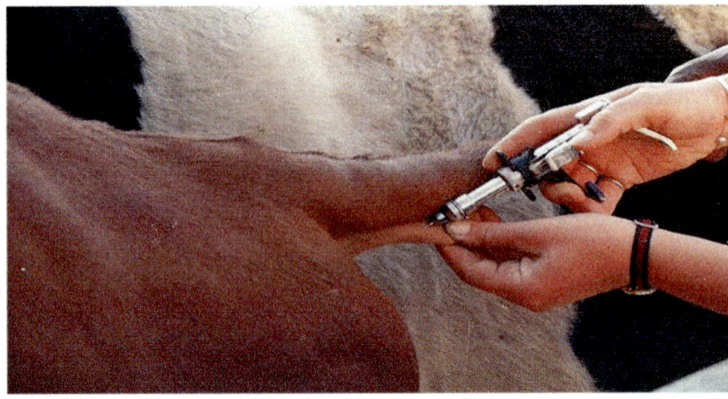

Fig. 9.4 Injecting tuberculin intradermally when using the caudal fold tuberculin test (photo by Dr. C. Daborn)

In cattle, there are three types of skin test: the single intradermal tuberculin test (SIT) and the single intradermal comparative cervical tuberculin test (CCT) that are done in the skin of the neck, and the caudal fold test (CFT) (Fig. 9.4). Depending on the specific situation, they may be used serially or as stand-alone tests. The caudal fold test is favored in the USA, Canada, Australia, and New Zealand, while the other two are routinely used in Europe, the UK, some of the African countries, and elsewhere. Generally, the SIT is used as the initial screening test, as it has the highest Se of the three. It is, however, important to select a specific test to be used in a country, dependent on the varying Se and Sp required for the environment, prevalence of the disease in the population, and the conditions under which the tests are to be conducted (Leslie and Hebert 1965; Awah-Ndukum et al. 2012).

The intensity of the reaction at the caudal fold and the neck sites differs as the sensitivity of the skin to tuberculin increases from the back toward the front and from the bottom to the top of a bovine. The skin of the hindquarters is three times less sensitive to tuberculin than that of the neck, and higher doses of PPD are used in the caudal fold test to compensate for this difference (OIE 2009; Schiller et al. 2010a). Even on the neck, although the Sp has not been determined, the tests have higher Se when they are done closer to the head and then have a higher probability of detecting reactors (Casal et al. 2014). Where the CFT is used, responder cattle are removed from infected herds, with or without using the CCT (Good and Duignan 2011).

The Se and Sp of a test are used to assess its accuracy as a diagnostic tool. The Se of a test is the proportion (as a %) of the number of diseased animals that give a positive test result, and the Sp is the proportion of non-diseased animals that give a negative test result (Martin 1984). Negative test results in diseased animals are referred to as "false negatives," whereas positive test results in non-diseased animals are "false positives" (Table 9.1).

The Se and Sp of the CCT are, respectively, between 68–95% and 96–99.9%, but they may vary substantially (Monaghan et al. 1994; Karolemeas et al. 2012). This low Se has the disadvantage of allowing the disease to persist in up to 29% of infected herds, thus allowing recurrent breakdowns, onward transmission of the infection to other herds, and spillover to local wildlife hosts to occur (Norby et al. 2004; Karolemeas et al. 2012). The choice of using either the single intradermal test (SIT) or the comparative cervical test (CCT) significantly influences the

Table 9.1 Interpretation criteria for TSTs (OIE 2009)

Test procedure	Diagnosis	Interpretation	Purpose of the test
SIT—increase in skin thickness			
≥4	Positive	Standard	Surveillance Confirmation of breakdowns
2–4	Inconclusive		
≤2	Negative		
CCT—difference between bovine and avian increase in skin thickness			
≥4	Positive	Standard	Surveillance Confirmation of breakdowns
2–4	Inconclusive		
≤2	Negative		
2–4	Positive	Strict	Infected herds High prevalence regions
>1	Positive	Severe	

interpretation of the results. Hence the outcome of the assessment and the decision as to which test is used should not just be a financial consideration; rather the suitability of a specific test should be based on local circumstances (Awah-Ndukum et al. 2012; Bezos et al. 2014). This is of particular importance in areas where large numbers of non-specific reactors occur (see below). The problem of false-positive tuberculin reactions due to significant cross-reactivity with atypical mycobacteria, reduces the diagnostic value of the SIT, and, depending on the types and levels of exposure to environmental mycobacteria, the proportion of false positives may be as high as 12% (Schiller et al. 2010a). To deal with this lack of Sp of the SITs, cattle can be retested with the CCT to enhance the Sp of the test and to reduce the number of false positives (Fischer et al. 2005; Good and Duignan 2011).

The lack of diagnostic accuracy associated with false positive and inconclusive reactions are the major problems with the TSTs. This is partially because tuberculin itself is a poorly defined cocktail of antigens that does not discriminate clearly between individuals infected with TB, and reacts to non-specific antigens, particularly those produced by other mycobacteria. Using the CCT 42–60 days after initial screening by SIT increases the Sp. However, testing has less Se if used in this way, as small reactions to bovine PPD, by the way the test is interpreted, are not classified as positive, and prior exposure to environmental mycobacteria may mask an *M. bovis* infection (Thom et al. 2008; Awah-Ndukum et al. 2012) resulting in many infected animals to be classified as negative.

Estimates of the Se and Sp of the TST vary substantially as the test results may be influenced by several factors at the time the test is performed (Table 9.2) (Martin 1984; Monaghan et al. 1994; Awah-Ndukum et al. 2012). Current TSTs thus have considerable limitations when used to test individual animals for which it is an inadequate procedure. The values for the ante- and postmortal Se and Sp of the various tests have been assessed by Nuñez-Garcia et al. (2018) and should be used as a guideline for selecting appropriate tests when developing control strategies.

The herd-level sensitivity [Se (Herd)] of a test depends on the individual level Se and Sp of the test, the number of animals tested in a herd, and the proportion of these animals that are actually infected. As the number of animals tested from a given herd

Table 9.2 Common causes of false-positive and false-negative responses in the TSTs for BTB

Test result	Reason for the inappropriate response and interpretation
False-negative	Early infection
	Anergy due to advanced, generalized *M. bovis* infection
	Temporary desensitization associated with short-interval, repeat testing
	Stress associated with poor nutrition, transportation, pregnancy, parturition
	Concurrent parasitic or viral infections
	Use of immunosuppressive drugs
	Tuberculin administered incorrectly, insufficient amounts of tuberculin, inadequate preparation of the injection site, inexperienced operators, inappropriate animal handling
	Low potency of tuberculin, expired product, product stored under inappropriate conditions, manufacturing errors
False-positive	Exposure to or infection with several non-tuberculous environmental mycobacteria
	Bacillus Calmette–Guérin (BCG) vaccination or vaccination against Johne's disease
	Passive immunity (in young calves)
	Skin tuberculosis caused by unidentified acid-fast mycobacteria
	Infection with other bacterial species
	Variation in the pressure applied to manual calipers

increases, the probability of detecting at least one infected animal (to declare the herd "diseased") is assumed to be greater than it is to identify an individual infected animal. However, testing a large number of animals to improve the Se increases the likelihood of obtaining false-positive results and reduces the Sp, particularly as the prevalence of the disease decreases in a herd or country (Anon 1994). The Se of the test in high-risk groups too may differ from those in the general population. By using different cut-off points when interpreting the reaction to a TST, the number of test positives may vary between the tests, and this may cause an inconsistency when determining the actual number of TB-infected animals (Anon 1994; Buddle et al. 2009).

In those countries where the SIT or CCT are used, interpretation of the tests is done in accordance with the guidelines of the OIE and, in the European countries, also the relevant EU trade directive (Directive 64/432/EEC). Because of specific local conditions, and other reasons, the OIE recommendations for the interpretation of the tests may have to be modified in specific countries, or regions of countries, to increase the Se and Sp of the tests to acceptable levels. There is no point, as an alternative, to just increase the potency of tuberculin, since there is then an increase in the number of cross-reactions with other mycobacteria, and thus, in the number of false positives (Ritchie 1959).

Cattle that were recently infected also do not react to any of the tests, since it takes from 3 to 6 weeks for them to develop a sufficient cell-mediated immune response that can be detected by the test (Good and Duignan 2011). Sensitized cattle react stronger during the early stages of the infection, and the response decreases over time as the disease progresses. The waning of the delayed cellular immune response causes this decline in responsiveness because midway through the course of the diseases, the immune response gradually switches to a predominantly humoral response. The extent of the immunological response may vary substantially within and between species, age cohorts, testing histories, and between sexes. Coinfections with liver fluke, Johne's disease (Álvarez et al. 2008), and bovine viral diarrhea virus also diminish the reaction (Byrne et al. 2018). Animals may be unresponsive, or anergic (Lepper et al. 1977), during the early stages of the infection, when lesions heal, and particularly in the advanced, generalized stages of the disease when they shed large numbers of *M. bovis* into the environment (Perla 1927).

It is generally accepted that the SIT is more Se but less Sp than both the CFT and the CCT, and in certain countries as many as 12% more cattle reacted to the SIT compared to the CCT (Anon 1994; Good and Duignan 2011). When applied to animals exposed to a high risk of infection, the SIT is effective in detecting preclinical cases (Buddle et al. 2009), and in many countries BTB was eradicated by rapidly removing tuberculin-positive reactors using this test. However, as the prevalence of the disease becomes lower, not all infected animals react to the injection of tuberculin, irrespective of whether the SIT or CCT is used. This phenomenon may be due to the presence of anergic reactors, animals that react equally to the bovine and avian PPD, animals in the advanced stages of the disease, lesions confined to single organs such as the udder, lesions that became inactive (latent), or peri-parturient cows (Ritchie 1959). Non-specific responses, that appear to be a worldwide problem, also complicate the caudal fold test.

Interpretation of the SIT According to the OIE-formulated guidelines for the interpretation of the tuberculin skin tests, a general approach when doing the SIT is to consider an increase of ≥4 mm as a positive reaction, whereas an increase of ≤2 mm, with no local signs, is defined as negative. An indeterminate reaction (IR) (inconclusive or suspicious) is recorded in the absence of local signs of a reaction, when the increase in skin thickness is between 2 and 4 mm (Vekemans et al. 1999).

Interpretation of the test is not only dependent on an increase in the thickness of the skin. The characteristics of the reaction at the site where the tuberculin has been injected should also be taken into consideration after visual assessment and by palpation. Non-specific reactions, such as those caused by the so-called skin lesions caused by environmental mycobacteria, result in the development of a hard, localized swelling. Those reacting to *M. bovis* usually have a diffuse, poorly circumscribed thickening of the skin characterized by the presence of edema, pain, heat, and, when severe, superficial necrosis of the skin in the central portion of the reaction site, distention of the lymphatics draining the area, and enlargement of the prescapular lymph node. A number of factors influence the Se and the outcome of the test (Table 9.2).

Depending on the required Se and objectives of the investigation, the skin tests can be interpreted according to standard, strict, or severe criteria (Table 9.1). Because of various factors, that may differ from country to country, the criteria for interpreting the skin reaction may differ and may have to be adapted for a specific

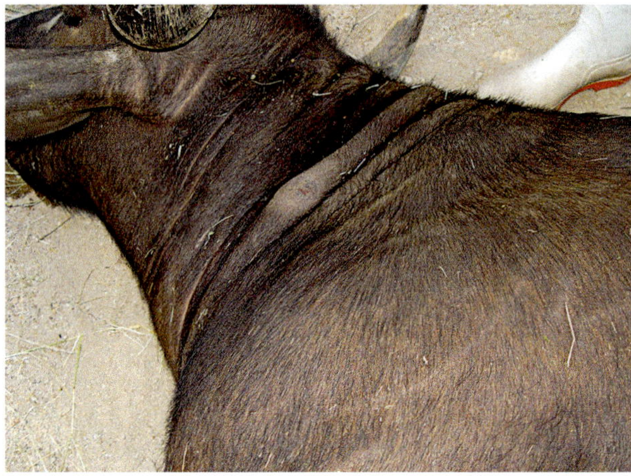

Fig. 9.5 The typical site in the middle of the neck for doing the single intradermal tuberculin test, manifesting a positive response

country. An example of this type of adaptation for different situations that can be used as a guideline in Africa, can be found in the South African BTB test manual (DAFF 2016). What is critical to remember though is that these criteria may have to be adapted over time based on the experience gained in each specific country as its campaign progresses. Good record keeping and monitoring the human resources and reagents are critical for quality control and consistency when interpreting the test results.

Single Intradermal Tuberculin Testing Technique Assessing the outcome of an SIT test in the skin of the neck involves measuring the skin thickness (after clipping the hair) using a spring-loaded caliper that will consistently exert the same pressure on the skin when measuring, and then injecting PPD-B tuberculin (0.1 ml of 2000 IU at 1 mg/ml) intradermally. Normally, the tuberculin is injected on one side of the neck, halfway between the juncture of the head and neck, and the skin fold in front of the shoulder, halfway between the top and bottom of the neck (Fig. 9.5). The injection site should have a normal skin texture and be without lumps (DAFF 2016). Tuberculin can be injected using a number of different types of syringe: the McLintock (UK) and the Dermojet (France) syringes are used most frequently. A small, pea-like swelling should be present at the site of the injection to confirm that the intradermal injection was successful. The skin thickness at the injection site is remeasured 72 h later, using the same caliper used to initially measure the skin thickness, to determine the increase in skin thickness. The site should also be palpated to assess the nature of the reaction.

The Single Intradermal Comparative Cervical Tuberculin Test (CCT) The CCT is applied by injecting avian and bovine PPD (0.1 ml of 2500 IU of each, at 1 mg/ml) 12–15 cm apart on the same side of the neck. The difference between the increase in skin thickness at the two sites after 72 h is used to determine the status of the animal (OIE 2009). Should the increase in the avian PPD be equal or greater than that of the bovine PPD, the animal is considered to be negative. The interpretation, should the bovine reaction exceed the avian reaction, is based on different cutoff points, referred to as standard, strict, or severe, to increase the likelihood of detecting more positive cases in a given situation (Table 9.1).

Serial Tuberculin Skin Testing To increase the specificity of a TST, sequential or serial testing may be used. An example of sequential testing is the use of the CFT or SIT to initially screen a herd. Then all responders, or those that inconclusively reacted to the bovine PPD, are tested with the CCT after an interval of 42–60 days. Only those animals that continue to react are then classified as positive. This sequential testing procedure greatly increases the specificity of the overall diagnostic procedure, although there is potential for a reduction in sensitivity because of the likelihood of false-negative results with the second test (Radunz and Lepper 1985).

Another approach in serial testing is to use the SIT or CCT as herd-screening tests, and then to subject all negative reactors to a second test, commonly the IFN-γ assay. Although both the TST and the IFN-γ assay detect a cell-mediated immune response, they are known to detect different sets of reactors that partially overlap (Wood et al. 1991). One or the other test will identify a certain, but different, percentage of the reactors. This process increases the Se of the overall diagnostic procedure, because animals reacting to either of the tests are considered to be positive, and they then in combination identify a larger number of infected animals. The process still remains inadequate, as the Se of the IFN-γ assay is only about 85% and its use does not contribute substantially to detecting all the positive reactors (Praud et al. 2015).

The application of the TST followed by a serological test, as an ancillary test, is a unique form of sequential testing. It appears that infected animals are stimulated by the initial intradermal test to increase antibody production, and they are then more likely to test positive with the subsequent serological test (Dowling and Schleehauf 1991). Using different serological techniques, up to 75% positive reactors out of a group of false negatives were detected by collecting blood for commercially available serological assays 15 days after the TST was done. This is a useful procedure when there is a need to identify anergic animals that are likely to be supershedders when the CCT with its low sensitivity is used, and in those cases where the INF-γ assay cannot be done due to resource limitations (Casal et al. 2014).

The Use of TSTs in Africa Because of limited financial resources and the cost of TSTs, and because the disease is not considered to be of importance to livestock or humans, few countries in Africa use tuberculin testing to assess the prevalence and distribution of BTB in their livestock population. For these reasons, TSTs on the continent are mostly used for research or limited surveillance purposes (Bedard et al. 1993).

There are a number of major impediments that limit the use of TSTs, particularly when testing in remote rural African areas. Doing the tests in these areas in which transhumant and extensive management practices prevail, is challenging. Under these circumstances, given the poor, or non-existing roads, the lack of infrastructure, and the reluctance of cattle owners to return their cattle to the testing site to evaluate the reactions, it becomes almost impossible to use tuberculin skin testing as a diagnostic tool. Further impediments include the limited availability of tuberculin, the lack of trained human resources to do the tests, and the lack of suitable handling facilities in which to restrain the animals for testing. It is possible that physiological and operational problems including stress associated with poor nutrition and exertion also cause a poor immunological response and render the tests less sensitive (Buddle et al. 2009). Concurrent parasitic infections also significantly undermine the reactivity to tuberculin in tropical Africa (Awah-Ndukum et al. 2012). The presence of external parasites and other parasitic conditions of the skin caused by *Demodex*, Filarioidea, and *Besnoitia* spp. further limit the ability to use the skin test (Martrenchar et al. 1993) because of the diffuse lesions that they cause in the skin that interfere with accurately measuring the increase in skin thickness.

The large number of non-specific reactors is a further major problem impacting the use of TSTs in Africa. In some instances as many as 50% of the animals tested are non-specific reactors (Carmichael 1937; Warren et al. 2006; Asiak et al. 2007; Michel 2008; Durnez et al. 2009; Müller et al. 2009; Katale et al. 2013; Zahran et al. 2014). To deal with these problems in some of the tests, cut-off points of ≥ 2 mm or ≥ 3 for CCT were used to limit the number of non-specific reactors (Müller et al. 2009; Muma et al. 2013). Because of the extent of non-specific reactors in the various countries, it has been recommended that the SIT should not be used as a preliminary screening test, but that the CCT should be routinely used. This, however, increases the cost of testing, which further limits the use of TSTs because of the inadequate financial resources available in most of the African countries.

The Use of TSTs in Wildlife in Africa For the purpose of control, wild animals are categorized as those animals that do not live under human supervision or control and do not have their phenotype selected by humans, captive wild animals, and feral animals. There is a long list of wild animals, either free-living or in captivity, that have been subjected to a variety of diagnostic procedures for BTB. However, only the intradermal test (both the SIT and CCT) and the IFN-γ assay used in African buffaloes (Michel et al. 2011) come close to being validated for the purpose (Cousins and Florisson 2005).

The CCT, done on the neck, was assessed in lions in the Kruger National Park (KNP), South Africa. By only using the reaction at the site where bovine PPD was injected, it detected 85% of positive animals (as confirmed by culture), and it was 5% more sensitive than the CCT. To obtain a sufficient reaction to the intradermal injection of both bovine and avian PPD in lions, doubling the dosage used in cattle, and an increase in skin thickness of ≥ 2 mm as the cut-off point, rendered the highest Se and Sp. The extensive infection of lions in the KNP with mycobacteria other than *M. bovis* is the most likely reason why the CCT was less Se, as it also influenced and masked the response to bovine tuberculin. Although these results can be used as a guideline for interpreting a TST in lions, the test has not been validated, and under different circumstances, the response may be different (Keet et al. 2010).

A substantial amount of work has been done in South Africa to develop a variety of tests for some of the more iconic species such as white (*Ceratotherium simum*) and black (*Diceros bicornis*) rhinoceroses, elephants, lions, and greater kudus, but most are not close to being validated for the purpose. Validation of tests for diagnostic purposes is a tedious process, and the methodology is contained in the OIE's Manual of Diagnostic Tests and Vaccines (OIE 2017). The OIE makes provisions for validation of tests in wildlife given the small numbers of animals that are dealt with and available for the validation process. The procedures for test validation in wildlife are contained in Chapter 3.6.7 of the Manual and can be accessed online (http://www.oie.int/standard-setting/terrestrial-manual/access-online/).

Given the likely importance of wildlife in the epidemiology of BTB in Africa, there is no doubt that some effort is needed to validate the available tests to allow assessment of the epidemiological role of wildlife in the various countries in which wildlife maintenance hosts have been identified. It is clear that not only wildlife maintenance hosts are important when dealing with BTB. The tests should also be available for spillover hosts, particularly those that are farmed with, those in zoological collections, and other free-ranging wildlife as they may yet be determined to play a significant role in the epidemiology of the disease, its control, and its final eradication from the continent.

From a practical perspective, developing diagnostic tests for wildlife species should focus on one-off, rapid, ante-mortal tests. The need to assess the reaction of the TSTs after a few days, prohibits their use in wildlife as a general diagnostic technique, because of their impracticality, the time and effort required to do the tests, and the cost of immobilizing animals for a second time (Fitzgerald and Kaneene 2012).

9.4 The Gamma Interferon (IFN-γ) Assay

The gamma interferon (IFN-γ) assay is one of the diagnostic tests for bovine tuberculosis approved by the OIE (Wood and Jones 2001; Gormley et al. 2006). The IFN-γ assay is based on the same cell-mediated immune response as the TSTs and measures the production of IFN-γ by sensitized lymphocytes exposed in vitro to PPD or similar cocktails of antigens. Compared to individual animal Se, the test also has a higher herd Se (Praud et al. 2015). The results vary accordingly when the test is used as a serial test to enhance Sp or done in parallel to enhance Se (Schiller et al. 2010a). Serial testing is advised in those areas where the prevalence of BTB is low, and frequent cross-reactions occur (Praud et al. 2015). Cross-reactions are caused by environmental mycobacteria other than *M. avium* that is the usual component of avian PPD used to detect non-specific, false-positive reactors (Gormley et al. 2013). Inclusion of the ESAT6/CFP10 cocktail antigen (Vordermeier et al. 2001; Pollock et al. 2003), PPD-N (*M. nonchromogenicum*), tuberculin derived from *Mycobacterium fortuitum*, Rv3615 (EspC), and Rv0287 (esxG) and ESAT6/CFP10 were immunogenic in the infected cattle and distinguished infected cattle from the non-infected NTM-exposed animals. Recombinant mycobacterial antigens (R-Mags) derived from NTM mycobacteria and heat shock proteins (HSP) may also be beneficial in attempting to more accurately identify true positive reactors and to reduce the number of false positives (Jenkins et al. 2018).

Briefly, the assay consists of two phases. Heparinized whole blood is incubated with different sets of antigens (normally bovine PPD and avian PPD for comparative purposes) and a control for about 18–24 h, and the amount of IFN-γ released by sensitized lymphocytes is assessed with a sandwich ELISA (Wood et al. 1991) or purified lymphocytes from peripheral blood, using the enzyme-linked immunospot (ELISPOT) (Maas et al. 2013). The test is easier to use than the skin test as it is necessary to immobilize cattle only once for the purpose of blood collection. The main problem with using the test is its cost (40–60 € per test), and the need to process the blood in a laboratory within at least 8 but not more than 30 h after collection because of the progressive decrease in secretion of IFN-γ by lymphocytes during the course of that time.

The IFN-γ assay has been approved for diagnostic and trade purposes in cattle by the OIE. It has been used as an ancillary test in Australia since the 1980s and in Europe from 2000 onward. It suffers from the same limitations as the SIT and the CCT, in that it lacks Sp and Se. In comparison to the SIT, the two tests detect different subpopulations of infected animals, and doing the two tests in parallel improves the Se and more positive reactors are detected than when used in isolation. The IFN-γ assay is more sensitive than the TSTs with a range of 96.2–98.1% (Wood et al. 1991; Wood and Jones 2001); it is also more Se than the CCT and is as specific (Schiller et al. 2010b).

The assay is commonly used as an ancillary test in most countries in which it is applied, either as a parallel or as a serial test. As with the TSTs, its specificity is influenced by a number of factors, including infection with nonpathogenic mycobacteria. It is also more specific (96.9%) in a TB-endemic zone compared to a TB-free zone (≤90.8%) (Palmer and Waters 2006; Anon 2007). The assay should not be used as a screening test for routine surveillance in low prevalence areas as it lacks sufficient Se, and with an increasing herd size, there is an increase in sensitization by environmental mycobacteria and consequently the number of non-specific reactors.

As with the SIT and the CCT, interpretation of the test remains somewhat subjective, and one should expect a degree of variation in interpretation depending on different operators, the variations in cut-offs used, the antigens used, varying test interpretation criteria, and epidemiological variables (Praud et al. 2015). Advantages of the use of IFN-γ in Egypt were confirmed. Here there were indications that using the test in parallel with the CCT was more effective, but that under their circumstances, financial constraints forced the use of it as a serial test (Abdellrazeq et al. 2014).

Disadvantages of the IFN-γ test include inadequate specificity, the logistical demands such as the time limit within which the test must be done, access to laboratory facilities and trained staff, the presence of non-specific reactions, the variation of potency of the PPDs used, and its high cost (Schiller et al. 2010a).

The advantages of using the IFN-γ assay include its increased sensitivity, being able to do more repeat testing, no need to visit the farm and handle the animals for a second time to read the test, and more objective procedures when reading the test results. Further advantages of the IFN-γ assay over TST include that, as there is no interference with the immune status of the host, there is no waiting period required between IFN-γ assays, thereby creating the opportunity for more rapid repeat testing. The test performance is also not substantially affected by the experience of field staff, and more objective testing schemes and interpretation of results are possible in comparison to the TSTs. Finally, the IFN-γ assay offers improved sensitivity and has the capability of detecting early infections (parallel use with TST therefore results in increased overall diagnostic sensitivity) (DAFF 2016; Schiller et al. 2010a; Wood and Jones 2001).

However, one significant drawback with regard to the use of the IFN-γ assay compared to the TSTs, is its high cost. With the exception of South Africa where the use of the IFN-γ assay is an integral part of the BTB control strategy in buffaloes (DAFF 2016), its high cost has led to the limited use of the test in Africa, and for this reason, it has never been used in many parts of Sub-Saharan Africa.

9.4.1 The Use of IFN-γ in Wildlife in Africa

Against the background of the known *M. bovis* infection in a range of wildlife species in Africa, few tests have been validated to diagnose the disease or to do surveillance to determine their role in the epidemiology of the disease. Reference in the section on tuberculin skin tests has been made to buffaloes and lions, but there are no published data about any of the other species. Some work has been done in a number of species to assess the use of the INF-γ assay or its modifications in an attempt to validate them according to the OIE guidelines.

In South Africa, the assay has been used fairly extensively for detecting BTB in free-ranging African buffaloes and lions. In buffaloes the IFN-γ assay and the CCT had a comparable Se and Sp (Raath et al. 1995; Grobler et al. 2002). Due to the misclassification as positive reactors (as much as 9.3%) due to presensitization with environmental mycobacteria in certain areas of South Africa, the assay was modified by the additional stimulation with crude proteins extracted from *M. fortuitum* cultures. When used as a stand-alone test for the detection of infected buffaloes, the results obtained appeared to have satisfactory Se under field conditions (Michel et al. 2011). The BOVIGAM® 1G assay had the highest Se, and when used with a number of ancillary tests, its Se approaches 100% (van der Heijden et al. 2016). Recent improvements increased the Sp of the assay by targeting specific antigens such as ESAT-6 and CFP-10. Currently different assays, such as BOVIGAM® EC and HP, are available commercially, and they appear to be more Se than the CCT. The modified QuantiFERON-TB Gold assay, a modified human assay that utilizes ESAT-6, CFP-10, and TB 7.7 (Parsons et al. 2011), has also been assessed, but it is less sensitive than the other tests that are now used (Goosen et al. 2014).

Recently a sensitive diagnostic gene expression assay was developed for lions allowing discrimination between *M. bovis*-infected and -uninfected lions. This test measures the amount of monokine induced by IFN-γ mRNA measuring the degree of sensitization by ESAT-6/CFP10 (Olivier et al. 2017).

A lot of work needs to be done, particularly in those African wildlife species that are vulnerable and face extinction. In Africa, rhinoceroses are on the verge of extinction, and since the more recent confirmation of *M. bovis* infections in this species (Espie et al. 2009), it has become even more important to develop a reliable ante-mortal diagnostic test for them. Recent work has gone a long way in adapting the IFN-γ assay for them by developing a rhino-specific IFN-γ, the RhIFN-γ (Morar et al. 2007). This test in BTB-negative white rhinos was 100% Sp (Morar et al. 2013), but further work is required to confirm its Se and Sp in BTB-infected rhinos.

9.5 Serology

The immune response to *M. bovis* infections is complex, and it varies substantially between species. A strong humoral response partially replaces the predominantly delayed cell-mediated immune response found during the early and mid-stages of the disease. Early detection of an antibody response is characteristically present only in a few species. In some of them, the response may be detectable as early as 10 days but usually from 3 to 4 weeks after infection allowing the use of ELISA as a reliable test in them. The low Sp of serology is probably the consequence of the characteristics of the humoral response in animals with tuberculosis. The initial characteristically strong cell-mediated immune response of BTB that is replaced by a humoral response influences the accuracy of tests based on both the cell-mediated and the humoral responses. The timeframe leading to this shift depends on many factors, including the species, its immunological status, and infection dose, and may be as long as 20 weeks in experimentally infected cattle. The antibody response in deer seems to develop earlier compared to cattle (Ritacco et al. 1991).

The use of multi-antigen combinations and cocktails of selected antigens led to the development of more refined serological tests such as multi-antigen print immunoassay (MAPIA), lateral flow rapid tests, the fluorescence polarization assay (FPA), multiplex plate systems, dual platform assays, chemiluminescent platforms, and improved ELISAs. Numerous issues such as the questions about Sp and its variability depending on the stage of the disease, remain. Other factors including immune compromised animals and coinfections with non-tuberculous mycobacteria must be addressed before these techniques will have the same Se and Sp as the cell-mediated immunity (CMI)-based tests and before they can be used with confidence as a replacement of the CMI-based diagnostic tests (reviewed by Schiller et al. 2010a).

The ELISAs generally have a lower Se and Sp than the TSTs and the IFN-γ assays, and there is a considerable variation in their efficacy, which is likely to improve during the later stages of the disease when the humoral response is the predominant immunological response (Nuñez-Garcia et al. 2018). Different animal species and mycobacterial strains respond to different antigens as they have distinct antigen reactivity profiles that affect the sensitivity of the tests (Bezos et al. 2014;

Lyashchenko et al. 2018). In goats, for instance, the Se was higher with the CCT and IFN-γ assay (83.7%, each) and the anamnestic ELISA (88.6%) compared to the Se of the standard ELISA (54.9%; Gutiérrez et al. 1998). The current application of the serological tests lies in using them in parallel with the TSTs and IFN-γ (Casal et al. 2014; Waters et al. 2017).

Using antibody detection as a diagnostic procedure would have the advantage of being cheap, allowing the testing of large numbers of specimens in a short time and standardization of the technique in different laboratories. Its use for the diagnosis of BTB has been investigated for a long time, but because of numerous unresolved issues and the general lack of Sp, very few serological tests are suitable for field use, and a substantial amount of work will still have to be done to allow their use for general diagnostic purposes.

9.5.1 *Serology in Wildlife*

Serology as a test for BTB has also been assessed in wildlife. The results, with few exceptions, have been similar to those in cattle. Serological tests though have been approved for diagnostic purposes in the USA for elephants and deer, and in the UK for badgers, but in general the results are too limited to determine the Se and Sp of the individual assays. The use of a test combination consisting of STAT-PAK, Enferplex, and *Mycobacterium* culture from tracheal lavage and swabs for BTB monitoring appears to have potential for general use for surveillance purposes (Brüns et al. 2017), but they require further refinement. There seems to be a strong relationship between serological responses and the presence of tuberculosis in African lions too, but the data are yet insufficient to determine the Se and Sp of the ElephantTB STAT-PAK® and VetTB® assays used (Miller et al. 2015).

9.6 Postmortal Examination (PME)

Of all the available postmortal diagnostic techniques, recognizing the lesions caused by an *M. bovis* infection in the various species remains one of the most important, and when correctly done, one of the most Se, Sp, and cost-effective tests with which to diagnose BTB (Rogers et al. 1980; Whipple et al. 1996; Liebana et al. 2008; Nuñez-Garcia et al. 2018). The ability to recognize and detect the lesions determines the success and quality of meat inspection in abattoirs and when examining incidental deaths as part of passive surveillance for BTB.

Both in countries where abattoir surveillance is used to monitor the absence of the disease, and in resource-poor countries where it is the only affordable method for BTB detection, it is thus critical for investigators to be fully acquainted with the physical appearance and distribution of the lesions and the inter- and intraspecific variation that may occur both in the macroscopical and histopathological appearance of the lesions and in their distribution. These may differ substantially, and unless examiners are aware of these characteristics, they may easily miss lesions in infected carcasses. If they do not know where to look, and what to look for, they will in all likelihood miss the lesions and the diagnosis. The least that can be expected of veterinarians and meat inspectors involved in this type of investigation is that they know where the relevant regional lymph nodes are located, and the variation in their normal appearance. They should also have an understanding of the inter- and intraspecific appearance and the variation in the lesions caused by *M. bovis* and other relevant mycobacteria that may infect cattle, small-stock and camels, and free-ranging African wildlife species.

The pathology of BTB is well documented in a number of older textbooks (Francis 1958; Jubb et al. 1993), and it appears that in those countries from which the disease has been eradicated, the lesions assume a different characteristic macroscopical appearance as they are mostly early lesions. In those countries, the advanced disease, with severe generalized lesions, and the full scope of the pathology of BTB is rarely seen. This should be kept in mind in African countries in which the disease has never been controlled, as the lesions and general manifestation of BTB may be substantially different (Domingo et al. 2014). This may have an influence on the ability to recognize the lesions and to detect diseased carcasses. It is important to monitor the appearance of lesions in different countries, as it will also give an indication of the nature of the disease and may influence control strategies. Further issues such as latent infections, the presence of no visible lesions (NVLs) in tested herds, and the distribution of lesions that can be used to improve the ease and accuracy of their detection should also be considered (Cassidy 2008).

Postmortal examination for diagnosing BTB is a powerful tool, if applied correctly. However, investigators must be adequately trained to recognize the lesions, and do their work with the necessary diligence and meticulous attention to detail, if the results are going to be reliable and of any value to the regulatory authorities, and the public at large. Regular monitoring and quality control of the processes should be part of the activities included in control programs.

9.6.1 The Pathology of BTB

Cattle Bovine TB manifests as a chronic, focal or multifocal, granulomatous caseous-necrotic inflammatory process that primarily localizes in the lungs, and their draining lymph nodes, but depending on the portal of entry, may occur in a number of other organs and lymph nodes (Domingo et al. 2014). The distribution of lesions differs according to various reports, and these differences may be a reflection of the different routes of infection by which cattle contract the disease. Generally, it was believed that most infections in cattle were contracted by inhalation, but it now appears that per os infection with localization in the palatine tonsils (in about 20% of cases) may play a significant role in the way in which BTB is contracted. Exudate, including the mycobacteria present in the lesions in the tonsils, drains into and causes lesions in the medial retropharyngeal lymph nodes, from where the bacteria disseminate to cause lesions in the lungs and elsewhere (Domingo et al. 2014).

Following infection, lesions develop in the organ in which it localizes and then extend to the regional lymph node. When granulomas in both these sites are visible, it is referred to as a complete primary complex; lesions in only one of the sites, usually the regional lymph node, are known as an incomplete primary complex. In one study, single lesion occurred in 66% of infected carcasses (Corner et al. 1990), but in others on average, 1.7 lymph nodes were affected in a group of 61 tuberculous animals (Liebana et al. 2008). Percentages of the occurrence and distribution of lesions differ. In carcasses examined by Corner et al. (1990), 29.4% of the lesions were found in the medial retropharyngeal nodes, 28.2% in the mediastinal nodes, 18.0% in the tracheobronchial nodes, 8.0% in the lungs, 2.9% in the mesenteric nodes, 2.4% in the parotid node, and 2.4% in the caudal cervical node. These sites should be the primary focus of inspection to detect infected animals with visible lesions. In some of the animals in which lesions occurred in the abdominal nodes, they were the only ones affected. Lesions are often found in the palatine tonsils (Neill et al. 2001), and they are common sites from which positive cultures can be obtained. They are often the location of non-visible lesions. The issue of no visible lesions (NVLs), i.e., those TST-positive animals in which no lesions can be detected, is an important one. As many as 10% of TST-positive reactors may contain no lesions, and 30% of TST-negative, in-contact animals in one study of a BTB-positive herd, contained NVLs detected by culture following postmortal examination (Gavier-Widén et al. 2009). It is critical to keep in mind that not being able to find lesions at necropsy does not mean that the animal is not infected with *M. bovis*.

Infection by inhalation normally causes the development of lesions in the lower respiratory tract where they tend to localize in the dorso-caudal portion of the caudal lung lobes, although, to a lesser extent, lesions may be present in all the other lobes of the lung (Liebana et al. 2008). In general, lesions are more commonly found in the thoracic cavity, then in the head, and lastly in the abdomen. When examined with care, pulmonary lesions may be found in up to 70% of animals, and 1.5 lobes on average contain lesions. It is normally easier to find small lesions in the regional lymph nodes compared to the lungs.

Tuberculous lesions on the serosal surfaces manifest typically as multiple, single to confluent clusters of sessile or pedunculated nodules resembling pearls and become markedly calcified (Figs. 9.6 and 9.7). Serosal lesions may proceed to develop a diffuse, caseous serositis with the development of large caseous plaques adherent to the thickened serosal surfaces (Neill et al. 2001).

Conceivably, single lesions could occur anywhere in the body, a situation that makes detection of all lesions a tedious process, and these abnormally located lesions are usually not detected during the course of routine meat inspection. Single lesions also occur unexpectedly in lymph nodes such as the sub-iliac, which is not one of the target nodes for routine meat inspection.

The lesion caused by an *M. bovis* infection is typically a focal granuloma, manifesting as irregular firm, white or yellowish nodules varying from 2 to 20 mm or more in diameter. They are characterized by the presence of an accumulation of

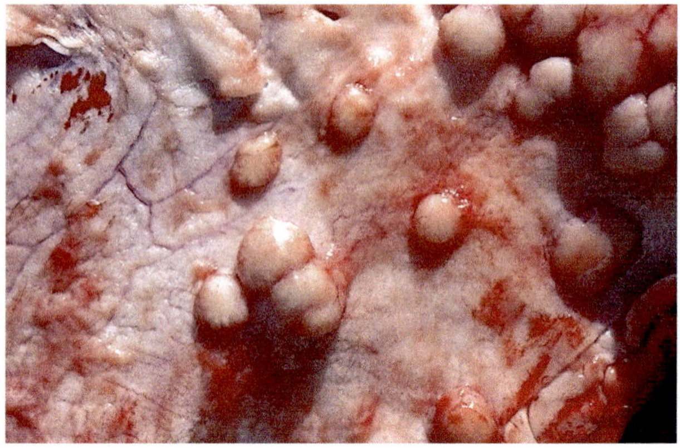

Fig. 9.6 Tuberculous peritonitis. Note the typical pearl-like granulomas that are often markedly calcified

Fig. 9.7 Sessile tuberculous granulomatous plaques on the pleural surface that can easily be confused with mesotheliomas

centrally located caseous necrotic material that may or may not become inspissated and calcified, and may undergo liquefaction usually in association with the presence of neutrophils. Liquefaction of the exudate creates a lesion that is easily confused with an abscess with a fluid purulent content, caused by a variety of pyogenic organisms. A rim of an active granulomatous response and an outer capsule surround the granuloma. Lesions can persist with or without progression, or they can heal and disappear. In cattle experimentally infected with *M. bovis*, the lesions are classified into four categories according to their stage of development (Cassidy et al. 1998; Wangoo et al. 2005). This then, although lesions following experimental infection may differ from those in natural cases, is the usual concept of the appearance of lesions expected in animals suffering from BTB. The lesions, however, may differ substantially in size and appearance in naturally infected cattle due to attributes that may be dependent on the breed of cattle, the age and gender of the animal, virulence of the mycobacterial strain, the immune status of the animal, and the stage of the disease (Domingo et al. 2014). In the advanced disease, lesions may vary substantially and are inconsistent in appearance with those in experimental animals or in the early stages of the disease.

A common characteristic of the lesions of BTB in the various species, including cattle, is the substantial variation that occurs between animals in their macro- and histopathological appearance (Neill et al. 2001; Ameni et al. 2007). In cattle, the lesions are characteristically yellowish-white nodules containing a dry gritty content, and may occupy the entire tissue space contained by the lymph node's capsule.

Lesions in the lymph nodes vary in appearance and may consist of ill-defined foci of necrosis with a predominantly neutrophil infiltrate, to clearly encapsulate granulomas with calcified caseous necrotic centers. Individual lesions are rarely severely fibrosed, and they lack the thick fibrous capsule that characterizes the so-called "typical tubercle." Occasionally lymph nodes with the appearance of large encapsulated abscesses containing a thick yellowish to greenish purulent exudate may be seen (Whipple et al. 1996).

Single lesions are commonly seen in the lung (Fig. 9.8), but they can be multiple and confluent. In generalized cases, multiple embolic granulomas may be scattered throughout the lungs. In many instances, a single, small granuloma is the only lesion seen, but in individual cases, multiple granulomas of up to 70 mm in diameter and scattered throughout the lungs may be present. In advanced cases, the lesions may obliterate large portions of the lungs, more often in the caudal lobes, and develop into a lobar pneumonia, but with retention of the interlobular septa (Buddle et al. 1994).

Local lesions may expand substantially over time and may destroy the normal structure of the organ. Thus, the expanding inflammatory process may destroy all the normal tissue of a lymph node and cause substantial irregular enlargement of individual nodes. In the lung, the lesions expand to affect whole lobules that may by extension, draining of the exudate via the airways, or dissemination of mycobacteria via the blood or the lymphatic vessels, spread to adjacent or distant lobules. In this way, large portions of the lungs may become affected. Multiple lesions may become confluent, and in advanced cases, the necrotic portions may undergo liquefaction and

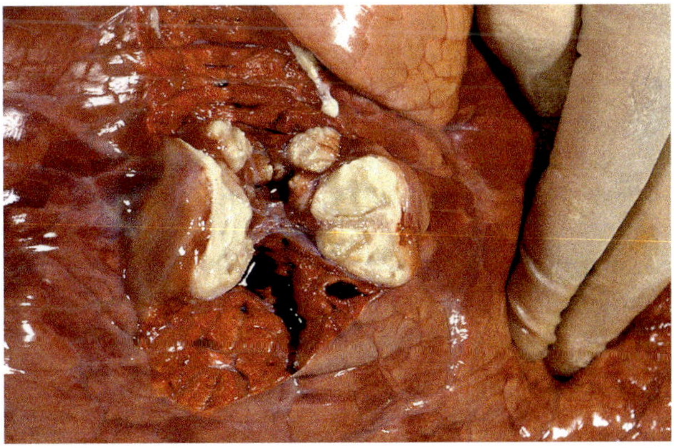

Fig. 9.8 A small cluster of focal, pyogranulomatous lesions in the lung resembling abscesses

cavitation. These cavities remain small and are limited to lobules because of the very rigid interlobular septa in cattle that persist and limit further expansion of the lesions (Domingo et al. 2014).

Depending on the immune status of an infected animal, dissemination of the infection may occur, either early during the course of the infection or at a later stage, a process that is, respectively, referred to as early or late, postprimary dissemination (or embolic spread) via the blood vascular or lymphatic systems. This type of dissemination usually results in the development of miliary granulomatous lesions in many organs and tissues throughout the body. The lesions in the lungs and on serosal surfaces of the thorax, the abdomen, and the pericardium are more commonly seen. Although they may initially be small, miliary nodules may become confluent and in this way develop into sizeable lesions. In generalized cases, lesions may also occur in the liver, kidneys, udder, and meninges (Neill et al. 2001).

For diagnostic purposes, lesions that macroscopically resemble those caused by mycobacterial infections should be confirmed histopathologically and/or by culture, as only a presumptive diagnosis can be made following the detection of the so-called "typical macroscopic lesions" (Liebana et al. 2008). There are numerous causes of disease that produce lesions with a similar macroscopical appearance to those caused by the mycobacteria. These include organisms of the *Nocardia/Streptomyces* spp. group, those causing actinomycosis and actinobacillosis, and *Staphylococcus*, *Dermatophilus*, *Corynebacterium*, and *Pseudomonas* spp. Additionally, hydatid cysts, squamous cell carcinoma, and fungal and foreign body granulomas produce lesions with a similar appearance. In certain instances, what appears to be just an abscess, can turn out to be a lesion caused by *M. bovis* of which the contents are liquefied. Confirming the diagnosis at least by histopathology, but preferably by culture, is of particular importance when dealing with single animals originating from herds with an unknown BTB history. Not making a diagnosis of BTB under those circumstances could have disastrous consequences.

Histopathologically, lesions develop from an initial accumulation of macrophages, epithelioid cells, neutrophils, and a few Langhans-type giant cells. In time, the central portion undergoes caseous necrosis following which the exudate becomes inspissated and develops varying degrees of mineralization. A typical active granulomatous inflammatory reaction, a peripheral cuff of lymphocytes, and finally a capsule that may not be fully developed, surround the lesion. Varying numbers of acid-fast mycobacteria may be detected in the cellular debris, and in the cytoplasm of the Langhans giant cells and epithelioid cells. Many of the lesions are paucibacillary and histologically contain few or no detectable mycobacteria. It must be again stressed that the lack of detectable acid-fast mycobacteria in histological sections is not a reason to rule out a diagnosis of tuberculosis (Domingo et al. 2014). In addition, only a presumptive diagnosis of BTB can be made on the basis of detecting acid-fast bacteria (AFB) histologically since it is not possible, based on their morphology, to identify either the species or the genus of the acid-fast organisms in histopathological sections (Corner 1994).

9.6.2 Small Stock and Camels

Camels Limited data are available about the disease in dromedary camels (*Camelus dromedarius*), and tuberculosis has only been reported on a few occasions in Egypt, Kenya, Somalia, the United Arab Emirates, Pakistan, India, and Australia (Kinne et al. 2006). In addition to *M. tuberculosis* and *M. bovis*, a range of non-tuberculous mycobacteria was isolated from lesions resembling those caused by *M. bovis* in camels (Kinne et al. 2006). Generally, lesions occur in the lungs, bronchial and mediastinal lymph nodes, pleura, and liver (Mason 1917; Elmossalami et al. 1971). In tuberculous carcasses about 87.8% contained single lesions. Of the lesions in 91 camels detected postmortally during post-slaughter examination, about 50% had lesions in the lungs in which they were located predominantly in the apical and cardiac lobes, while lesions in the lymph nodes were most commonly present in the mesenteric nodes and less often in the mediastinal nodes. Another study reported that 57.5% of the lesions occurred in the thoracic lymph nodes and the lungs, 27.2% in the lymph nodes of the head, and 15% in the mesenteric lymph nodes (Beyi et al. 2014). Of the camels from which mycobacteria were isolated in one study, only two contained *M. bovis*, while the rest were infected with mycobacteria other than the *M. tuberculosis* complex (MTC; Mamo et al. 2011). The characteristic histopathological features of tuberculosis in camels resemble those seen in other animals, including the presence of acid-fast, rod-shaped bacteria (Kinne et al. 2006).

Sheep Sheep are considered, perhaps erroneously, not to be important in the epidemiology of BTB, and they are considered to be less susceptible to infection with *M. bovis* than both cattle and goats. However, cases of BTB have been diagnosed in New Zealand and Spain, and in a number of African countries (Davidson et al. 1981; Muñoz-Mendoza et al. 2016). The lesions caused by *M. bovis* in sheep closely resemble those in cattle and vary from isolated mineralized granulomas in the lymph nodes to extensive, soft, caseous granulomas in the dorsal areas of the lung and some of its regional lymph nodes (Davidson et al. 1981). The lesions vary in size from <5 cm in diameter to large nodules that presented as extensive diffuse purulent, caseous and encapsulated, partially calcified granulomas that occupied portions or the entire organ. Histopathologically, the lesions varied according to their stage of development from focal granulomas, which as they increased in size, developed central caseous necrosis with mineralization in association with the typical granulomatous inflammatory reaction that included the presence of epithelioid and Langhans giant cells. Varying numbers of acid-fast organisms were present in lesions in early and advanced stages of their development (Muñoz-Mendoza et al. 2016).

Goats *Mycobacterium caprae*, a recently added member of the MTC, is the usual but not exclusive cause of tuberculosis in goats (Sahraoui et al. 2011; Sanchez et al. 2011). Generally, small, firm, yellowish nodules characterize the lesions seen in goats. In the lymph nodes, these nodular lesions usually become large, and they may coalesce and occupy large portions of the node. In the lungs, the nodules are prominent, large, rounded, and occur within lobules and contain a pasty yellowish-white content. In advanced cases, a tuberculous pneumonia characterized by progression to a caseous lobular pneumonia develops, while the rest of the parenchyma contains miliary granulomas of different sizes that also extend to the serosal surfaces. Advanced lesions in the lungs tend to undergo cavitation caused by the influx of neutrophils and the presence of large numbers of acid-fast bacteria. The development of cavities is accompanied by a change in the cellular components, particularly neutrophils, the ratio of different T-cell types, and liquefaction of the exudate that then drains into adjacent bronchi (Sanchez et al. 2011).

Histopathologically, the reaction is proliferative and contains small, calcified, central areas of caseous necrosis surrounded by a granulomatous inflammatory reaction containing epithelioid and Langhans giant cells, and then a layer of lymphocytes next to the outer surrounding connective tissue capsule. Few or no acid-fast bacilli are present in the necrotic debris or cell cytoplasm, but large numbers are seen in the liquefied exudate of advanced lesions. At the edge of these lesions in the lungs, large areas contain a serous intra-alveolar inflammatory exudate (Seva et al. 2000).

Pigs Tuberculosis in domestic pigs is now rare in countries that have successfully implemented tuberculosis control programs. The lesions differ according to cause (either *M. tuberculosis*, *M. bovis*, *M. avium*, or *M. intracellulare*) or the route of infection. Lesions caused by *M. tuberculosis*, *M. avium*, and *M. intracellulare* remain localized and cause small granulomas varying in diameter from 2 to 10 mm. They are usually limited to the lymph nodes of the intestinal tract, head, and neck, although generalization may occur. The lesions are non-encapsulated, lardaceous and proliferative, and rarely develop necrosis or calcification.

The lesions caused by the different mycobacteria causing atypical tuberculous lymphadenitis cannot be differentiated histopathologically or immunohistochemically, and their identity should be confirmed by culture (Thorel et al. 2001). The lesions caused by *M. bovis* may become generalized and have a caseo-calcareous appearance resembling the lesions seen in cattle. Dissemination of the infection, usually from the intestinal tract and its nodes, causes miliary tubercular lesions in a number of organs including the lungs, liver, and spleen. The lesions in the lungs may become extensive with consolidation often of the cranial lobes. Skeletal lesions, particularly of the axial skeleton, are commonly seen, but extension to the serosal surfaces is rare (Jubb et al. 1993).

9.6.3 The Pathology of BTB in African Wildlife

The limited available data about the situation in Africa, and the relevance of wildlife reservoirs in other parts of the world, suggest that wildlife in Africa may become increasingly important in the epidemiology of BTB. As a result it is important to determine their BTB status if control programs are to be successful in those countries in which wildlife reservoirs occur (Tweddle and Livingstone 1994; Ryan et al. 2006; Renwick et al. 2007; Munyeme et al. 2010; Hang'ombe et al. 2012).

Of the wildlife species in which the disease has been detected in Africa, all were first diagnosed with BTB during postmortal examination, and this situation is expected to remain in the future. What is of importance in this respect is that there is a marked intra- and interspecific variation in the appearance of the lesions and in their distribution in the various species. The lesions in some species closely resemble those in cattle, but in others (particularly lions, leopards, and hyenas), they are distinctly different and bear no resemblance to the tuberculous granulomas generally associated with mycobacterial infections.

The pathology of the disease is inadequately described in all the African wildlife species in which BTB has been diagnosed. However, the differences in the appearance and distribution of the lesions in predators, antelopes, omnivores, and primates give an indication of the variation that can be expected, and investigators should not be misled by what has been described for long, as the "typical" tuberculous granuloma when investigating the presence of BTB in wildlife species. A further complication of detecting wildlife infected with *M. bovis* is the occurrence of BTB in those species that are not often encountered because of their small size and nocturnal habits, such as rats (*Rattus norvegicus*) (Little et al. 1982) and hedgehogs (*Erinaceus europaeus*) (Lugton et al. 1995) that have been found to be infected with *M. bovis*. Others have no visible lesions, and any one of them may play a role in the dissemination of the infection in ecosystems (Gavier-Widén et al. 2009).

The description of the lesions in the following species serves as an example of the diversity of lesions that may be encountered in African species of wildlife, and it is presented in some detail to guide further investigation of the disease in wildlife on the continent, and in the different African countries where they may occur.

African Buffaloes In African buffaloes, the lesions often resemble those seen in cattle in which they are characterized by containing a fibrous capsule, marked caseation, and some calcification. The various early developmental stages of the granulomas can similarly be classified as they are in cattle (Wangoo et al. 2005; Laisse et al. 2011). In buffaloes, some of the lesions are atypical and have a neoplastic, fibro-lardaceous appearance (Fig. 9.9), some contain branching interdigitating areas of caseous necrosis lacking calcification and a detectable capsule (Fig. 9.10), and others appear as well-encapsulated aggregates of an inspissated caseous necrotic debris (Fig. 9.11) (Guilbride et al. 1963; Woodford 1982). Pus may be present in some of the lesions in buffaloes, and it is invariably of a lighter color than that seen in tuberculous lesions in cattle. The caseous necrotic centers of the granulomas may become liquefied both in the lungs (Fig. 9.12) and in the lymph nodes. In advanced cases of the disease, the carcasses are cachectic.

The distribution of the lesions is quite consistent in that in most buffaloes with tuberculosis, lesions occur in the lungs (50% of cases had only pulmonary lesions), while in the largest proportion of infected buffaloes, lesions occurred in the bronchial and/or mediastinal lymph nodes. Rarely, lesions occurred exclusively in the retropharyngeal and cervical lymph nodes (Woodford 1982) or in the palatine

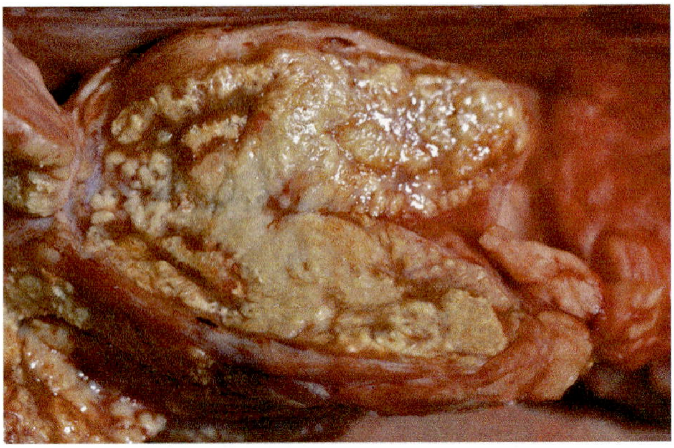

Fig. 9.9 Buffalo: lardaceous appearance of the granulomatous lesion seen on the cut surface of a mediastinal lymph node. Note the absence of necrosis and a distinct capsule

Fig. 9.10 Focal lesion in a lymph node of a buffalo manifesting interdigitating areas of necrosis, branching early fibrosis, and the lack of a distinct capsule (formalin-fixed specimen)

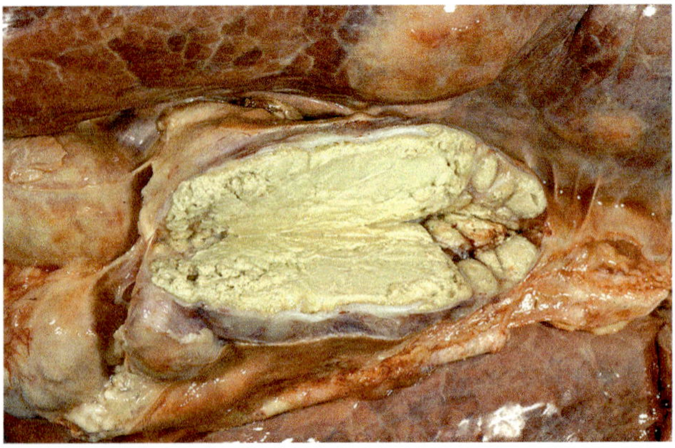

Fig. 9.11 Enlarged and encapsulated, inspissated, caseo-necrotic exudate in the mediastinal node of a buffalo

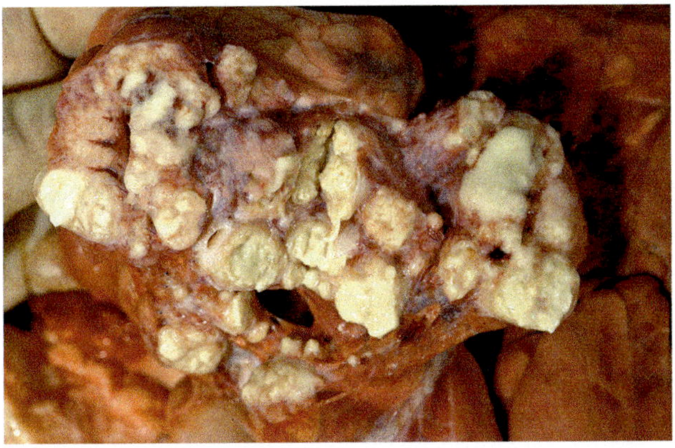

Fig. 9.12 Liquefied exudate in a multifocal pyogranulomatous pneumonia in an African buffalo

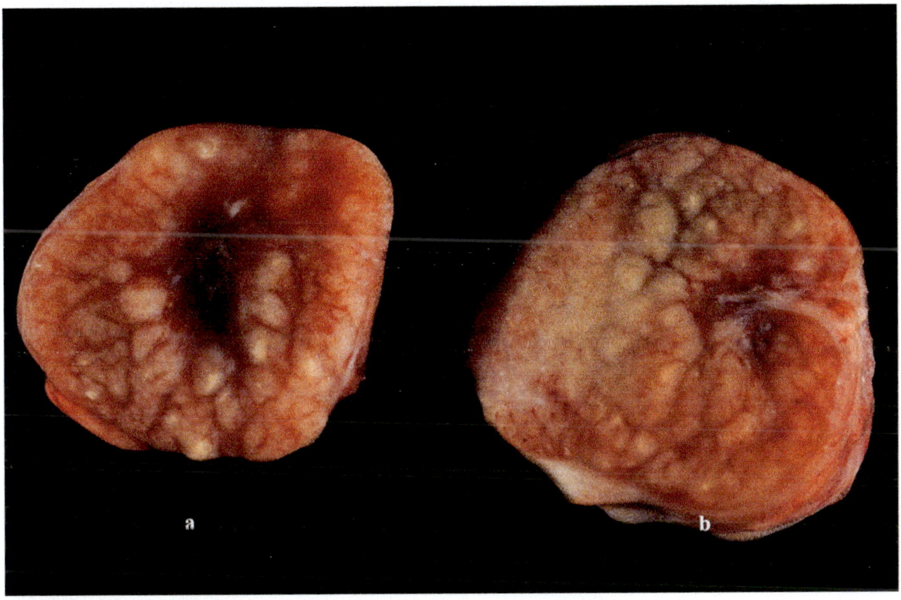

Fig. 9.13 Palatine tonsils. (**a**) Normal tonsil containing a few crypt abscesses. (**b**) This tonsil contains a poorly defined lesion partially effacing the normal structure of the tonsil. The granuloma blends into the normal tissue and is not encapsulated

tonsils. Lesions in the tonsils may be difficult to detect, and they are most easily located by palpating the organ as they cause an increased consistency of the affected parts of the tonsils. The granulomas in the tonsils are usually not encapsulated, and the reaction extends imperceptibly into the surrounding normal tissue and sometimes into the adjacent skeletal muscle (Fig. 9.13). On cursory examination, they may be confused with cysts and crypt abscesses that commonly occur in the tonsils in buffaloes (Kriek et al. 1994).

Pulmonary lesions may occur throughout the lungs, but they are most often found in the dorso-caudal aspects of the caudal lobes; often, only single lesions are detected, and they may be quite small and are then most easily detected by thorough palpation of the pulmonary parenchyma. The lesions vary in appearance from small almost undetectable, encapsulated, discrete, fibro-caseous granulomas to numerous coalescing foci (Fig. 9.14) of up to 30 cm in diameter. In many instances, they are distributed multifocally in the caudal portion of the lung where they may displace large portions of the lungs (Fig. 9.15), sometimes to the extent that more than 60% of the functional tissue has been destroyed. Lesions in the lung too may have a smooth lardaceous appearance (Fig. 9.16) and may vary in color; sometimes they may have an orangey appearance dependent on the characteristics of the strain of *M. bovis* that infected the animal. The granulomas commonly undergo limited to extensive caseous necrosis, and the necrotic exudate may become partially liquefied (Keet et al. 1994). Other organs and tissues are rarely affected, but when they are, the lesions

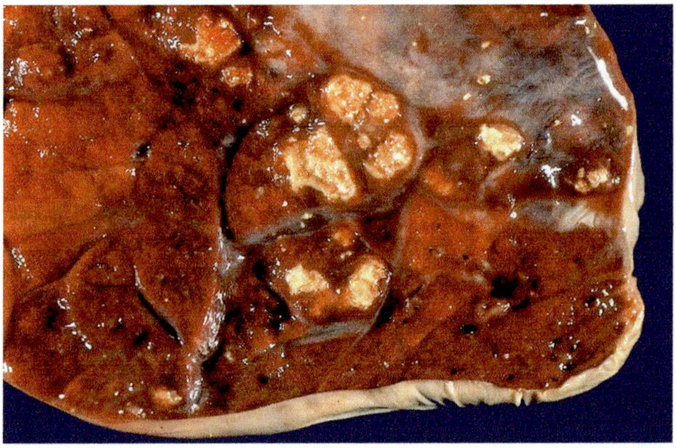

Fig. 9.14 Multifocal granulomatous pneumonia with a lobular distribution

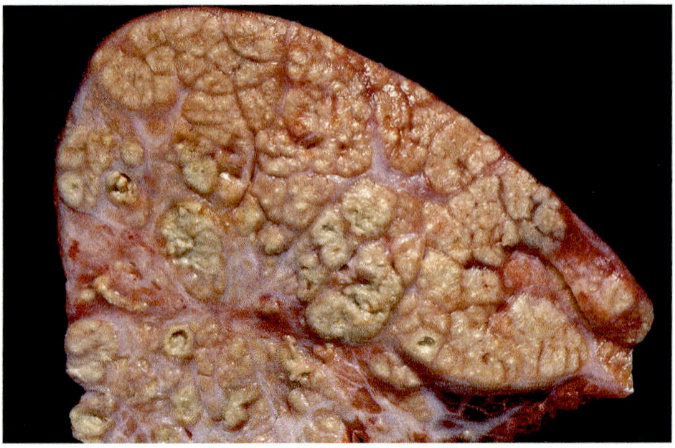

Fig. 9.15 Buffalo lung with an extensive lobar distribution of the lesions. Note the retention of the lobular structure

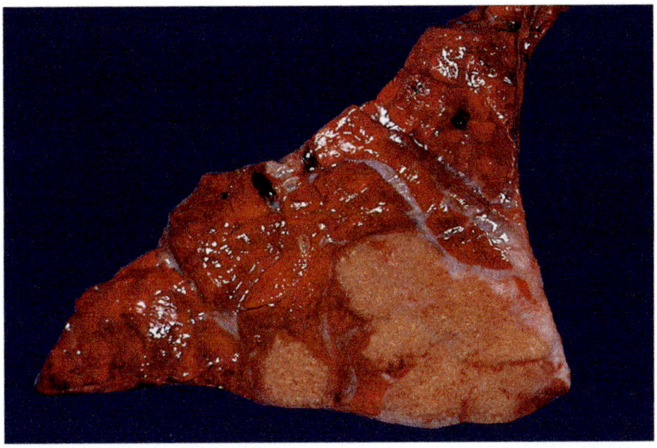

Fig. 9.16 Focal, locally extensive, lardaceous, granulomatous lesion in the lung. Note the orangey tinge of the affected tissue and the lack of the usual features associated with lesions caused by *M. bovis*

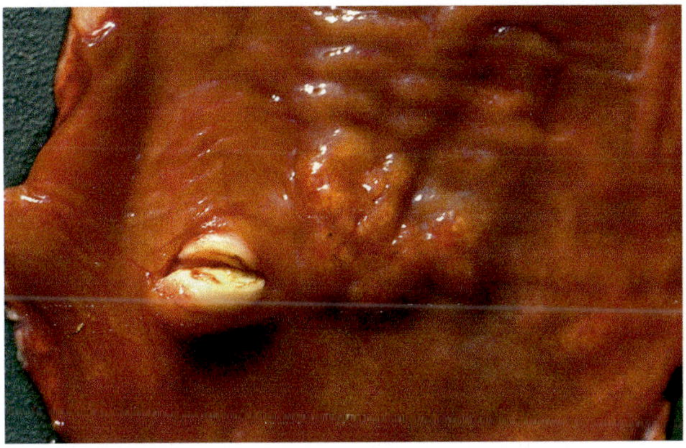

Fig. 9.17 Granulomatous enteritis causing slight, irregular, linear, elevation of the mucosa, containing an inspissated, partially calcified content when cut into

have the characteristic appearance of those seen in lymph nodes, the palatine tonsils, and the lungs. *Elaeophora sagitta* also causes localized parasitic granulomas in the dorso-caudal portions of the lungs, that are easily confused with those of BTB, thus emphasizing the need to confirm the cause of the lesion histopathologically.

Intestinal lesions are rarely seen in buffaloes, except in advanced cases in which they develop following expectoration and swallowing exudate containing large numbers of bacteria from advanced lesions in the lungs. The intestinal lesions are then more commonly seen in the distal portion of the small intestine, and at the ileocecal junction. Here they have a linear configuration and are mostly located within the mucosa and submucosa where their presence causes the development of small irregular ulcers (Fig. 9.17). Where these lesions occurred, the draining mesenteric nodes were always affected (Kriek et al. 1994).

Animals with generalized tuberculosis also manifested a granulomatous peritonitis, while a granulomatous pleuritis and pericarditis occurred in a quarter of the buffaloes with generalized BTB. Individual buffaloes rarely contain lesions only in

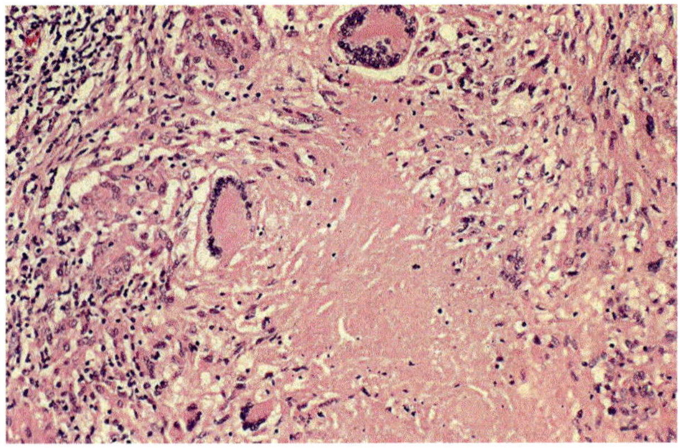

Fig. 9.18 Histopathological characteristics of a typical tuberculous granuloma. Note the central caseous necrosis and the presence of a number of Langhans giant cells, epithelioid cells, loosely arranged lymphocytes, and early fibrosis at the periphery of the lesion

the liver, spleen, or uterus and in the gastro-hepatic nodes. Peripheral lymph nodes are rarely affected but they do develop lesions in advanced cases of the disease.

The histopathological appearance of the lesions in African buffaloes differs little from that seen in cattle, the inflammatory reaction being characterized by a granulomatous reaction, the presence of epithelioid cells, Langhans giant cells, central caseous necrosis with some calcification, and a pronounced peripheral fibrous capsule (Fig. 9.18). Not all lesions are encapsulated. In some of the lesions, masses of neutrophils are present in the center of the necrotic portions, and these undergo liquefaction as a consequence of this inflammatory reaction. The numbers of acid-fast bacteria seen in sections differ substantially; in some instances, numerous bacteria are scattered throughout the inflammatory reaction, while in others they are very sparse, and in a significant percentage of the cases, acid-fast organisms are not detectable (Woodford 1982).

Greater Kudus The lesions in the different organs are well demarcated, encapsulated, firm, and irregular, and have a caseous necrotic and granulomatous appearance on cut surface. In less severely affected animals, lesions are predominantly present in the lymph nodes of the head but without sufficient enlargement to cause visible swelling below the ears or in the parotid area. The nodes of the neck are less often involved, and lesions occur inconsistently in different lymph nodes, including the caudal cervical node. Multifocal, granulomatous, mediastinal lymphadenitis, and lesions in the lungs that are then limited to a few caseo-necrotic granulomas scattered throughout the pulmonary parenchyma, may occur in these early cases (Thorburn and Thomas 1940).

The irregular, and often prominent, swelling in the parotid area is a characteristic feature of the disease in some kudus with tuberculosis. When cut into, they contain cavities filled with a watery, turbid, whitish, purulent exudate enveloped by a thin,

Fig. 9.19 The typical fluid purulent appearance of the exudate surrounding exiting structures on the side of the neck and in the parotid region of a kudu

well-vascularized capsule (Fig. 9.19). The pressure exerted by the large volumes of accumulated exudate causes it to extend between the anatomic structures of the neck (Weber and Van Hooven 1992). Ventral to the ears (where the parotid salivary gland and lymph nodes are situated), large, poorly encapsulated, caseo-necrotic masses replace the normal tissue. In many instances, these lesions drain through a cutaneous sinus (Thorburn and Thomas 1940).

In advanced cases, the disease becomes generalized. All the lymph nodes of the head, neck, thoracic cavity, as well as some of the caudal cervical, mesenteric, and the hepatic nodes, are often involved. The bronchial nodes are usually markedly enlarged (up to 10 cm in diameter), and their normal structure is completely replaced by a severe, granulomatous inflammatory reaction that also contains large masses of caseo-necrotic debris. Lymphatics draining into a lymph node are often affected by a multifocal, granulomatous lymphangitis, the individual granulomas in the lymphatics being up to 2 mm in diameter. Corneal opacity and a fibrinous exudate in the anterior chamber of both eyes may be present indicating spread to the ocular structures in some of the advanced cases (Keet et al. 2001).

The lungs in advanced cases are severely affected, and the physical appearance of the granulomas varies according to their locality. Those in the cranial lobes are larger and more numerous than in the caudal lobes and replace most of the normal parenchyma forming extensive areas of consolidation. The granulomas in the cranial lobes are clustered, even-sized (30–40 mm in diameter), ovoid, oblong or triangular, and well encapsulated (Fig. 9.20). They contain a smooth and inspissated caseo-necrotic exudate that may be partially liquefied and poorly calcified. Similar granulomas occur in the caudal portions of the lungs, but as a few isolated masses scattered randomly throughout the lobes. In addition, miliary tubercles appearing as pearly areas of granulomatous inflammation of about 3–5 mm in diameter can be scattered throughout the caudal lobes. These small granulomas tend to be clustered, thereby imparting a mulberry-like appearance to the lesion. A diffuse granulomatous pleuritis that predominantly involves the affected portions of the cranial lobes, results in fibrous adhesions to the parietal pleura (Fig. 9.21) (Keet et al. 2001).

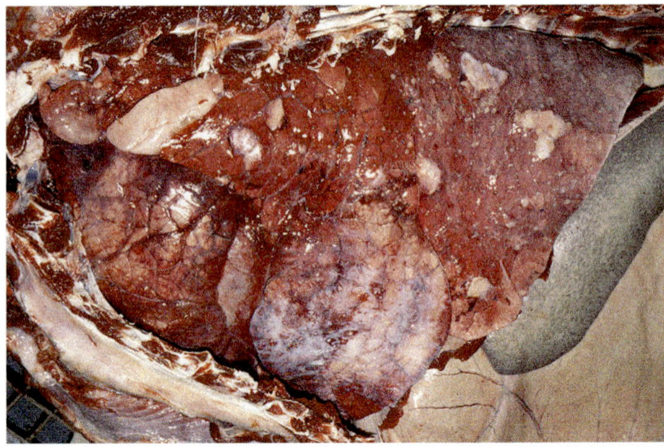

Fig. 9.20 Lung of a kudu manifesting the typical, well-demarcated, and often angular, granulomatous lesions seen in this species

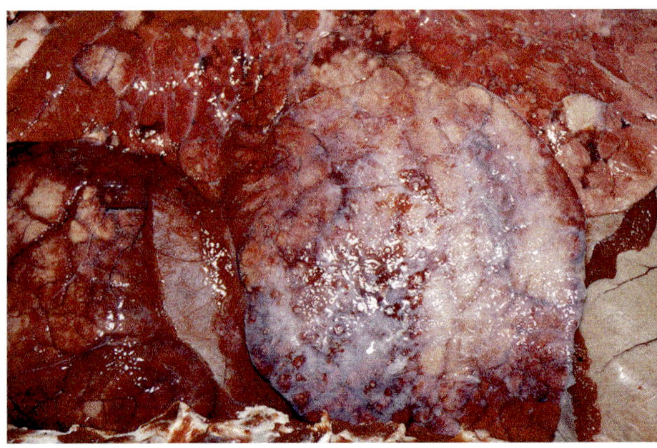

Fig. 9.21 Diffuse, chronic pleuritis with extensive fibrosis on the apical lobe of the lung in a kudu

Histologically, the lesions are encapsulated and contain extensive areas of caseous necrotic debris surrounded by a typical granulomatous inflammatory reaction containing numerous epithelioid cells, Langhans giant cells, and lymphocytes. The lesions are further characterized by the presence of large numbers of neutrophils particularly in the caseous necrotic debris. Even very early lesions contain aggregates of neutrophils. Few acid-fast bacteria are present in the cytoplasm of Langhans giant cells and in the exudate (Keet et al. 2001).

African Lions The lesions caused by *M. bovis* in lions are substantially different in their macroscopical and histological appearance compared to the lesions seen in other wildlife species in which this infection occurs (Keet et al. 1998). Histologically, homogenous infiltrates of macrophages, the lack of necrosis, calcification, and the absence of Langhans giant cells characterize the lesions. Normally mycobacteria are absent from these lesions.

In the lungs, the inflammatory reaction is initially limited to the interstitial tissue, while a mixed macrophage and neutrophil exudate may be present in the

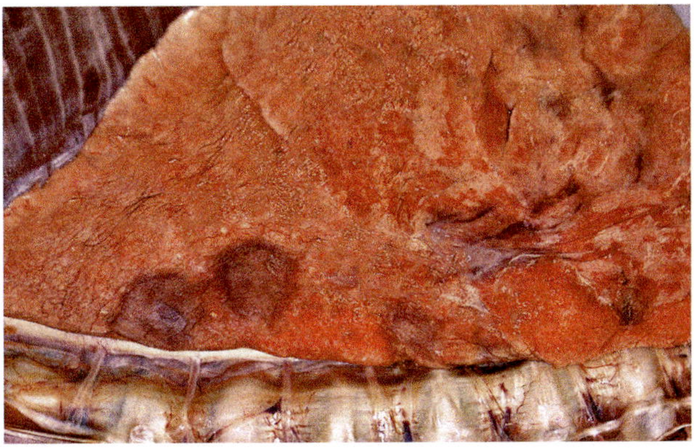

Fig. 9.22 Lion lung with two defined, focal, raised foci characteristic of the lesions in the early stages of the disease

alveoli. The areas of granulomatous inflammation are interspersed with foci of an acute inflammatory reaction containing fibrin, a neutrophil exudate, and a marked inflammatory edema. The alveolar walls are thickened and infiltrated by a mixed inflammatory exudate consisting of macrophages, fibroblasts, and proliferating type 2 pneumocytes. Marked bronchiectasis is seen within which there is a copious, mucopurulent, inflammatory exudate in which large numbers of mycobacteria can be seen following staining with the Ziehl-Neelsen (ZN) staining technique.

The initial macroscopical lesions in the lungs are nondescript, raised foci of a few centimeters in diameter that bulge slightly above the pleural surface and fade into the surrounding tissue, and are most easily detected following collapse of the lungs after the thoracic cavity is opened (Fig. 9.22). In more advanced cases, these lesions become confluent and affect large portions of the lung (Fig. 9.23). The lesions, when cut into, are ill defined and contain a semifluid mucopurulent exudate (Fig. 9.24) that originates from the dilated portions of the bronchi that are the consequences of bronchiectasis (Fig. 9.25). The parenchyma surrounding the dilated bronchi contains an inconspicuous granulomatous inflammatory reaction seen as a collar of pale, tan-colored tissue that blends into the surrounding normal parenchyma.

The lymph nodes of these animals may be enlarged, but it is not possible to detect lesions macroscopically because of the nature of the inflammatory reaction that lacks the normal features of a mycobacterial infection. In many instances, the lymph nodes may also contain large cystic spaces that extend throughout the node, but these lesions too, contrary to some anecdotal opinions, have no diagnostic significance when dealing with tuberculosis.

Lions suffering from BTB are emaciated in the terminal stage of the disease and often have large cutaneous granulomas and arthritis caused by the infection. The elbow joints may be substantially enlarged by a chronic arthritis, and contain numerous fibrin globules as part of the exudate (Fig. 9.26). Hypertrophic osteopathy may be seen in the long bones of some of the infected animals in association with joint lesions. Occasionally, panophthalmitis may be seen as a consequence of the infection (Fig. 9.27). In many of the terminal cases of BTB, renal amyloidosis may

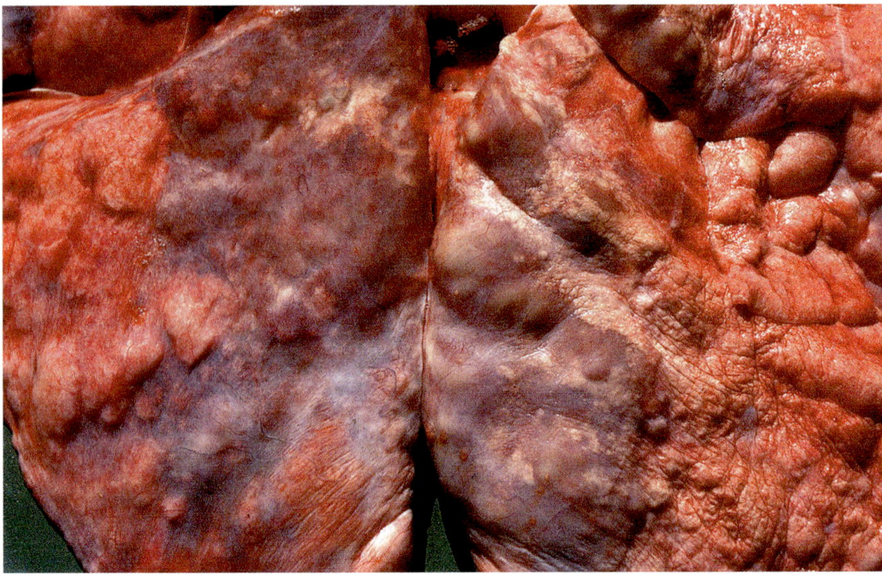

Fig. 9.23 The atypical lesions in the lung of a lion in the advanced stage of the disease resembling and extensive interstitial pneumonia

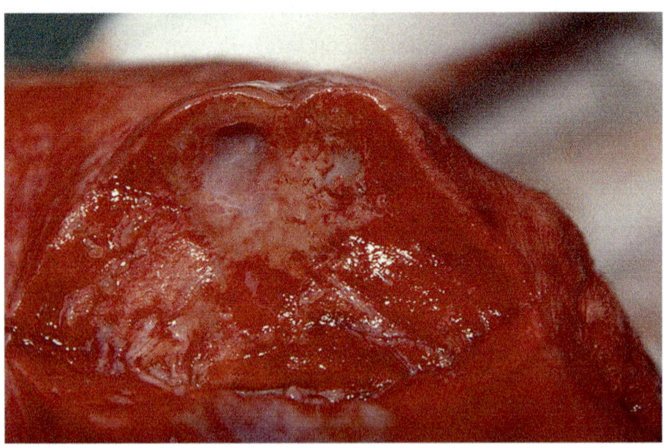

Fig. 9.24 Cavities caused by extensive bronchiectasis in the lung of a lion. These bronchi contain copious amounts of a mucopurulent exudate

be detected. This lesion may or may not be directly related to the mycobacterial infection, since it is also seen in free-ranging lions that do not suffer from the disease.

Lesions in leopards and hyenas are very similar to those seen in lions. In cheetahs, the appearance of the lesions in a single animal was recorded, and they differed markedly from those seen in lions. Macroscopically, the only lesions occurred in the lungs as extensive areas of necrosis in a multifocal to confluent granulomatous

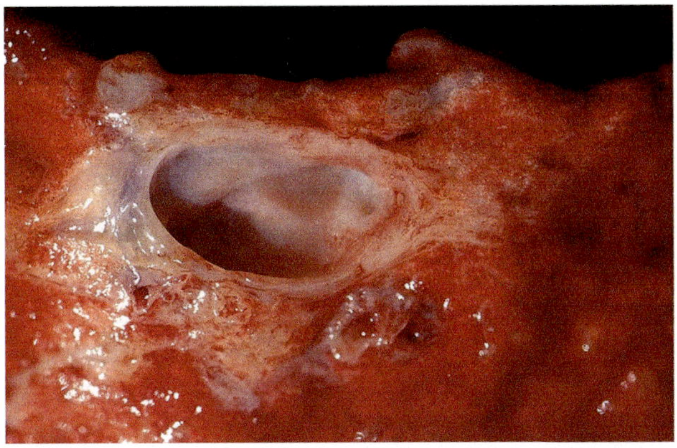

Fig. 9.25 Cavity caused by bronchiectasis in a lion's lung from which the exudate has been washed off to reveal the dilated bronchus

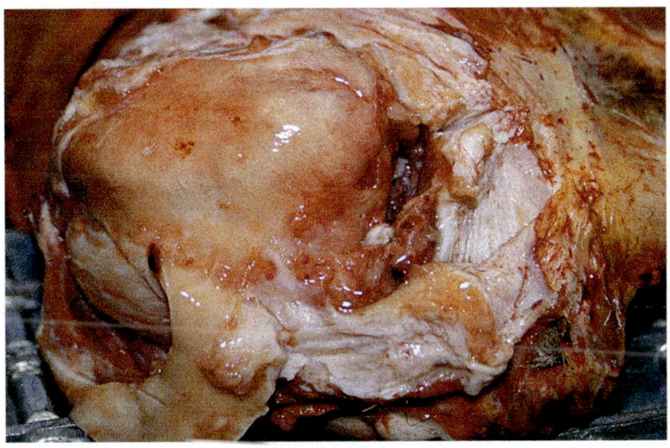

Fig. 9.26 Severe chronic, deforming arthritis caused by an *M. bovis* infection in a lion

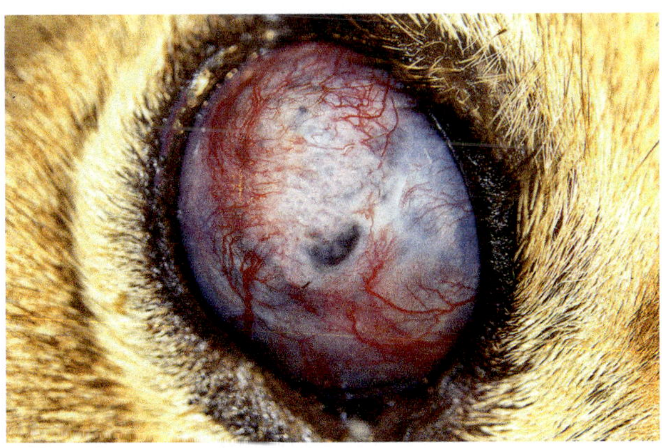

Fig. 9.27 Severe chronic granulomatous panophthalmitis in a lion

pneumonia containing a fluid, purulent exudate. Histologically, the lesions were characterized by the presence of multiple aggregates of neutrophils, occasional foci of coagulative necrosis, limited fibrin exudation, and limited calcification of a poorly encapsulated expanding granulomatous inflammatory reaction. The alveoli surrounding the inflammatory response contained macrophages and epithelioid cells, and the associated terminal bronchioles were plugged with a necrotic exudate containing many neutrophils. Scattered acid-fast bacteria occurred in the exudate and intra-cytoplasmically in macrophages at the periphery of the lesions (Keet et al. 1996).

Chacma Baboon The macroscopic lesions are typical tuberculous granulomas that may occur throughout the body. The distribution of the lesions suggests that there is early postprimary embolic spread following the initial infection. Lesions appear consistently in the mesenteric lymph node, spleen, and lungs (Fig. 9.28). The lymph nodes of the head and neck are inconsistently affected, as are others such as the mammary, inguinal, and axillary lymph nodes. In individual cases, lesions may also be present in the liver and kidneys. Tuberculous granulomas may occur in some of the vertebra.

The lesions are commonly well-circumscribed with a multifocal to confluent distribution. Their consistency varies from solid in those with fibrosis, to soft when the content liquefies. The initial lesions in the lungs have a miliary distribution reflecting a typical embolic origin. In time, they become larger and confluent and the exudate then may become liquefied, and cavities in the lung tissue may develop (Fig. 9.29).

Histologically, the lesions are characterized by areas of central caseation, containing aggregates of necrotic neutrophils within a multifocal to coalescing granulomatous pneumonia. Infiltrates of macrophages, epithelioid cells, and Langhans giant cells, lymphocytes, and plasma cells surround the necrotic centers of the granulomas. The lesions are poorly encapsulated, and a miliary spread can be seen. Liquefied exudate within granulomas may drain into adjacent bronchioles and

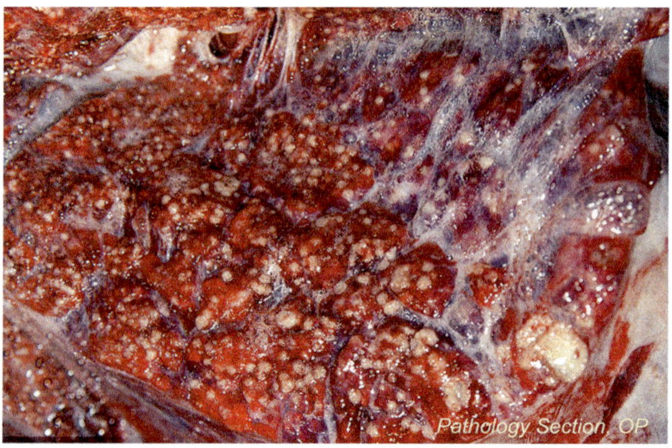

Fig. 9.28 Extensive, miliary, granulomatous pneumonia in a baboon following late, postprimary dissemination of the infection

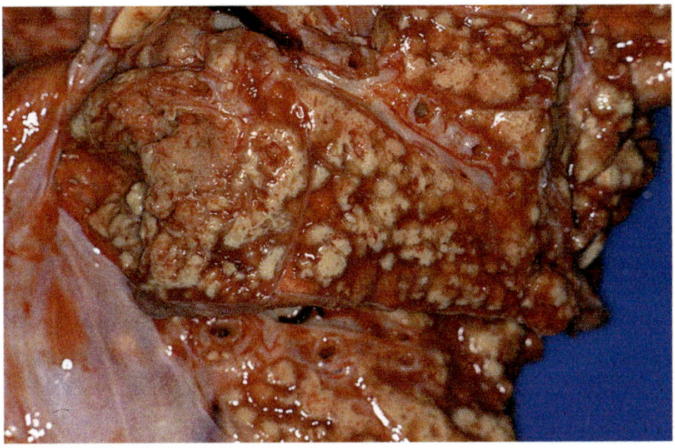

Fig. 9.29 Multifocal to confluent, granulomatous pneumonia in a baboon. Note the liquefaction of the exudate on the left, and collapse of the tissue resulting in cavitation

bronchi in which the mucosa contained a multifocal caseous granulomatous inflammation causing ulceration in areas (Keet et al. 1996). Granulomatous lesions may also be present in the liver, spleen, and lymph nodes. Many acid-fast bacteria occur within the lesions (Keet et al. 2000).

Warthogs Woodford (1982) described the lesions in warthogs infected with *M. bovis*. They were seen as calcified abscesses in the submaxillary lymph nodes and lungs. The lesions in the lungs may be extensive and consist of a caseo-calcific consolidation of the entire lung due to coalescing masses of small granulomas with an embolic distribution. In generalized cases, lesions may also be seen on serosal surfaces, and there may be dissemination to many lymph nodes. Both *M. bovis* and atypical mycobacteria may be isolated from lymph nodes with lesions.

Black Rhinoceros The lesions described in one case were limited and presented as multifocal, firm and irregular granulomas varying in diameter from 1 to 6 cm in the dorso-cranial portions of the lung lobes. They were encapsulated and contained a creamy necro-caseous exudate in which numerous acid-fast mycobacteria were detected in smears of the exudate (Miller et al. 2017).

Based on the appearance of the lesions in the various species, it must be clear that one should expect a substantial variation in their macroscopical appearance and distribution. It should be easy to make a presumptive diagnosis in advanced cases and in those with single or only a few large lesions. The challenge lies in detecting the lesions when they are single, small, or not visible and when they occur in sites in which they normally are not expected. The likelihood of missing the diagnosis with a cursory examination emphasizes the need to do a complete necropsy, should the intention be to use it for surveillance purposes. The implications of ignoring lesions in wildlife species and considering them to be "just abscesses," as has been done in the past, may have important implications as the disease then spreads imperceptibly throughout the ecosystem for many years before it is finally diagnosed.

9.6.4 *Detection of Lesions Caused by* M. bovis *in Livestock in Abattoirs*

There are different reasons for detecting lesions caused by *M. bovis* infection in domestic and wild animals. These include abattoir examination for the soundness of meat for human consumption, surveillance purposes for the control of BTB, and research. A single procedure for reliably detecting lesions that would satisfy the demands of each of these activities does not exist. Each of the different activities requires a different level of accuracy, and thus more or less attention to detail when doing the examinations. If you do not want to do what is required, or do not have the inclination to do an appropriate examination fit for the purpose of the investigation, do not bother to do it, because you will be wasting your time. Different protocols have been designed for each of these activities, and they will be dealt with in the following sections.

Meat Inspection in Abattoirs Routine abattoir inspection is based on procedures legislated by animal health authorities and/or the quarantine division of the Ministry of Agriculture in individual countries (Murray 1986; Kaneene et al. 2006; Awah-Ndukum et al. 2012; Koro et al. 2013). Inspection is intended to be conducted by trained meat inspectors with the main objective of ensuring that meat and organs are fit for human consumption. This includes detecting the presence of BTB that will, depending on the extent and distribution of the lesions, prompt rejection of affected organs or tissues, or result in total condemnation of the carcass for human consumption (Etter et al. 2006). The procedure varies somewhat between countries. In some, a specific set of lymph nodes must be incised and the cut surface inspected for the presence of lesions. In others, the nodes are merely palpated and assessed visually, a procedure that may detect 16% fewer lesions, and is thus substantially less Se (Corner 1994). Further inspection involves visual examination and palpation of organs, such as the liver and kidneys, and palpation and in situ incision of the tracheobronchial, mediastinal, and caudal cervical lymph nodes and the lungs. Other lymph nodes and organs are incised if lesions are detected in one of these tissues. In some countries, the mesenteric lymph nodes and lymph nodes of the head (Awah-Ndukum et al. 2010) and supra-mammary lymph nodes (Ngandolo et al. 2009) must also be routinely inspected.

By following the protocol for routine meat inspection, 43% of tuberculous animals may be missed (Corner 1994), while Bekele and Belay (2011) and Aylate et al. (2013), respectively, found that as many as 90.5 and 87% of cattle with tuberculous lesions, were missed. Stärk et al. (2014) similarly found that the success of meat inspection in general in detecting lesions is highly correlated with the presence of clinical and/or pathological signs in affected animals and that early or subclinical cases are likely to be missed at slaughter. The impact of the use of visual-only inspection instead of incising the lymph nodes and tissues was negligible for most notifiable diseases and conditions with the exception of detectable cases of tuberculosis. In addition to the specific methodology used, the success in detecting lesions during the course of abattoir inspection will also depend on the competency of the examiner, and the time spent on examining the animal. One investigation reported that the mean time spent to inspect a carcass (1.2 ± 0.4 min) by the routine method was virtually five times shorter than the detailed method (5.8 ± 1.9 min) (Bekele and Belay 2011), hence the difference in the successful detection of lesions when they are present.

Meat inspection at abattoirs is the main test method used in the final stages of eradication programs and for documenting freedom from BTB in developed countries from which the disease in cattle has been eradicated. The lack of the sensitivity of using routine meat inspection as a means of surveillance in the advanced stages of eradication schemes, although it may be much less expensive than sustaining the use of TSTs, is thus not without risk. This is particularly applicable when the issue of not being able to detect cases with non-visible lesions with current technology, is taken into consideration (Gavier-Widén et al. 2009).

Detailed Postmortem Inspection Under certain circumstances, for instance, for research purposes and the reliable detection of lesions in positive reactors, because of the low sensitivity of routine meat inspection to detect BTB, it is necessary to increase the sensitivity of postmortal detection of BTB-positive carcasses. The sensitivity of the investigation can be improved substantially by simply examining six pairs of lymph nodes, a process that is sufficient to detect 95% of cattle with macroscopic lesions (Corner et al. 1990). The specificity of macroscopic examination by the parallel use of histopathology can also enhance the Se of the investigation to 94–95% (Rohonczy et al. 1996).

Because of a long list of tissues, organs, and anatomical sites affected by BTB, the results obtained vary greatly in scope from study to study (Corner 1994). Commonly, the following lymph nodes are sliced at a thickness of 2 mm: parotid, retropharyngeal, mediastinal, tracheobronchial, mesenteric, submaxillary, iliac, precrural, prescapular, supra-mammary, inguinal, ischiatic, portal, and sternal. Organs including the lungs, liver, kidneys, mammary gland, intestines, heart, and meninges are thoroughly examined visually and by palpation and are incised at thicknesses varying from 2 to 20 mm (Asseged et al. 2004; Biffa et al. 2010). The cut surfaces are examined under a bright light source for the presence of TB-like lesions. Sectioning and the detection of small lesions are easier after overnight refrigeration of the lymph nodes. To increase the sensitivity of the detailed examination, aseptic specimens from all lymph nodes should be submitted for culture, according to the protocol used (Corner 1994).

Meat Inspection in Africa With the exception of the situation in a few countries, meat inspection at abattoirs in Africa is in a parlous state (Bekele and Belay 2011). The purpose of meat inspection is to focus on the removal of grossly abnormal meat, organs, and related products from the food chain, thus preventing the distribution of infected products that could cause disease in humans. It should also assist in the detection of specific diseases to allow tracing-back infections to their source for control purposes (Biadglegne et al. 2013; Habarugira et al. 2014). Of these, the removal of grossly abnormal products for human consumption is probably the easiest to accomplish and is overemphasized in Africa.

Meat inspection protocols currently utilized in abattoirs in Africa are generally insufficient to detect the majority of visible TB lesions. Factors contributing to the low sensitivity include the lack of adequately trained personnel, heavy workload, poor physical facilities with inadequate lighting, and the lack of diligence of the inspectors (Awah-Ndukum et al. 2010; Biffa et al. 2010). Furthermore, as the examination is performed under variable conditions, inconsistencies of interpretation are difficult to avoid. Corruption, to limit the impact of the loss incurred by condemned carcasses, has also been highlighted as a limiting factor in applying the existing regulations (Sulieman and Hamid 2002).

Standardization of abattoir inspection protocols (in line with international sanitary requirements), enhanced training, quality control measure to assess the standard of meat inspection, and raising public awareness are essential and cost-effective measures that should be implemented to improve the situation (Biffa et al. 2010) and to increase the accuracy of diagnosing BTB in abattoirs.

The specificity of meat inspection as a means to detect BTB in Africa is also affected by a plethora of non-specific TST reactions caused by environmental mycobacteria and the presence of lesions caused by parasitic infections and bacterial or mycotic granulomas and abscesses caused by pyogenic bacteria, which may be macroscopically indistinguishable from tuberculous granulomas (Liebana et al. 2008). A number of examples are quoted to demonstrate the extent of the problem. A study in Mali revealed that 72% of tuberculous lesions detected during routine meat inspection were due to pathogens other than *M. bovis* (Maas et al. 2013). In another report, only 11.2% of BTB-like lesions in Tanzania were positive for *M. bovis* (Mwakapuja et al. 2013). Various pathogens including the pulmonary form of nocardiosis (in Sudan) (Awad 1962), granulomas caused by parasites such as liver flukes, actinomycotic granulomas (Berrada 1993), and fungi (Kuria and Gathogo 2013) may be confused with the lesions of BTB. In addition, NTMs (up to 44.8% in certain studies) cause similar lesions to those caused by *M. bovis* that may result in incorrect diagnoses in the absence of the ability to confirm the diagnosis, at least by histopathology, and, preferably by culture (Berg et al. 2009; Sahraoui et al. 2009; Biffa et al. 2010).

The limitations of using abattoir diagnostic data as a means of surveillance for BTB in Africa must be kept in mind by regulatory authorities and policy makers. Using these data also have an effect on the time it takes to detect an infection in a herd as the estimated median time for detection by abattoir surveillance of a herd after the disease was introduced was 5.75 years compared to 1.99 years when a TST was used as a routine diagnostic procedure (Fischer et al. 2005). However, in Africa, as monitoring of BTB using bacteriology and TSTs is time-consuming and prohibitively expensive, and the laboratories are ill-equipped, it is inevitable that data obtained from meat inspection will be the only way in which some useful information about the extent of BTB in the continent's livestock population can be gathered. To do this, the process will have to be exploited to its full potential, always keeping in mind that the process is flawed (Atiadeve et al. 2014).

9.7 Microscopic Detection of *M. bovis*

The use of microscopy to detect acid-fast bacteria (AFB) is a rapid and inexpensive, but inconclusive, method to tentatively diagnose mycobacterial infections. The presence of AFBs justifies a presumptive diagnosis of tuberculosis, especially when clinical signs and typical lesions are present. Microscopic examination can be done in two main ways: firstly, by examining smears from mucus obtained from the respiratory tract, or smears prepared from exudate obtained from lesions resembling those caused by *M. bovis* and, secondly, by detecting AFBs in histological sections of tissues containing the suspect lesions.

Smear Examination Conventionally, smears for TB diagnosis are stained with the Ziehl–Neelsen (ZN) technique that allows detection of acid-fast bacteria (AFB) (Fig. 9.30), but auramine O/rhodamine staining and immunohistochemistry are also used (Varello et al. 2008), the latter of which is the most sensitive (Watrelot-Virieux et al. 2006). The smears, except those made from respiratory mucus, should be made from exudate obtained from visible tuberculous lesions. In the case of histopathology, tissue specimens from a portion of the lesion bordering normal tissue, should be fixed in 10% buffered formalin for processing. The biggest limitation of using this technique is that the specific mycobacterial genus cannot be determined, and mycobacteria can also morphologically be confused with *Rhodococcus equi* that is also acid-fast. Generally, smear examination is very specific, but it has a low Se because smears from fresh specimens require up to 10,000 bacteria/ml for smear positivity, and not all cases with tuberculosis are shedders (Adu-Bobi et al. 2009). For ZN-stained smears the Se and Sp were, respectively, 33.9 and 100% (Varello et al. 2008), while Damina et al. (2011) recorded a Se of 82.7%. Not all investigators reported the same results, as in Burkina Faso, Tarnagda et al. (2014) detected AFBs in only 37.3% of smears prepared from 102 animals with BTB-like lesions.

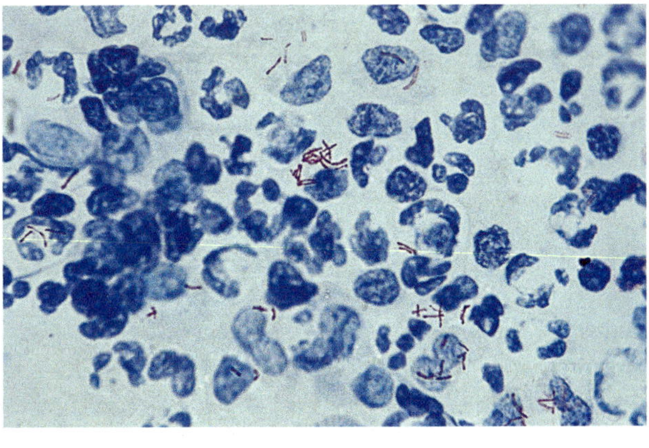

Fig. 9.30 Microscopical appearance of a positive, Ziehl-Neelsen-stained smear of exudate collected from a tuberculous lesion containing scattered, and a small cluster of acid-fast mycobacteria (1000× magnification)

Smears from nasal or respiratory tract mucus are not often used in animals, but they have been used, for instance, to diagnose TB in elephants (Mikota et al. 2001) and recently in lions (Miller et al. 2015). Smear examination is not considered to be a reliable diagnostic technique and it has a low Se and Sp. In wildlife, the use of microscopy has been recommended for surveys and game meat inspection, particularly when combined with macroscopic postmortal examination (Maas et al. 2013).

Histopathology The examination of histological sections for diagnosing BTB is not a stand-alone test. It should be used to confirm a macroscopical presumptive diagnosis based on the presence of lesions resembling those expected in animals suffering from BTB. In histopathological sections, the diagnosis is based on the presence of the typical appearance of the lesion, in addition to the presence of AFBs, or those stained with auramine O, or by immunocytochemistry. Depending on the species of animal, and the nature of the lesion, very few bacteria may be present (paucibacillary cases), or they may not be visible at all, probably because of their limited numbers in the exudate. Not being able to detect AFBs histologically is not a reason to rule out *M. bovis* as the cause of the lesion, as is the case with smears, when there are too few bacteria in the lesions to allow their detection. This is a known phenomenon in many species, including wildlife, and particularly in African buffaloes, where a large percentage of cases with lesions do not contain detectable bacteria histologically.

The Se and Sp of the technique tend to differ in various published studies. Varello et al. (2008) found a high Se (93.4%) and Sp (92.3%) for histopathology, but it is clear that postmortal inspection, histopathology, and culture do not detect the same sets of positive animals, although there is an overlap between the results of the individual tests. Using these tests in series or in parallel increases the Se (Rohonczy et al. 1996). The use of PCR to identify the different MTC species on histological sections also increases the sensitivity of the investigation in those cases where histopathology reveals typical tuberculous lesions and the presence of mycobacteria, but the specimens were culture negative (Miller et al. 2002).

9.8 Bacterial Culture and Typing Procedures

Culture is considered to be the diagnostic gold standard for the detection and confirmation of BTB in livestock and wildlife (Liebana et al. 2008; Gavier-Widén et al. 2009). Although it is considered as such, it too is neither 100% Se nor 100% Sp, and this adds to the difficulties of eradicating BTB. The sensitivity of culture to detect mycobacteria is influenced by a number of factors, including the collection of appropriate specimens, the time taken before the specimens are processed for culturing, the way in which specimens are processed, and the culture media used (Gormley et al. 2014). Using conventional culturing techniques delay making a final diagnosis because of the slow growth of many of the mycobacteria, as it may take up to 15 weeks or more before some of the bacteria become detectable and available for species identification and typing (Corner et al. 2012). As is the case in postmortal examination of animals for the detection of lesions, the inability to culture mycobacteria from submitted specimens does not imply that the specific animal did not contain viable mycobacteria of the MTC, particularly when there were cases with early infection and in the preclinical stage of the disease, or latent carriers (Gormley et al. 2014).

That bacterial culture is not 100% sensitive, is due to a number of factors, and they should be kept in mind when collecting and processing specimens for culture (Ramos et al. 2015). Tissue specimens for diagnostic culture should contain visible lesions if there is to be any likelihood of mycobacteria being cultured. If the intention is to detect non-visible lesions, adherence to specific protocols outlining the range of tissues to collect and to maximize the possibility of detecting mycobacteria should be strictly followed and adapted according to the species dealt with. Some of the causes of failure to culture the organisms include:

- The type of culture medium and O_2 concentration during culturing: Three types of media are commonly used in Africa for culturing mycobacteria. These are egg-based media such as Lowenstein–Jensen (LJ) or Stonebrink's (SB), agar-based media such as Middlebrook (7H10 and 7H11) for primary isolation, and liquid media such as Middlebrook 7H9. All these media contain malachite green, a selective dye required for the growth of mycobacteria in culture (Buxton and Fraser 1977). Details of the techniques are available in guidelines available from the OIE and in various other publications (O'Brien et al. 2008; Mohamed et al. 2011). None of these media has 100% Sp, and hence attention to detail, including specimen collection and decontamination procedures in particular, are important processes that should be addressed, also in the future when developing improved culturing systems (Gormley et al. 2014).
- The use of automated liquid culture systems such as the BACTEC 460 and 12B, BACTEC MGIT 960, and VersaTREK systems is an improvement on solid media, as it cuts down the average time for first detecting growth to as short as 15.8 days (Gormley et al. 2014). This technology is probably beyond the reach of most of the African countries because of cost and the lack of infrastructure and adequately trained human resources.
- There are two important factors that determine growth of *M. bovis* on culture. Glycerol, which is used as the primary carbon source in culture media used for mycobacterial culture, promotes the growth of several species of mycobacteria including *M. tuberculosis*, but inhibits the growth of *M. bovis*. It is thus critical to use media with an alternative carbon source such as pyruvic acid, to enhance the growth of *M. bovis* on culture (Collins and Grange 1983). Furthermore, contrary to most mycobacteria that are aerobic organisms, *M. bovis* is microaerophilic, and the correct environmental O_2 concentration should be maintained when they are cultured (Addo et al. 2007).
- Depending on the culture medium used, and the pre-culture processing, too few organisms may be present in the specimens to allow positive culture (Anon 1994; Collins et al. 1994).

- Using an incorrect protocol for the purpose of the investigation when collecting specimens for culture from individuals and groups of animals (Crawshaw et al. 2008).
- Not collecting a sufficient amount of tissue, organs, or exudate for culture. This applies particularly to milk, where about 100 ml from each quarter is recommended to optimize the likelihood of culturing the organism for typing purposes (Kleeberg 1984; Ben Kahla et al. 2011).
- The way in which specimens are collected, stored, and transported and the time involved in getting the specimens to the diagnostic laboratory. Tissues should be collected aseptically and processed immediately for culture. Samples stored at 4–6°C must reach the laboratory as soon as possible and must be processed within 24–48 h after collection; otherwise, they should be frozen for storage and transport. For optimal detection of mycobacteria, the sediment from centrifuged milk must be examined immediately (within 24–48 h of collection) or otherwise frozen.
- Decontamination, a process intended to selectively eliminate non-mycobacterial flora and enhance the isolation and detectability of mycobacteria by preventing other bacterial overgrowth, is another factor that influences the isolation of mycobacteria from specimens. The decontamination process may significantly decrease the number of viable mycobacteria, and up to 80–90% may die (Krasnow and Wayne 1969; Anon 1994). The type of decontaminant and the concentration at which it is used, are critical. Decontaminants include detergents such as 0.375–0.75% hexadecylpyridinium chloride (HPC), alkalis (2–4% sodium hydroxide), and acids (5% oxalic acid).

9.8.1 Species Identification and Typing

Culture on its own is a mechanism of detecting *M. bovis* in specimens collected from animals suspected to be suffering from BTB. Colony morphology does not allow accurate species identification, although the characteristic growth pattern and morphology of the colonies provide sufficient information to make a presumptive diagnosis of *M. bovis*.

A series of tests based on phenotypic characteristics and biochemical properties was traditionally used during the past decades to identify the various species of the MTC. The process is slow, cumbersome and inaccurate, and results are often ambiguous (Ramos et al. 2014; Ramos et al. 2015). These tests cannot be done in many laboratories, and they have largely been replaced by a number of sophisticated molecular genotyping techniques allowing linage and strain identification (Fig. 9.31). Identifying the individual strains is critical for implementing molecular epidemiological studies to determine the origin and spread of the disease to allow the design of more focused control measures and eradication campaigns (Collins 1999; Courcoul et al. 2014; Drewe and Smith 2014; Gormley et al. 2014).

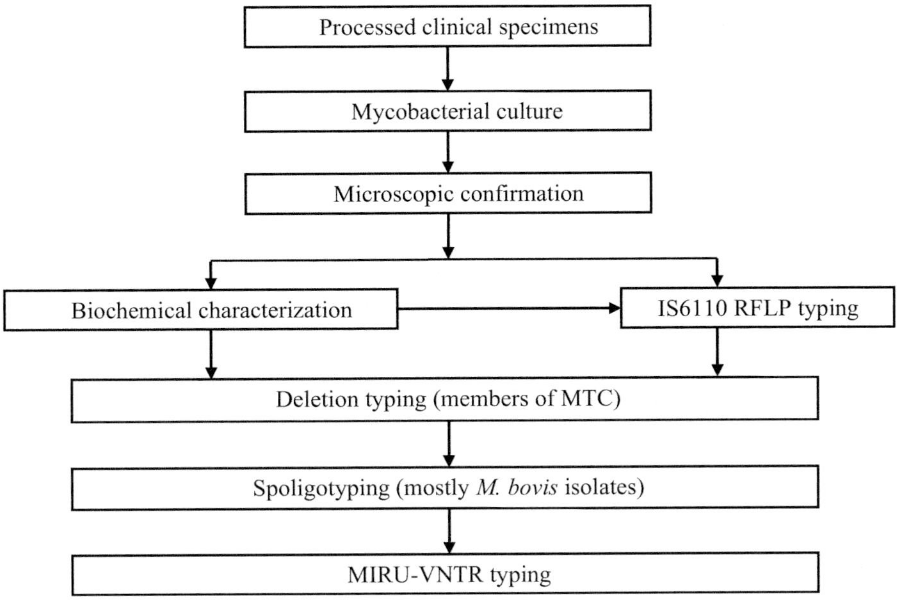

Fig. 9.31 Schematic flowchart showing mycobacterial identification techniques and common molecular diagnosis steps

Recent developments also include the use of different molecular techniques for typing and strain identification of *M. bovis* to characterize and untangle the molecular epidemiology of the disease to more adequately control its spread and to determine and control the origin of the infection. Few countries in Africa have the financial and human resources to implement the use of these tests. However, because of increasing collaboration with foreign laboratories, some data have become available. This information can be used to determine the source of infections and patterns of spread, enabling more focused approaches to controlling the infection (Cadmus et al. 2011; Ramadan et al. 2012). The details of these matters are addressed in Chap. 8 (molecular epidemiology) and in some of the country reports.

Despite the extensive genetic homology (99.9% genomic sequence homology and identical 16S rRNA sequences) of the members of the MTC, individual species, lineages, and strains can be differentiated by comparing differences in the sequences in the remaining 0.1% of their genomic structure. These differences are the consequence of variations in single nucleotide polymorphism (SNP) and large sequence polymorphism (LSPs), also known as regions of difference (RDs) (Brosch et al. 2002; Garnier et al. 2003). These characteristics are the basis of the various molecular diagnostic techniques.

During the past few decades, PCR-based assays, DNA mapping, and whole genome sequencing have been widely used for species and strain identification of mycobacteria and to elucidate genetic diversity, their evolution, pathogenicity, drug resistance, and the geospatial distribution of *M. bovis* in parts of Africa.

Much of the current research is focused on developing more sensitive and specific tests that are inexpensive and rapid. They should also be appropriate for use particularly in resource-poor countries, and should be suitable for use in the field in rural areas where there is a lack of infrastructure, electricity, and a cold chain.

Two categories of tests utilizing genetic markers exist: those that use whole genome techniques and those that are based on partial genome techniques. The partial genome techniques include restriction fragment length polymorphism (RFLP) analysis, spoligotyping, variable number of tandem repeat typing (VNTR), IS6110-ampliprinting (MPTR), random amplified polymorphic deoxyribonucleic acid analysis (RADP), RD typing (RD), and single nucleotide polymorphism (SNP) typing. Whole genome typing allows the use of the microarray technique for comparing particular strains to sequenced reference strains (Gormley et al. 2014).

The first step following culture is to distinguish the MTC mycobacteria from non-tuberculous mycobacteria, and this can only be done by the use of nucleic acid amplification methods such as PCR and spoligotyping. Both these tests are technically challenging, time-consuming, and expensive and, although highly discriminatory, are prone to homoplasy or convergent evolution, where the same genetic profile could be obtained in distinct MTC strains that are phylogenetically unrelated, thus confounding strain classification and phylogenetic inference (Hauer et al. 2016). Given the limitations of many of the current techniques, and the cost and technical ability required to use them, only some have been used in Africa, and it appears that there are specific problems in various countries limiting their use. The following typing tests are more often used in Africa:

IS6110 Restriction Fragment Length Polymorphism (RFLP) This typing test has for long been used as the gold standard for differentiating *M. tuberculosis* strains (van Embden et al. 1993), but it has a limited value for typing *M. bovis* in which certain strains, such as those found in Europe, have one or only a few copies of IS6110 (Smith 2012). In Africa too about 90% of isolates in Madagascar (Razanamparany et al. 2006) and 83% in Cameroon (Njanpop-Lafourcade et al. 2001), possess only a single copy of IS6110, while 44% of isolates in South Africa contain only two copies (Michel 2008). The resolution, because of the low number of IS6110 copies in *M. bovis* strains, is too low for it to be of value. In addition, the test is time-consuming, labor-intensive, and technically demanding and requires large amounts of high-quality DNA (Deriemer and Daley 2004). Despite these limitations, the use of IS6110 RFLP typing remains essential in Africa as it is used to screen AFB-positive bacteria from cultures to differentiate between members of the MTC and NTMs or other related mycobacterial genera (Rigouts et al. 1996; Michel et al. 2008; Malama et al. 2014).

Spoligotyping (Spacer Oligonucleotide Typing) Spoligotyping involves PCR amplification of highly polymorphic repetitive units in a particular genomic region, known as the direct repeat (DR) locus, which are only present in members of the MTC. The DR locus belongs to the clustered, regularly interspaced, short palindromic repeat (CRISPR) family of repetitive DNA (Skuce and Neill 2001). The DR region in MTC could be associated with selective advantages and may be of specific biological importance to the pathogen (van Embden et al. 1993).

The DR locus includes a series of well-conserved, virtually identical, 36 bp DRs separated by unique, non-repetitive DNA spacer sequences of 34–41 bp length (the direct variant repeat (DVR) units). The spacers usually occur only once (occasionally twice) and are separated by one or several DRs and other spacers (Boer et al. 2004). The DR region may contain over 60 DVR units; however, only 43 of the spacer units (37 spacer sequences of *M. tuberculosis* H37Rv and 6 of *M. bovis* BCG P3) are used in standard spoligotyping of the MTC. The DR region is polymorphic because of the loss (deletion) of a single or multiple spacers, a condition that cannot be repaired by recombination, and, hence, is lost from a lineage forever (Smith et al. 2006) and thus remains a stable marker that does not change over time.

Spoligotyping detects the presence or absence of these 43 spacers of known sequence in the DR region by hybridization of PCR-amplified spacer DNA to a set of immobilized oligonucleotides, representing each of the unique spacer DNA sequences (Boer et al. 2004). The spacers are arranged on the blot in the order in which they appear in the H37Rv DR sequence (spacers 1–19, 22–32, and 37–43) and the sequence of *M. bovis* BCG P3 (spacers 20, 21, and 33–36). Polymorphism among spoligotypes of MTC is based on the presence or absence (loss of hybridization signal) of one or more spacers, compared to the complete 43-spacer set of *M. tuberculosis* H37Rv (Kamerbeek et al. 1997). The simplicity and robustness of this procedure and the ease and accuracy of presenting results in highly diverse portable numerical values make spoligotyping a widely used tool in clinical laboratories and for application in molecular epidemiology, evolutionary, and population genetic studies (Zumárraga et al. 1999).

Spoligotyping also has serious limitations. Its main disadvantage is that all genetic polymorphisms are restricted to a single genomic region, the DR locus, which limits its resolution. Furthermore, spoligotype patterns can be homoplasic, i.e., identical spoligotype patterns can be found in epidemiologically unrelated strains (Milián-Suazo et al. 2002), nullifying the assumption that strains with the same spoligotype patterns are identical by descent. However, homoplasic spoligotype patterns can be rare (Smith et al. 2006), and could be differentiated, when present, by variations in their VNTR patterns and other molecular markers. Therefore, from a practical point of view, spoligotype patterns are useful in the diagnosis of TB under African condition, since all MTC strains from humans and animals can be differentiated (Munyeme et al. 2009). This is important because it is becoming clear that animals that appear to be affected by *M. bovis* may actually be infected by *M. tuberculosis* or *M. africanum* (Sanou et al. 2014). The extent to which *M. tuberculosis* can occur in lymph nodes of cattle with BTB-like lesions is surprising; as in a study in the Eastern Cape, South Africa, 96.9% of isolates from 182 MTC-positive lymph nodes were strains of *M. tuberculosis* (Bhembe et al. 2017). This has implications for disease management (as MTC stains differ in their drug resistance profiles) and for establishing the source of infection. The advantage of combining spoligotyping with MIRU-VNTR (Weniger et al. 2010) and Hain®Genotype MTBC was demonstrated in Cameroon where it was determined that *M. bovis* strain Afr1 was the dominant clonal complex with 97 unique genotypes. It also determined that a third of infected animals with multiple lesions had multi-strain infections. The data generated by the combination of these techniques allow characterizing stable and unstable endemic infections in different parts of the country, thus enabling the implementation of suitable control measures for the specific situation (Egbe et al. 2017).

Spoligotyping has several attributes that make it particularly suitable for resource-poor African countries, and it is the most popular molecular technique in Africa for the following reasons:

- It can detect and type MTC directly in a range of clinical samples (Deriemer and Daley 2004; Kamerbeek et al. 1997),
- It is fast and relatively easy to perform and requires only a small amount of DNA (Deriemer and Daley 2004; Zumárraga et al. 1999).
- Spoligotyping patterns can easily be compared with results from other countries or regions by using the freely accessible international database (http://www.Mbovis.org) (Smith and Upton 2012). This database holds information and authoritative names for spoligotype patterns of all animal-adapted MTC strains (and *M. africanum*), in which RD9 is deleted.

Region-of-Difference (RD) Deletion Analysis RD variability is attributed to deletion or truncation of specific genomic elements due to replication slippage of DNA polymerase (Brosch et al. 2002), and has a nonrandom distribution, with a tendency to aggregate in a specific region (Rao et al. 2005). These may be missing genes that encode virulence factors such as the organism's development of resistance to drugs and adaptation to environment and host, the presence or absence of which could give clues to clonal lineage of various mycobacterial isolates in specific geographic regions (Rao et al. 2005). *Mycobacterium tuberculosis* lacks 16 RDs, labeled as RD1 to RD16 that occur in other members of the MTC. Some are deleted from specific members of the MTC, and their absence allows distinguishing between the various species of the MTC (Parsons et al. 2002). Such comparative analysis of deleted distinct regions from the MTC genomic structure, provides a basis for differentiation and understanding the evolutionary lineage/relationship of members of the group. For example, Parsons et al. (2002) were able to accurately differentiate 88 well-characterized isolates of *M. tuberculosis*, *M. africanum*, *M. microti*, *M. bovis*, and *M. bovis*-BCG using PCR-based RD analysis of only six regions (RD1, RD3, RD5, RD9, RD10, and RD11).

In Africa, RD analysis is frequently used to differentiate *M. bovis* (which lacks RD4) from other members of the MTC and to differentiate *M. tuberculosis* from animal-derived strains of the MTC, including *M. africanum*. Additionally, RD analysis provides an understanding of the lineages of the current population of mycobacterial pathogens, and in addition to its diagnostic use, RD analysis could be a useful tool in epidemiological and evolutionary studies of mycobacterial species (Gordon et al. 1999; Brosch et al. 2002). Currently RD is considered to be the gold standard for genotyping mycobacteria in the MTC (Faksri et al. 2016).

Multilocus Variable-Number Tandem Repeat Analysis (MLVA) Typing MLVA is a PCR-based molecular typing technique that exploits natural variations in tandem repeat (TR) DNA sequences, also known as variable number tandem repeat (VNTR) loci, that are found in multiple loci in the genome of a variety of organisms (Ablordey et al. 2005). In MLVA, specific primers target the flanking sequences of these VNTR regions (Diguimbaye-Djaibé et al. 2006), the variations of which generate PCR amplicons of different fragment length; such length polymorphism forms the basis for the generation of diverse alleles. VNTRs used in mycobacterial studies are known as mycobacterial interspersed repetitive units (MIRUs), Queen's University Belfast (QUB), exact tandem repeats (ETRs), or Mtub (*M. tuberculosis*) (Diguimbaye-Djaibé et al. 2006). Although there are subtle differences between the loci identified by these approaches, there is an overlap, and some of the chromosomal regions are known as ETR, MIRU, and/or VNTR loci (Smith et al. 2006). More than 40 of these loci have been investigated so far. Nearly half of them produced no discrimination, and the remaining loci gave varying degrees of discrimination among *M. bovis* isolates depending on the panel of VNTRs used (Collins 2011). When small numbers of VNTR loci are used, the level of discrimination achieved is not good. Larger panels and other sets of polymorphic repeats (MIRU-VNTR) involving 12 loci and 15 or 24 loci subsets, are frequently used, but this causes confusion, as the results of the different combinations cannot be compared. Of 29 loci combinations (QUB, ETR, MIRU, and Mtub) in South Africa, only 13 loci were stable and polymorphic, 14 were monomorphic, and 2 could not be amplified (Hlokwe et al. 2013), and of 26 MIRU-VNTR loci from 32 *M. bovis* strains in Burkina Faso, 9 loci were monomorphic (Sanou et al. 2014). In Chad, five loci were highly discriminative, two were moderately so, three were poor, and six loci showed no polymorphism at all (Hilty et al. 2005) (Table 9.3). The technique has also been used to type and determine the origin of *M. bovis* strains in Ethiopia (Biffa et al. 2014) and the molecular epidemiology of the disease in Nigeria (Jenkins et al. 2011). These reports emphasize the importance of evaluating the discriminatory power of different loci for each country or geographical region. It is thus important that the loci employed in individual countries should be those that will provide the biggest likelihood to discriminate the characteristic clonal complexes of its *M. bovis* population. The strain genetic variability in a country depends on the dominant existing clonal complexes, and the best results will be rendered by a combination of typing methods (Hauer et al. 2016).

Using the web-based server (http://www.miru-vntrplus.org/) makes it easier to determine the lineages of MTC strains. Furthermore, comparison between strains can also be made by comparing information maintained on the server including copy numbers of 24 MIRU loci and other molecular features. Additionally, the web-based server allows standardized nomenclature of different MIRU genotypes for ease of scientific communication (Hilty et al. 2005; Allix-Béguec et al. 2008). However, in order to adequately exploit what the server has to offer, African diagnosticians and researchers will have to align their selection of loci combinations for diagnostic purposes, as the combinations used in different countries in Africa are significantly different from the 24 loci set contained in the server.

Table 9.3 Molecular techniques to characterize *M. bovis* and other members of the MTC in Africa

Country	Technique	Target locus/ molecule	Objective	No. of isolates	No. of genotypes	Reference
Algeria	Spoligotyping	DR	Characterization of MTC	89	23[a]	Sahraoui et al. (2009)
	VNTR	ETR-AE	Characterization of *M. bovis*		35	
Burkina Faso	Spoligotyping	DR	Characterization of *M. bovis*	33	12	Sanou et al. (2014)
	MIRU-VNTR	26 loci	Characterization of *M. bovis*	32	24	
Cameroon	RFLP	IS6110	Characterization of *M. bovis*	18	3	Njanpop-Lafourcade et al. (2001)
Chad	RFLP	IS6110	Characterization of *M. bovis*	15	6	Diguimbaye-Djaibé et al. (2006)
	Spoligotyping	DR	Characterization of *M. bovis*	55	12	
Ethiopia	Multiplex PCR	RD4	Differentiate *M. bovis* from other members of MTC	1	1	Firdessa et al. (2012)
	Spoligotyping	DR	Characterization of *M. bovis*	27	4	
	VNTR	24 loci	Characterization of *M. bovis*	27	5	
	Spoligotyping	DR	Characterization of *M. bovis*	45	12	Biffa et al. (2010)
	MIRU-VNTR	24 loci	Characterization of *M. bovis*	58	19	Biffa et al. (2014)
Madagascar	RFLP	IS6110	Speciation of *M. bovis* isolates	180	9	Razanamparany et al. (2006)
		RD	Differentiate *M. bovis* from other members of MTC		20	
	Spoligotyping	DR	Characterization of *M. bovis* isolates		12	
Mali	Spoligotyping	DR	Characterization of *M. bovis*	20	7	Müller et al. (2008)
Nigeria	Spoligotyping	DR	Characterization of M. bovis	178	34	Cadmus et al. (2011)
	VNTR	ETR A-F	Characterization of *M. bovis*	1[b]	40	
	Deletion typing	RD	Differentiate AFB	142[c]	3	Damina et al. (2011)
	Spoligotyping	DR	Characterization of *M. bovis*	13	8	Jenkins et al. (2011)
	VNTR	16 Loci	Characterization of *M. bovis*	49	30	

South Africa	RFLP	IS6110	Characterization of *M. bovis*	49	16	Michel et al. (2008)
	Spoligotyping	DR	Characterization of *M. bovis*	50	12	
	VNTR	ETR A-F	Characterization of *M. bovis*	42	13	
Zambia	Spoligotyping	DR	Characterization of *M. bovis*	27	1	Malama et al. (2014)
	MIRU-VNTR	9 loci	Characterization of *M. bovis*		16	
	Spoligotyping	DR	Characterization of *M. bovis*	31	10	Munyeme et al. (2009)
				52	3	Hang'ombe et al. (2012)
	MIRU-VNTR	26 loci			9	

[a]*M. caprae* (SB1451)
[b]SB0944
[c]*M. bovis* (107), *M. tuberculosis* (6), *M. africanum* (2), others (27) not typed

Lateral Flow Immunochromatographic Device The recently developed rapid lateral-flow immunochromatographic device (LFD) to confirm the identity of *M. bovis* is less sensitive than the qPCR methods. Its development is an attempt to create a rapid, inexpensive test. Currently the LFD results vary from inconsistent, as the test does not detect all spoligotypes, particularly because of *M. caprae* that also causes bovine tuberculosis and contain spoligotypes not detected by LFD, to almost 100% concurrence with other tests. The intention for the test is not to replace spoligotyping but to aid in the detection of, *M. bovis*-positive cultures that are not detected by Ziehl-Neelsen staining (Stewart et al. 2017).

9.8.2 Diagnosis of BTB in Wildlife

The presence of BTB in wildlife in the same ecosystem as that of cattle has become an international complication when attempting to eradicate BTB from cattle. There is increasing recognition of the role that different species of wildlife play in the epidemiology of the disease and in sustaining the infection by becoming maintenance hosts. Africa is richly endowed with a wide variety of wildlife species, most of which are probably susceptible to infection with members of the MTC. Some reference has been made to the diagnosis of the infection in wildlife in other sections of this chapter. Little attention, however, has until recently been given to the diagnostics of BTB in African wildlife, and authorities dealing with BTB face an enormous challenge to cope with the issues given the number of species and distribution of them throughout Africa. Against the background of the history of the role of wildlife in developed countries and elsewhere, dealing with wildlife, in the process of eradicating BTB will in all likelihood be a mammoth task.

Maas et al. (2013) recently comprehensively reviewed the problems associated primarily with the ante-mortal diagnosis of tuberculosis in wildlife. One of the main impediments of making a correct diagnosis in these species, is the lack of specific and sensitive diagnostic tests that have been validated and are specific and sensitive enough to satisfy the demands of the various reasons for testing them for the presence of the disease. These include surveillance, monitoring, diagnosis in individual animals, and certification of the BTB status for various purposes including those of import and export.

It has become a habit, and a bad one at that, to use the current ante- and postmortal tests, without validation, in a variety of wildlife species, and to rely on the results based on the extrapolation of cut-off points and responses in various other species. This practice cannot be condoned, although certain importing and exporting countries seem to accept these test results. Allowing this type of testing and accepting the results are high-risk practices, and it emphasizes the need to validate the tests, at least for those wildlife species that are traded in, and for vulnerable species facing extinction.

9.9 Conclusion

It is clear that many challenges remain in the quest to develop diagnostic tests of sufficient Se and Sp that are affordable and rapid, and that will allow resource-limited countries to develop protocols to deal with a disease that is increasingly becoming more complex to control. Some of these factors include the major differences in the epidemiology of the disease in different countries and coinfections that influence the application and interpretation of the available tests. It is critical that combinations of applicable diagnostic tests are available for use in individual countries to enable effective control strategies to be designed.

It is inevitable that the lack of adequate information in resource-poor countries will lead to uncoordinated efforts and the failure of campaigns (Marcotty et al. 2009). Hopefully, soon, there will be ante- and postmortal diagnostic tests that have sufficient Se and Sp and are rapid and affordable to assist resource-limited countries to contribute to the international eradication of tuberculosis, including zoonotic TB, in humans. Until such time, the uncontrolled presence of BTB in under-resourced countries will remain an international threat and an impediment to the global drive to eradicate the disease, because of their inability to adequately diagnose, and thus to control BTB in livestock and wildlife.

References

Abdellrazeq GS, Elnaggar MM, Osman HS et al (2014) Prevalence of bovine tuberculosis in Egyptian cattle and the standardization of the interferon-gamma assay as an ancillary test. Transbound Emerg Dis 63(5):497–507

Ablordey A, Swings J, Hubans C et al (2005) Multilocus variable-number tandem repeat typing of *Mycobacterium ulcerans*. J Clin Microbiol 43(4):1546–1551

Addo KK, Owusu-Darko K, Yeboah-Manu D et al (2007) Mycobacterial species causing pulmonary tuberculosis at the Korle Bu Teaching Hospital, Accra, Ghana. Ghana Med J 41(2):52–57

Adu-Bobi NAK, Mak-Mensah EE, Achel DG (2009) Preliminary investigation of bovine tuberculosis in suspected beef from a metropolitan abattoir in Ghana with Ziehl-Neelsen microscopy. Pak J Biol Sci 12(17):1222–1225

Allix-Béguec C, Harmsen D, Weniger T et al (2008) Evaluation and user-strategy of MIRU-VNTRplus, a multifunctional database for online analysis of genotyping data and phylogenetic identification of *Mycobacterium tuberculosis* complex isolates. J Clin Microbiol 46 (8):2692–2699

Álvarez J, de Juan L, Bezos J et al (2008) Interference of paratuberculosis with the diagnosis of tuberculosis in a goat flock with a natural mixed infection. Vet Microbiol 128:72–80

Ameni G, Aseffa A, Engers H (2007) High prevalence and increased severity of pathology of bovine tuberculosis in Holsteins compared to Zebu breeds under field cattle husbandry in central Ethiopia. Clin Vaccine Immunol 14(10):1356–1361

Anon (1994) Livestock disease eradication evaluation of the cooperative state–federal bovine tuberculosis eradication program. Committee on bovine tuberculosis. National Academy Press, Washington

Anon (2007) Bovine TB: the scientific evidence. Final report of the independent scientific group on cattle TB. Department of Environment, Food and Rural Affairs (DEFRA), London

Asiak IE, Ohare OB, Emikpe BO et al (2007) The use of ELISA in the detection of bovine tuberculosis in slaughtered trade cattle and sedentary herds in south west Nigeria. J Anim Vet Adv 6(7):883–886

Asseged B, Woldesenbet Z, Yimer E et al (2004) Evaluation of abattoir inspection for the diagnosis of *Mycobacterium bovis* infection in cattle at Addis Ababa Abattoir. Trop Anim Health Prod 36:537–546

Atiadeve SK, Gyamfi OK, Mak-Mensah E et al (2014) Slaughter surveillance for tuberculosis among cattle in three metropolitan abattoirs in Ghana. J Vet Med Anim Health 6(7):198–207

Awad FI (1962) Studies on type-determination and epidemiology of tuberculosis among cattle in Sudan. Zentralbl Veterinarmed B 9(9):890–898

Awah-Ndukum J, Kudi AC, Bradley G et al (2010) Prevalence of bovine tuberculosis in abattoirs of the littoral and western highland regions of Cameroon: a cause for public health concern. Vet Med Int. https://doi.org/10.4061/2010/495015

Awah-Ndukum J, Kudi AC, Bah GS et al (2012) Bovine tuberculosis in cattle in the highlands of Cameroon: seroprevalence estimates and rates of tuberculin skin test reactors at modified cut-offs. Vet Med Int. https://doi.org/10.1155/2012/798502

Aylate A, Shah SN, Aleme H et al (2013) Bovine tuberculosis: prevalence and diagnostic efficacy of routine meat inspection procedure in Woldiya municipality abattoir north Wollo zone, Ethiopia. Trop Anim Health Prod 45(3):855–864

Bedard BG, Martin SW, Chinombo D (1993) A prevalence study of bovine tuberculosis and brucellosis in Malawi. Prev Vet Med 16:193–205

Bekele M, Belay I (2011) Evaluation of routine meat inspection procedure to detect bovine tuberculosis suggestive lesions in Jimma municipal abattoir, south West Ethiopia. Glob Vet 6 (2):172–179

Ben Kahla I, Boschiroli ML, Souissi F et al (2011) Isolation and molecular characterisation of *M. bovis* from raw milk in Tunisia. Afr Health Sci 11(S1):S2–S5

Berg S, Firdessa R, Habtamu M et al (2009) The burden of mycobacterial disease in Ethiopian cattle: implications for public health. PLoS One 4(4):e5068. https://doi.org/10.1371/journal.pone.0005068

Berrada J (1993) *Mycobacterium bovis* infection in cattle in Morocco: preparation and evaluation of chemical extracts for use in detection of immune responses. PhD Thesis, Iowa State University

Beyi AF, Gezahegne KZ, Mussa A et al (2014) Prevalence of bovine tuberculosis in dromedary camels and awareness of pastoralists about its zoonotic importance in Eastern Ethiopia. J Vet Med Anim Health 6(4):109–115

Bezos J, Álvarez J, Romero B et al (2012) Tuberculosis in goats: assessment of current in vivo cell-mediated and antibody-based diagnostic assays. Vet J 191:161–165

Bezos J, Casal C, Romero B et al (2014) Current ante-mortem techniques for diagnosis of bovine tuberculosis. Res Vet Sci 97:S44–S52

Bhembe NL, Jaja IF, Nwodo UU et al (2017) Prevalence of tuberculous lymphadenitis in slaughtered cattle in Eastern Cape, South Africa. Int J Infect Dis 61:27–37

Biadglegne F, Tesfaye W, Sack U et al (2013) Tuberculous lymphadenitis in northern Ethiopia: in a public health and microbiological perspectives. PLoS One 8(12):e81918. https://doi.org/10.1371/journal.pone.0081918

Biffa D, Bogale A, Skjerve E (2010) Diagnostic efficiency of abattoir meat inspection service in Ethiopia to detect carcasses infected with *Mycobacterium bovis*: implications for public health. BMC Public Health 10(1):462

Biffa D, Johansen TB, Godfroid J et al (2014) Multi-locus variable-number tandem repeat analysis (MLVA) reveals heterogeneity of *Mycobacterium bovis* strains and multiple genotype infections of cattle in Ethiopia. Infect Genet Evol 23:13–19

Boer MD, Zanden AV, van Soolingen D (2004) Simultaneous detection and typing of *Mycobacterium tuberculosis* complex bacteria. Spoligotyping and TB. Isogen Lifesciences, The Netherlands

Bonsu OA, Laing E, Akanmori BD (2000) Prevalence of tuberculosis in cattle in the Dangme-West district of Ghana, public health implications. Acta Tropica 76:9–14

Brosch R, Gordon SV, Marmiesse M et al (2002) A new evolutionary scenario for the *Mycobacterium tuberculosis* Complex. Proc Natl Acad Sci USA 99(6):3684–3689

Brüns AC, Tanner M, Williams MC et al (2017) Diagnosis and implications of *Mycobacterium bovis* infection in banded mongooses (*Mungos mungo*) in the Kruger National Park, South Africa. J Wildl Dis 53(1):19–29

Brush EF (1898) The association of human and bovine tuberculosis. Wynkoop Hallenbeck Crawford Co. Printers, New York, p 140

Buddle BM, Aldwell FE, Pfeffer A et al (1994) Experimental *Mycobacterium bovis* infection of cattle: effect of dose of *M. bovis* and pregnancy on immune responses and distribution of lesions. N Z Vet J 42(5):167–172

Buddle BM, de Lisle GW, Pfeffer A (1995) Immunological responses and protection against *M. bovis* in calves vaccinated with a low dose of BCG. Vaccine 13(12):1123–1130

Buddle BM, Livingstone PG, de Lisle GW (2009) Advances in ante-mortem diagnosis of tuberculosis in cattle. N Z Vet J 57(4):173–180

Buxton A, Fraser G (1977) *Mycobacterium*. In: Animal microbiology, vol I. Blackwell, Oxford, pp 229–235

Byrne AW, Graham J, Brown C et al (2018) Modelling the variation in skin-test tuberculin reactions, post-mortem lesion counts and case pathology in tuberculosis-exposed cattle: effects of animal characteristics, histories and co-infection. Transbound Emerg Dis 65(3):844–858

Cadmus SIB, Gordon SV, Hewinson RG (2011) Exploring the use of molecular epidemiology to track bovine tuberculosis in Nigeria: an overview from 2002 to 2004. Vet Microbiol 151:133–138

Carmichael J (1937) A brief note on tuberculosis in Tropical Africa with special reference to Uganda. J Cardiovasc Pharmacol Ther 50:383–385

Casal C, Díez-Guerrier A, Álvarez J et al (2014) Strategic use of serology for the diagnosis of bovine tuberculosis after intradermal skin testing. Vet Microbiol 170(3–4):342–351

Cassidy JP (2008) The pathology of bovine tuberculosis: time for an audit. Vet J 176:263–264

Cassidy JP, Bryson DG, Pollock JM et al (1998) Early lesion formation in cattle experimentally infected with *M. bovis*. J Comp Pathol 119:27–44

Collins DM (1999) Molecular epidemiology: *Mycobacterium bovis*. In: Rutledge C, Dale J (eds) Mycobacteria–molecular biology and virulence. Blackwell, Boston, MA, pp 123–135

Collins DM (2011) Advances in molecular diagnostics for *Mycobacterium bovis*. Vet Microbiol 151:2–7

Collins CH, Grange JM (1983) The bovine tubercle bacillus: a review. J Appl Bacteriol 55:13–29

Collins DM, Radford AJ, de Lisle GW et al (1994) Diagnosis and epidemiology of bovine tuberculosis using molecular biological approaches. Vet Microbiol 40:83–94

Corner LA (1994) Post mortem diagnosis of *Mycobacterium bovis* infection in cattle. Vet Microbiol 40(1-2):53–63

Corner LA, Melville L, McCubbin K et al (1990) Efficiency of inspection procedures for the detection of tuberculous lesions in cattle. Aust Vet J 67:389–392

Corner LAL, Gormley E, Pfeiffer DU (2012) Primary isolation of *Mycobacterium bovis* from bovine tissues: conditions for maximizing the number of positive cultures. Vet Microbiol 156:162–171

Courcoul A, Moyen JL, Brugere L et al (2014) Estimation of sensitivity and specificity of bacteriology, histopathology and PCR for the confirmatory diagnosis of bovine tuberculosis using latent class analysis. PLoS One 9(3):e90334

Cousins DV, Florisson N (2005) A review of tests available for use in the diagnosis of tuberculosis in non-bovine species. Rev Sci Tech Off Int Epiz 24(3):1039

Crawshaw TR, Griffiths IB, Clifton-Hadley RS (2008) Comparison of a standard and a detailed postmortem protocol for detecting *Mycobacterium bovis* in badgers. Vet Rec 163(16):473–477

DAFF (2016) Bovine tuberculosis scheme manual (interim). Department of Agriculture, Forestry and Fisheries, Republic of South Africa, pp 80

Damina MS, Owoludun OA, Chukwukere S et al (2011) The use of deletion analysis in the detection of *M. bovis*, *M. tuberculosis* and *M. africanum* among slaughtered cattle in Plateau State, north central Nigeria. Nig Vet J 32(1):9–15

Davidson RM, Alley MR, Beatson NS (1981) Tuberculosis in a flock of sheep. N Z Vet J 29 (1–2):1–2

Deriemer K, Daley CL (2004) The molecular epidemiology of tuberculosis. In: Madkour MM (ed) Tuberculosis. Springer, Berlin, pp 57–74

de Vos V, Bengis RG, Kriek NPJ et al (2001) The epidemiology of tuberculosis in free-ranging African buffalo (*Syncerus caffer*) in the Kruger National Park, South Africa. Onderstepoort J Vet Res 68:119–130

Diguimbaye-Djaibé C, Hilty M, Ngandolo R et al (2006) *Mycobacterium bovis* isolates from tuberculous lesions in Chadian zebu carcasses. Emerg Infect Dis 12(5):769–771

Di Marco V, Mazzone P, Capucchio MT et al (2012) Epidemiological significance of the domestic black pig (*Sus scrofa*) in maintenance of bovine tuberculosis in Sicily. J Clin Microbiol 50 (4):1209–1218

Doherty ML, Bassett HF, Quinn PJ et al (1996) A sequential study of the bovine tuberculin reaction. Immunology 87(1):9

Domingo M, Vidal E, Marco A (2014) Pathology of bovine tuberculosis. Res Vet Sci 97:S20–S29

Dowling LA, Schleehauf SM (1991) Specific antibody responses to *Mycobacterium bovis* in infected cattle analysed with six mycobacterial antigens in enzyme-linked immunosorbent assays. Res Vet Sci 50(2):157–161

Downs SH, Parry JE, Upton PA et al (2017) Methodology and preliminary results of a systematic literature review of ante-mortem and post-mortem diagnostic tests for bovine tuberculosis. Prev Vet Med 153:117–126

Drewe JA, Smith NH (2014) Molecular epidemiology of *M. bovis*. In: Thoen CO, Steele JH, Kaneene JB (eds) Zoonotic tuberculosis: *M. bovis* and other pathogenic mycobacteria, 3rd edn. Wiley, Chichester, pp 79–88

Durnez L, Sadiki H, Katakweba A et al (2009) The prevalence of *Mycobacterium bovis*-infection and atypical mycobacterioses in cattle in and around Morogoro, Tanzania. Trop Anim Health Prod 41:1653–1659

Egbe NF, Muwonge A, Ndip L et al (2017) Molecular epidemiology of *Mycobacterium bovis* in Cameroon. Sci Rep 7(1):4652

Elmossalami E, Siam MA, El Sergany M (1971) Studies on tuberculous-like lesions in slaughtered camels. Zbl Vet Med B 18:253–261

Espie IW, Hlokwe TM, van Pittius NCG et al (2009) Pulmonary infection due to *Mycobacterium bovis* in a black rhinoceros (*Diceros bicornis minor*) in South Africa. J Wildl Dis 45 (4):1187–1193

Etter EMC, Ameni G, Roger FLM (2006) Tuberculosis risk assessment in Ethiopia: safety of meat from cattle slaughtered in abattoirs. In: Proceedings of the 11th international symposium on veterinary epidemiology and economics (ISVEE), Cairns, Australia. www.sciquest.org.nz.

Faksri K, Xia E, Tan JH et al (2016) *In silico* region of difference (RD) analysis of *Mycobacterium tuberculosis* complex from sequence reads using RD-Analyzer. BMC Genomics 17(1):847

Firdessa R, Tschopp R, Wubete A et al (2012) High prevalence of bovine tuberculosis in dairy cattle in central Ethiopia: implications for the dairy industry and public health. PLoS One 7(12): e52851

Fischer EAJ, van Roermunda HJW, Hemerik L et al (2005) Evaluation of surveillance strategies for bovine tuberculosis (*Mycobacterium bovis*) using an individual based epidemiological model. Prev Vet Med 67:283–301

Fitzgerald SD, Kaneene JB (2012) Wildlife reservoirs of bovine tuberculosis worldwide: hosts, pathology, surveillance, and control. Vet Pathol 50(3):488–499

Francis J (1958) Tuberculosis in animals and man. A study in comparative pathology

Gallagher J, Macadam I, Sayer J et al (1972) Pulmonary tuberculosis in free-living lechwe antelope in Zambia. Trop Anim Health Prod 4:204–213

Garnier T, Eiglmeier K, Camus JC et al (2003) The complete genome sequence of *Mycobacterium bovis*. Proc Natl Acad Sci USA 100(13):7877–7882

Gavier-Widén D, Cooke MM, Gallagher J et al (2009) A review of infection of wildlife hosts with *Mycobacterium bovis* and the diagnostic difficulties of the 'no visible lesion' presentation. N Z Vet J 57(3):122–131

Good M, Duignan A (2011) Perspectives on the history of bovine TB and the role of tuberculin in bovine TB eradication. Vet Med Int 2011:410470. p11

Goosen WJ, Miller MA, Chegou NN et al (2014) Agreement between assays of cell-mediated immunity utilizing *Mycobacterium bovis*-specific antigens for the diagnosis of tuberculosis in African buffaloes (*Syncerus caffer*). Vet Immunol Immunopathol 160(1-2):133–138

Gordon SV, Brosch R, Billault A et al (1999) Identification of variable regions in the genomes of tubercle bacilli using bacterial artificial chromosome arrays. Mol Microbiol 32(3):643–655

Gormley E, Doyle MB, Fitzsimons T et al (2006) Diagnosis of *M. bovis* infection in cattle by use of the gamma-interferon (Bovigam®) assay. Vet Microbiol 112:171–179

Gormley E, Doyle M, Duignan A et al (2013) Identification of risk factors associated with disclosure of false positive bovine tuberculosis reactors using the gamma-interferon (IFNγ) assay. Vet Res 44(1):117

Gormley E, Corner LAL, Costello E et al (2014) Bacteriological diagnosis and molecular strain typing of *Mycobacterium bovis* and *Mycobacterium caprae*. Res Vet Sci 97:S30–S43

Grobler DG, Michel AL, De Klerk LM et al (2002) The gamma-interferon test: its usefulness in a bovine tuberculosis survey in African buffaloes (*Syncerus caffer*) in the Kruger National Park. Onderstepoort J Vet Res 69(3):221

Guilbride PDL, Rollinson DHL, McAnulty EG et al (1963) Tuberculosis in the free living African (cape) buffalo (*Syncerus caffer caffer* Sparrman). J Comp Pathol Ther 73:337–348

Gutiérrez M, Tellechea J, Marín JFG (1998) Evaluation of cellular and serological diagnostic tests for the detection of *Mycobacterium bovis*-infected goats. Vet Microbiol 62:281–290

Habarugira G, Rukelibuga J, Nanyingi MO et al (2014) Bovine tuberculosis in Rwanda: prevalence and economic impact evaluation by meat inspection at Société des Abattoirs de Nyabugogo-Nyabugogo Abattoir, Kigali. J S Afr Vet Assoc 85(1):1062. https://doi.org/10.4102/jsava.v85li1.1062

Hang'ombe MB, Munyeme M, Nakajima C et al (2012) *Mycobacterium bovis* infection at the interface between domestic and wild animals in Zambia. BMC Vet Res 8:221

Hauer A, Michelet L, De Cruz K et al (2016) MIRU-VNTR allelic variability depends on *Mycobacterium bovis* clonal group identity. Infect Genet Evol 45:165–169

Hilty M, Diguimbaye C, Schelling E et al (2005) Evaluation of the discriminatory power of variable number tandem repeat (VNTR) typing of *M. bovis* strains. Vet Microbiol 109:217–222

Hines N, Payeur JB, Hoffman LJ (2006) Comparison of the recovery of *Mycobacterium bovis* isolates using the BACTEC MGIT 960 system, BACTEC 460 system, and Middlebrook 7H10 and 7H11 solid media. J Vet Diagn Invest 18:243–250

Hlokwe TM, van Helden P, Michel A (2013) Evaluation of the discriminatory power of variable number of tandem repeat typing of *M. bovis* isolates from southern Africa. Transbound Emerg Dis 60(Suppl 1):111–120

Jenkins AO, Cadmus SIB, Venter EH et al (2011) Molecular epidemiology of human and animal tuberculosis in Ibadan, Southwestern Nigeria. Vet Microbiol 151:139–147

Jenkins AO, Gormley E, Gcebe N et al (2018) Cross reactive immune responses in cattle arising from exposure to *Mycobacterium bovis* and non-tuberculous mycobacteria. Prevent Vet Med 152:16–22

Jolles AE, Cooper DV, Levin SA (2005) Hidden effects of chronic tuberculosis in African buffalo. Ecology 86(9):2258–2264

Jubb KVF, Kennedy PC, Palmer N (1993) Pathology of domestic animals, vol 2. Academic, London, pp 641–652

Kamerbeek J, Schouls L, Kolk AM et al (1997) Simultaneous detection and strain differentiation of *Mycobacterium tuberculosis* for diagnosis and epidemiology. J Clin Microbiol 35(4):907–914
Kaneene JB, Miller RA, Meyer RM (2006) Abattoir surveillance: the U.S. experience. Vet Microbiol 112:273–282
Karolemeas K, de la Rua-Domenech R, Cooper R et al (2012) Estimation of the relative sensitivity of the comparative tuberculin skin test in tuberculous cattle herds subjected to depopulation. PLoS One 7(8):e43217
Katale BZ, Mbugi EV, Karimuribo ED et al (2013) Prevalence and risk factors for infection of bovine tuberculosis in indigenous cattle in the Serengeti ecosystem, Tanzania. BMC Vet Res 9 (1):267
Keet DF, Kriek NP, Huchzermeyer H et al (1994) Advanced tuberculosis in an African buffalo (*Syncerus caffer* Sparrman). J S Afr Vet Assoc 65:79–83
Keet DF, Kriek NP, Penrith ML et al (1996) Tuberculosis in buffaloes (*Syncerus caffer*) in the Kruger National Park: spread of the disease to other species. Onderstepoort J Vet Res 63:239–244
Keet DF, Kriek NPJ, Penrith ML et al (1998) Tuberculosis in free-ranging lions in the Kruger National Park. In: Proceedings of ARC-Onderstepoort OIE international congress on anthrax, brucellosis, contagious bovine pleura-pneumonia, clostridial and mycobacterial diseases, with WHO co-sponsorship. Bergen-Dal, Kruger National Park, South Africa
Keet DF, Kriek NPJ, Bengis RG et al (2000) The rise and fall of tuberculosis in a free-ranging chacma baboon troop in the Kruger National Park. Onderstepoort J Vet Res 67(2):115
Keet DF, Kriek NPJ, Bengis RG (2001) Tuberculosis in kudus (*Tragelaphus strepsiceros*) in the Kruger National Park. Onderstepoort J Vet Res 68(3):225
Keet DF, Michel AL, Bengis RG et al (2010) Intradermal tuberculin testing of wild African lions (*Panthera leo*) naturally exposed to infection with *Mycobacterium bovis*. Vet Microbiol 144 (3–4):384–391
Kinne J, Johnson B, Jahans KL et al (2006) Camel tuberculosis—a case report. Trop Anim Health Prod 38:207–213
Kleeberg HH (1984) Human tuberculosis of bovine origin in relation to public health. Rev Sci Tech Off Int Epiz 3(1):11–32
Koro FK, Foko E, Ngatchou AF et al (2013) First insight into the current prevalence of bovine tuberculosis in cattle slaughtered in Cameroon: the case of main abattoirs of Yaoundé and Douala. Br Microbiol Res J 3(3):272–279
Krasnow I, Wayne LG (1969) Comparison of methods for tuberculosis bacteriology. Appl Microbiol 18(5):915–917
Kriek N, Bengis R, de Vos V et al (1994) The pathology of tuberculosis in buffalo in the Kruger National Park. In: Wildlife ranching: a celebration of diversity: Proceedings of the 3rd international wildlife ranching symposium, October 1992, Pretoria, South Africa, p 170
Kuria JKN, Gathogo SM (2013) Concomitant fungal and *Mycobacterium bovis* infections in beef cattle in Kenya. Onderstepoort J Vet Res 80(1):1–4
Laval G, Ameni G (2004) Prevalence of bovine tuberculosis in zebu cattle under traditional animal husbandry in Boji district of western Ethiopia. Rev Med Vet (Toulouse) 155(10):494–499
Lepper A, Pearson CW, Corner LA (1977) Anergy to tuberculin in beef cattle. Aust Vet J 53 (5):214–216
Leslie LW, Hebert CN (1965) The use of dilute tuberculins for testing cattle. Br Vet J 121 (9):427–436
Laisse CJM, Gavier-Widén D, Ramis R et al (2011) Characterization of tuberculous lesions in naturally infected African buffalo (*Syncerus caffer*). J Vet Diagn Invest 23:1022–1027
Liebana E, Johnson L, Gough J et al (2008) Pathology of naturally occurring bovine tuberculosis in England and Wales. Vet J 176:354–360
Little TWA, Swan C, Thompson HV et al (1982) Bovine tuberculosis in domestic and wild mammals in an area of Dorset. III. The prevalence of tuberculosis in mammals other than badgers and cattle. Epidemiol Infect 89(2):225–234

Lugton IW, Johnstone AC, Morris RS (1995) *Mycobacterium bovis* infection in New Zealand hedgehogs (*Erinaceus europaeus*). N Z Vet J43(7):342–345
Lyashchenko KP, Gortázar C, Miller MA et al (2018) Spectrum of antibody profiles in tuberculous elephants, cervids, and cattle. Vet Microbiol 214:89–92
Maas M, Michel AL, Rutten VPMG (2013) Facts and dilemmas in diagnosis of tuberculosis in wildlife. Comp Immunol Microbiol Infect Dis 36:269–285
Malama S, Johansen TB, Muma JB et al (2014) Characterization of *M. bovis* from humans and cattle in Namwala District, Zambia. Vet Med Int. https://doi.org/10.1155/2014/187842
Mamo G, Bayleyegn G, Tessema TS et al (2011) Pathology of camel tuberculosis and molecular characterization of its causative agents in pastoral regions of Ethiopia. PLoS One 6(1):e15862
Marcotty T, Matthys F, Godfroid J et al (2009) Zoonotic tuberculosis and brucellosis in Africa: neglected zoonoses or minor public-health issues? The outcomes of a multi-disciplinary workshop. Ann Trop Med Parasitol 103(5):401–411
Martin SW (1984) Estimating disease prevalence and the interpretation of screening test results. Prev Vet Med 2:463–472
Martrenchar A, Njanpop BM, Yaya A et al (1993) Problems associated with tuberculosis and brucellosis skin-test methods in northern Cameroon. Prev Vet Med 15:221–229
Mason FE (1917) Tuberculosis in camels. J Comp Pathol Ther 30:80–84
Menin Á, Fleith R, Reck C et al (2013) Asymptomatic cattle naturally infected with *M. bovis* present exacerbated tissue pathology and bacterial dissemination. PLoS One 8(1):e53884. https://doi.org/10.1371/journal.pone.0053884
Michel AL (2008) Tuberculosis in wild and domestic animals in South Africa. PhD thesis, Universiteit Utrecht, The Netherlands
Michel AL, Hlokwe TM, Coetzee ML et al (2008) High *M. bovis* genetic diversity in a low prevalence setting. Vet Microbiol 126:151–159
Michel AL, Cooper D, Jooste J et al (2011) Approaches towards optimising the gamma interferon assay for diagnosing *Mycobacterium bovis* infection in African buffalo (*Syncerus caffer*). Prev Vet Med 98(2-3):142–151
Mikota SK, Peddie L, Peddie J et al (2001) Epidemiology and diagnosis of *Mycobacterium tuberculosis* in captive Asian elephants (*Elephas maximus*). J Zoo Wildl Med 32(1):1–16
Milián-Suazo F, Banda-Ruíz V, Ramírez-Casillas C et al (2002) Genotyping of *M. bovis* by geographic location within Mexico. Prev Vet Med 55:255–264
Miller JM, Jenny AL, Payeur JB (2002) Polymerase chain reaction detection of *Mycobacterium tuberculosis* complex and *Mycobacterium avium* organisms in formalin-fixed tissues from culture-negative ruminants. Vet Microbiol 87(1):15–23
Miller M, Buss P, Hofmeyr J et al (2015) Antemortem diagnosis of *Mycobacterium bovis* infection in free-ranging African lions (*Panthera leo*) and implications for transmission. J Wildl Dis 51 (2):493–497
Miller MA, Buss PE, van Helden PD et al (2017) *Mycobacterium bovis* in a free-ranging black rhinoceros, Kruger National Park, South Africa, 2016. Emerg Infect Dis 23(3):557
Mohamed M, Moussa LM, Mohamed KF et al (2011) BACTEC MGIT 960™ system for screening of *Mycobacterium tuberculosis* complex among cattle. Afr J Biotechnol 10(63):13919–13923
Monaghan ML, Doherty ML, Collins JD et al (1994) The tuberculin test. Vet Microbiol 40:111–124
Morar D, Tijhaar E, Negrea A et al (2007) Cloning, sequencing and expression of white rhinoceros (*Ceratotherium simum*) interferon-gamma (IFN-γ) and the production of rhinoceros IFN-γ specific antibodies. Vet Immunol Immunopathol 115(1–2):146–154
Morar D, Schreuder J, Mény M et al (2013) Towards establishing a rhinoceros-specific interferon-gamma (IFN-γ) assay for diagnosis of tuberculosis. Transbound Emerg Dis:60(s1):60–60 (s1):66
Müller B, Steiner B, Bonfoh B et al (2008) Molecular characterization of *M. bovis* isolated from cattle slaughtered at the Bamako abattoir in Mali. BMC Vet Res 4:26
Müller B, Vounatsou P, Ngandolo BNR et al (2009) Bayesian receiver operating characteristic estimation of multiple tests for diagnosis of bovine tuberculosis in Chadian cattle. PLoS One 4 (12):e8215. https://doi.org/10.1371/journal.pone.0008215

Muma B, Syakalima M, Munyeme M et al (2013) Bovine tuberculosis and brucellosis in traditionally managed livestock in selected districts of southern province of Zambia. Vet Med Int 2013: ID 730367. https://doi.org/10.1155/2013/730367

Muñoz-Mendoza M, Romero B, Cerro AD et al (2016) Sheep as a potential source of bovine TB: epidemiology, pathology and evaluation of diagnostic techniques. Transbound Emerg Dis 63 (6):635–646

Munyeme M, Rigouts L, Shamputa IC et al (2009) Isolation and characterization of *M. bovis* strains from indigenous Zambian cattle using Spacer Oligonucleotide typing technique. BMC Microbiol 9:144. https://doi.org/10.1186/1471-2180-9-144

Munyeme M, Muma JB, Siamudaala VM et al (2010) Tuberculosis in Kafue lechwe antelopes (*Kobus leche kafuensis*) of the Kafue Basin in Zambia. Prev Vet Med 95:305–308

Murray G (1986) Ante-mortem and post-mortem meat inspection: an Australian Inspection Service perspective. Aust Vet J 63(7):211–215

Mwakapuja RS, Makondo ZE, Malakalinga J et al (2013) Molecular characterization of *M. bovis* isolates from pastoral livestock at Mikumi-Selous ecosystem in the eastern Tanzania. Tuberculosis 93:668–674

Neill SD, Bryson DG, Pollock JM (2001) Pathogenesis of tuberculosis in cattle. Tuberculosis 81 (1):79–86

Ngandolo BNR, Müller B, Diguimbaye-Djaïbe C et al (2009) Comparative assessment of fluorescence polarization and tuberculin skin testing for the diagnosis of bovine tuberculosis in Chadian cattle. Prev Vet Med 89:81–89

Njanpop-Lafourcade BM, Inwald J, Ostyn A et al (2001) Molecular typing of *M. bovis* isolates from Cameroon. J Clin Microbiol 39(1):222–227

Norby B, Bartlett PC, Fitzgerald SD et al (2004) The sensitivity of gross necropsy, caudal fold and comparative cervical tests for the diagnosis of bovine tuberculosis. J Vet Diagn Invest 16 (2):126–131

Nuñez-Garcia J, Downs SH, Parry JE et al (2018) Meta-analyses of the sensitivity and specificity of ante-mortem and post-mortem diagnostic tests for bovine tuberculosis in the UK and Ireland. Prev Vet Med 153:94–107

O'Brien DJ, Schmitt SM, Berry DE et al (2008) Estimating the true prevalence of *Mycobacterium bovis* in free-ranging elk in Michigan. J Wildl Dis 44(4):802–810

OIE (2009) Bovine tuberculosis. In: Terrestrial manual. Chapter 2.4.7, p 16

OIE (2017) Manual of diagnostic tests and vaccines for terrestrial animals. http://www.oie.int/standard-setting/terrestrial-manual/access-online/

Olivier TT, Viljoen IM, Hofmeyr J et al (2017) Development of a gene expression assay for the diagnosis of *Mycobacterium bovis* infection in African lions (*Panthera leo*). Transbound Emerg Dis 64(3):774–781

Palmer MV, Waters WR (2006) Advances in bovine tuberculosis diagnosis and pathogenesis: what policy makers need to know. Vet Microbiol 112:181–190

Parsons LM, Brosch R, Cole ST et al (2002) Rapid and simple approach for identification of *Mycobacterium tuberculosis* complex isolates by PCR-based genomic deletion analysis. J Clin Microbiol 40(7):2339–2345

Parsons SD, Cooper D, McCall AJ et al (2011) Modification of the QuantiFERON-TB Gold (In-Tube) assay for the diagnosis of *Mycobacterium bovis* infection in African buffaloes (*Syncerus caffer*). Vet Immunol Immunopathol 142(1–2):113–118

Perla D (1927) Experimental epidemiology of tuberculosis. J Exp Med 45(2):209–226

Pollock JM, McNair J, Bassett H et al (2003) Specific delayed-type hypersensitivity responses to ESAT-6 identify tuberculosis-infected cattle. J Clin Microbiol 41(5):1856–1860

Praud A, Boschiroli ML, Meyer L et al (2015) Assessment of the sensitivity of the gamma-interferon test and the single intradermal comparative cervical test for the diagnosis of bovine tuberculosis under field conditions. Epidemiol Infect 143(1):157–166

Raath JP, Bengis RG, Bush M et al (1995) Diagnosis of tuberculosis due to *Mycobacterium bovis* in the African Buffalo (*Syncerus caffer*) in the Kruger National Park. In: Griffin F, de Lisle G (eds) Tuberculosis in wildlife and domestic animals. University of Otago Press, Dunedin, pp 313–315

Radunz BL, Lepper AW (1985) Suppression of skin reactivity to bovine tuberculin in repeat tests. Aust Vet J 62(6):191–194

Ramadan HH, El-Gohary AHN, Mohamed AA et al (2012) Detection of *Mycobacterium bovis* and *Mycobacterium tuberculosis* from clinical samples by conventional and molecular techniques in Egypt. Glob Vet 9(6):648–654

Ramos DF, Tavares L, da Silva PE (2014) Molecular typing of *Mycobacterium bovis* isolates: a review. Braz J Microbiol 45(2):365–372

Ramos DF, Silva PEA, Dellagostin OA (2015) Diagnosis of bovine tuberculosis: review of main techniques. Braz J Biol 75(4):830–837

Rao KR, Kauser F, Srinivas S et al (2005) Analysis of genomic downsizing on the basis of region-of-difference polymorphism profiling of *Mycobacterium tuberculosis* patient isolates reveals geographic partitioning. J Clin Microbiol 43(12):5978–5982

Razanamparany VR, Quirin R, Rapaoliarijaona A et al (2006) Usefulness of restriction fragment length polymorphism and spoligotyping for epidemiological studies of *M. bovis* in Madagascar: description of new genotypes. Vet Microbiol 114:115–122

Renwick R, White PCL, Bengis RG (2007) Bovine tuberculosis in southern African wildlife: a multi-species host–pathogen system. Epidemiol Infect 135:529–540

Rigouts L, Maregeyat B, Traore H et al (1996) Use of DNA restriction fragment typing in the differentiation of *Mycobacterium tuberculosis* complex isolates from animals and humans in Burundi. Tuber Lung Dis 77:264–268

Ritacco V, López B, De Kantor IN et al (1991) Reciprocal cellular and humoral immune responses in bovine tuberculosis. Res Vet Sci 50:365–367

Ritchie JN (1959) Tuberculosis. In: Stableforth AW, Galloway IA (eds) Diseases due to bacteria, vol 2. Butterworths, London, UK, pp 713–744

Rogers RJ, Donald BA, Schultz K (1980) The distribution of *Mycobacterium bovis* in Queensland cattle herds with observations on the laboratory diagnosis of tuberculosis. Aust Vet J 56 (11):542–546

Rohonczy EB, Balachandran AV, Dukes TW et al (1996) A comparison of gross pathology, histopathology, and mycobacterial culture for the diagnosis of tuberculosis in elk (*Cervus elaphus*). Can J Vet Res 60(2):108

Ryan TJ, Livingstone PG, Ramsey DSL et al (2006) Advances in understanding disease epidemiology and implications for control and eradication of tuberculosis in livestock: the experience from New Zealand. Vet Microbiol 112:211–219

Sahraoui N, Müller B, Guetarni D (2009) Molecular characterization of *M. bovis* strains isolated from cattle slaughtered at two abattoirs in Algeria. BMC Vet Res 5:4

Sahraoui N, Muller B, Mamache B et al (2011) Tuberculosis in cattle and goats in the north of Algeria. Vet Res 4(4):100–103

Sanchez J, Tomás L, Ortega N et al (2011) Microscopical and immunological features of tuberculoid granulomata and cavitary pulmonary tuberculosis in naturally infected goats. J Comp Pathol 145(2-3):107–117

Sanou A, Tarnagda Z, Kanyala E et al (2014) *Mycobacterium bovis* in Burkina Faso: epidemiologic and genetic links between human and cattle isolates. PLoS Negl Trop Dis 8(10):e3142. https://doi.org/10.1371/journal.pntd.0003142

Schiller I, Oesch B, Vordermeier HM et al (2010a) Bovine tuberculosis: a review of current and emerging diagnostic techniques in view of their relevance for disease control and eradication. Transbound Emerg Dis 57(4):205–220

Schiller I, Vordermeier HM, Waters WR et al (2010b) Bovine tuberculosis: effect of the tuberculin skin test on in vitro interferon gamma responses. Vet Immunol Immunopathol 136:1–11

Seva J, Hernández D, Bernabé A et al (2000) Immunophenotypical characterization of the lymphocyte infiltrate in caprine pulmonary tuberculosis. J Comp Pathol 123(2–3):96–103

Skuce RA, Neill SD (2001) Molecular epidemiology of *Mycobacterium bovis*: exploiting molecular data. Tuberculosis 81(1–2):169–175

Smith NH (2012) The global distribution and phylogeography of *M. bovis* clonal complexes. Infect Genet Evol 12:857–865

Smith NH, Upton P (2012) Naming spoligotype patterns for the RD9-deleted lineage of the *Mycobacterium tuberculosis* complex: www.Mbovis.org. Infect Genet Evol 12:873–876

Smith NH, Gordon SV, de la Rua-Domenech R et al (2006) Bottlenecks and broomsticks: the molecular evolution of *Mycobacterium bovis*. Nat Rev Microbiol 4:670–681

Stärk KDC, Alonso S, Dadios N et al (2014) Strengths and weaknesses of meat inspection as a contribution to animal health and welfare surveillance. Food Control 39:154–162

Stewart LD, McCallan L, McNair J et al (2017) Multilaboratory evaluation of a novel lateral flow immunochromatographic assay for confirming isolation of *Mycobacterium* bovis from veterinary diagnostic specimens. J Clin Microbiol 55(12):3411–3425

Sulieman MS, Hamid ME (2002) Identification of acid fast bacteria from caseous lesions in cattle in Sudan. Zoonoses Public Health 49:415–418

Tarnagda Z, Kanyala E, Zingué D et al (2014) Prevalence of tuberculosis in bovine carcasses in two slaughterhouses of Burkina Faso. Int J Microbiol Immunol Res 2(6):92–100

Thom M, Morgan JH, Hope JC et al (2004) The effect of repeated tuberculin skin testing of cattle on immune responses and disease following experimental infection with *M. bovis*. Vet Immunol Immunopathol 102:399–412

Thom M, Howard C, Villarreal-Ramos B et al (2008) Consequence of prior exposure to environmental mycobacteria on BCG vaccination and diagnosis of tuberculosis infection. Tubercle 88:324–334

Thorburn JA, Thomas AD (1940) Tuberculosis in the Cape kudu. J S Afr Vet Med Assoc 11:3–10

Thorel M-F, Huchzermeyer HF, Michel AL (2001) *Mycobacterium avium* and *Mycobacterium intracellulare* infection in mammals. Rev Sci Tech Off Int Epiz 20(1):204–218

Tschopp R, Schelling E, Hattendorf J et al (2010) Repeated cross-sectional skin testing for bovine tuberculosis in cattle kept in a traditional husbandry system in Ethiopia. Vet Rec 167:250–256

Tweddle NE, Livingstone P (1994) Bovine tuberculosis control and eradication programs in Australia and New Zealand. Vet Microbiol 40:23–39

Van der Heijden EMDL, Jenkins AO, Cooper DV et al (2016) Field application of immunoassays for the detection of *Mycobacterium bovis* infection in the African buffalo (*Syncerus caffer*). Vet Immunol Immunopathol 169:68–73

Van Embden JDA, Cave MD, Crawford JT et al (1993) Strain identification of *Mycobacterium tuberculosis* by DNA fingerprinting: recommendations for a standardized methodology. J Clin Microbiol 31(2):406–409

Van Pinxteren LAH, Ravn P, Agger EM et al (2000) Diagnosis of tuberculosis based on the two specific antigens: ESAT-6 and CFP10. Clin Diagn Lab Immunol 7(2):155–160

van Soolingen D, De Haas PEW, Haagsma J et al (1994) Use of various genetic markers in differentiation of *M. bovis* strains from animals and humans and for studying epidemiology of bovine tuberculosis. J Clin Microbiol 32(10):2425–2433

Varello K, Pezzolato M, Mascarino D et al (2008) Comparison of histologic techniques for the diagnosis of bovine tuberculosis in the framework of eradication programs. J Vet Diagn Invest 20(2):164–169

Vekemans M, Cartoux M, Diagbouga S et al (1999) Potential source of human exposure to *M. bovis* in Burkina Faso, in the context of the HIV epidemic. Clin Microbiol Infect 5(10):617–621

Vordermeier HM, Whelan A, Cockle PJ et al (2001) Use of synthetic peptides derived from the antigens ESAT-6 and CFP-10 for differential diagnosis of bovine tuberculosis in cattle. Clin Diagn Lab Immunol 8:571–578

Wangoo A, Johnson L, Gough J et al (2005) Advanced granulomatous lesions in *Mycobacterium bovis*-infected cattle are associated with increased expression of type I procollagen, γδ (WC1+) T cells and CD 68+ cells. J Comp Pathol 133(4):223–234

Warren RM, van Pittius NCG, Barnard M et al (2006) Differentiation of *Mycobacterium tuberculosis* Complex by PCR amplification of genomic regions of difference. Int J Tuberc Lung Dis 10 (7):818–822

Waters WR, Vordermeier HM, Rhodes S et al (2017) Potential for rapid antibody detection to identify tuberculous cattle with non-reactive tuberculin skin test results. BMC Vet Res 13 (1):164

Watrelot-Virieux D, Drevon-Gaillot E, Toussaint Y et al (2006) Comparison of three diagnostic detection methods for tuberculosis in French cattle. Zoonoses Public Health 53(7):321–325

Weber A, Van Hooven W (1992) Tuberculosis of the parotid salivary gland in a kudu *Tragelaphus strepsiceros*. Koedoe 35:119–122

Weniger T, Krawczyk J, Supply P et al (2010) MIRU-VNTRplus: a web tool for polyphasic genotyping of *Mycobacterium tuberculosis* complex bacteria. Nucleic Acids Res 38 (Suppl):326–331

Whelan AO, Clifford D, Upadhyay B et al (2010) Development of a skin test for bovine tuberculosis for differentiating infected from vaccinated animals. J Clin Microbiol 48(9):3176–3181

Whipple DL, Bolin CA, Miller JM (1996) Distribution of lesions in cattle infected with *Mycobacterium bovis*. J Vet Diagn Invest 8(3):351–354

Wood PR, Jones SL (2001) BOVIGAMTM: an in vitro cellular diagnostic test for bovine tuberculosis. Tuberculosis 81(1):147–155

Wood PR, Corner LA, Rothel JS et al (1991) Field comparison of the interferon-gamma assay and the intradermal tuberculin test for the diagnosis of bovine tuberculosis. Aust Vet J 68:286–290

Woodford MH (1982) Tuberculosis in wildlife in the Ruwenzori National Park Uganda (Part I). Trop Anim Health Prod 14:81–88

Zahran RN, El Behiry A, Marzouk E et al (2014) Comparison of LCD array and IS6110-PCR with conventional techniques for detection of *Mycobacterium bovis* isolated from Egyptian cattle and buffaloes. Int J Mycobacteriol 3:197–204

Zumárraga MJ, Martin C, Samper S et al (1999) Usefulness of spoligotyping in molecular epidemiology of *Mycobacterium bovis*-related infections in South America. J Clin Microbiol 37(2):296–303

Chapter 10
The Control of Bovine Tuberculosis in Africa

Asseged B. Dibaba and Nicolaas P. J. Kriek

10.1 Background

There are differing opinions about the need to control bovine tuberculosis (BTB), caused by *Mycobacterium bovis*, in cattle in many parts of the world. In the UK, for instance, the opinion has recently been expressed that the cost of attempting eradication is not justified given the low prevalence of the disease and the availability of pasteurized milk to all its inhabitants that protects them against the threat of contracting zoonotic TB by consuming *M. bovis*-containing milk from diseased cows (Torgerson and Torgerson 2009). Zoonotic TB is considered to be primarily a food-borne disease, and it is probably the major reason for controlling BTB internationally, but transmission of the infection may, under certain circumstances, also take place by inhalation of infected droplets, although this route seems to be of minor importance.

In most African countries, but for different reasons, BTB is also considered to be of minor importance, and veterinary regulatory authorities in many of the countries do not list it as a notifiable disease. The argument, based on the reasons given that may justify this attitude in the UK, does not apply to most of the African countries. The problem in Africa is that BTB occurs at a high prevalence in many of the countries although it may be in a patchy fashion, but most of the countries have insufficient data to assess the situation in cattle and in humans to allow them to make an informed decision about controlling the disease. A large proportion of Africa's inhabitants do not have access to pasteurized milk, or prefer to drink raw or

A. B. Dibaba
Department of Pathobiology, College of Veterinary Medicine, Tuskegee University, Tuskegee, AL, USA
e-mail: adibaba@tuskegee.edu

N. P. J. Kriek (✉)
Department of Paraclinical Sciences, Faculty of Veterinary Science, University of Pretoria, Onderstepoort, South Africa

A. B. Dibaba et al. (eds.), *Tuberculosis in Animals: An African Perspective*,
https://doi.org/10.1007/978-3-030-18690-6_10

fermented milk, and are thus fully at risk of contracting zoonotic TB. The negative impact of the policy of not controlling the disease must also be seen against the background of BTB being listed as a List B disease, considered to be of socioeconomic or public health (zoonotic) importance within countries, and of significance to the international trade of animals and animal products, by the World Organization for Animal Health (OIE) and its 174 member countries.

The eradication or the ability to limit the disease to very low prevalence levels is important for those countries that manage to control the disease. They rely on the multilateral trade policy framework based on the General Agreement on Tariffs and Trade (GATT), the Agreement on the Application of Sanitary and Phytosanitary Measures (SPS), and the Agreement on Technical Barriers to Trade (TBT) of the World Trade Organization (WTO) to facilitate safe trade and to avoid unnecessary trade barriers. The WTO mandated the OIE as the international standard-setting organization to compile the international standards, guidelines, and recommendations related to global animal health with the main purpose of facilitating international trade in terrestrial and aquatic animals and their products. The application of SPS measures avoids the introduction of pathogens via international trade in animals and animal products while at the same time preventing countries from setting up unjustified sanitary barriers to inhibit trade (Brückner 2009).

Because of these reasons, the developed nations instituted campaigns and sustain them to control and eradicate *M. bovis* from their national cattle herd. The success of these programs has been mixed; some of the European countries managed to eradicate the disease, but even in the developed world, some experience difficulties to contain and eradicate BTB, usually because of the presence of a wildlife maintenance host in the same ecosystem. What is important is that even after reducing the prevalence of BTB, and following the introduction of pasteurization with the consequent reduction in the risk of contracting zoonotic TB, the disease continues to cause production losses in cattle when poorly controlled.

Designing and applying control programs for BTB pose various challenges, many of them dependent on the diversity of people and ecological zones on the continent. Africa is often spoken of as if it were a single country. It is the second largest of the continents, and its 54 states are located in tropical and subtropical ecozones that include tropical, moist and dry forests, subtropical humid and dry forests, subtropical steppe, tropical and subtropical mountain systems, and deserts (FAO 2018). The composition of its people is as diverse, with different social norms, standards, age-old traditions, and development that vary from fairly advanced to some of the poorest countries on the globe. This influences the way in which people live and conduct their farming practices and assess the value of the livestock that they own. Cattle in certain cultures are not often slaughtered but are kept for prestige, milk, draft power, dowry, and savings to offset crop failure. The ethnic groups of Africa number in the thousands, each generally with its own language (or dialect of a language) and culture. The major ethnolinguistic groups include the Afro-Asiatic, Khoisan, Niger-Congo, and Nilo-Saharan populations. This diversity of factors makes it impossible to construct a single protocol for the control and eradication of BTB from the continent, since many of them have an impact on the epidemiology of the disease and the way in which it can be practically controlled (Zinsstag et al. 2008).

The ability to control the disease, in the final analysis, is conceivably mostly dependent on the stage of development of a particular country and the availability of competent human and sufficient financial resources that can be allocated for the control of *M. bovis* infections. It is likely that in the future this will be the main factor that will determine the success of attempts to control or eradicate the disease from various countries on the African continent. Currently, the information available about the distribution and prevalence of bovine and zoonotic TB is, with a few exceptions, limited, and the figures quoted are often speculative and based on fragmented and incomplete datasets. The general lack of information and the false assumption that BTB is not a problem, either in livestock or in humans in Africa, are often used to justify the decision not to control the disease. An additional argument is that under extensive animal husbandry systems prevailing in most of Africa where animals constantly live in the open, close contact between animals is reduced, the spread of *M. bovis* tends to be slow, and the prevalence of BTB remains at a low endemic level (Benkirane 1998). This attitude is probably also the cause of the general apathy of cattle owners to deal with BTB (and any other disease) that does not cause regular visible losses. Consequently, veterinary activities on the continent focus on the rapidly fatal, economically important, trans-boundary diseases such as rinderpest (in the past), CBPP, and East Coast fever (Carmichael 1937; Awad 1962).

The internationally accepted method of successfully controlling and eradicating BTB is based on the test-and-slaughter approach following the detection usually of positive animals by the tuberculin skin test that is based on the assessment of the local dermal immunological response to the intradermal injection of PPD. This appears to be a simple matter, but it is an extremely tedious and expensive process. In the USA, reducing the prevalence to 0.001% in the national herd took 50 years at a cost of US$450 million (Gilsdorf et al. 2006). The successful control of the disease is also dependent on having access to adequate and sustainable funding, adequate veterinary diagnostic laboratories, and competent human resources, such as veterinary technicians and veterinarians, and acceptance by the policymakers, stockowners, and consumers of dairy products that the control of BTB is important and to the benefit of society as a whole.

This chapter deals with the various issues and the challenges to devise practical control programs for Africa that can be sustained with limited financial resources that would benefit the health and well-being of its inhabitants and the productivity of its livestock.

10.2 Reasons for Controlling BTB

The presence of BTB is not only a livestock issue, but it has an, often unquantified, impact on human health, wildlife, income generated by international trade in livestock and livestock products, tourism and its various related activities, ecosystems, and the national economy. Depending on the status of BTB in domestic and wild animals, countries at various stages of development have different reasons for attempting to eradicate BTB. In Australia and Ireland, for example, an animal market restriction was the driving force (Collins 2006). Thus, although BTB control enhances productivity, the desire for disease freedom for trade purposes is the primary driving force for eradication in many developed countries. In line with this, many countries and international organizations such as the OIE, have the requirement for countries to provide scientific evidence of disease status. This is likely to put additional pressure on countries for disease eradication and for continued disease surveillance, particularly those that do not do any surveillance for the presence of BTB and its zoonotic implication, as is the case in many of the African countries.

There are three principal reasons, addressing both animal and human health, for controlling BTB in cattle:

- The financial loss and societal impact caused by the decreased productivity (meat and milk), reproductivity, and replacement costs of BTB-infected animals, in addition to the inability to participate in the lucrative international trade in livestock and dairy products.
- The zoonotic health risk for humans.
- The risk of infecting other livestock and wildlife with the consequent impact on ecosystems and the development of additional maintenance hosts for *M. bovis*.

Many of these matters are dealt with in other chapters in this book in more detail, and only the highlights of each will be given as background for the rationale of controlling BTB.

10.2.1 Economic Importance of Bovine Tuberculosis

Bovine TB is a chronic debilitating disease, characterized by a progressive loss of body condition, reduced milk production, lower reproductive rates, and losses caused by reduced feed utilization, a decrease in the average productive age, a reduced market value due to poor body condition, condemnation of carcasses or portions of carcasses at abattoirs of BTB-infected cattle, and the additional processing costs following condemnation at abattoirs. The presence of BTB in the national herd also prevents participation in international trade in cattle, milk, and milk-derived products. There is an attempt by the WTO to reduce trade restrictions, such as using certain diseases like BTB as trade barriers, to increase globalization and the integration of markets with decreasing border control, as is the case in Western Africa. All these efforts will put additional strain on countries with limited resources to manage livestock diseases.

There are few publications detailing the actual cost of the consequences of a BTB infection in a country and the benefits that accrue when the disease is dealt with successfully. Some data are available with emphasis on the cost/benefit ratios of eradication programs. A good example of the accrued benefits of an eradication program is in the USA where the control campaign is considered to have been both an economic and animal health success. When the cooperative State-Federal Eradication Program was initiated in 1917 in the USA, about 5% of the cattle population was tuberculous, and about 50,000 carcasses were annually condemned in abattoirs because of BTB (Ranney 1960). From its initiation until 1959, the focus was on single animal testing and, from then on, on abattoir surveillance. It takes a long time for the prevalence to decline, and in the USA, it took 40 years for it to decline from 1% of the cattle population to 0.1%. The benefits obtained by this reduction are the costs not experienced because of the reduction in the number of infected animals, the so-called foregone costs. These losses include on-farm losses (reduced milk production, decreased reproductive capacity, culling, and replacement costs) and slaughter costs in terms of condemnation of BTB-positive cattle detected at the abattoir. The calculated benefit amounts to about US$2 billion per annum. In the event that the control scheme was not introduced in 1917, and depending on the herd prevalence, the prevalence of BTB in the national herd could have increased to between 30 and 60%. Depending on a number of variables, the estimated economic return on investment of the eradication program in the USA varies from US$13 to US$55 billion (Gilsdorf et al. 2006). In Canada, the cost/benefit ratio was calculated to be 1:33 following the implementation of the herd program. In other countries, such as Ireland, the benefit exceeded the costs by 85%, while in the UK the costs exceeded the benefit, probably because of the complication of the presence of a wildlife maintenance host in the ecosystem that sustains the infection in the cattle population almost indefinitely or until such time that an effective vaccine is produced (Power and Watts 1987).

With respect to a more detailed analysis of the costs, according to some estimates, the overall productive efficiency of BTB-infected cows may be reduced by 10–25% and the milk production by 10–12%; and sterility of tuberculous cows increases by 5–10% due to the involvement of the reproductive organs (DAFF 2016). There is a decrease of the average weight of cattle with detectable BTB-like lesions of 14 kg compared to healthy carcasses in Niger, while in Madagascar, Malagasy zebu carcasses with gross visible lesions weighed 3.1–9.7 kg less than those without lesions (Boukary et al. 2012). In Morocco the annual direct and indirect losses due to BTB were estimated at $44,260,411 (Berrada 1993). Ejeh et al. (2014) estimated the direct economic loss attributable to BTB in cattle slaughtered in Makurdi abattoirs, Nigeria. Out of 61,654 cattle slaughtered from 2008 to 2012, 1172 (1.9%) were positive for BTB. Although there was no record of whole carcass condemnation due to BTB lesions, a total of 1935 organs, weighing 3046.5 kg and valued at 2.9×10^6 Naira (1.8×10^4 US$), were condemned during the study period. In Cameroon, where a total of 466,816 slaughtered cattle were inspected over a 9-year period (1995–2003), tuberculous lesions were detected in 0.2–0.8% of the animals inspected (depending on the area). Approximately 48.8–81.5% of carcass condemned in the abattoirs included in the study was due to BTB (Awah-Ndukum et al. 2010).

It is probably impossible to quantify the economic losses attributable to BTB in Africa at this stage, but there is no reason to believe that the nature and extent of the losses will be different to those in other countries across the globe. An additional factor on the continent is that the disease also causes an indirect loss in agricultural productivity because of the reduction in animal traction power caused by the debilitation that is characteristic of cattle suffering from BTB. Some data exist, and in Central Ethiopia it was estimated that the milk yield was reduced by 5–13%, the number of services per conception increased from 1.25 to 2.02, and the number of milking days was reduced from 328 days in BTB-negative to 294 days in BTB-positive cows (Ameni et al. 2010).

A common argument, which may be a gross generalization, against the implementation of BTB eradication schemes in Africa, is that a large proportion of the cattle herd is managed under extensive, free-ranging conditions that limit the spread of the diseases. Consequently, should BTB be present, its prevalence will be low and remain so, and it thus does not have a significant impact on the animals or pose a public health risk. The situation, however, is changing rapidly. Intensive livestock improvement schemes that include the importation of improved European breeds of cattle are being implemented in several African countries. An increasing risk too lies in the rapid urbanization driven by poverty and public neglect across the whole of Africa and the establishment of small-scale dairy farms in peri-urban and urban areas to satisfy the increasing demand for milk. These farming practices differ markedly from the extensive transhumant rural management systems, and it is inevitable under these circumstances that there will be a substantial increase in the prevalence of BTB. A good example of this pattern is found in Nigeria where the prevalence of BTB increased from 0.3% in 1976 to 7.3% in 2003, primarily due to the intensification of animal husbandry, the importation of foreign breeds of cattle, and the lack of preventive measures to control BTB (Ofukwu et al. 2008). These changes, that are likely to accelerate with the expected increasing urbanization in Africa, are causes for concern, and it is imperative that there should be a reappraisal of the general approach that the disease in Africa is not important enough to warrant its control.

10.2.2 Public Health Importance of BTB

During the 1900s zoonotic tuberculosis was highly prevalent in some of the countries of the developed world such as in the UK and Germany. It is unlikely that it will ever be known how many people actually died from *M. bovis* infection during the course of time. It is clear though that it was a major problem at times and that it had a substantial impact on human health. In the early 1900s in Great Britain, the proportion of human cases of TB caused by *M. bovis* varied from 5 to 30% (Cousins 2001), and in 1917, zoonotic *M. bovis* infections were responsible for 15,000 deaths in the USA (Palmer 2013). Nevertheless, the nationwide BTB eradication programs and the pasteurization of milk successfully decreased the incidence of *M. bovis* infection in humans in these countries to negligible levels (Palmer and Waters 2011). There is no doubt though that human TB caused by *M. bovis* still occurs in developed countries (Collins and Grange 1983; Grange and Collins 1987) but that it is of minor importance.

The protection of consumers from contracting milk-borne zoonotic diseases, such as tuberculosis, brucellosis, and others, is the main reason for the introduction and the general use of pasteurization. This process primarily, and not the eradication of BTB from cattle, led to the drastic reduction of the occurrence of zoonotic tuberculosis in humans in those countries where processed milk is consumed. But, even in those developed countries where the prevalence of BTB is low, there is still a significant risk of *M. bovis* infection particularly with on-farm consumption of unpasteurized cows' milk, the sale of unpasteurized milk and dairy products, and exposure to infectious aerosols from tuberculous animals and their carcasses (de la Rua-Domenech 2006).

With the exception of South Africa and Namibia where the commercial dairy industry is on par with those in the developed world, and where pasteurized milk is available to almost the entire population, the situation in Africa is unclear and complex. That pasteurized milk is widely available in South Africa and Namibia is the likely reason that in those areas in which TB is rife, no or very few cases of *M. bovis* infection have been detected in humans (see Chap. 3 for further information).

The information for the rest of the continent is limited and fragmented. In Africa, routine laboratory human TB diagnostics do not differentiate between *M. bovis* and *M. tuberculosis* infections, and it is not possible to evaluate the contribution of *M. bovis* to the current TB epidemic in humans. Where the situation was investigated, zoonotic TB occurred, albeit in varying, and in many instances, at a low prevalence. For instance in Tanzania, where BTB is widespread in livestock, *M. bovis* has been diagnosed in humans (Roug et al. 2014), and in Ethiopia it was isolated from pastoralists (Gumi et al. 2012) and from the consumers and producers of dairy products (Kidane et al. 2002). Generally, researchers, quoting the results of various studies to substantiate this view, conclude that *M. bovis*, also in Africa, does not contribute significantly to the number of human TB cases. One such example is that only four sputum samples collected from 984 tuberculous patients yielded *M. bovis* (Firdessa et al. 2013). Similar results were obtained in Ethiopia (Berg et al. 2015), which is globally ranked seventh in respect to its TB burden, with close to 400 TB cases per 100,000 of the population (WHO 2013). Several other African studies, however, confirmed human TB caused by *M. bovis* (Cleaveland et al. 2005; Kazwala et al. 2001; Kidane et al. 2002; Oloya et al. 2008). In countries such as Ethiopia, where 80% of the population owns livestock (CSA 2011) and 82% of the milk supplied to consumers is unpasteurized (Ameni and Erkihun 2007), it is unlikely that zoonotic TB will not be present. In addition, in central Ethiopia, 83% of farmers keep their shoats inside their house at night (Tschopp et al. 2011), and because of this there is a continuous risk of aerosol exposure to *M. bovis* even in rural areas due to this close association between humans and their livestock.

To resolve the uncertainties about the prevalence and distribution of the disease, studies on TB due to *M. bovis* in both cattle and humans should be accorded high priority, particularly in those communities across Africa that live in known BTB hot spots where the occurrence of zoonotic infections can be expected. It is important again to emphasize that in events where BTB establishes itself in dairy herds, the disease poses a serious public health risk, especially for children, through consumption of *M. bovis*-containing milk and where there is a strong physical bond between humans and their livestock in the pastoral African husbandry settings with exposure to airborne *M. bovis*-containing aerosols.

Playing down the importance and potential effect of zoonotic tuberculosis in Africa may have serious consequences in the future if the disease in cattle is left uncontrolled. The risk of becoming infected with zoonotic TB is dependent on the prevalence of the disease in cattle. With a low TB prevalence, *M. bovis* is often considered to be the most common cause of extra-pulmonary TB (EPTB) in humans that is mainly acquired from drinking milk from tuberculous cattle (Dankner and Davis 2000). However, when TB is highly prevalent in the general population (as in Ethiopia), the majority of EPTB cases are caused by *M. tuberculosis* (Kidane et al. 2002). It is thus necessary to understand the epidemiological role of the pattern of occurrence and prevalence of TB (LoBue et al. 2003) before drawing any firm conclusions from the few isolated studies that are available and are used to create the impression of the limited importance of BTB and zoonotic TB in Africa. These influence policymakers and opportunistically convince them that there is no need to control the infection in either humans or in animals (Beyene et al. 2009; Byarugaba et al. 2009).

10.2.3 The Role of Other Domestic and Wild Animals

Domesticated Animals *Mycobacterium bovis* is known to infect a large number of other domestic and wild animals. The list of domestic species in which BTB has been diagnosed includes goats (Hiko and Agga 2011), sheep (Opuda-Asibo 1995), camels (Elmossalami et al. 1971), and pigs (Muwonge et al. 2012).

Pastoralists and smallholder crop–livestock farming communities in Africa commonly practice mixed livestock rearing systems, and close contact between cattle and other livestock species is common. When BTB is highly prevalent in cattle, the disease is likely to spread to other species when intermingling occurs. The detection of identical *M. bovis* strains in cattle, pigs, and humans along a cattle corridor in Uganda corroborates this assumption (Muwonge et al. 2012). A similar situation exists in Kenya where the prevalence of BTB in camels supplying milk to Nairobi was up to 4.48% (Lamuka et al. 2018). With a few exceptions, the infection in all domesticated species throughout Africa has been epidemiologically linked with infected cattle.

No link between BTB infection in cattle herds and flocks of small ruminants in southern Ethiopia could be established (Tschopp et al. 2011), and the risk of spread between species appears to be limited under those specific circumstances. It is clear, however, that the possibility of becoming infected is area-dependent. In the Somali pastoralist areas of southeastern Ethiopia, a BTB prevalence of 2, 0.4, and 0.2%, respectively, in cattle, camels, and goats and a herd/flock prevalence of 14.3, 3.1, and 2.9% were recorded (Gumi et al. 2012). It is expected, in general, that owing to the low prevalence of BTB in extensive animal husbandry systems, the mixing of various domestic species does not seem to play a significant role in the epidemiology of BTB because of the limited direct contact between species. This suggests that minimum intervention, but with the implementation of biosecurity measures and segregation, would be sufficient to effectively limit the risk of interspecies transmission.

Wildlife During the course of time, it became clear that *M. bovis*, being a multihost pathogen, has the ability to infect a range of animals, including wildlife in the same ecosystem, thus creating an epidemiologically complex situation in which bi- or multidirectional transmission of the disease can occur. It also became clear, should one of the infected species become a maintenance host, that it is impossible with the current management techniques at our disposal to control and eradicate the infection (Santos et al. 2015). This has been the experience in countries with wildlife reservoirs such as possums (*Trichosurus vulpecula*) in New Zealand, Eurasian badgers (*Meles meles*) in the UK and Ireland, cervids in North America, and wild ungulates, mostly wild boar (*Sus scrofa*), and red deer (*Cervus elaphus*) on Continental Europe.

The complexity at the interface varies in that in certain countries a two-host situation exists, while in others a multihost system in which multiple infected species and more than one maintenance host may exist. The role of each of these wildlife species in the overall dynamics of the disease is different due to differences in the persistence of the bacteria in each of the hosts, and behavioral differences (Pilosof et al. 2017). In these situations susceptible species communities are often composed of both domestic and wildlife hosts that include species other than the main reservoir species. The environment itself might contribute to the complexity by maintaining viable mycobacteria of the *M. tuberculosis* complex (MTC) group in water or soil, thus further complicating the epidemiology of the disease at these interfaces (Gortázar et al. 2015).

Transmission of the infection in ecosystems can be direct or indirect. Direct transmission requires close contact and is expected to play a major role in intraspecific transmission, whereas close contact between individuals of different species is usually rare and indirect routes such as predation and environmental contamination are more likely to be important. The way in which transmission takes place is difficult to determine, but environmental contamination of watering and feeding areas may play a major role in the transmission of the disease such as between white-tailed deer and cattle in North America, badgers and cattle in the UK and Ireland, and wild ungulates and cattle in the Iberian Peninsula. In these countries, in particular, it has been impossible to eradicate BTB from cattle, even after the schemes have been in operation for a hundred years or more (Santos et al. 2015).

Although it has been known for almost a century that various species of African wildlife are susceptible to, and become infected with *M. bovis* (Paine and Martinaglia 1929) (see Chap. 5), their role in the epidemiology and control of the disease has largely been ignored by regulatory authorities and researchers. It is only during the last couple of decades, and against the background of the role of wildlife maintenance hosts in sustaining *M. bovis* infections in cattle in some of the developed countries, that their potential role in the epidemiology of the disease and its control was recognized and prompted a reappraisal of control measures by some of the African countries. This situation currently prevails in a number of eastern and southern African countries known for their diversity of wildlife. In these areas, African buffaloes (*Syncerus caffer*) in South Africa, Uganda, Tanzania, Mozambique, and Zimbabwe (Woodford 1982; Bengis et al. 2001; Kalema-Zikusoka et al. 2005), greater kudus (*Tragelaphus strepsiceros*) in South Africa and Tanzania, and Kafue lechwe (*Kobus leche kafuensis*) on the Kafue flats in Zambia (Gallagher et al. 1972) are known to be maintenance hosts. The risk of transmission from wildlife is particularly high at the wildlife–livestock interface in these countries where bi-directional transmission of the disease has been confirmed. In South Africa in the Kruger National Park (KNP) and Hluhluwe–iMfolozi Park, infection was introduced into the ecosystem through direct contact between cattle and buffaloes. The molecular characterization of 189 *M. bovis* isolates from the two parks showed that the respective epidemics were each caused by a single, but unrelated, *M. bovis* genotype (Michel et al. 2009). What is of major concern under these circumstances is that bovine tuberculosis and other diseases can spread between buffalo populations across national parks, community land, and countries and thus pose a risk to animal and human health in surrounding wildlife and farming areas (Caron et al. 2016).

Limited information about the role of wildlife in the epidemiology of BTB in Africa exists, but prevalence rates of bovine tuberculosis of up to 50% at herd level are known to occur in Zambia where cattle and Kafue lechwe share grazing and water (Admassu et al. 2015). In South Africa too, transmission of the infection from infected buffaloes to cattle of subsistence farmers at the wildlife–livestock interface has been established, and it is clear that infected wildlife in the conservation areas constitute a risk factor for bovine tuberculosis infection of neighboring cattle, even when the livestock and wildlife are separated by well-maintained disease-control fences. There are growing concerns about the increasing spatial distribution of BTB in South Africa and its spread to an increasing number of wildlife species (Musoke et al. 2015). There is also evidence of clonal expansion of some ancestral strains and of coinfections with two or three *M. bovis* strains on some of the South African cattle and game farms, which suggested independent introductions of the infection from epidemiologically unrelated sources (Hlokwe et al. 2014). The common element in many of these wildlife hosts appears to be close family groups that spread the disease within the species, and environmental or feed contamination that result in the dissemination of the disease to other wildlife species (Fitzgerald and Kaneene 2013).

It is important to keep in mind that environmental factors also play a role in the dissemination of the disease in these extended and complex ecosystems. In certain infected areas, 55.8% of the water points tested positive for MTC in mud samples, while 8.9% of them were positive in the case of water samples. A higher percentage of MTC-positive samples were found at the smallest waterholes and where cachectic animals, assumed to be in the advanced stage of the disease and actively shedding large numbers of mycobacteria, were identified (Barasona et al. 2017).

There have been a few attempts more recently to detect BTB in wildlife species in other areas of Africa. No infection was found in Ethiopia (Tschopp et al. 2010) or in the Okavango Delta of Northern Botswana (Jori et al. 2013). The extent of the infection in wildlife in most of the African countries is currently unknown, and the importance of the infection in them remains a matter of conjecture. Though cattle remain the major source of infection for other domestic and wild animals (and humans) in many parts of Africa, the role of wildlife should not be ignored because of the lack of information to support such a decision.

Against the background of the known role of infected wildlife hosts in sustaining the infection and interfering with its control and eradication, surveillance programs in various African countries should include wildlife to allow the design of successful strategies to contain the disease. The situation is bound to be different for each individual country, and even within countries, and extrapolation of data between countries may prove to be costly and without benefit. Increasing degrees of diversity should be expected to increase the complexity of the problem, and these factors should be kept in mind when formulating disease-control strategies to manage cattle in ecosystems characterized by seasonally limited resources and intense wildlife–livestock interactions (Sintayehu et al. 2017).

Experience with the presence of wildlife maintenance hosts of BTB in many countries has shown that while *M. bovis* can be more easily controlled when the disease is limited to livestock species, it is impossible to eradicate once it has spread into ecosystems with established free-ranging wildlife maintenance hosts (Miller and Sweeney 2013). Controlling BTB in wildlife is a major challenge and at this stage almost impossible. Decreasing population densities of wildlife reservoirs of BTB and improving biosecurity to prevent interaction between wild and domestic animals, help to control the lateral spread of the disease, but these methods alone are insufficient to eradicate the disease. Other methods such as focal depopulation of wildlife species, if those species are not endangered or highly valued by the public, are additional strategies that can be employed. It is for these reasons that the South African authorities are attempting to control the movement of African buffaloes in the wildlife ranching sector across the country and from known BTB-infected conservation areas in an attempt to limit the spread of the disease. Against the background of the extensive spread of the disease to other wildlife species in the country, only focusing on buffaloes may not be sufficient to limit the role of BTB-infected wildlife.

There are two examples of countries that managed to or are in the process of eradicating BTB in the presence of wildlife hosts: Australia and New Zealand. Their successes contrast vividly with those of the UK in which the infection is increasing both in the wildlife hosts and in cattle herds. Regardless of the methods employed, the control of bovine tuberculosis once it establishes itself in a wildlife maintenance host population is generally a long-term commitment (Fitzgerald and Kaneene 2013). Given the experience in those countries in which the disease has been complicated by the presence of a wildlife reservoir, there is no doubt that the African regulatory authorities, when designing control programs without adequate surveillance for the presence of the disease in wildlife, will, at their peril, ignore the role of wildlife in the epidemiology of BTB. The reasons for the successful control of the disease while dealing with wildlife maintenance hosts are discussed in detail by Livingstone et al. (2015), and the principles applied may be used as guidelines in countries with similar epidemiological patterns of the disease.

10.2.4 The Challenges of Dealing with BTB in Africa

There is no quick and easy way to control and eradicate BTB. The test-and-slaughter method used by the developed countries in accordance with the recommendations of the OIE forms the basis of all the successfully completed programs that resulted in reducing the prevalence of the disease to very low levels or in its eradication. Conducting such a campaign is a cumbersome process requiring a substantial investment of financial resources, full cooperation of the farming community, access to adequate diagnostic facilities, and competent veterinary human resources. This is where the problem lies with most of the African countries should they consider to attempt to control and eradicate BTB. It is recognized that these programs are exceedingly expensive and thus not easily affordable, and cannot be applied in many parts of the world.

Most African countries do not have the resources to implement and sustain the conventional test-and-slaughter program, and doing nothing remains the status quo (Grange et al. 1994). However, because of the increasing urbanization and need for milk, and the increasing small-scale farming in urban and peri-urban areas, in addition to the intensification caused by the implementation of livestock improvement schemes in several African countries, the dynamics and epidemiology of BTB are rapidly changing on the continent, and the problem of BTB is likely to increase in magnitude. There is a perception by most of the veterinary authorities in Africa that although they are aware that BTB and zoonotic TB are present, they are not important. With the limited financial resources and usually inadequate number of veterinary personnel to manage a wide array of animal diseases under unfavorable conditions, they have no option but to focus on more economically important transboundary diseases and those that are rapidly fatal and cause many acute deaths in domestic animals and humans (Awad 1962). Most of the countries do not participate in the international trade with livestock and their products, and their lack of participation in these markets is a further reason why they are not too concerned about the presence of the disease and its limitation on participating in international trade in livestock and their products.

The consequences of doing nothing have been addressed and modeled a number of times. One such study modeled the consequences of various BTB control strategies in Tanzania, including the prospect of doing nothing. The introduction of one infected animal into a herd of 40 animals where no control is applied, will cause the proportion of infected animals to increase to a maximum of 11.3% after

10 years, with an average of 6.7% (Roug et al. 2014). Entertaining this option spells eventual disaster for human and animal health, and control programs in Africa with the current free cross-border movements of cattle, such as in Western Africa according to their trade agreement. Unlimited cross-border movement may become an increasing practice because of the political efforts to create free trading across Africa (BBC News 2015).

To resolve these issues, as there is no alternative, an adaption and phased implementation of permutations of the current test-and-slaughter programs should be considered. There is a number of matters that must be addressed: detection, surveillance, risk factor analysis, mapping and calculating the rate of spread, increasing the knowledge and attitude of both farmers and policymakers, improving the competency of veterinary and abattoir officials, and reducing human exposure by reducing the consumption of raw milk, and reducing the direct contact between humans and potentially infected livestock. It has further been suggested that population-based molecular markers of infectious pathogens be used to assist with developing targeted control strategies in resource-limited settings. These methods focus on the unique host population-driven configurations of pathogen genotype clustering, cluster-size distribution, diversity and spatial distribution to infer contact activity, potential transmission events, and most importantly to identify hot spots of disease on which to focus when setting priority areas in which to commence a control program (Egbe et al. 2017). These data should allow a structured approach to allocate limited resources to deal with the problem, rather than to do nothing.

Following surveillance, and even for surveillance purposes, countries or regions can be subdivided into zones or compartments allowing the control of BTB on a piecemeal and geographic basis. This allows regulatory authorities to design, implement, and manage BTB control programs that have different objectives as a consequence of varying epidemiological, economic, social, administrative, and legal factors. Based on these concepts, BTB can be eradicated from parts of a state or country, to allow subpopulations of animals that are kept under different management or husbandry practices to reach disease-free status independent of the disease status of others. This approach is not without its challenges since the authorities will have to prevent inter- and intraspecific transmission of the infection from neighboring areas. Dealing with wildlife maintenance hosts is another matter and, at this stage, is exceedingly difficult if not impossible. Until the epidemiology and ecology of BTB in the maintenance hosts are understood, and the way in which transmission occurs from the wildlife host to livestock, devising successful control programs remains in the realms of impossibility (Livingstone et al. 2015).

In Africa, it is also necessary too to deal with the lack of the political will to implement the program against the background of poverty, cultural beliefs and customs, illiteracy, high population growth, urbanization, the lack of public awareness of the zoonotic implications of BTB, the ongoing social disruption and displacement of people and their livestock within and between countries. The biggest problem, perhaps, is the lack of information about the disease that does not allow the authorities to adequately assess its importance, nor to develop a control program that will be executable given the limitations that exist in most of the countries.

The role of livestock in the social structure of many tribes in Africa creates marked resistance to the implementation of eradication programs particularly in those instances where cattle do not appear to be visibly ill. As livestock are highly valued by pastoralists, culling an animal suffering from BTB that does not cause acute clinical signs is perceived as a draconian measure and is experienced as a devastating blow to individual farmers with both emotional and financial implications (Drewe et al. 2014). The remoteness of many of the cattle herds in rural areas that are not accessible by road, and do not have the infrastructure to handle cattle for testing purposes, is a major impediment, as is the requirement for animals after 72 hours to return to a gathering point for the test to be evaluated. Most livestock keepers do not have the monetary means to privately finance their participation in these programs and will not be able to participate without substantial governmental support (Roug et al. 2014). With the exception of few of the countries, no African government is able to do so without substantial international agency and NGO financial support.

10.3 Bovine TB Control Strategies

From an African perspective and keeping the various limitations in mind, there is a number of permutations that could be considered, singly or in combination, to control the disease, as well as to limit the risk of zoonotic infections. These include applying or modifying the test-and-slaughter approach, application of biosecurity measures, increasing pasteurization, and, hopefully in the not too distant future, vaccination.

10.3.1 Test-and-Slaughter (TS) Policy

Conventionally, BTB control programs depend on the diagnosis of BTB in cattle through whole-herd testing, slaughtering of test-positive cattle, and quarantine of infected herds. This process is repeated until the disease has finally been eradicated. In many countries, this program initially focused on a voluntary herd testing scheme, followed by area testing, in which all cattle in a specified area were subjected to tuberculin testing at intervals of less than a year (Morris et al. 1994). The program in the USA was initially confined to herds of purebred cattle under the Accredited Herd Plan in which a certificate was issued to herd owners when all cattle in the herd passed repeated tuberculin tests over a period of at least 1 year (Ranney 1960). When a sufficient number of herds were accredited, herd testing was extended into area testing, in which all cattle owners in a county allowed their herds to be tuberculin tested. Cattle positive to the tuberculin test were shipped to slaughter, and herds from which infected animals were removed were placed under movement restriction. The herds were then retested after prescribed periods, until no further reactors were detected or, depending on the number of infected animals in a herd, were totally depopulated. The testing was repeated sequentially at specified intervals until a herd, an area, or a state had no infected animals and could be declared BTB-free.

Focusing on the early detection and slaughter of clinical cases of BTB to remove the sources of infection from within the herd, thereby decreasing the likelihood of in-contact animals to be readily infected, is an important approach. Quarantining an infected herd prevents disease spread beyond that herd. Once the test-and-slaughter programs commenced, the incidence of clinical cases of BTB rapidly declined, reversing the year-on-year increase in the incidence of bovine TB and halting its geographical expansion. In the USA, when the national campaign began in 1917, about 5% of the US cattle population was tuberculous. By 1940, only 0.5% of cattle reacted to the skin test (Martin 1994). It thus became clear that BTB could be eradicated by the use of the tuberculin test to identify infected herds and to clear those herds from infection. As a result, several countries commenced BTB eradication programs, with more or less similar success.

In the 1930s, an estimated 15–20% of the national cattle herd in the UK was infected with *M. bovis* (Hope and Villarreal-Ramos 2008). The application of the test-and-slaughter policy in high-incidence areas in the 1950s, followed by the nationwide application of control measures in the 1960s, reduced the incidence of BTB to 0.5% in the national herd in 1979. In Spain, the prevalence of TB in cattle was similarly reduced from 1.4 to 0.3% between 1996 and 2005 (Naranjo et al. 2008), although further reduction has proven difficult (Martín-Hernando et al. 2007). In Denmark (Magnus 1966) and Switzerland (Fritsche et al. 2004), where nationwide, centrally directed eradication programs began in the early 1930s, BTB in cattle was eradicated as long ago as in the 1950s.

The BTB eradication programs in many of the developed countries are regarded as being among the most effective control measures ever mounted against any bacterial disease (Grange et al. 1994). After its initiation about 100 years ago, routine, whole-herd tuberculin testing is still being performed in the UK (and many other countries) at intervals of 1, 2, 3, or 4 years, depending on the BTB risk in the area (Reilly and Courtenay 2007). Iran is a good example of the consequences of interrupting an eradication program. In 1967, it implemented the European style test-and-slaughter program that was then compulsory for modern and semi-modern cattle farms and for traditional farms located at their perimeters. Due to various sociopolitical events, the program was interrupted between 1978 and 1981. The 4-year lapse resulted in an increase in the prevalence of BTB from 0.4% in 1977 to 2.7% in 1984 (Tadayon et al. 2013).

A prerequisite of the test-and-slaughter program and the subsequent control of the disease is that, unless executed meticulously and attention given to the various risk factors that favor recurrence, the disease commonly returns. In one instance, about one-third of the *M. bovis*-infected herds were still positive after the disease was deemed to be eradicated (Roswurm and Ranney 1973) mainly as a consequence of the incorrect performance and interpretation of the tuberculin skin test (Morris et al. 1994). In all the countries that used this form of control, cases of TB detected at slaughter are followed up, and thorough epidemiologic investigations are conducted to determine the source and potential further dissemination of the infection (Wahlström and Englund 2006; Drewe et al. 2014; McCluskey et al. 2014). To limit the risk of recurrence, and because of the low sensitivity and specificity when used as an individual-animal test, depopulation of herds with a very high prevalence is favored over test-and-slaughter for the eradication of BTB (Martin 1994).

10.3.1.1 The Current Status of BTB Control Programs in Africa

Bovine TB is poorly controlled in Africa, where economic and sociopolitical factors largely determine the degree to which the disease can be controlled in individual countries. The extent of the current control programs in Africa allows the interventions in various countries to be roughly grouped into three categories.

Control Programs on Par with those in the Developed World South Africa is one among a few countries in Africa in which a comprehensive control program has been implemented and sustained for some time. Regulations to control the disease were included in the Cape Act (No. 16 of 1906) that replaced the Animal Diseases Act of 1893 and made provision for extensive tuberculin testing and the payment of partial compensation for cattle destroyed that were diagnosed with BTB. These measures were later incorporated in the subsequent Union of South Africa's Disease of Stock Act in 1911. To deal with the increasing prevalence of the disease in commercial dairy herds, the South African state veterinary authorities in 1969 introduced a comprehensive National Tuberculosis Scheme under the Animal Diseases and Parasites Act (Act 13 of 1956) targeting BTB in commercial cattle herds. The campaign resulted in the reduction of the prevalence of BTB from 11.9% in 1971 to 0.4% in 1995 (DAFF 2017). The regulations dealing with BTB were later incorporated in the Animal Diseases Act (Act 35 of 1984) with the aim of eradicating BTB from the national cattle herd. However, in spite of the application of control measures and provisions for compulsory notification of a diagnosis of BTB in herds, compulsory testing of herds in which the disease was suspected to be present, and application of control measures such as quarantine, slaughtering of infected animals, INH (Isoniazid, also known as isonicotinylhydrazide) treatment, and the disinfection of premises to contain and eradicate the disease, the disease was never eradicated. Following South Africa's independence in 1994, and the provincialization of veterinary services, the quality of the control decreased because of a reduction in funding and the variable application of the norms and standards contained in the regulations under the Act by the different provinces. Currently, BTB is present at an unknown prevalence in all nine provinces of South Africa (Michel et al. 2008). Act 7 of 2002 (the Animal Health Act) was promulgated more recently and replaced Act 35 of 1984. In addition, because the breeding and sale of African buffaloes, a local wildlife maintenance host of BTB, became a particularly lucrative business, the Directorate of Animal Health of the Department of Agriculture, Forestry and Fisheries currently strictly controls all buffalo movements in the country (DAFF 2017). These measures are contained in the veterinary procedural notice for buffalo disease risk management in South Africa (also referred to as the Buffalo Veterinary Procedural Notice) regulating the keeping of African buffaloes in South Africa (DAFF 2017).

A comprehensive draft manual outlining the various current programs to deal with the disease in South Africa has been accepted for implementation (DAFF 2016). Participation in the program is voluntary, except that when a herd is found to be infected, it is placed under quarantine until such time that the disease has been eradicated from it. A deficiency in the control program is that it still focuses primarily on commercial farming enterprises; cattle in the informal sector, comprising about 50% of cattle in the country, are only tested on an ad hoc basis. In addition, in terms of wildlife, it only addresses the issue of BTB in buffaloes, while a number of other wildlife species, and greater kudus in particular, are infected and may act as maintenance hosts. Kudus that are known to have been infected for about a century have the ability to cross fences and other barriers with ease, may act as maintenance hosts, and can transmit the disease to other species as has happened on a number of occasions. More than 20 different wildlife species are now known to be infected in South Africa, while only the movement of buffaloes is regulated.

Apart from Namibia and Botswana, which both have lucrative meat export markets, and successfully control BTB, the Kingdom of Morocco is the only other country that is relatively well prepared to control BTB. With technical assistance from the FAO, Morocco has strengthened its organizational and technical capacity to conduct surveillance and to control BTB. Laboratory equipment and reagents for BTB diagnosis were supplied, and technical personnel and field veterinarians received adequate training on laboratory and field diagnostics of the disease in cattle. Following that, a national BTB survey was conducted. Advocacy to enhance farmer awareness and to build trust are key elements in the list of their project objectives. With the herd prevalence at a staggering 32%, Morocco is now in a position to launch an effective BTB control/eradication program (Berrada 2006).

Second-Tier Countries These include Algeria (Sahraoui et al. 2009), Egypt (Zahran et al. 2014), Madagascar (Quirin et al. 2001), Mozambique (Moiane et al. 2014), Tanzania (Roug et al. 2014), and Tunisia (Ben Kahla et al. 2011). In these countries, BTB is a notifiable disease, and it is officially controlled by the test-and-slaughter method, but only in registered dairy herds. Unfortunately, animals found positive are not routinely sent for slaughter because of a lack of funds to compensate the owners for their losses following the condemnation of carcasses in abattoirs. Individual animals and herds are not identified, and the movement of cattle cannot be controlled, with the consequence that the programs have a negligible impact on the prevalence and spread of the disease. In, at least, Mozambique and Tanzania, wildlife reservoirs of the infection, which make control even more complex, exist.

Third-Tier Countries The remaining African countries, notably Nigeria (Ejeh et al. 2014), Cameroon (Awah-Ndukum et al. 2012), Burkina Faso (Coulibaly and Yameogo 2000), Niger (Boukary et al. 2012), Uganda (Bernard et al. (2005), Zambia (Malama et al. 2013), and many others, have a statutory policy for the control of BTB, which is either not implemented or very poorly applied. For example, in Nigeria, the control of BTB is the responsibility of the Federal Ministry of Agriculture and Rural Development, which administers the Animal Diseases (Control) Decree of 1988 (Abubakar et al. 2011), but because of political and socioeconomic constraints, and a lack of awareness, it is inadequately implemented. The only meaningful control measure applied in these countries is limited to meat inspection in the abattoirs in which tuberculous carcasses are fully or partially condemned depending on the extent of the BTB-like lesion detected during inspection. Positive cases are not followed up and no epidemiological investigations are conducted. The detection of infected carcasses at abattoir inspection is more of a public health measure than an attempt to control BTB in the national cattle herd. Owners are also not compensated for carcasses that are condemned, with the result that carcasses with generalized BTB are often declared fit for consumption (Quirin et al. 2001; Asiimwe et al. 2009). In many of these countries, existing meat and milk inspection practices are totally inadequate to protect consumers from consuming *M. bovis*-infected meat and milk (Ismaila et al. 2015). The general level of competency is reflected by the observation that "these reports were based on slaughterhouse scanty records written by unqualified and nonchalant/uncommitted meat inspectors. . ." (Okoro et al. 2014).

10.3.1.2 Legislation and Policy

In addition to knowing what the BTB status of a country is, including its distribution, prevalence, and an understanding of the epidemiology of the disease, there must be a sound policy and effective legislation to control it. Weak enforcement of a veterinary policy is usually the consequence of a lack of capacity, poor advocacy, and the lack of sound evidence of the nature and importance of BTB. Awareness of the potentially harmful effects is vitally important, not only for the communities that face daily exposure to them but also for the policy actors and key decision-makers that play a role in preventing the emergence and spread of BTB as it relates to human health, agriculture, and development (Okello and Welburn 2014).

Control programs should be compiled against the background of the specific patterns of the disease in each country, in which the situation may vary substantially and requires information to better inform future testing choices, private biosecurity investments, and vector population control. Unless there is strict enforcement of, and meticulous adherence to, the rules and regulations of all the components, programs are bound to fail, becoming a process of merely going through the motions, and destined to become a complete waste of human and financial resources without any benefit to society.

The lack of policymakers to institute control mechanisms for BTB is often caused by a lack of an understanding of the disease and not knowing what the economic impact is of its presence. There is a disparity in the way in which policymakers and veterinarians approach the control of a disease. Economists focus on increasing human welfare, irrespective of whether the disease is reduced or the farmers' welfare is improved. They focus on how the disease affects total economic welfare, considering a variety of economic relationships, the effect of changing incentives, and the implementation of new policies, including animal health policies. For economists, animals are important but are simply part of the production environment. These issues must be addressed by the veterinary authorities if policymakers are to be convinced that there is a need to allocate resources for the control of BTB (Jarvis and Valdes-Donoso 2018).

An integrated, national approach is considered to be essential to effectively control diseases such as BTB in a country, and the successful implementation of BTB control programs depends on the availability of a legal framework allowing enforcement of the control measures (Cousins 2001). It is imperative to understand that the effective control of BTB relies on a high level of competence, requiring continuing education and periodic quality control. The central management should provide detailed technical and administrative guidelines and protocols, and adequate supervision and quality control while conducting the program. The central coordinating authority should also set standards and formulate definitions and rules that are practical at all levels to allow the program to be executed.

This implies that the state veterinary services of each country should establish the objectives and goals of BTB control programs according to its specific situation and limitations. These goals and objectives should include the control and activity levels required, the financial needs, and the level of training for field, abattoir, and laboratory personnel, and the responsibility of each stakeholder. In addition, high-quality administrative support is critical for BTB control programs, including the availability of legal procedures to enforce the prescribed activities, particularly the performance of the intradermal tests that is probably one of the most critical steps in the control process, and that supervised slaughtering and meat inspection are correctly done (Moda 2006). These aspects are often insufficiently monitored and prone to result in major differences in the efficiency of the procedures and the reliability of their results that will inevitably determine the degree of success and the cost-effectiveness of the program.

10.3.2 Designing Tuberculin Testing and Control Programs

By the nature of the disease, there is no quick and easy way in which to deal with BTB that will allow its control and eradication from African countries. A number of basic principles apply when using test-and-slaughter programs to control BTB in a country.

Knowledge of the extent, the spread, and the epidemiological features of BTB are essential for designing specific control strategies tailored for the needs of the specific country. In the event that it is not possible to immediately launch a nationwide eradication program, an alternative route could be followed. This would entail focusing on those areas and farming enterprises in which the prevalence of BTB is high and of importance. The first step should thus be to conduct a national survey on which to base these decisions before commencing the program. If this is not possible, the focus should be on those areas that are conducive to the establishment, maintenance, and spread of the disease. It is then advisable to categorize herds and locations based on the BTB risks that they pose and to focus on those that are deemed to be of the greatest importance. The high-risk enterprises include intensive dairy farms and urban and peri-urban smallholder farms, and only once BTB control has been achieved in them, to move on to the known BTB-infected pastoral herds (Asseged et al. 2014).

In many countries, it may not be practicable to test the entire national cattle population. In these countries, the biggest challenge will be to achieve BTB eradication in pastoral areas. Activities in the remote rural areas are hampered by an inadequate road transport system and poor infrastructure, making it difficult, if not impossible, to restrain the animals to do the test. There is also a general apathy of the pastoral stockowners to participate in the testing program and, without compensation, the removal of animals suffering from a disease that they have never heard of or of which they do not comprehend the impact on cattle and humans. Even in South Africa with its more structured approach, it is difficult to convince rural subsistence farmers to return their cattle to a gathering point after three days for assessment and interpretation of the test results.

It is critical too to understand that merely testing cattle and removing reactors do not constitute a control or eradication program. This process includes numerous other activities based on the One Health approach and keeping in mind that eradicating BTB from the cattle herds is probably not the main objective for dealing with the disease but that the primary objective is to control zoonotic TB and to contribute to the eradication of TB in humans.

10.3.2.1 Variations of the Test-and-Slaughter Program

The BTB control and eradication programs implemented successfully in the developed countries were all based on the test-and-slaughter approach and related activities. Conducting this type of campaign in developed countries with a sophisticated infrastructure and commercially oriented farms remains a cumbersome process. It is, however, facilitated by the existence of adequate animal identification systems, functional diagnostic laboratories, competent meat inspectors at abattoirs, and the ability to trace the origin of reactors, to limit uncontrolled movement of infected animals and those from infected herds. These practices enable their regulatory veterinary services to conduct these campaigns with relative ease and the confidence that the system will work because of the commitment of stockowners to cooperate in the eradication of the disease. One could use their programs as templates for Africa, but given the many limitations that exist on the continent, it may be prudent to use a template designed by an African country with extensive experience in conducting an eradication scheme. South Africa has been controlling BTB from the beginning of the 1900s and actively from 1969, and its current program framework can be used to give some guidance to other African countries when designing their control and eradication programs. (For more detail on the history of the South African programs, see Chap. 19.)

The points made in the Introduction in the South African BTB Eradication Manual are pertinent to other countries in Africa, and the following extract from it will provide some indication of the logic behind the different programs and how they are integrated (DAFF 2016):

> The Bovine Tuberculosis Scheme was officially introduced by the Division of Veterinary Services during 1969 with the purpose of eradicating the disease in the Republic of South Africa.
>
> At the initial stage it was already realized that the need for testing for tuberculosis (TB) in herds varied and that it was largely determined by the type of farming animal husbandry being conducted. In order to incorporate as many stock owners in the scheme as possible, the testing procedures and the various testing programmes were adapted through the years. Such modifications also necessitated changes in the administrative processes and consequent adaptations to the manual.
>
> The various testing programmes as set out in the manual should be regarded as appropriate means of achieving the final objective of the programmes: the total eradication of bTB. In order to reach this goal, thorough planning and careful use of available funds and manpower would be required. Extensive knowledge of the scheme as presented in this manual is a prerequisite for sensible planning and execution of the programme.
>
> The scheme begins with comprehensive guidance to stock owners who wish to join and for his herd to be assigned to the correct testing programme. The previous accreditation programme has been discontinued and replaced by the present scheme. Thus, stud and dairy herds should be incorporated into the Maintenance (old Annual Diagnostic) Programme as many of these require a declaration for the sale of stud animals or milk. Where there is no need for a TB-free declaration or where such a declaration is not of much value to the owner, the incorporation should be into the Surveillance Programme and not the Maintenance Programme. A single negative herd test under the Surveillance Programme makes it possible to test many more herds to track down bovine TB and thereby achieving the final objective sooner.

The various programs are dependent on the use of the tuberculin tests and structured to aid the detection, control, combating, and eventual eradication of BTB. For this purpose, DAFF use the following programs to control BTB in South Africa:

1. Maintenance herd program.
2. Surveillance herd program.
3. Diagnostic testing program (individual animals).
4. Infected herd program.

In South Africa, with the exception of the infected herd program, it is voluntary to join a program. However, if infection is established at the first or later tests, the herd is incorporated in the infected herd program, and although the owner voluntarily participated in the other programs, he/she cannot withdraw in an attempt to avoid further tests and the slaughter of positive reactors. Tests and further action in these herds for the eradication of the infection are compulsory and enforced in terms of the Animal Diseases Act, 1984 (Act No. 35 of 1984).

Only the broad outlines of the various programs will be provided, and the relevant detail, including the test procedures, is available in the detailed TB Control Manual (DAFF 2016).

Maintenance Herd Program This program incorporates all herds that require negative certification. Identification of individual animals in these herds is not required, but the owner may identify his cattle at his own cost. Following the first test, cattle with suspect reactions to the tuberculin test are identified permanently for further tests. Should there be cattle that react positively, the herd is incorporated in the infected herd program in which all the animals are then individually identified by ear tags.

New herds, in which case the initial tests can be done by state officials free of charge, that enter the program must undergo two consecutive negative tests at an interval of not less than 3 months and not more than 6 months, after which a declaration is issued that states that the animals tested negative for tuberculosis on the specific dates of the test. Thereafter all animals of all sexes above the age of 12 months are retested every 2 years.

It is the onus of the stockowner to keep his herd free from tuberculosis during the interim period by not allowing it to become infected, for example, by purchasing cattle with an unknown BTB status or by contact with known or suspected BTB-infected animals.

Surveillance program This program is structured to establish the prevalence of tuberculosis in a certain area and is conducted as a survey at the state's expense and executed by government officials. Tests in this program should preferably be undertaken on an organized basis, such as testing a whole municipal area until the tuberculosis status of the whole municipality is eventually established.

This program can also accommodate stockowners, at their own expense, that do not want to join the maintenance program but are anxious to determine the tuberculosis status of their herds. For this purpose, all cattle of all sexes older than 18 months belonging to all the owners on a specific farm (considered to be an epidemiological unit) are subjected to the single intradermal tuberculin test.

In this program, a herd is only tested once except when suspect reactors are detected in which case the whole herd is again tested after 3 months, or the owner may elect to incorporate his herd in the maintenance program. Cattle with suspect reactions to the tuberculin test are identified with ear tags for further tests. Should positive reactors be found with the tuberculin test, the herd is incorporated in the infected herd program.

Following a single reactor-free surveillance test, stockowners should be advised and motivated to keep their herds free from tuberculosis by means of good management practices, such as purchasing cattle from herds in the maintenance program or having cattle tested for BTB prior to purchasing them and bringing them onto the property.

Diagnostic Herd Program The diagnostic herd program deals with cattle destined for export and for cattle that are at a quarantine station following importation. They are subject to a comparative test with bovine and avian tuberculin done by the officer in charge of the quarantine station. Cattle for export are usually tested at the owner's expense—regardless of whether the test is conducted by a private or a state veterinarian.

These tests are also conducted when an owner wishes to test one or more of the cattle in a herd that they suspect to be suffering from BTB. Private or state veterinarians do these tests at the stockowner's expense. Positive reactors found during the test automatically result in the herd being treated in accordance with the requirements of the infected herd program.

Infected Herd Program All herds in which a BTB infection has been diagnosed either during meat inspection, milk examination, postmortem examination, or clinically, but especially when positive tuberculin tests have been conducted, are regarded as infected and are placed under quarantine. The necessary steps are then taken to eradicate the infection in the herd and then to keep the herd BTB-free. The manual provides a detailed exposition of the way in which infected herds must be dealt with but stresses that it must be regarded as a guideline as circumstances in each herd will differ and that each herd must thus be treated on its own merits. The activities required in the infected herd program include:

1. Duties of the farmers.
2. Duties of the state veterinarians.
3. Control over the BTB-positive herds.
4. Testing procedures.
5. Dealing with positive reactors.
6. Cleaning and disinfection.
7. Branding of reactors.
8. Quarantine.
9. Permits.
10. Arrangement for dealing with reactors: immediate slaughtering, trial slaughtering, and postponed slaughtering.
11. Milk for human consumption.
12. Clinical cases.

10.3.2.2 Surveillance

Surveillance is a critical initial step in initiating control and eventual eradication of BTB in any country and, even in resource-poor countries, should be executed with as much precision as can be afforded. There are recognized standard procedures for livestock that are mainly dependent on the use of tuberculin testing or abattoir records (with its known limitations).

The process in wildlife is much more complex and usually consists of three independent processes: passive scanning surveillance on hunted wildlife; passive surveillance on animals found dead (road kills), moribund, or with abnormal behavior; and active surveillance on hunted or culled animals. Surveillance, and the reliability of the results, in wildlife species is further complicated by unknown geographic dispersion of populations, unknown population densities, difficulties in observing and collecting animals, and the nonrandom selection of the most accessible animals leading to potential sampling biases. These activities may be supplemented by active surveillance based on ante-mortal blood testing collected when animals are immobilized for various reasons. The detection of tuberculosis-like lesions in the field has a low sensitivity, and the prevalence of BTB in wildlife species is usually underestimated or not detected (Fitzgerald and Kaneene 2013; Rivière et al. 2015).

10.3.2.3 Movement Restriction and Biosecurity

The imposition of movement restrictions on tuberculin reactors assumes that cattle-to-cattle transmission of BTB is of critical importance. Movement restriction to prevent disease dissemination is an essential component of any BTB control strategy, and it can be employed to broadly control the spread of the disease in countries where resources are limited and comprehensive control programs cannot be implemented. There are good examples that this strategy can be used successfully. In the USA, biosecurity and movement restrictions which commenced as early as 1900 to prevent both the entry of infected animals from Europe and their local dissemination played a pivotal role in eliminating BTB on a regional basis (Good and Duignan 2011). Mathematical models similarly highlighted the value of movement restrictions. As an example, in Great Britain, the R_0 for between-herd spread of BTB in cattle is 1.1 (Cox et al. 2005). That means there would be 11 new infected herds for each 100 existing infected herds, emphasizing that restricting movement of infected or high-risk herds is critical for the control of BTB.

To facilitate the initiation of control programs, attempts should be made to perform pre-movement testing of all animals from herds in high-risk areas (except when they go directly to slaughter). This would aid in the identification of pockets of infection that may exist and prevent the dissemination of large numbers of infected cattle. Hence as a strategy, when initiating the control of BTB in African countries, cattle movement should be regulated by zoning the specific countries or regions into low- and high-risk areas and by prohibiting animal movement from the high- to the low-risk areas (Bourne 2007). This should have a beneficial effect in Africa where there are high (urban/peri-urban)- and low (pastoral)-risk areas and intensive, high- and extensive, low-risk herds. Given the insufficient collaboration at regional and national levels, the lack of quarantine and border control, and the substantial illegal movement across national borders, this strategy is currently bound to be almost impossible to implement and manage unless there is a drastic change of mind and commitment to control BTB and a changed approach to the control of BTB in those regions (FAO 2012).

Test-and-slaughter is not only a practical approach but also is more economical than other BTB control measures (Roswurm and Ranney 1973). Nevertheless, by applying the broad principles of biosecurity, it would be possible to reduce the risk of cattle becoming infected. It is critical that the policy of test-and-slaughter be vigorously applied at all levels and that BTB-positive animals are quarantined and slaughtered as there are many instances where reactors are knowingly sold on the open market, contributing to the further dissemination of the infection (Berrada 1993). Imposing these measures will inevitably require a degree of discipline and compliance that is currently absent from many African countries.

10.3.2.4 Public Awareness

The prevailing ignorance and reluctance to acknowledge the importance of BTB and zoonotic TB is one of the biggest challenges in Africa that will have to be addressed before embarking on control and eradication campaigns if there is to be any hope of success. There is a well-documented lack of knowledge across the continent of the presence and the role of BTB in the health and welfare of its human and animal populations (Munyeme et al. 2010; Tamiru et al. 2013; Tebug et al. 2014; Kidane et al. 2015; Kazoora et al. 2016).

Confinement, poor hygiene, and poor replacement schemes increase the risk of BTB. Therefore, farmers' education, regarding the nature of BTB, is critical, since most sanitary measures enforced by law are likely to fail if the people do not understand and support them. Hence, commonsense cleanliness of cow sheds, floors, feeding, and water troughs, and maintaining a closed herd at all times are important activities if BTB is to be controlled.

Public awareness campaigns and sensitization of farmers and the general public are key components in the overall strategy when attempting to control zoonotic TB. Specific measures include general hygiene and boiling of milk before consumption. A simulation model in Tanzania (Roug et al. 2014) indicated that boiling of milk substantially contributed to reducing human exposure, especially early in the control regimen. Therefore, regulatory authorities in different countries should focus on the implementation of pasteurization or some form of heat treatment of milk to control zoonotic TB even when they lack the ability to control the disease in cattle.

10.4 Vaccination

The Independent Scientific Committee in Great Britain concluded that the development of a vaccine for cattle would be the best option for the long-term control of BTB (Bourne 2007). Access to an effective vaccine and more reliable diagnostic tests, particularly in wildlife (Wilson et al. 2011; Buddle et al. 2011), are required to effectively control BTB (Buddle et al. 1995). In developing countries, vaccination, once there is an effective vaccine, could be used to reduce the prevalence of BTB to a point where implementation of the test-and-slaughter strategy would be economically feasible (Milian-Suazo et al. 2003). It should perhaps be stressed that vaccination in itself would be insufficient to eradicate BTB, but would merely be an additional mechanism to reduce the prevalence of the disease, thus allowing easier elimination by the test-and-slaughter approach.

The existing vaccine, BCG, is an attenuated strain derived from serial passage of *M. bovis* on glycerol-soaked potato slices by Calmette and Guerin in 1921 for use as a live vaccine to prevent the establishment of tuberculous lesions in *M. bovis*-challenged calves. Its effectiveness in cattle is disappointing and inconsistent (Berggren 1977; Berggren 1981; Griffin 2000; McNair et al. 2007; Hope and Villarreal-Ramos 2008), and it induces skin reactivity following vaccination that does not allow differentiation between naturally infected cattle and vaccinates (Moodie 1977). The level of protection varies significantly across studies (Waters et al. 2014), it being protective in some populations, but not others (Martin 1994). In Malawi, BCG vaccination to control the disease was attempted for a period of 8 years, but its use was unsuccessful and discontinued, and eventually there were no significant differences in the prevalence of BTB in the vaccinated and unvaccinated cattle (Ellwood and Waddington 1972; Berggren 1981).

The development of an effective vaccine for different species has been a challenging and elusive process, and the ability to develop a new vaccine with an effect superior to that of BCG seems to be as difficult if not more so than the ongoing search for effective therapeutic compounds with which to treat the disease. Although the current vaccine candidates may reduce the development and size of lesions and/or bacterial loads of infected organs, there is still no vaccine candidate that can induce sterile immunity against BTB (Waters et al. 2012).

One of the major problems with the current vaccines is that they induce an immunological reaction that makes it impossible to differentiate between naturally infected animals and vaccinates. Currently, any vaccination strategy that incorporates BCG is not recommended in countries where control or trade based on such testing are in operation (OIE 2009). An additional future challenge for the development of an effective vaccine and its use in cattle and wildlife species thus is the development of a diagnostic test that can differentiate between infected and vaccinated animals (the so-called DIVA test). Vordermeier et al. (2014) reviewed the recent advances in the development of new vaccines and the differentiation between vaccinated and naturally infected cattle.

Funding around the globe involved in the control and eradication of BTB is currently focused on the development of wildlife vaccines. As is the case in cattle, the protective effect of BCG vaccination differs substantially in various wildlife species in which it was tested. Vaccination significantly reduced the incidence of positive responses to a serological test that is correlated with the extent and severity of BTB infection in badgers (Wilson et al. 2011), and in wild boar oral re-vaccination with BCG yielded a strong protective response against challenge with a field strain of *M. bovis* (Gortazar et al. 2014). The vaccine efficacy exceeded 95% in possums (Tompkins et al. 2009), but in a controlled study, BCG vaccination had no protective effect in African buffaloes under semi-free-ranging conditions

(Michel 2008). In buffaloes, modeling indicated that annual vaccination of more than 70% of the population would be needed to reduce the herd prevalence to less than 1%, assuming a lifelong protection. It was hence concluded that, although vaccination may provide a means of controlling BTB, it cannot be used as a stand-alone strategy and should be combined with other control measures if eradication is the objective (Cross and Getz 2006).

10.5 Chemotherapy

Treatment with INH was used on a limited scale in the past. Its use is not recommended since treatment of food animals with drugs used in the treatment of human tuberculosis creates the problem of the presence of drug residues in meat, milk, and other consumable products, thus aiding the development of drug-resistant strains of pathogenic mycobacteria. Because of the potential for the development of drug resistance (FAO 2012), and the long period (4–12 months of daily dosing) needed for chemotherapy in livestock for BTB (Dean et al. 2008), its use is contraindicated from a public health perspective (Martin 1994).

10.6 Conclusion

Controlling and eradicating BTB, and eradicating zoonotic TB, are not easy tasks. Given the nature of the disease, and the known difficulties of eradicating it under certain environmental conditions, particularly in the presence of wildlife maintenance hosts, control and eradication programs require adequate financial resources, commitment, and perseverance over a matter of decades. Whether the African countries will be able to do this within the context of the roadmap for the global eradication of TB (WHO 2017) is a moot point and an objective that at this stage appears not to be attainable. International human and animal health agencies, and NGOs involved in the eradication of the disease in humans, should take cognizance of these challenges and limitations, and they will have to provide substantial support to the countries on the African continent should they want to have any hope of globally eradicating the disease in humans successfully.

The following actions are considered to be critical when countries initiate control programs (Admassu et al. 2015):

1. Public health information campaigns are needed to raise community awareness about the risk of TB transmission through consumption of raw/undercooked milk and meat.
2. Organize regular capacity building through in-service training for both professionals (technical training) and zoonotic diseases awareness training for nonprofessional personnel working in abattoirs.

3. Standardization of abattoir inspection protocols, enhanced training and proficiency testing of meat inspections, and raising public awareness are recommended as essential and cost-effective interventions to improve meat inspection services, with subsequent protection of consumers' health.
4. Due attention should be given to this economic important disease by the government, and to redress the efforts to minimize its occurrence together with health professionals.
5. Improved interaction between the livestock owners and medical and veterinary personnel is a perquisite for the investigation of the zoonotic importance of *M. bovis*, and further investigations are required for minimizing its devastating effect in animals and humans.

References

Abubakar UB, Ameh JI, Abdulkadir IA et al (2011) Bovine tuberculosis in Nigeria: a review. Vet Res 4(1):24–27

Admassu B, Kebede E, Shite A (2015) Review on bovine tuberculosis. Eur J Biol Sci 7(4):169–185

Ameni G, Erkihun A (2007) Bovine tuberculosis on small-scale dairy farms in Adama town, Central Ethiopia, and farmer awareness of the disease. Rev Sci Tech Off Int Epiz 26:711–719

Ameni G, Bekele S, Tolosa T (2010) Preliminary study on the impact of bovine tuberculosis on the reproductive efficiency and productivity of Holstein dairy cows in Central Ethiopia. BAHPA 58:222–226

Asiimwe BB, Asiimwe J, Kallenius G et al (2009) Molecular characterization of 511 *Mycobacterium bovis* isolates from cattle carcasses at a city slaughterhouse in Uganda. Vet Rec 164:655–658

Asseged B, Tameru B, Habtemariam T (2014) Status and control of bovine tuberculosis in Ethiopia. In: Thoen CO, Steel JH, Kaneene JB (eds) Zoonotic tuberculosis: *Mycobacterium bovis* and other pathogenic mycobacteria, 3rd edn. Wiley-Blackwell, Chichester, pp 109–132

Awad FI (1962) Studies on type-determination and epidemiology of tuberculosis among cattle in Sudan. Zentralbl Veterinarmed B 9(9):890–898

Awah-Ndukum J, Kudi AC, Bradley G et al (2010) Prevalence of bovine tuberculosis in abattoirs of the littoral and western highland regions of Cameroon: a cause for public health concern. Vet Med Int 2010:495015. https://doi.org/10.4061/2010/495015

Awah-Ndukum J, Kudi AC, Bradley GI et al (2012) Prevalence of bovine tuberculosis in cattle in the highlands of Cameroon based on the detection of lesions in slaughtered cattle and tuberculin skin tests of live cattle. Vet Med 57(2):59–76

Barasona JA, Vicente J, Diez-Delgado I et al (2017) Environmental presence of *Mycobacterium tuberculosis* complex in aggregation points at the wildlife/livestock interface. Trans Emerg Dis 64(4):1148–1158

BBC News (2015) Africa creates TFTA—Cape to Cairo free-trade zone. http://www.bbc.com/news/world-africa-33076917

Ben Kahla I, Boschiroli ML, Souissi F et al (2011) Isolation and molecular characterisation of *M. bovis* from raw milk in Tunisia. Afr Health Sci 11(S1):S2–S5

Bengis RG, Keet DF, Michel AL et al (2001) Tuberculosis, caused by *Mycobacterium bovis*, in a kudu (*Tragelaphus strepsiceros*) from a commercial game farm in the Malelane area of the Mpumalanga Province, South Africa. Onderstepoort J Vet Res 68:239–241

Benkirane A (1998) Bovine tuberculosis in Africa. Wild Anim Rev 90(1):54–56

Berg S, Schelling E, Hailu E et al (2015) Investigation of the high rates of extrapulmonary tuberculosis in Ethiopia reveals no single driving factor and minimal evidence for zoonotic transmission of *M. bovis* infection. BMC Infect Dis 15:112. https://doi.org/10.1186/s12879-015-0846-7

Berggren SA (1977) Incidence of tuberculosis in BCG vaccinated and control cattle in relation to age distribution in Malawi. Br Vet J 133:490–494

Berggren SA (1981) Field experiment with BCG vaccine in Malawi. Br Vet J 137:88–94

Bernard F, Vincent C, Matthieu L et al (2005) Tuberculosis and brucellosis prevalence survey on dairy cattle in Mbarara milk basin (Uganda). Prev Vet Med 67:267–281

Berrada J (1993) *Mycobacterium bovis* infection in cattle in Morocco: preparation and evaluation of chemical extracts for use in detection of immune responses. PhD thesis, Iowa State University

Berrada J (2006) Capacity building for surveillance and control of tuberculosis. In: FAO animal production and health proceedings. FAO/WHO/OIE Expert and Technical Consultation, Rome, pp 49–53

Beyene D, Bergval I, Hailu S et al (2009) Identification and genotyping of the etiological agent of tuberculous lymphadenitis in Ethiopia. J Infect Dev Ctries 3(6):412–419

Boukary AR, Thys E, Rigouts L et al (2012) Risk factors associated with bovine tuberculosis and molecular characterization of *M. bovis* strains in urban settings in Niger. Transbound Emerg Dis 59(6):490–502

Bourne J (2007) Bovine TB: the scientific evidence. A science base for a sustainable policy to control TB in cattle: an epidemiological investigation into bovine tuberculosis. Final report of the independent scientific group on cattle TB. Department of Environment, Food and Rural Affairs (DEFRA), London

Brückner CK (2009) The role of the world organisation for animal health (OIE) to facilitate the international trade in animals and animal products. Onderstepoort J Vet Res 76:141–146

Buddle BM, de Lisle GW, Pfeiffer A et al (1995) Immunological responses and protection against *M. bovis* in calves vaccinated with a low dose of BCG. Vaccine 13(12):1123–1130

Buddle BMD, Wedlock N, Denis M et al (2011) Update on vaccination of cattle and wildlife populations against tuberculosis. Vet Microbiol 151:14–22

Byarugaba EF, Sylvain ME et al (2009) Pulmonary TB and *M. bovis*, Uganda. Emerg Infect Dis 15 (1):124–125

Carmichael J (1937) A brief note on tuberculosis in tropical Africa with special reference to Uganda. J Comp Pathol 50:383–385

Caron A, Cornelis D, Foggin C et al (2016) African buffalo movement and zoonotic disease risk across transfrontier conservation areas, southern Africa. Emerg Infect Dis 22(2):277

Cleaveland S, Mlengeya T, Kazwala RR et al (2005) Tuberculosis in Tanzanian wildlife. J Wildl Dis 41(2):446–453

Collins JD (2006) Tuberculosis in cattle: strategic planning for the future. Vet Microbiol 112:369–381

Collins CH, Grange JM (1983) The bovine tubercle bacillus: a review. J Appl Bacteriol 55:13–29

Coulibaly ND, Yameogo KR (2000) Prevalence and control of zoonotic diseases: collaboration between public health workers and veterinarians in Burkina Faso. Acta Trop 76:53–57

Cousins D (2001) *Mycobacterium bovis* infection and control in domestic livestock. Rev Sci Tech Off Int Epiz 20(1):71–85

Cox DR, Donnelly CA, Bourne FJ et al (2005) Simple model for tuberculosis in cattle and badgers. Proc Natl Acad Sci U S A 102(49):17593

Cross PC, Getz WM (2006) Assessing vaccination as a control strategy in an ongoing epidemic: bovine tuberculosis in African buffalo. Ecol Model 196:494–504

CSA (2011) Report on Livestock and Livestock Characteristics. Federal Democratic Republic of Ethiopia, Central Statistical Agency (CSA). Agricultural Sample Survey 2010/11 (2003 E.C.), Volume II. Statistical Bulletin 505

DAFF (2016) Bovine tuberculosis manual. Forestry and Fisheries, Republic of South Africa, Department of Agriculture. http://www.nda.agric.za/vetweb/pamphlets&Information/Policy/Tuberculosis%20in%20Cattle%20Interim%20Manual%20for%20the%20Veterinarian%20&%20AHT%20-%20Sept2....pdf

DAFF (2017) Veterinary procedural notice for buffalo disease risk management in South Africa. Forestry and Fisheries, Republic of South Africa, Department of Agriculture. http://www.daff.gov.za/vetweb/pamphlets&Information/Policy/Buffalo%20Disease%20Risk%20Management%20VPN_Signed%202017-02-17.pdf

Dankner WM, Davis CE (2000) *Mycobacterium bovis* as a significant cause of tuberculosis in children residing along the United States-Mexico border in the Baja California region. Pediatrics 105:e79

De la Rua-Domenech R (2006) Human *Mycobacterium bovis* infection in the United Kingdom: incidence, risks, control measures and review of the zoonotic aspects of bovine tuberculosis. Tuberculosis (Edinb) 86(2):77–109

Dean GS, Rhodes SG, Coad M et al (2008) Isoniazid treatment of *Mycobacterium bovis* in cattle as a model for human tuberculosis. Tuberculosis (Edinb) 88:586–594

Drewe JA, Pfeiffer DU, Kaneene JB (2014) Epidemiology of *M. bovis*. In: Thoen CO, Steele JH, Kaneene JB (eds) Zoonotic tuberculosis: *M. bovis* and other pathogenic mycobacteria, 3rd edn. Wiley, New York, pp 63–77

Egbe NF, Muwonge A, Ndip L et al (2017) Molecular epidemiology of *Mycobacterium bovis* in Cameroon. Sci Rep 7(1):4652

Ejeh EF, Raji MA, Bello M et al (2014) Prevalence and direct economic losses from bovine tuberculosis in Makurdi, Nigeria. Vet Med Int. 2014:904861. https://doi.org/10.1155/2014/904861

Ellwood DC, Waddington FG (1972) A second experiment to challenge the resistance to tuberculosis in BCG vaccinated cattle in Malawi. Br Vet J 128:619–626

Elmossalami E, Siam MA, El Sergany M (1971) Studies on tuberculous-like lesions in slaughtered camels. Zoonoses Public Health 18:253–261

FAO (2012) Bovine tuberculosis at the animal-human-ecosystem interface. In: Transboundary animal diseases bulletin. No. 40—FAO animal production and health division. FAO-EMPRES, Rome, Italy

FAO (2018) Ecological zones in Africa, Chapter 12. http://www.fao.org/docrep/004/y1997e/y1997e0h.htm

Firdessa R, Berg S, Hailu E et al (2013) Mycobacterial lineages causing pulmonary and extrapulmonary tuberculosis, Ethiopia. Emerg Infect Dis 19(3):460–463

Fitzgerald SD, Kaneene JB (2013) Wildlife reservoirs of bovine tuberculosis worldwide: hosts, pathology, surveillance, and control. Vet Path 50(3):488–499

Fritsche A, Engel R, Buhl D et al (2004) *Mycobacterium bovis* tuberculosis: from animal to man and back. Int J Tuberc Lung Dis 8(7):903–904

Gallagher J, Macadam I, Sayer J et al (1972) Pulmonary tuberculosis in free-living lechwe antelope in Zambia. Trop Anim Health Prod 4:204–213

Gilsdorf MJ, Ebel ED, Disney TW et al (2006) Benefit and cost assessment of the US bovine tuberculosis eradication program. In: Thoen JMCO (ed) *Mycobacterium bovis* infection in animals and humans, 2nd edn. Iowa State University Press, Ames, IA, pp 89–99

Good M, Duignan A (2011) Perspectives on the history of bovine TB and the role of tuberculin in bovine TB eradication. Vet Med Int. 2011:410470. https://doi.org/10.4061/2011/410470

Gortázar C, Beltrán-Beck B, Garrido JM et al (2014) Oral re-vaccination of Eurasian wild boar with *M. bovis* BCG yields a strong protective response against challenge with a field strain. BMC Vet Res 10:96

Gortázar C, Che Amat A, O'Brien DJ (2015) Open questions and recent advances in the control of a multi-host infectious disease: animal tuberculosis. Mammal Rev 45(3):160–175

Grange JM, Collins CH (1987) Bovine tubercle bacilli and disease in animals and man. Epidemiol Infect 99(2):221–234

Grange JM, Daborn C, Cosivi O (1994) HIV-related tuberculosis due to *Mycobacterium bovis*. Eur Respir J 7:1564–1566

Griffin JF (2000) Veterinary tuberculosis vaccine development. Clin Infect Dis 30(Suppl 3):S223–S228

Gumi B, Schelling E, Firdessa R et al (2012) Low prevalence of bovine tuberculosis in Somali pastoral livestock, Southeast Ethiopia. Trop Anim Health Prod 44(7):1445–1450. https://doi.org/10.1007/s11250-012-0085-5

Hiko A, Agga GE (2011) First-time detection of mycobacterium species from goats in Ethiopia. Trop Anim Health Prod 43:133–139

Hlokwe TM, Van Helden P, Michel AL (2014) Evidence of increasing intra and inter-species transmission of Mycobacterium bovis in South Africa: are we losing the battle? Prev Vet Med 115(1–2):10–17

Hope JC, Villarreal-Ramos B (2008) Bovine TB and the development of new vaccines. Comp Immunol Microbiol Infect Dis 31:77–100

Ismaila UG, Rahman HA, Saliluddin SM (2015) Knowledge on bovine tuberculosis among abattoir Workers in Gusau, Zamfara state, Nigeria. Int J Public Health Clin Sci 2(3):45–58

Jarvis LS, Valdes-Donoso P (2018) A selective review of the economic analysis of animal health management. J Agric Econ 69(1):201–225

Jori F, Mokospasetso M, Etter E et al (2013) Preliminary assessment of bovine tuberculosis at the livestock/wildlife interface in two protected areas of northern Botswana. Transbound Emerg Dis 60(Suppl. 1):28–36

Kalema-Zikusoka G, Bengis RG, Michel AL et al (2005) A preliminary investigation of tuberculosis and other diseases in African buffalo (*Syncerus caffer*) in queen Elizabeth National Park, Uganda. Onderstepoort J Vet Res 72(2):145–151

Kazoora HB, Majalija S, Kiwanuka N et al (2016) Knowledge, attitudes and practices regarding risk to human infection due to *Mycobacterium bovis* among cattle farming communities in western Uganda. Zoonoses Public Health 63(8):616–623

Kazwala RR, Daborn CJ, Sharp JM et al (2001) Isolation of *Mycobacterium bovis* from human cases of cervical adenitis in Tanzania: a cause for concern? Int J Tuberc Lung Dis 5:87–91

Kidane D, Olobo JO, Habte A et al (2002) Identification of the causative organism of tuberculous lymphadenitis in Ethiopia by PCR. J Clin Microbiol 40(11):4230–4234

Kidane AH, Sifer D, Aklilu M et al (2015) Knowledge, attitude and practice towards human and bovine tuberculosis among high school students in Addis Ababa, Ethiopia. Int J Livest Res 5:1–11

Lamuka PO, Njeruh FM, Gitao GC et al (2018) Prevalence of bovine and avian tuberculosis in camel herds and associated public health risk factors in Isiolo County, Kenya. Trop Anim Health Prod 50(5):937–945. https://doi.org/10.1007/s11250-017 1486-1482

Livingstone PG, Hancox N, Nugent G et al (2015) Toward eradication: the effect of *Mycobacterium bovis* infection in wildlife on the evolution and future direction of bovine tuberculosis management in New Zealand. N Z Vet J 63(Suppl. 1):4–18

LoBue PA, Betacourt W, Peter C et al (2003) Epidemiology of *Mycobacterium bovis* disease in San Diego County, 1994–2000. Int J Tuberc Lung Dis 7(2):180–185

Magnus K (1966) Epidemiological basis of tuberculosis eradication: 3. Risk of pulmonary tuberculosis after human and bovine infection. Bull World Health Organ 35:483–508

Malama S, Muma JB, Godfroid J (2013) A review of tuberculosis at the wildlife-livestock-human interface in Zambia. Infect Dis Poverty 2:13. http://www.idpjournal.com/content/2/1/13

Martin SW (1994) Livestock disease eradication: evaluation of the cooperative state-federal bovine tuberculosis eradication program/committee on bovine tuberculosis, board on agriculture, National Research Council. National Academy Press, Washington, D.C.

Martín-Hernando MP, Höfle U, Vicente J et al (2007) Lesions associated with *Mycobacterium tuberculosis* complex infection in the European wild boar. Tubercle 87:360–367

McCluskey B, Lombard J, Strunk S et al (2014) *Mycobacterium bovis* in California dairies: a case series of 2002–2013 outbreaks. Prev Vet Med 115:205–216

McNair J, Welsh MD, Pollock JM (2007) The immunology of bovine tuberculosis and progression toward improved disease control strategies. Vaccine 25:5504–5511

Michel AL (2008) Tuberculosis in wild and domestic animals in South Africa. PhD thesis, Universiteit Utrecht, The Netherlands

Michel AL, Hlokwe TM, Coetzee ML et al (2008) High *M. bovis* genetic diversity in a low prevalence setting. Vet Microbiol 126:151–159
Michel AL, Coetzee ML, Keet DF et al (2009) Molecular epidemiology of *Mycobacterium bovis* isolates from free-ranging wildlife in south African game reserves. Vet Microbiol 133 (4):335–343
Milian-Suazo F, Anaya-Escalera AM, Gallegos-Torres RM (2003) A review of *M. bovis* BCG protection against TB in cattle and other animal species. Prev Vet Med 58:1–13
Miller RS, Sweeney SJ (2013) *Mycobacterium bovis* (bovine tuberculosis) infection in north American wildlife: current status and opportunities for mitigation of risks of further infection in wildlife populations. Epidemiol Infect 141:1357–1370
Moda G (2006) Non-technical constraints to eradication: the Italian experience. Vet Microbiol 112:253–258
Moiane I, Machado A, Santos N et al (2014) Prevalence of bovine tuberculosis and risk factor assessment in cattle in rural livestock areas of Govuro District in the southeast of Mozambique. PLoS One 9(3):e91527. https://doi.org/10.1371/journal.pone.0091527
Moodie PA (1977) Tuberculosis reactions in BCG-vaccinated cattle. Br Vet J 133:642–645
Morris RS, Pfeiffer DU, Jackson R (1994) The epidemiology of *Mycobacterium bovis* infections. Vet Microbiol 40:153–177
Munyeme M, Muma JB, Munang'andu HM et al (2010) Cattle owners' awareness of bovine tuberculosis in high and low prevalence settings of the wildlife-livestock interface areas in Zambia. BMC Vet Res 6:21. http://www.biomedcentral.com/1746-6148/6/21
Musoke J, Hlokwe T, Marcotty T et al (2015) Spillover of *Mycobacterium bovis* from wildlife to livestock, South Africa. Emerg Infect Dis 21(3):448
Muwonge A, Johansen TB, Vigdis E et al (2012) *Mycobacterium bovis* infections in slaughter pigs in Mubende district, Uganda: a public health concern. BMC Vet Res 8:168. https://doi.org/10.1186/1746-6148
Naranjo V, Gortazar C, Vicente J et al (2008) Evidence of the role of European wild boar as a reservoir of *Mycobacterium tuberculosis* complex. Vet Microbiol 127:1–9
Ofukwu RA, Oboegbulem SI, Akwuobu CA (2008) Zoonotic *Mycobacterium species* in fresh cow milk and fresh skimmed, unpasteurised market milk (nono) in Makurdi, Nigeria: implications for public health. J Anim Plant Sci 1(1):21–25
OIE (2009) Bovine Tuberculosis. In: World Organization for Animal Health: terrestrial manual, chaps. 2, 4, 7, p 16
Okello AL, Welburn SC (2014) The importance of veterinary policy in preventing the emergence and re-emergence of zoonotic disease: examining the case of human African trypanosomiasis in Uganda. Front Public Health 2:218
Okoro OJ, Anosa GN, Oboegbulem SI et al (2014) Comparative assessment of postmortem inspection and immunochromatographic techniques for the detection of bovine tuberculosis in slaughter cattle in Nigeria. Trop Anim Health Prod 46(5):831–836
Oloya J, Opuda-Asibo J, Kazwala R et al (2008) Mycobacteria causing human cervical lymphadenitis in pastoral communities in the Karamoja region of Uganda. Epidemiol Infect 136:636–643
Opuda-Asibo J (1995) Regional and country status reports, Uganda. In: Thoen CO, Steel JH (eds) *Mycobacterium bovis* infection in animals and humans, 1st edn. Iowa State University Press, Ames, IA, pp 299–303
Paine R, Martinaglia G (1929) Tuberculosis in wild buck living under natural conditions. J Comp Pathol Therap 42(1):1–8
Palmer MV (2013) *Mycobacterium bovis*: characteristics of wildlife reservoir hosts. Transbound Emerg Dis 60(Suppl 1):1–13
Palmer MV, Waters WR (2011) Bovine tuberculosis and the establishment of an eradication program in the United States: role of veterinarians. Vet Med Int 2011:816345. https://doi.org/10.4061/2011/816345

Pilosof S, Greenbaum G, Krasnov BR et al (2017) Asymmetric disease dynamics in multihost interconnected networks. J Theor Biol 430:237–244

Power AP, Watts BGA (1987) The badger control policy: an economic assessment. Ministry of Agriculture, Forestry and Food, London

Quirin R, Rasolofo V, Andriambololona R et al (2001) Validity of intradermal tuberculin testing for the screening of bovine tuberculosis in Madagascar. Onderstepoort J Vet Res 68(3):231

Ranney F (1960) Bovine tuberculosis eradication. Dis Chest 39:150–157

Reilly LA, Courtenay O (2007) Husbandry practices, badger sett density and habitat composition as risk factors for transient and persistent bovine tuberculosis on UK cattle farms. Prev Vet Med 80:129–142

Rivière J, Le Strat Y, Dufour B et al (2015) Sensitivity of bovine tuberculosis surveillance in wildlife in France: a scenario tree approach. PLoS One 10(10):e0141884

Roswurm JD, Ranney AF (1973) Sharpening the attack on bovine tuberculosis. Am J Public Health 63(10):884–886

Roug A, Perez A, Mazet JAK et al (2014) Comparison of intervention methods for reducing human exposure to *M. bovis* through milk in pastoralist households of Tanzania. Prev Vet Med 115:157–165

Sahraoui N, Müller B, Guetarni D et al (2009) Molecular characterization of *M. bovis* strains isolated from cattle slaughtered at two abattoirs in Algeria. BMC Vet Res 5:4. https://doi.org/10.1186/1746-6148-5-4

Santos N, Santos C, Valente T (2015) Widespread environmental contamination with *Mycobacterium tuberculosis* complex revealed by a molecular detection protocol. PLoS One 10(11):e0142079

Sintayehu DW, Heitkönig IM, Prins HH et al (2017) Effect of host diversity and species assemblage composition on bovine tuberculosis (bTB) risk in Ethiopian cattle. Parasitology 144(6):783–792

Tadayon K, Mosavari N, Feizabadi MM (2013) An epidemiological perspective on bovine tuberculosis spotlighting facts and dilemmas in Iran, a historically zebu-dominant farming country. IJMM 5(1):1

Tamiru F, Hailemariam M, Terfa W (2013) Preliminary study on prevalence of bovine tuberculosis in cattle owned by tuberculosis positive and negative farmers and assessment of zoonotic awareness in ambo and toke Kutaye districts, Ethiopia. J Vet Med Anim Health 5(10):288–295

Tebug S, Njunga GR, Chagunda MG et al (2014) Risk, knowledge and preventive measures of smallholder dairy farmers in northern Malawi with regard to zoonotic brucellosis and bovine tuberculosis. Onderstepoort J Vet Res 81(1):1–6

Tompkins DM, Ramsey DSL, Cross ML et al (2009) Oral vaccination reduces the incidence of tuberculosis in free-living brushtail possums. Proc R Soc Lond (Biol) 276:2987–2995

Torgerson PR, Torgerson DJ (2009) Public health and bovine tuberculosis: what's all the fuss about? Trends Microbiol 18(2):67–72

Tschopp R, Schelling E, Hattendorf J et al (2010) Repeated cross-sectional skin testing for bovine tuberculosis in cattle kept in a traditional husbandry system in Ethiopia. Vet Rec 167:250–256

Tschopp R, Bobosha K, Aseffa A et al (2011) Bovine tuberculosis at a cattle-small ruminant-human interface in Meskan, Gurage region, Central Ethiopia. BMC Infect Dis 11(1):318

Vordermeier HM, de Val BP, Buddle BM et al (2014) Vaccination of domestic animals against tuberculosis: review of progress and contributions to the field of the TBSTEP project. Res Vet Sci 97:S53–S60

Wahlström H, Englund L (2006) Adopting control principles in a novel setting. Vet Microbiol 112:265–271

Waters WR, Palmer MV, Buddle BM et al (2012) Bovine tuberculosis vaccine research: historical perspectives and recent advances. Vaccine 30:2611–2622

Waters WR, Maggioli MF, McGill JL et al (2014) Relevance of bovine tuberculosis research to the understanding of human disease: historical perspectives, approaches, and immunologic mechanisms. Vet Immunol Immunopathol 159:113–132

WHO (2013) WHO country cooperation strategy 2012–2015 Ethiopia. WHO Regional Office for Africa, Brazzaville, Republic of Congo

WHO (2017) Roadmap for zoonotic tuberculosis

Wilson GJ, Carter SP, Delahay RJ (2011) Advances and prospects for management of TB transmission between badgers and cattle. Vet Microbiol 151:43–50

Woodford MH (1982) Tuberculosis in wildlife in the Ruwenzori National Park, Uganda (part II). Trop Anim Health Prod 14(3):155–160

Zahran RN, El Behiry A, Marzouk E et al (2014) Comparison of LCD array and IS6110-PCR with conventional techniques for detection of *Mycobacterium bovis* isolated from Egyptian cattle and buffaloes. Int J Mycobacteriol 3:197–204

Zinsstag J, Schelling E, Roth F et al (2008) Economics of bovine tuberculosis. In: Thoen CO, Steele JH, Gilsdorf MJ (eds) *Mycobacterium bovis* infection in animals and humans, 2nd edn. Wiley, New York, pp 68–83

Part III
Country Reports

Chapter 11
Bovine Tuberculosis: Status, Epidemiology, and Public Health Implications in Burkina Faso

Adama Sanou

11.1 Introduction

Burkina Faso is a resource-poor country situated in the heart of West Africa. It is surrounded by six countries and shares its northern and western borders with Mali, its northeastern border with Niger, its southeastern border with Benin, and its southern border with Ghana, Togo, and Côte d'Ivoire. As a Sahelian country, Burkina Faso's economy mainly depends on agriculture and livestock farming. Although dairy farms, with the introduction of exotic breeds, have more recently been established around urban centers, the extensive, transhumant type of cattle farming remains the main form of livestock farming in the country.

The country's livestock population was recently estimated at about 8.7 million cattle, 8.7 million sheep, and 13 million goats. The administrative region of Sahel, one of Burkina Faso's 13 administrative regions, is a predominantly pastoral area and has the largest number of cattle of all the regions. The Fulani pastoralists (also known as the Peulh), one of more than 60 ethnic groups in Burkina Faso, practice an extensive transhumant husbandry system in the Sahel region where it is their main socioeconomic activity. They live in close and permanent contact with their animals and consume raw or curdled raw milk.

Several important animal diseases are prevalent in Burkina Faso, including bovine tuberculosis (BTB) that occupies an important place (Coulibaly and Yameogo 2000). The presence of TB creates both animal and public health, and economic problems for Burkina Faso. Economically, the financial losses attributable to BTB are related to decreased livestock productivity (reduced milk production,

A. Sanou (✉)
Bacteriology Laboratory of Centre MURAZ, Department of Biomedical Sciences, Centre MURAZ, Bobo Dioulasso, Burkina Faso

Research and Training Unit in Science and Technology, Polytechnic University of Bobo-Dioulasso, Bobo-Dioulasso, Burkina Faso

A. B. Dibaba et al. (eds.), *Tuberculosis in Animals: An African Perspective*,
https://doi.org/10.1007/978-3-030-18690-6_11

Table 11.1 Bovine tuberculosis surveys conducted in Burkina Faso (Letroteur 1952; Sere 1966; Gidel et al. 1969a, b; Rey et al. 1986; Delafosse et al. 1995; Vekemans et al. 1999; Traoré et al. 2004; Boussini et al. 2012; Tarnagda et al. 2014)

Study period	Location	Diagnostic method	Sample size	No. positive (%)
1951	Bobo-Dioulasso	TST	390	39 (1)
1952	Bobo-Dioulasso	TST	111	14 (12.6)
1950–1961	Bobo-Dioulasso	AI	99,409	13,303 (13.4)
1964	Ouagadougou	TST	143	19 (13.3)
Unspecified	Ouagadougou	TST	480	73 (15.2)
1965–1968	Bobo-Dioulasso	AI[a]	82,961	7971 (9.6)
1967–1968	Dédougou	TST	2432	69 (2.8)
1967–1968	Dori	TST	2231	133 (6)
Unspecified	Department of Sahel	TST	2229	≈ 87 (3.9)
1994	Bobo-Dioulasso	AI[a]	2700	100 (3.7)
1996	Bobo-Dioulasso	AI[a]	199	38 (19)
Unspecified	Bobo-Dioulasso	TST	174	23 (13)
1997	Whole country	AI	135,822	177 (0.13)
2001–2002	Ouagadougou	TST	325	90 (27.7)
2004–2005	Ouagadougou	TST	874[b]	26 (3)
2004–2005	Ouagadougou	TST	214[c]	11 (5.1)
2004–2005	Ouagadougou	TST	332[d]	49 (14.8)
2011	Ouagadougou, Bobo-Dioulasso	AI[a]	1499	102 (6.8)

TST tuberculin skin test, *AI* abattoir inspection
[a]Confirmed by culture
[b]Modern system
[c]Extensive system
[d]Intra-urban system

weight loss, etc.), slaughterhouse seizures/condemnations, and the added cost of processing tuberculous carcasses. Bovine TB also constitutes an obstacle to access the lucrative international markets because of the restrictions on trade with livestock and animal products from BTB-infected countries.

During the course of recent decades, BTB was regularly diagnosed in Burkina Faso (Table 11.1), and it remains a major animal health problem.

Based on the presence of tuberculous lesions, from which mycobacteria were isolated, observed during meat inspection in slaughterhouses in the two largest cities in Burkina Faso (i.e., Bobo-Dioulasso and Ouagadougou), its prevalence in 2014 was estimated to be 6.8% (Tarnagda et al. 2014). This observation has a wider implication because it indicates that BTB is widespread in Burkina Faso, as cattle slaughtered in these two slaughterhouses not only originate from neighboring villages but also from the other main cattle production areas in the country (Sanou et al. 2014). These two abattoirs receive beef cattle from all over the country and provide meat to the two largest cities in Burkina Faso, and processes carcasses for export.

11.2 Epidemiology of BTB Caused by *M. bovis* in Burkina Faso

Evidence of the presence of BTB in Burkina Faso dates back to the colonial period. There are often differences in the prevalence detected by tuberculin skin test surveillance and data obtained from abattoirs because the source of animals differs for the two datasets.

In Bobo-Dioulasso, based on the single cervical intradermal test (SIT), the prevalence was estimated to be 10% in 1951 ($n = 390$) and 12.6% in 1952 ($n = 111$) (Sere 1966). A large survey conducted in 1969 in Dori and Dédougou using SIT reflected a prevalence of 6.0 and 2.8%, respectively (Gidel et al. 1969b). Follow-up studies conducted in Bobo-Dioulasso in 1999 confirmed the persistence of BTB with an estimated prevalence of 13.0% (Vekemans et al. 1999). There seems to be significant differences in the prevalence of BTB depending on the farming system. According to a study conducted in an intra-urban dairy farm in Ouagadougou in 2004, the BTB prevalence was estimated to be 27.7% (Traoré et al. 2004). In intra-urban and suburban farms in Ouagadougou representing three different beef production systems (extensive, an enhanced system, including semi-intensive and intensive farming, and an intra-urban-intensive breeding variant), the BTB prevalence in 2012 was 5.1, 3.0, and 14.8%, respectively, for animals originating from the three systems (Boussini et al. 2012).

Because of the lack of funding to conduct intradermal tuberculin tests, BTB is usually diagnosed in Burkina Faso by detecting suspected tuberculous lesions during meat inspection in abattoirs. Trained animal health professionals (meat inspectors) conduct these inspections, and the data that they gather have been the subject of several published reports (Sere 1966; Gidel et al. 1969a). Based on routine meat inspections conducted at Bobo-Dioulasso slaughterhouse in 1995 and 1999, the BTB prevalence was, respectively, 3.7 and 19% (Delafosse et al. 1995; Vekemans et al. 1999).

Several factors appear to influence the prevalence and distribution of BTB in Burkina Faso. These include husbandry systems, the age and sex of the cattle, and uncontrolled trans-boundary movement of livestock from neighboring BTB-infected countries (Gidel et al. 1969b; Tarnagda et al. 2014; Regnoult 1963). The prevalence of BTB is higher in farming systems where animals live in close contact with one another, where large numbers of cattle are housed in a confined airspace and where there is inadequate health monitoring (Vekemans et al. 1999; Boussini et al. 2012), thus partly explaining why the Sahelian regions generally have a higher prevalence of BTB (Gidel et al. 1969b). The association between a longer lifespan of an animal and the increased risk of having BTB (Vekemans et al. 1999) and the animal's sex is reflected by the prevalence in females of 6.7% compared to 2.9% in males and 17% in 2-year-old animals compared to 33.7% in those older than 6 years (Boussini et al. 2012; Tarnagda et al. 2014).

Burkina Faso is located in the heart of West Africa, and it is a transit hub for the movement of people and their livestock to and from its six neighboring countries

Table 11.2 *Mycobacterium bovis* spoligotype signatures isolated from humans and animals in Burkina Faso (Sanou et al. 2014)

SB numbers[a]	Host species	Geographical location	Clonal complex (%)[b]
SB2282	Cattle	Bobo-Dioulasso	Af5 (3%)
SB2283	Human	Solenzo	Af5 (3%)
SB1398	Cattle	Bobo-Dioulasso	Af5 (6%)
SB2284	Human	Bobo-Dioulasso	Af5 (3%)
SB2285	Cattle	Bobo-Dioulasso	Af5 (3%)
SB0857	Cattle	Ouagadougou	Af1 (3%)
SB0300	Cattle, human	Bobo-Dioulasso, Ouagadougou	Af1 (9%)
SB2286	Cattle	Ouagadougou	Af1 (9%)
SB2287	Cattle	Ouagadougou	Af1 (3%)
SB0944	Cattle, human	Bobo-Dioulasso, Ouagadougou, Koupéla	Af1 (52%)
SB1439	Cattle	Ouagadougou	Af1 (3%)
SB2288	Human	Ouagadougou	Af1 (3%)
Total			Af5 (18.2%)
			Af1 (81.8%)

[a]SB number = name of spoligotype based on *M.bovis*.org database nomenclature
[b]Af1 = African 1 clonal complex; Af5 = putative African 5 clonal complex

(Sere 1966; Sanou et al. 2014). Thus, cross-border cattle movement, particularly from Mali and Niger, appears to significantly influence the epidemiology of BTB, and it is likely to have caused an increase in the prevalence of BTB in Burkina Faso. This assumption is supported by data from local abattoirs in the Nouna sector that is adjacent to the BTB-infected areas of Mopti, Ségou, and Macina in Mali (Sanou et al. 2014; Regnoult 1963).

Mycobacterium bovis is the major causative agent of BTB and is responsible for 89–97% of cases in cattle in Burkina Faso (Letroteur 1952; Sere 1966). There are many gaps in the understanding of the epidemiology of BTB in Burkina Faso, but a detailed molecular epidemiological investigation of the types of human and bovine *M. bovis* isolates in 2014 shed some light on the origin of the infection in the country. The strains of *M. bovis* in Burkina Faso are characterized by a considerable genetic diversity, and at least 12 *M. bovis* spoligotypes, divided into two subgroups, were identified (Table 11.2).

All *M. bovis* strains in this study belonged either to the Af1 clonal complex, found in other countries of West-Central Africa, such as Mali, Cameroon, Nigeria, Chad, and Niger, or to the putative Af5 clonal complex that has also been described in Mali. In Burkina Faso, the putative Af5 strains comprise about 18.2% of the isolates and are geographically localized in the western region (Bobo-Dioulasso and Solenzo), an area bordering Mali (Fig. 11.1). The presence of common or related genotypes in Burkina Faso and Mali is probably the consequence of transhumance between these countries, and of the passage, through Burkina Faso, of Malian

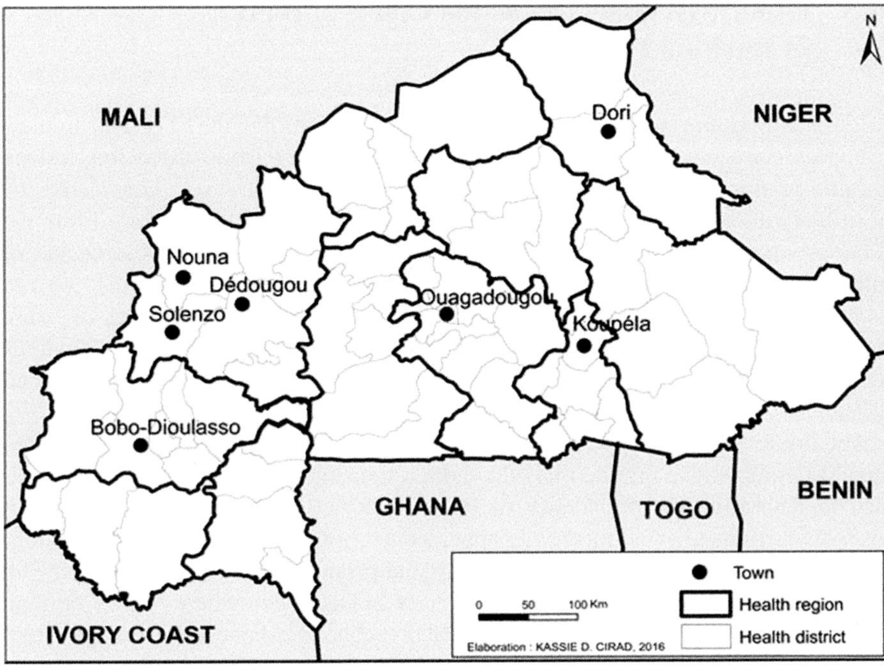

Fig. 11.1 Map of Burkina Faso showing major cities and regional and international borders

livestock to Ghana and Nigeria in the south. Spoligotype signatures belonging to the putative Af5 clonal complex have been reported in Europe, and it is likely that the putative Af5 clonal complex was introduced into this region of Africa from Europe, probably via North Africa (Müller et al. 2008; Haddad et al. 2001; Sahraoui et al. 2009). The larger group of isolates, comprising about 81.8% of the total national *M. bovis* isolates, belongs to the Af1 clonal complex, also including the SB0944 spoligotype pattern, a dominant strain that accounts for almost half of all the isolates. This group is characterized by the lack of spacer 30 and is also the most common type in other West-Central African Sahelian countries such as Mali, Nigeria, Chad, Niger, and Cameroon (Boukary et al. 2012; Müller et al. 2009). In the Af1 clonal complex, the SB0944 spoligotype signature is considered the most recent common ancestor (progenitor). It is the most common pattern in this group and comprises 40% of isolates in Chad, 46.1% in Nigeria, and 62.7% in Cameroon. It is also the most abundant type (52%) in Burkina Faso (Sanou et al. 2014; Müller et al. 2008). The spread of the Af1 clonal complex over this large area of West-Central Africa is probably caused by the long-distance, cross-border migration of livestock, a feature of the transhumant production system mainly practiced by the Fulani pastoralists in the Sahel (Müller et al. 2008).

11.3 Other Mycobacteria as the Cause of BTB in Burkina Faso

In addition to *M. bovis*, other mycobacterial species (including the MTC and atypical mycobacterial species) have also occasionally been isolated from tuberculous lesions in cattle in Burkina Faso. In several instances, these species were considered the most likely etiological agents involved in the development of the lesions. From the MTC group, *M. tuberculosis* and *M. africanum*, the main etiological agents of human TB, were frequently isolated. In one study (Tarnagda et al. 2014), 9.6 and 6.4%, respectively, of TB bacilli isolated from tuberculous carcasses were *M. africanum* and *M. tuberculosis*. Several other studies reported similar findings but at a lower prevalence (Gidel et al. 1969b; Delafosse et al. 1995; Vekemans et al. 1999).

The involvement of atypical mycobacteria in cattle infections was first suspected when the comparative intradermal tuberculin test, using avian and bovine PPDs, was used to determine the prevalence of BTB (Gidel et al. 1969b). The number of positive reactions to avian tuberculin appeared to correlate with the extent of poultry production in the study areas. Accordingly, the proportion of reactors to *M. avium* PPD was low in Dori (<1%) compared to 3.9% in Dédougou where poultry farming was more common. A number of atypical mycobacteria have been isolated from carcasses of the reactors in Burkina Faso (Sanou et al. 2014) including *M. aquae*, *M. batteyi*, *M. bortelense*, and *M. kansasii*; *M. farcinogenes* has not yet been detected (Gidel et al. 1969a; Delafosse et al. 1995).

11.4 TB Due to *M. bovis* in Other Animal Species

Up until now, *M. bovis* has been almost exclusively isolated from cattle in this country (Delafosse et al. 1995; Vekemans et al. 1999). Given that *M. bovis* is the major causative agent of BTB in a broad spectrum of animal hosts, domestic and wild, it should be anticipated that because of Burkina Faso's huge faunal population (both wild and domestic), *M. bovis* should also occur in species other than cattle. Because of a lack of funding, other species, including wildlife, received very little attention with regard to their TB status. There are no published reports of the presence of BTB in wildlife, and the available data on other domestic species are scant and incomplete. BTB was detected in pigs and horses slaughtered in Bobo-Dioulasso abattoir with a prevalence of 2.2 and 0.3%, respectively. Tuberculosis was also diagnosed in sheep, and it is suspected to occur in a number of other domesticated species, but without confirmation (Gidel et al. 1969a).

11.5 Public Heath Implication of *M. bovis*

Mycobacterium bovis is a potential health hazard, both for animals and humans, and the risk of becoming infected depends on the route of transmission. In Burkina Faso, aerosol transmission is strongly facilitated by the close and prolonged contact between diseased animals and pastoralists (Gidel et al. 1969b), such as the Fulani, who have an increased risk of contracting BTB due to their livestock-associated lifestyle. The Fulani ethnic group has a higher rate of the pulmonary form of TB compared to populations that are not as exposed to *M. bovis*-infected cattle (Vekemans et al. 1999).

The oral route of transmission following consumption of infected milk and meat products also presents a considerable risk of *M. bovis* transmission from animals to humans and vice versa (Vekemans et al. 1999; Tarnagda et al. 2014). In Burkina Faso, the zoonotic risk associated with the consumption of meat and dairy products from diseased animals appears to be important. *Mycobacterium bovis* was isolated from nearly 22% of dairy products collected from small markets and industrial dairy farms in Bobo-Dioulasso (Vekemans et al. 1999). The risk of contracting BTB appears to be particularly high in the Sahelian regions where milk is usually consumed raw and where the BTB infection rates in cattle are typically high, especially in older cows. The risk of contracting TB by humans is three times higher in villages with high BTB infection rates than in BTB-free villages (Rey et al. 1986). It seems therefore that the high TB prevalence in young children may be related to the prevailing high prevalence of BTB in the local cattle population, suggesting that *M. bovis* infections may be more prevalent in these populations than anticipated.

Determining the actual zoonotic role of *M. bovis* in Burkina Faso is difficult particularly because of the lack of adequate diagnostic facilities and techniques, and the true prevalence of *M. bovis* infection in the human population remains unknown. The risk of contracting zoonotic BTB in Burkina Faso will remain until such time that adequate surveillance, good hygienic practices (including pasteurization of milk and cooking of meat), and other good public health practices are instituted.

11.6 Control of BTB in Burkina Faso

In developed countries, BTB control programs based on a test-and-slaughter strategy and milk pasteurization significantly reduced the incidence of BTB in cattle and humans. In developing countries such as Burkina Faso, the situation is problematic, and the disease is mostly inadequately controlled because of the lack of financial resources and clear policies, and poor application of control strategies.

In Burkina Faso, two directorates (Directorate of National Laboratory of Livestock and the Directorate of Animal Health) are mandated to control BTB by two decrees (KITI No. AN VII-0113 /FP/AGRI-EL portant réglement de la Police Zoo-Sanitaire au Burkina du 22 Novembre 1989 and KITI no. AN VII 114 FP-

AGRI-EL portant réglementation de la santé publique vétérinaire du 22 Novembre 1989). Although BTB is listed as a notifiable, communicable disease, it is not currently actively controlled by the Epidemiological Surveillance Network (Réseau de Surveillance Épidemiologique, RESUREP) of the Ministry of Animal Resources.

The Regional Directorates of Animal Health are responsible for monitoring serious livestock diseases (including BTB) in each of the 13 regions of Burkina Faso. Under the supervision of Regional Livestock Laboratories, herd testing (including SIT for BTB) is conducted annually in different regions of the country as part of a general disease surveillance program. Bovine TB test-positive animals are slaughtered under veterinary supervision but without compensation. These animals are also allowed to be slaughtered for consumption (without any veterinary supervision), or they may be sold on the open market. Voluntary BTB diagnostic tests are sometimes performed on privately owned dairy cattle when required by international buyers. Thus control of milk-borne BTB is subject to voluntary control by individual owners, with the support of their trading partners.

In addition to the surveillance programs, there are veterinary control posts at the entry points along the borders of Burkina Faso that monitor a number of serious infectious diseases. Veterinary officers at these control posts alert RESUREP when suspected cases of BTB are encountered, and the regional officers and the RESUREP then conduct further investigations and institute appropriate control procedures.

To effectively deal with BTB in Burkina Faso in the future, the following measures should be considered:

1. Tuberculin testing. Systematically use SIT to detect infected herds, and apply the appropriate control measures (slaughter of infected animals, disinfection of premises, etc.).
2. Adequate funding. It is necessary that funding be allocated for financial compensation of farmers who lose their animals, and further financial support of the program.
3. Quarantine facilities. As a country in the heart of West-Central Africa, Burkina Faso could benefit from quarantining animals coming from other countries that are known to be BTB-infected.
4. Managing the zoonotic risk. Although efforts are being made to limit human exposure along the meat chain based on abattoir inspection procedures, few measures are taken to limit the consumption of BTB-infected milk and other dairy products (Vekemans et al. 1999). The ministries of Health and of Animal Resources will benefit by cooperating to deal with the public health risk posed by BTB (Coulibaly and Yameogo 2000). For effective control of zoonotic TB, efforts must be made to enhance the diagnostic capabilities and BTB surveillance in both humans and animals through strong collaboration between the two mandated public institutions. In addition to SIT and slaughter schemes that are necessary to eradicate BTB from cattle herds, specific measures such as pasteurization of milk and other dairy products and improved meat inspection protocols in slaughterhouses should be put in place to further limit the risk of transmission.

5. Public awareness campaigns. These activities are an integral part of the fight against BTB. It is important that breeders, butchers, traders, consumers, and the general public have adequate knowledge about the way in which *M. bovis* is transmitted and of the sources of infection and their zoonotic risk (Coulibaly and Yameogo 2000).

Acknowledgments The author is grateful to Dr. Daouda Kassie, researcher in geography of health at the French Agricultural Research Centre for International Development (CIRAD), for the map of Burkina Faso.

References

Boukary AR, Thys E, Rigouts L et al (2012) Risk factors associated with bovine tuberculosis and molecular characterization of *Mycobacterium bovis* strains in urban settings in Niger. Transbound Emerg Dis 59:490–502

Boussini H, Traore A, Tamboura HH et al (2012) Prevalence of tuberculosis and brucellosis in intra-urban and peri-urban dairy cattle farms in Ouagadougou, Burkina Faso. Rev Sci Tech 31:943–951

Coulibaly ND, Yameogo KR (2000) Prevalence and control of zoonotic diseases: collaboration between public health workers and veterinarians in Burkina Faso. Acta Trop 76:53–57

Delafosse A, Traore A, Kone B (1995) Isolation of pathogenic *Mycobacterium* strains in cattle slaughtered in the abattoir of Bobo-Dioulasso, Burkina Faso. Rev Elev Med Vet Pays Trop 48:301–306

Gidel R, Albert JP, Lefevre M et al (1969a) Mycobacteria of animal origin isolated by the Muraz Center from 1965 to 1968: technics of isolation and identification; results. Rev Elev Med Vet Pays Trop 22:495–508

Gidel R, Albert JP, Retif M (1969b) Survey of bovine tuberculosis by means of tuberculin test in various regions of West Africa (Upper Volta and Ivory Coast): results and general considerations. Rev Elev Med Vet Pays Trop 22:337–355

Haddad N, Ostyn A, Karoui C et al (2001) Spoligotype diversity of *Mycobacterium bovis* strains isolated in France from 1979 to 2000. J Clin Microbiol 39:3623–3632

Letroteur R (1952) Quelques observations sur la tuberculose bovine à Bobo-Dioulasso. Bull Serv Elev Industr Anim Afr Occid Franc 5:25–32

Müller B, Steiner B, Bonfoh B et al (2008) Molecular characterisation of *Mycobacterium bovis* isolated from cattle slaughtered at the Bamako abattoir in Mali. BMC Vet Res 4:26

Müller B, Hilty M, Berg S et al (2009) African 1, an epidemiologically important clonal complex of *Mycobacterium bovis* dominant in Mali, Nigeria, Cameroon, and Chad. J Bacteriol 19:1951–1960

Regnoult MG (1963) Animal tuberculosis in west African territories of French expression. Rev Pathol Gen Physiol Clin 63:1093–1115

Rey JL, Villon A, Saliou P et al (1986) Tuberculosis infection in a cattle-breeding region in Sahelian Africa. Ann Soc Belg Med Trop 66:235–243

Sahraoui N, Muller B, Guetarni D et al (2009) Molecular characterization of *Mycobacterium bovis* strains isolated from cattle slaughtered at two abattoirs in Algeria. BMC Vet Res 5:4

Sanou A, Tarnagda Z, Kanyala E et al (2014) *Mycobacterium bovis* in Burkina Faso: Epidemiologic and genetic links between human and cattle isolates. PLoS Negl Trop Dis 8:e3142

Sere A (1966) La Tuberculose bovine en Haute-Volta. Imprimerie de Centre Camilli et Fournie

Tarnagda Z, Kanyala E, Zingué D et al (2014) Prevalence of Tuberculosis spp. species in bovine carcasses in two slaughterhouses of Burkina Faso. Int Microbiol Immunol Res 2:92–100

Traoré A, Tamboura HH, Bayala B et al (2004) Prévalence globale des pathologies majeures liées la production laitière bovine en système d'élevage intra-urbain à Hamdallaye (Ouagadougou). Biotechnologie agronomie société et environnement (Centre de recherches agronomiques Gembloux; Faculté universitaire des sciences agronomiques de Gembloux) 8:3–8

Vekemans M, Cartoux M, Diagbouga S et al (1999) Potential source of human exposure to *Mycobacterium bovis* in Burkina Faso, in the context of the HIV epidemic. Clin Microbiol Infect 5:617–621

Chapter 12
The Status of Bovine Tuberculosis in Cameroon

Julius Awah-Ndukum, Nkongho Franklyn Egbe, and Victor Ngu-Ngwa

12.1 Introduction

Human and animal tuberculosis (TB) are well-controlled diseases in most developed countries. However, the rapid population growth, widespread poverty, regional conflict, migration between countries, inappropriate application of disease control measures, and the HIV/AIDS epidemic continue to boost the number of cases and the negative effects of human and animal TB in most of Africa, including Cameroon (WHO 2009). Bovine TB (BTB) is prevalent in cattle in many African countries (Ayele et al. 2004; Zinsstag et al. 2006), but the exact prevalence of BTB in cattle and the extent of zoonotic TB due to *M. bovis* on the continent are largely unknown. It is also generally unknown which of the *M. bovis* strains circulate in animals in Africa, the number of existing *M. bovis* maintenance hosts, and the role that they play in maintaining the infection and contribute to the spread of the infection. However, historical and clinical data, and the sporadic use of the intradermal tuberculin tests, the single cervical (SIT), and the comparative cervical (CCT) tests (Awah-Ndukum et al. 2012b), have clearly established the endemic nature of BTB in cattle in Cameroon (Awah-Ndukum et al. 2012a, b; Egbe et al. 2016). There is, however, a lack of understanding of the magnitude of the infection, and its distribution.

Effective application of the test-and-slaughter policy, the basis of national BTB control programs in developed countries, is not yet practicable in many developing countries because of logistical, political, and financial constraints. Evaluation and application of effective alternative strategies that are technically feasible and

J. Awah-Ndukum (✉) · V. Ngu-Ngwa
School of Veterinary Medicine and Sciences, University of Ngaoundere, Ngaoundere, Cameroon

N. F. Egbe
Tuberculosis Reference Laboratory Bamenda, Bamenda, Cameroon

A. B. Dibaba et al. (eds.), *Tuberculosis in Animals: An African Perspective*,
https://doi.org/10.1007/978-3-030-18690-6_12

economically viable under these circumstances, should be the primary objective in preventing the widespread occurrence of the disease in Africa.

Though BTB is widespread in Cameroon (Awah-Ndukum et al. 2012a, b; Egbe et al. 2016), it remains a neglected zoonotic disease, and its presumptive diagnosis in animals is based mostly on detecting the characteristic macroscopic lesions found at slaughter, and during meat inspection in abattoirs. In many communities in Cameroon, a number of practices increase the risk of contracting zoonotic BTB. In this country, it is common for humans and their animals to share an unhygienic microenvironment and water sources, especially during the dry season. In addition, the food preferences and eating habits of many Cameroonians, which include ingesting fresh animal products such as raw milk and meat, predispose them to contracting zoonotic TB. Other factors that may contribute to zoonotic BTB in Cameroon include the poor implementation of existing disease control legislation, poor monitoring and notification of the presence of the disease, and the lack of collaboration between the different public sector services responsible for controlling zoonotic diseases. In addition, inadequately trained veterinary and medical professionals, a poor diagnostic capability, the lack of public awareness about zoonotic TB, animal husbandry practices that predispose to the occurrence of the disease and human exposure to the infection, further complicate the matter (Ayele et al. 2004; AU/IBAR 2006; Kelly et al. 2016).

Because of the close human-livestock interaction in livestock-rearing communities in Cameroon, opportunities exist for the transmission of *M. bovis* to humans. Although an estimated median of 2.8% (0–37.7%) of human TB cases in Africa is due to *M. bovis*, significantly higher prevalences have been reported in some communities, and a similar trend may be expected in Cameroon (Müller et al. 2013).

12.2 Bovine Tuberculosis in Animals in Cameroon

Cameroon's human population of over 20 million live in five agroecological zones (AEZ), namely: the (1) Sudano-Sahel (in the North and Far North regions), (2) Guinea High Savannah (in the Adamawa Region), (3) Western Highlands (in the Northwest and West regions), (4) mono-modal humid forests (in the Southwest and Littoral regions), and (5) bimodal humid forests (in the Center, East and South regions). Over 70% of the population lives in rural areas, and their livelihood depends on crop farming and livestock rearing (Tanya 2004).

The national livestock population comprises 6 million cattle (mainly indigenous, multipurpose zebu breeds), 4 million sheep, 4.6 million goats, 0.2 million horses, and about 1.8 million pigs (FAOSTAT 2014). There is also a wide range of wildlife, including predators, wild hogs, wild ruminants, and rodents inhabiting the different AEZs. The majority of livestock in Cameroon are traditionally and extensively managed, and the farmers are dependent on limited ranges and feed that are more limited during the dryer seasons and the AEZ in which they live. There is extensive transhumant movement of livestock during the dry season to cope with the feed shortage. The dominant pastoral production and extensive husbandry systems increase their vulnerability to many diseases.

Contact of livestock with wildlife (during seasonal migration) and their congregation during grazing in the field, veterinary interventions, and in livestock markets provide ample opportunity for the transmission of infectious diseases. These diseases have a huge impact on animal productivity and wellbeing; they also threaten public health because of their zoonotic transmission to humans (including BTB), with the poor being particularly vulnerable. Therefore, improved surveillance for BTB and accurate estimations of its magnitude and distribution in cattle are essential to devise appropriate intervention strategies in Cameroon.

Information on the occurrence of BTB in Cameroon is scant especially before the year 2000. Available data extracted from FAO-OIE-WHO Animal Health Yearbooks (1992–1997) and the OIE (1997–2001) confirm that BTB is prevalent in the four sub-Saharan geopolitical regions. However, a definitive regional prevalence could not be estimated due to inadequate data from several key countries such as Cameroon in the Central African region. The disease status in Cameroon is often only documented as "low and sporadic," "disease reported," "disease suspected but not confirmed," and "serological evidence with no clinical disease." No disease information was available for animals other than cattle in those reports.

Although BTB in cattle occurs widely in Cameroon, there is no structured surveillance for its presence and distribution, and information about its occurrence is mainly dependent on the detection of macroscopic lesions during the postmortal examination of carcasses and meat inspection in abattoirs (Awah-Ndukum et al. 2012b; Egbe et al. 2016). Following the provisional diagnosis of BTB during these events and immunologic reactions to ancillary tests, direct smear microscopy of Ziehl-Neelsen-stained smears of the exudate of suspect lesions, culture (on solid/liquid media), and the results of molecular diagnostic techniques are used to confirm the presence and identification of the *Mycobacterium* spp. isolated from the lesions (Awah-Ndukum et al. 2013; Koro-Koro et al. 2013; Egbe et al. 2016). Intradermal tuberculin tests, immunochromatographic (lateral-flow-based rapid test), and IFN-γ assays (Bronsvoort et al. unpublished) (Awah-Ndukum et al. 2012a, b) are only occasionally used to diagnose the disease in cattle in Cameroon. In contrast to several reports from other African countries that documented BTB in other domestic animals such as pigs, small ruminants, horses, dogs, and wildlife (Cadmus et al. 2009; Hiko and Agga 2011; Hlokwe et al. 2011; Katale et al. 2012), BTB has not been detected in these animals in Cameroon.

Estimates of the national prevalence of BTB in cattle are based on opportunistic samples obtained from accessible segments of the cattle population. The distribution of the disease appears to be very patchy, and, depending on the sampling and detection methods, the prevalence varies widely from <1% to >50% (Table 12.1). Establishing the true extent of the disease will require representative samples of the total population from across the country. More often too, different prevalence estimates are reported from the same study population because of differing interpretations of the tests used, and there is a need to standardize the application and interpretation of diagnostic tests used in the country (Awah-Ndukum et al. 2012a).

Table 12.1 Current prevalence of bovine tuberculosis in cattle in Cameroon

Locations	Number of animals	Prevalence (%) according to diagnostic techniques used	Reference (test interpretation standard used)
North and Far North regions	890	10.6[a]; 2.7[b]	Martrenchar et al. (1993) (Ministry of Agriculture; France; Order: 07/11/90)
Northwest region	2492	3[a] (extensive farms); 13[a] (Ranches)	Merlin and Tsangueu (1985) (OIE standard cut-off point)
West Region (experimental farm)	142	14.8[a] (42% Zebu; 9.02% crosses)	Nfi and Ndi (1997) (OIE standard cut-off point)
Adamawa Region (experimental farm)	1395	1.4[a] (Zebu cattle); 2.8[a] (Holstein and their crosses)	Tanya et al. (1985) (OIE standard cut-off point)
Northwest Region	166 (48% exotic, 28% zebu, 20% crosses)	26[a]	Muchaal (2002) (Canadian Food Inspection Agency field protocol)
Douala abattoir	385,784 [1995–2003]	0.82[c]	Awah-Ndukum et al. (2010) (PM detection of TB lesions)
Bamenda abattoir	45,737 [1995–2003]	0.18[c]	
Bamenda abattoir	33,835 [2006–2008]	0.6[c]	
Dschang abattoir	1460 [2006–2008]	4.2[c]	
Bamenda abattoir	39 (zebu)	31[d]	Awah-Ndukum et al. (2010) (OIE and WHO standards)
		51[d,e]	
Bamenda abattoir	90 (zebu)	60[f]	Awah-Ndukum et al. (2010) (Manufacture protocol)
Bamenda abattoir	163 (zebu)	42.33[c]; >38[d,e]	Egbe et al. (2016) (OIE and WHO standards)
Bamenda abattoir	1129	3.99[c]; 2.75[d,e]	Egbe et al. (2016) (PM detection of TB lesions; OIE and WHO standards)
Maroua abattoir	175 (zebu)	27.42[c]; 68.57[f]; 16.0[d]	Awah-Ndukum et al. (unpublished)
	85 (zebu)	33.0[c]; 25.88[a]; 21.18[b]; 70.59[f]; 22.35[d]	(OIE standard cut-off point)
Maroua abattoir	122	14.75[c]; 13.1[d,e]	Egbe et al. (2016) (PM detection of TB lesions; OIE and WHO standards)
Garoua abattoir	160	23.75[c]; 21.3[e,d]	Egbe et al. (2016) (PM detection of TB lesions; OIE and WHO standards)
Ngaoundere abattoir	935	11.33[c]; 7.70[e,d]	Egbe et al. (2016) (PM detection of TB lesions; OIE and WHO standards)

(continued)

Table 12.1 (continued)

Locations	Number of animals	Prevalence (%) according to diagnostic techniques used	Reference (test interpretation standard used)
Yaounde abattoir	319,475 [2006–2012]	0.16[c]	Awah-Ndukum et al. (unpublished)
			(PM detection of TB lesions)
Yaounde and Douala	16,316	1.03[c]; 0.60[d]; 0.49[d,e]	Koro-Koro et al. (2013)
Yaounde abattoir	9127	0.81[c]	(PM detection of TB lesions; OIE and WHO standards)
Douala abattoir	7189	1.30[c]	
WHC and ADP	2853 [2009]	12.21[a]; 4.67[b]	Awah-Ndukum et al. 2012b (OIE standard cut-off point)
WHC	2126	14.03[a]; 5.38[b]	
Adamawa (Vina)	727	6.89[a]; 2.57[b]	
Bamenda abattoir	129,165 [1994–2010]	0.46[c]	
WHC and ADP	2853 [2009]	10.33[a]; 7.48[b]	Awah-Ndukum et al. (2012a)
WHC	2126	5.86[a]; 8.63[b]	
Adamawa (Vina)	727	11.86[a]; 4.10[b]	(OIE standard cut-off point and Manufacturer's protocol)
WHC and ADP	2853 [2009]	20.18[a]; 5.97[b]	Awah-Ndukum et al. (unpublished)
WHC	2126	11.18[a]; 6.97[b]	(OIE standard cut-off point and Manufacturer's protocol)
Adamawa (Vina)	727	23.25[a]; 3.05[b]	
WHC and ADP	1381 [2010]	18.35[a]; 7.41[b]	Awah-Ndukum et al. (2012a)
WHC and ADP	807	13.14[a]; 3.59[b]; 37.17[f]	
		23.42[a]; 6.31[b]; 43.24[f]	(OIE standard cut-off point and Manufacturer's protocol)
WHC	444 (zebu/exotic)	0.59[a]; 0.43[b]; 29.75[f]	
Adamawa (Vina)	363 (zebu)	24.68[a]; 9.89[b]	
WHC	1018 (zebu/exotic)		

[a]SIT: single intradermal tuberculin skin test
[b]CCT: single intradermal comparative cervical tuberculin skin test
[c]Slaughter/meat inspection
[d]AFS: acid-fast staining
[e]Liquid/solid culture of lesions
[f]Anti-bovine tuberculosis antibody assay (AntiBTBAb)
Year given in square bracket refers to duration of prevalence study
WHC and ADP, highlands of Cameroon (Western Highlands and Adamawa Plateau); *PM* postmortem

Bovine TB has been detected throughout the course of the year and the monthly prevalence ranges from 0.3 to 0.8% (Fig. 12.1). The observed increase in BTB in recent years seems to be associated with an increase in the number of cattle

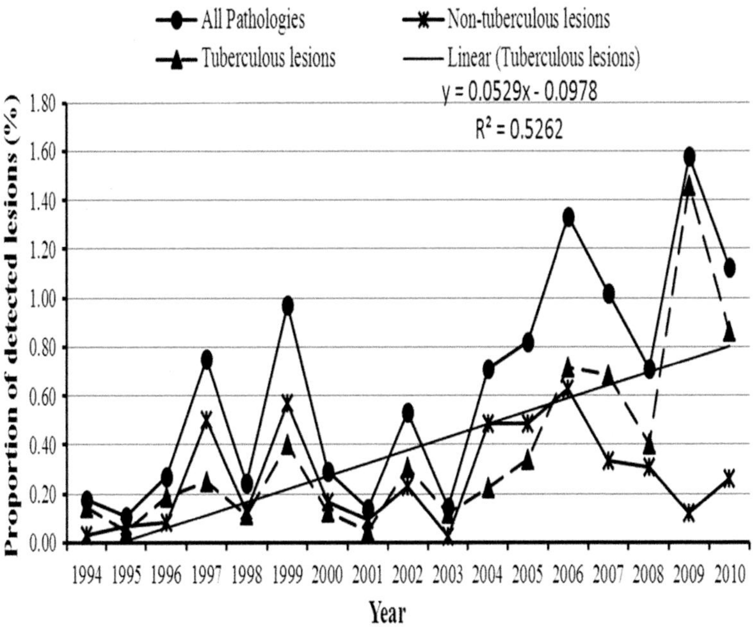

Fig. 12.1 Annual prevalence of tuberculous and non-tuberculous lesions recorded in slaughtered cattle at the Bamenda municipal abattoir, Cameroon

slaughtered and an increased efficiency of meat inspection (Awah-Ndukum et al. 2012b; Egbe et al. 2016). This increasing trend does not necessarily reflect an actual increase of the disease, but rather an improved diagnostic capability and better meat inspection procedures.

Molecular epidemiological data of BTB in Cameroon reveal a marked heterogeneity of strains with over 46 spoligotypes circulating in the country (Fig. 12.2). Mycobacteriological culture and spoligotyping confirm that though *M. bovis* is the principal etiological agent of BTB in Cameroon (Egbe et al. 2016; Awah-Ndukum et al. 2013), there is often a mixed and widely diverse infection of *M. bovis* and other *Mycobacterium* spp., including *M. tuberculosis*, *M. gordonae*, *M. phlei*, *M. fortuitum*, *M. mucogenicum*, and *M. scrofulaecum* (Egbe et al. 2016). The extent of their contribution to the disease burden is unknown, and confirmation of the cause of tuberculous lesions by mycobacteriological and molecular diagnostic techniques is essential to establish their importance.

For effective surveillance of BTB in Cameroon, meat inspection should be supported by tracing-back suspicious cases to the herds of origin to allow appropriate measures such as isolation and slaughter, and restriction of movement to other herds, to be instituted to limit further spread of the infection.

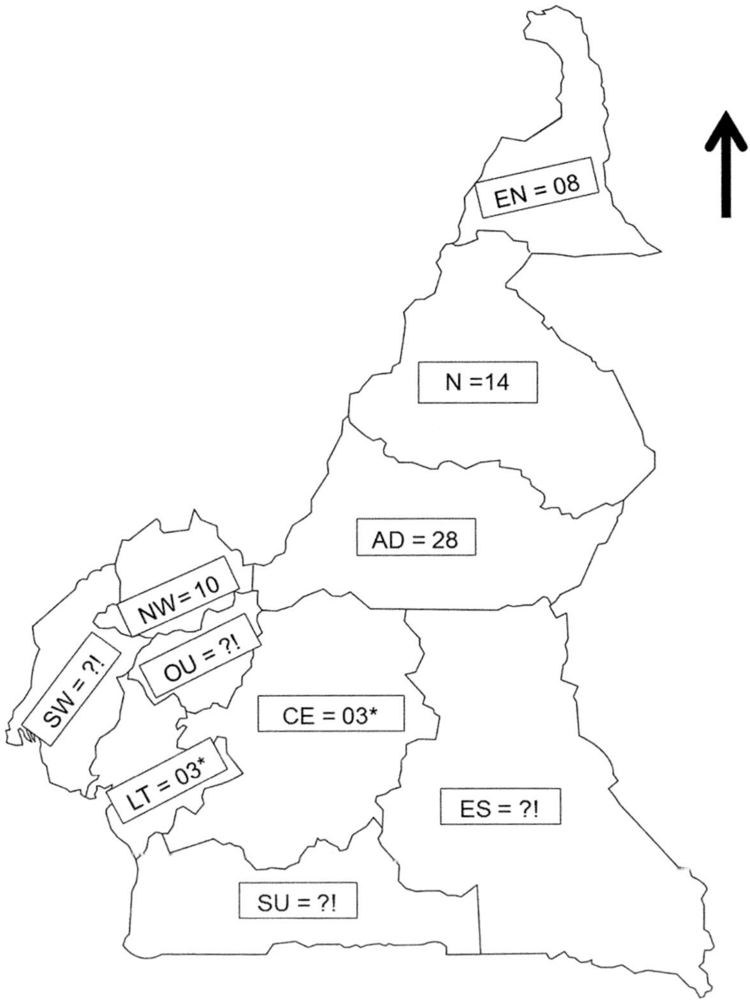

Fig. 12.2 Geographical distribution of the 46 *M. bovis* spoligotypes isolated in Cameroon. (1) Regions of Cameroon: EN (Far North); N (North); AD (Adamawa); NW (North West); OU (West); SW (South West); LT (Littoral); CE (Centre); SU (South); ES (East); (2) AD = SB0300, SB0893, SB0954, SB0955, SB1027, SB1099, SB1104, SB1418, SB1462, SB2313, SB2316, SB2317, SB2318, SB2319, SB2320, SB2321, SB2323, SB2327, SB2328, SB2330, SB2331, SB0944, SB0951, SB0953, SB1025, SB1460, SB2033, SB2162, SB2324; (3) EN = SB0944, SB0951, SB1460, SB1459, SB2325, SB2332, SB2333, SB0952; (4) N = SB0944, SB0951, SB1025, SB1460, SB2033, SB2324, SB1459, SB0120, SB2329, SB2334, SB2035, SB0952, SB1461, SB1463; (5) NW = SB0944, SB0953, SB2162, SB1026, SB2035, SB2161, SB2163, SB2164, SB2314, SB2315); (6) CE and LT = SB2033, SB2035, SB1419; (7) ?!: No data available (SW, OU, SU and ES are not major cattle producing areas); (8) *: cattle slaughtered in LT and CE mainly originate from AD, N, NW; (9) Spoligotype patterns were named according to www.Mbovis.org International database (Smith and Upton 2012). (Sources: Egbe et al. 2017; Awah-Ndukum et al. 2013; Koro-Koro et al. 2013; Njanpop-Lafourcade et al. 2001)

12.3 Evaluation of Diagnostic Tests Under Local Conditions

Based on the substantial variation in the results reported for the same groups of cattle by different investigators, it appears that there is a marked inconsistency in the application of criteria used for the interpretation of the test results. This implies that the performance of diagnostic tests and interpretation of results at various cut-off values have serious implications for assessing the actual status of BTB in Cameroon.

Since the available diagnostic tests were developed in different settings with different breeds of cattle, it was deemed necessary to assess the accuracy of various diagnostic techniques under Cameroonian conditions. As an example, performing the CCT in Maroua and using $\geq$2 mm, $\geq$3 mm, and $\geq$ 4 mm as cut-off points and the presence of visible BTB lesion as a reference test, the estimated sensitivities were, respectively, 78.5%, 67.8%, and 57.1%, and the specificities were, respectively, 85.9%, 94.7%, and 96.5% (Awah-Ndukum et al. unpublished data). Using the SIT at cut-off points of $\geq$3 mm and $\geq$ 4 mm, the respective sensitivity was 82.1% and 71.4%, and the specificity was 91.2% and 96.5%. When the presence of TB lesion in addition to Ziehl-Neelsen smear microscopy was used as the reference test to define the disease status, a sensitivity, respectively, of 100%, 89.4%, and 73.6% and a specificity of 83.3%, 92.4%, and 93.9% at $\geq$2 mm, $\geq$3 mm, and $\geq$ 4 mm cut-off points were obtained. Furthermore, receiver operating characteristic (ROC) analysis showed better performance at the $\geq$3-mm cut-off (over 91%) compared to the $\geq$4-mm cut-off point (84%) suggesting that in Cameroon a stricter interpretation of the specific tuberculin skin tests would detect more BTB-positive cattle. Bronsvoort et al. (unpublished data) similarly obtained better values for the sensitivity and specificity of the CCT at a cut-off value of $\geq$3 mm compared to $\geq$4 mm. Awah-Ndukum et al. (2012a) previously reported that irrespective of the tuberculin test cut-off values, there is a strong association between seroprevalence using the lateral flow rapid-based (immune-chromatographic) assay and the tuberculin test results.

The detection of lesions that are consistent with those caused by *M. bovis* during postmortal examination provides most of the information on which the prevalence and distribution of BTB in Cameroon are based. This technique too is flawed; there is ample evidence that under-recording and under-detection are very common, and the information provided is most likely a marked underestimation of the actual situation. Animals demonstrating poor health and diminished productivity are customarily the ones removed from herds and slaughtered for meat production. As they are usually disproportionately old-aged, abattoir inspection may not provide a true estimate of BTB in the local cattle population as the prevalence of BTB is expected to be higher in this group of animals.

Interpretation and the detection of TB-like lesions in slaughtered cattle can be difficult. Routine meat inspection generally only detects lesions in about 50% of carcasses containing tuberculous lesions. These lesions may resemble abscesses (with yellowish pus) or manifest as firm, yellowish, nodular lesions (often "gritty" on cutting), and are commonly detected in the lungs and associated lymph nodes (over 60%), followed by lymph nodes of the head, mesenteric lymph nodes, and the liver. The granulomatous lesions may easily be confused with parasitic granulomas, non-specific inflammatory reactions (Corner 1994; Shitaye et al. 2006; Edwards et al. 1997), and lesions caused by *Nocardia*, *Corynebacterium*, and other pyogranuloma-causing organisms (Grist 2008), and wrongly diagnosed as tuberculous unless confirmed by culture. Nonetheless, in slaughtered cattle, TB lesions were 3–5 times more prevalent than similar lesions caused by a different etiological agent.

12.4 The Epidemiology of BTB in Cattle in Cameroon

There are indications that the prevalence of BTB in Cameroon differs substantially between its regions. The reasons for these differences are not well understood and need further investigation. Generally, the gathering of animals at communal sites, mixing of different age groups (adult and aging), stressors such as adverse environmental factors, and long trekking to grazing and drinking spots are some of the major factors that appear to influence the prevalence and distribution of BTB in cattle in the country. Further factors that may affect its prevalence include the lack of application of quarantine measures; uncontrolled animal movement and smuggling of live animals from neighboring countries like Nigeria, Chad, and the Central African Republic where BTB is widespread (Cadmus et al. 2006; Diguimbaye et al. 2006a; Müller et al. 2009); and the mass influx of refugees and their livestock (displacement of human and animals resources) due to social unrest and civil wars in neighboring countries.

In Cameroon, many conditions favor the occurrence and spread of BTB in cattle (Awah-Ndukum et al. 2014; Kelly et al. 2016). For example, over 84% of cattle live to a very old age, and over 70% of cattle are kept in moderate to large herds, and in traditional, extensive (38%) or semi-extensive (58%) management systems. Many cattle trek at least 5 km daily for grazing and drinking (60%), and there are ample inter-herd intermingling and animal-animal contact (99%) when individual herds of the same or different owners come together (Table 12.2).

There are perceptions of apparent resistance of some of the indigenous African breeds in Cameroon to contracting BTB (Inangolet et al. 2008; Ameni et al. 2006; Oloya et al. 2006), although the prevalence of BTB in the predominantly zebu breeds of Cameroonian cattle still appears to be high. A lower prevalence of BTB was recorded in the Namchi, Gudali, and White Fulani breeds, compared to the Red Bororo cattle, suggesting that some of the local breeds may have a degree of innate resistance or tolerance to the disease. There are similar differences between the Gudali and Fulani breeds (Egbe et al. 2016). The reasons for the perceived resistance to BTB of local breeds may be more complex and include the possibility of reduced virulence and infectiousness of the causative agent, perhaps because of *M. bovis* being exposed to the effects of the harsh tropical climate (Oloya et al. 2006).

The burden of BTB is higher in the Northern Regions than in the Northwest (Fig. 12.1; Table 12.1), the odds of detecting gross TB lesions in carcasses being 4 times higher in cattle originating from the Northern Region compared to those

Table 12.2 Herd management and practices (values are in percentages)

Variable	Contact with other herd ($n = 318$: Yes = 315, No = 3)	Average daily trekking distance ($n = 316$: ADP = 208, WHC = 108)			Husbandry system ($n = 347$: AD = 228, WHC = 119)			Reasons for removing cattle ($n = 295$: ADP = 199, WHC = 96)		
		Long (>10 km)	Moderate (5 km - ≥ 10 km)	Short (≤ 5 km)	Extensive	Semi-intensive	Intensive	Old age	Income generation	Poor productivity
Highland region										
ADP	100.0	0.0	69.2	30.3	14.0	82.0	3.9	87.9	6.5	40.7
WHC	96.8	24.1	17.6	59.3	84.0	12.6	3.4	77.1	40.6	45.8
Total	99.1	8.2	51.58	40.2	38.0	58.2	3.7	84.4	17.6	42.4
Occupation										
Breeder	98.9	9.2	53.6	36.0	38.1	57.9	4.0	84.9	17.9	39.8
Butcher	100	50.0	0.0	50.0	66.7	0.0	33.3	50.0	50.0	50.0
"Buyem sellem"	100	0.0	0.0	75.0	71.3	28.6	0.0	66.7	50.0	50.0
Herdsmen	100	0.0	46.9	53.1	32.2	66.1	1.7	86.1	8.3	58.3
Educational level										
None	99.5	7.0	58.1	34.9	29.8	53.9	0.4	89.4	11.8	37.6
Primary	98.9	9.9	50.5	39.6	39.5	54.1	5.5	79.3	23.9	38.0
Secondary	96.0	14.8	22.2	63.0	50.0	38.2	11.8	50.0	50.0	50.0
Postsecondary	100	100	0.0	0.0	40.0	60.0	0.0	60.0	40.0	60.0
Not indicated	100	0.0	0.0	100	28.6	42.9	28.6	0.0	100.0	0.0

n total number of respondents, *ADP* Adamawa Plateau, *WHC* Western Highlands (Awah-Ndukum et al. unpublished; Awah-Ndukum et al. 2014)

from the Northwest Region (Egbe et al. 2016). A number of factors may affect this distribution, such as the uncontrolled movement of cattle in the Northern Regions, including those from neighboring countries that are BTB-infected. Elsewhere, such as in the Western Highlands and Far Northern Regions, continuous close contact between animals due to increasing animal and human population densities, and limited grazing, influence the prevalence. A lower prevalence has been recorded in regions with lower population densities, abundant natural pasture, and low herd-herd (animal-animal) contact (Awah-Ndukum et al. unpublished data).

Though intensive husbandry practices create favorable conditions for BTB transmission by providing opportunities for closer and prolonged contact between animals than those in extensive management systems (Ayele et al. 2004; Inangolet et al. 2008), in Cameroon, more BTB-positive reactions to tuberculin tests were recorded in animals kept under semi-extensive management compared to the other systems. In the extensive and semi-extensive systems, increased transmission of BTB can result from close contact, particularly during periods of drought, between animals sharing common grazing, water, and salt licks. Additionally, the intermingling of animals from different regions at cattle markets or during veterinary interventions may also enhance transmission of the infection (Ayele et al. 2004).

An important feature of BTB in Cameroon is the higher prevalence of the disease recorded in beef compared to dairy cattle, and in small compared to large herds (Fig. 12.3). This seems to result from fewer sources of *M. bovis*, or the absence of diseased cattle in dairy herds, and reduced contact between animals in the large herds that are usually managed in extensive farming systems. The characteristics of the traditional pastoral systems in the tropics, such as close contact between animals in shared microenvironments, as animals gather at the few watering points and salt-licks, or congregate under trees or shaded areas to protect them from the intense tropical ambient temperatures, mimic conditions created by intensive management systems (Ayele et al. 2004). These conditions often create increased opportunities for nose-to-nose and mouth-to-mouth contact between cattle, and enhance transmission of the disease.

Another important factor that favors the spread of BTB in the traditional pastoral husbandry system, is the high mobility of cattle herds, crisscrossing the regions during transhumance. Environmental stress associated with this movement and the ensuing mixing of transhumant and semi-intensive resident herds in the village, wildlife, and other domestic species create ideal conditions for exposure to and widespread transmission of *M. bovis*.

Although a wide diversity of wildlife and a vast livestock-wildlife interface exist in Cameroon, no data are available about the occurrence of BTB in its wildlife. Neither is there any information about the disease in the other economically important domestic species such as pigs, sheep, and goats. Several of these species, including wildlife, have been reported to serve as maintenance hosts for BTB in other countries (Courtenay et al. 2006). The need for a comprehensive investigation of the status of BTB in Cameroon, including establishing specific herd- and animal-level risk factors and identifying the maintenance hosts of the infection, and its zoonotic importance cannot be overemphasized.

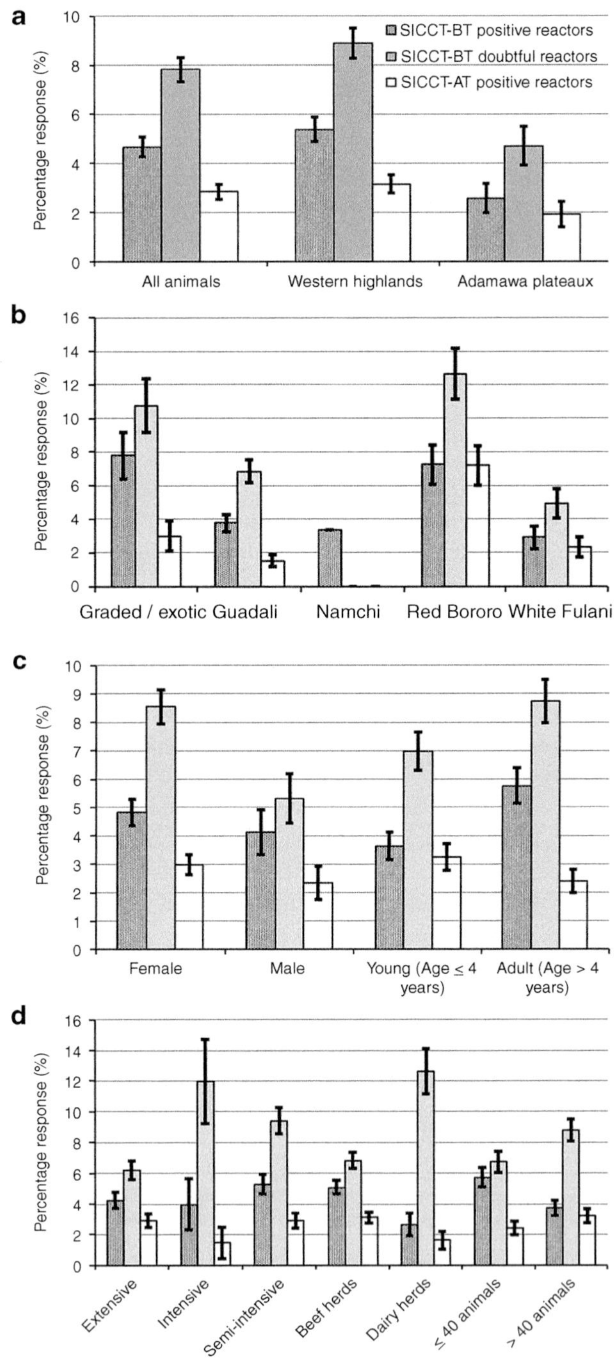

Fig. 12.3 Single intradermal comparative cervical tuberculin skin test (SICCT) responses according to (**a**) study location, (**b**) breed, (**c**) sex and age group, (**d**) management systems, and herd sizes (Awah-Ndukum et al. unpublished data)

12.5 Control of BTB in Cameroon

Although BTB has enormous public health significance, it is a neglected zoonosis in Cameroon. Governmental resources are inadequate for monitoring animal diseases including the zoonoses, and the private sector lacks the necessary capacity and resources to assume or to share the responsibility.

Implementation of existing legislature governing BTB is poor. The existing control programs are poorly applied, and the control of BTB is mainly dependent on the regulation of animal movements and by the postmortal examination of carcasses for the presence of tuberculous lesions. Further constraints include the poor execution of a number of activities: regular tuberculin testing and the removal of positive reactors, strict meat inspection and removal of suspected lesions, tracing tuberculous cases to farms or areas of origin, and restricting the movements of infected animals. There is also an absence of the application of broad hygienic and biosecurity measures reflecting deep-seated problems with respect to inter-institutional veterinary and medical collaboration causing them to not best serve the needs of society.

There are attitudinal problems by stock owners too in controlling the disease. Many cattle owners endeavor to increase their livestock numbers (increasing the size of their "living banks") but are unaware of the negative impact of BTB on animal productivity and health such as economic loss, lower production, and poor animal health (Table 12.3). Not many stockmen ($\approx$30%) can recognize the presence of BTB in their or adjacent herds, and most cattle owners (>86%) report that they do not implement known control measures to deal with the disease in their communities that are predominantly rural in "seminatural" or "semi-wild" areas (Table 12.4). They do, however, accept condemnations at abattoirs as an acceptable governmental measure to remove BTB-infected animals from the food chain.

Tuberculin skin testing and the elimination of infected animals (test-and-slaughter) that have been used effectively in other parts of the world (Good 2006; Pavlik 2006) are not practicable in Cameroon because of the lack of compensation when infected animals are slaughtered. Instead, testing and segregation of positive reactors, with phased slaughtering of the infected animals, could be economically and technically achievable as an alternative to the conventional test-and-slaughter method (WHO 1994). In the interim, there is a need to intensify meat inspection for the detection of BTB, to maintain reliable abattoir records, and to validate the various diagnostic tests used for screening live cattle for the presence of BTB in Cameroon.

There is an urgent need for a multidisciplinary approach in Cameroon, based on the "One Health" philosophy (Kahn et al. 2007; Vallat 2009), to control BTB. Specific activities in this regard include enhancing public awareness through continuing education of cattle owners, medical and veterinary professionals, and the general public on the potential risk of BTB, proper food (animal products) handling, good husbandry practices, personal hygiene, and maintaining a healthy environment. Restricting movements of BTB-infected herds, cooperative efforts by veterinary and medical personnel to maximize TB detection rates, engaging the populations at risk, and good health surveillance systems are essential activities that should be implemented to ensure effective control of the disease in humans and animals.

Table 12.3 Factors affecting meat/milk consumption habit of cattle owners (values are in percentages)

Variable	Milk is used for consumption ($n = 295$: ADP = 196, WHC = 99)	Fresh milk consumption ($n = 475$: ADP = 296, WHC = 179)		Meat consumption ($n = 477$: ADP = 297, WHC = 180)	
		Boiled or raw	Raw milk	Raw meat	Suya and Kilishi
Highland region					
ADP	93.4	82.8a	66.9a	16.8a	97.3a
WHC	92.9	79.6b	50.8b	14.3a	50.5b
Total	93.2	81.6	60.8	15.9	79.4
Occupation					
Cattle breeder	95.2	88.3c	71.1c	13.2b	81.3c
Butcher	100.0	62.6d	37.4d	15.5b	79.1de
"Buyem sellem"	80.0	68.6d	31.4d	5.7b	60.0e
Herdsmen	82.0	95.0ce	75.0c	31.7c	85.0d
Education (school level)					
None	96.0	90.5f	79.7e	22.0d	83.9d
Primary school	88.6	74.8g	47.4f	9.9e	79.6g
Secondary school	86.4	68.3g	38.0g	13.3e	73.3h
Postsecondary	100	83.3	16.7	0.0	66.7
Not indicated	100	73.1g	46.2f	11.5e	53.8j

n = total number of respondents, *ADP* Adamawa Plateau, *WHC* Western Highlands, a–k: different letters in the same column are significantly different ($P < 0.05$) (Awah-Ndukum et al. unpublished data; Awah-Ndukum et al. 2014)

12.6 Current Status of Human Tuberculosis in Cameroon

Human tuberculosis (TB) is highly prevalent in Cameroon, and it has a current annual incidence of over 200 cases per 100,000 population (WHO 2009) (Table 12.5). The unfavorable local socioeconomic conditions, the close association of TB with the HIV epidemic, and the widespread mycobacterial resistance to drugs used in the treatment of TB hamper its control (Kuaban et al. 2000a). Because of the close association between the re-emergence of TB and the emerging HIV/AIDS epidemic in Cameroon (WHO 2009) (Table 12.5), the seroprevalence of HIV in TB patients serves as an accurate indicator of the prevalence of HIV in the general population (Noeske et al. 2004). In addition to HIV/AIDS, other factors such as poverty, malnutrition, stress, and smoking are important risk factors predisposing to contracting the disease (Pešut et al. 2008).

Table 12.4 Degree of interaction of cattle handlers with their cattle and their knowledge of zoonotic BTB and its modes of transmission (values are in percentages)

Variable	Proportion of respondents ($n = 489$)		Contact between cattle handlers and animals ($n = 489$: ADP = 302, WHC = 187)		Know BTB is zoonotic ($n = 477$)	Mode of transmission of BTB to humans ($n = 477$: ADP = 298, WHC = 179)		
		Own other livestock spp.	Daily contact with cattle herds	≥ 1-day contact with cattle herds per week		Knows milk is a vehicle	Knows raw meat is a vehicle	Knows inhalation route (aerosol)
Highland region								
ADP	61.8	8.3a	87.8a	12.3a	51.7a	14.4a	43.6a	11.4a
WHC	38.2	73.8b	86.1a	13.9a	62.0b	14.0a	46.4a	19.5b
Total	100	33.3	87.1	12.9	55.6	14.3	44.7	14.5
Occupation								
Cattle breeder	57.7	35.5c	83.7b	16.3bd	53.1 cd	19.3b	43.6bc	16.4c
Butcher	22.5	28.2d	97.3c	2.7c	63.0c	7.4c	55.6b	12.0d
"Buyem sellem"	7.6	64.9e	75.7b	24.3b	67.6c	5.9c	35.3c	8.8d
Herdsmen	12.3	13.3d	91.7c	13.3d	46.7d	8.3c	36.7c	13.3d

n = total number of respondents, *ADP* Adamawa Plateau, *WHC* Western Highlands, a–k: different letters in the same column are significantly different ($P < 0.05$) (Awah-Ndukum et al. unpublished data; Awah-Ndukum et al. 2014)

Table 12.5 Estimated burden of human TB in Cameroon, 2000–2007

Parameter	2000	2001	2002	2003	2004	2005	2006	2007
Incidence[a] rates of TB (per 100,000 population)	168	181	194	202	204	202	197	192
Prevalence[a] rates of TB (per 100,000 population)	228	241	240	227	228	2103	201	195
Incidence rates of TB/HIV-positive number (per 100,000 population)	77	83	88	91	91	89	86	83
Prevalence rates of TB/HIV-positive number (per 100,000 population)				20	20	16	15	41
Prevalence of HIV in incident TB cases of all ages[b] (%)				–	31	26	15	43

[a]Incidence and prevalence estimates include TB in people with HIV
[b]Prevalence of HIV in incident TB cases of all ages [adapted from Global TB control, surveillance, planning, and financing (WHO 2009)]

The prevalence of human TB in Cameroon is still increasing, especially in the economically active age group (21–40 years) and in immunocompromised individuals, such as those suffering from HIV/AIDS (WHO 2009; Noeske et al. 2004; Ane-Anyangwe et al. 2006). About 40% of all TB patients in the country are also HIV-positive, and the prevalence of HIV both in the general adult population and in adults with TB is steadily increasing. More recent studies of HIV/AIDS in TB cases of all ages in the general population indicated an even higher prevalence (ranging from 31 to 43%) (WHO 2009). Due to the magnitude and the increase of the two infections in Cameroon, it is expected that even after the HIV epidemic in the general population reached its peak and starts to decline, TB/HIV coinfection would probably continue to occur since TB can be contracted at any time during the course of an HIV infection (Noeske et al. 2004).

12.7 Zoonotic (*M. bovis*) TB in Humans in Cameroon

Tuberculosis (TB) is globally a leading cause of human death by a single causal infectious agent. The extent of human TB caused by *M. bovis* is unknown, but it seems to account for only a small percentage of the cases of TB reported in humans. Thus, in the developed countries, *M. bovis* accounts for less than 0.5–7.2% of the cases of human TB, while in developing countries, 10–15% or more are estimated to be caused by the infection (de la Rua-Domenech 2006). Reliable data are unavailable about the number of human cases of TB caused by *M. bovis* in Cameroon, due to the lack of attention to the problem by policy makers, and the limited number of available diagnostic facilities able to identify *M. bovis*. In most of the rural areas, for TB, the main diagnostic approach is the detection of acid-fast bacilli (AFB) in the sputum of suspected cases by direct microscopy, and only urban reference laboratories attempt to isolate the organism. These referral laboratories, however, make little or no attempt to identify the isolates to the strain/species level, thus probably missing many *M. bovis* cases. Their argument is that the same drug regimen would be used to treat all TB cases, although resistance of *M. bovis* to some of the first-line anti-TB drugs has been reported worldwide (Diguimbaye et al. 2006b; Gibson et al. 2004), including in Cameroon (Kuaban et al. 2000a, b). Currently, the level of public awareness of the zoonotic risk of BTB, and the lack of animal and human health surveillance programs to control TB, is very limited; this is in spite of an established epidemiologic association between tuberculin-positive cattle and human TB in several countries, such as in Ethiopia and Zambia (Regassa et al. 2008).

Because of a number of practices that may enhance the transmission of the disease, there is concern that zoonotic *M. bovis* infections may be more prevalent in Cameroon than anticipated. Given the close association between livestock and humans because of prevailing farming practices in Cameroon, the high prevalence of BTB in the indigenous cattle could contribute to an increase in the prevalence of human TB. Opportunities exist for the transmission of *M. bovis* to humans because of the very close human-livestock interaction (Table 12.4) favoring aerosol transmission, and the habit of consuming unpasteurized milk (Thoen et al. 2006) and raw meat (Table 12.3). Other people at risk, such as butchers, abattoir workers, and those with low levels of education, appear not to be well informed about the risk posed by and the modes of transmission of zoonotic TB. Of TB patients surveyed in Bamenda (Awah-Ndukum et al. 2014), only 17.3% were aware of zoonotic TB, and Kelly et al. (2016) too reported a low level of awareness by both dairy farmers and cattle owners/pastoralists that consuming milk could cause zoonotic TB. A large segment of the population (32.1%) drink fresh milk (pasteurized or unpasteurized), 19.8% drink unpasteurized milk, 2.5% eat raw meat, and 61.3% eat *Suya* that is meat briefly roasted over hot charcoal or fire. The practice of pooling milk from cows of several owners could further increase the risk to a larger number of people consuming the raw product (Awah-Ndukum et al. 2014). Under these conditions, particularly those suffering from HIV/AIDS (Noeske et al. 2004) are at risk (Berg et al. 2009).

PCR-based genomic deletion analyses showed evidence of *M. bovis* in human and *M. tuberculosis* in cattle samples (Awah-Ndukum et al. 2011) suggesting possible animal-to-human and human-to-animal transmission cycles. Molecular analyses confirmed the presence of *M. bovis* in humans with pulmonary TB in the Western and Northwestern regions of Cameroon (Niobe-Eyangoh et al. 2003; Egbe et al. 2017). This indicates the possibility of human-to-human transmission of *M. bovis* by the respiratory route. Similarly, a range of mycobacteria, including *M. tuberculosis* and several mycobacteria other than tubercle (MOTT) bacilli, has been isolated from cattle (Egbe et al. 2016).

The epidemiologic link between TB in humans and BTB in cattle, with the possibility of a cattle-human-cattle cyclical transmission, is of serious concern in Cameroon due to issues related to drug resistance. *Mycobacterium bovis* strains isolated from human TB cases also have different drug susceptibility profiles compared to *M. tuberculosis* strains (Diguimbaye et al. 2006b; Gibson et al. 2004).

The contribution of BTB to the overall TB morbidity and mortality in Cameroon thus needs a broad investigative approach. This would involve establishing the epidemiology of the disease in cattle and human populations, identification of TB-causing agents and their respective sources, maintenance hosts, possible routes of transmission, and associated risk factors.

The public health threat of *M. bovis* in Cameroon therefore requires the urgent attention of veterinary and medical professionals, biomedical and ecological experts, social workers, and policy makers within the context of the "One Health" approach.

12.8 Conclusion

The prevalence of BTB in Cameroon is high and it poses a major zoonotic threat to the human population. Most herdsmen and livestock handlers, small-scale farmers, nomads, and wage laborers in Cameroon are poorly educated about the presence and the consequences of BTB. Currently, urban and peri-urban livestock farming is growing fast, bringing animals closer to the large urban populations, thus exposing more and more people to the zoonotic risk of the disease. These changes are occurring in the absence of any meaningful BTB control policy, including restriction of cattle movement within the country, and from neighboring countries. Due to the significant public health implications, prompt action to control BTB in both animals and humans in Cameroon is required.

Establishing basic biosecurity measures (such as farm and personal hygiene, good husbandry practices, movement control, regular health inspection, and proper food handling) and public education programs to raise awareness about zoonotic TB are the activities that need immediate attention. For the long-term control of BTB in animals, assessment of the full extent of the problem in Cameroon is required. Therefore, a comprehensive molecular epidemiological study is needed to provide accurate data about cattle-cattle transmission and the role of wildlife (and domestic animals) reservoirs in the maintenance and transmission of TB in animals and humans. Only once the full extent of the problem has been determined, can the most appropriate and cost-effective control measures for Cameroon be planned and implemented.

References

Ameni G, Aseffa A, Engers H et al (2006) Cattle husbandry in Ethiopia is a predominant factor affecting the pathology of bovine tuberculosis and gamma interferon responses to mycobacterial antigens. Clin Vacc Immunol 13:1030–1036

Ane-Anyangwe IN, Akenji TN, Mbacham WF et al (2006) Seasonal variation and prevalence of tuberculosis among health seekers in the South Western Cameroon. East Afr Med J 83:588–595

AU/IBAR (2006) Pan African Animal Health Yearbook 2006. African Union/Inter-African Bureau for Animal Resources Nairobi, Kenya

Ayele WY, Neill SD, Zinsstag J et al (2004) Bovine tuberculosis: an old disease but a new threat to Africa. Int J Tuberc Lung Dis 8:924–937

Awah-Ndukum J, Kudi AC, Bradley G (2010) Prevalence of bovine tuberculosis in abattoirs of the Littoral and Western highland regions of Cameroon: a cause for public health concern. Vet Med Int 2010:8. https://doi.org/10.4061/2010/495015

Awah-Ndukum J, Kudi AC, Bradley G et al (2011) Preliminary report on the zoonotic significance of tuberculosis in cattle in Cameroon. In: Kofer J, Schobesberger H (eds) Animal health and sustainable livestock production. Proceedings of the 15th International Congress on animal hygiene organized by the International Society of animal hygiene, Vienna, Austria. Austria, 3–7 Jul 2011, Tribun EU 1, pp 193–195

Awah-Ndukum J, Kudi AC, Bah GS et al (2012a) Bovine tuberculosis in cattle in the highlands of Cameroon: seroprevalence estimates and rates of tuberculin skin test reactors at modified cut-offs. Vet Med Int 2012(Article ID 798502):13. https://doi.org/10.1155/2012/798502

Awah-Ndukum J, Kudi AC Bradley B et al (2012b) Prevalence of bovine tuberculosis in cattle in the highlands of Cameroon based on the detection of lesions in slaughtered cattle and tuberculin skin tests of live cattle. Vet Medicina 57:59–76

Awah-Ndukum J, Kudi AC, Bradley G et al (2013) Molecular genotyping of *Mycobacterium bovis* isolated from cattle tissues in the North West region of Cameroon. Trop Anim Health Prod 45:829–836

Awah-Ndukum J, Kudi AC, Bah CS (2014) Risk factors analysis and implications for public health of bovine tuberculosis in the highlands of Cameroon. Bull Anim Health Prod Afr 62:353–376

Berg S, Firdessa R, Habtamu M et al (2009) The burden of mycobacterial disease in Ethiopian cattle: Implications for public health. PLoS One 4(4):e5068

Cadmus S, Palmer S, Okker M et al (2006) Molecular analysis of human and bovine tubercle bacilli from a local setting in Nigeria. J Clin Microbiol 44:29–34

Cadmus SI, Adesokan HK, Jenkins AO et al (2009) *Mycobacterium bovis* and *M. tuberculosis* in goats, Nigeria. Emerg Infect Dis 15:2066–2067

Corner LA (1994) Post mortem diagnosis of *Mycobacterium bovis* infection in cattle. Vet Microbiol 40:53–63

Courtenay O, Reilly LA, Sweeney FP et al (2006) Is *Mycobacterium bovis* in the environment important for the persistence of bovine tuberculosis? Biol Lett 2:460–462

de la Rua-Domenech R (2006) Human *Mycobacterium bovis* infection in the United Kingdom: Incidence, risks, control measures and review of the zoonotic aspects of bovine tuberculosis. Tuberculosis 86:77–109

Diguimbaye DC, Hilty H, Ngandolo R et al (2006a) *Mycobacterium bovis* isolates from tuberculous lesions in Chadian Zebu carcasses. Emerg Infect Dis 12:769–771

Diguimbaye DC, Hilty M, Ngandolo R et al (2006b) Molecular characterization and drug resistance testing of *M. tuberculosis* isolates from Chad. J Clin Microbiol 44:1575–1577

Edwards DS, Johnston AM, Mead GC (1997) Meat inspection: an overview of present practices and future trends. Vet J 154:135–147

Egbe NF, Muwonge A, Ndip L et al (2016) Abattoir-based estimates of mycobacterial infections in Cameroon. Sci Rep 6:24320. https://doi.org/10.1038/srep24320

Egbe NF, Muwonge A, Ndip L, Kelly RF, Sander M, Tanya V, Ngu Ngwa V, Handel IG, Novak A, Ngandalo R, Mazeri S, Morgan KL, Asuquo A, Bronsvoort BMdeC (2017) Molecular epidemiology of Mycobacterium bovis in Cameroon. Sci Rep 7:4652. https://doi.org/10.1038/s41598-017-04230-6

FAOSTAT (2014) Cameroon livestock population. http://www.fao.org/faostat/en

Gibson AL, Hewinson G, Goodchild T et al (2004) Molecular epidemiology of disease due to *Mycobacterium bovis* in humans in the United Kingdom. J Clin Microbiol 42:431–434

Good M (2006) Bovine tuberculosis eradication in Ireland. Ir Vet J 59:154–162

Grist A (2008) Bovine Meat Inspection—Anatomy, physiology and disease conditions, 2nd edn. Nottingham University Press, Nottingham, p 296

Hiko A, Agga G (2011) First-time detection of *Mycobacterium* species from goats in Ethiopia. Trop Anim Health Prod 43:133–139

Hlokwe TM, Jenkins AO, Streicher EM et al (2011) Molecular characterization of *Mycobacterium bovis* isolated from African buffaloes (*Syncerus caffer*) in Hluhluwe-iMfolozi Park in KwaZulu-Natal, South Africa. Onderstepoort J Vet Res 78:232

Inangolet F, Demelash B, Oloya J et al (2008) A cross-sectional study of BTB in the transhumant and agro-pastoral cattle herds in the border areas of Katakwi and Moroto districts, Uganda. Trop Anim Health Prod 40:501–508

Kahn LH, Kaplan B, Steele JH (2007) Confronting zoonoses through closer collaboration between medicine and veterinary medicine (as 'one medicine'). Vet Italia 43:5–19

Katale BZ, Mbugi EV, Kendal S et al (2012) Bovine tuberculosis at the human-livestock-wildlife interface: Is it a public health problem in Tanzania? A review. Onderstepoort J Vet Res 79:84–97

Kelly RF, Hamman SM, Morgan KL et al (2016) Knowledge of bovine tuberculosis, cattle husbandry and dairy practices amongst pastoralists and small-scale dairy farmers in Cameroon. PLoS One 11:e0146538. https://doi.org/10.1371/journal.pone.0146538

Koro-Koro F, Bouba FE, Ngatchou AFS et al (2013) First insight into the current prevalence of bovine tuberculosis in cattle slaughtered in Cameroon: the case of main abattoirs of Yaoundé and Douala. Br Microbiol Res J 3:272–279

Kuaban C, Bercion R, Jifon G et al (2000a) Acquired anti-tuberculosis drug resistance in Yaounde, Cameroon. Int J Tuberc Lung Dis 4:427–432

Kuaban C, Bercion R, Noeske J et al (2000b) Anti-tuberculosis drug resistance in the West Province of Cameroon. Int J Tuberc Lung Dis 4:356–360

Martrenchar A, Njanpop BM, Yaya A et al (1993) Problems associated with tuberculosis and brucellosis skin-test methods in northern Cameroon. Prev Vet Med 15:221–229

Merlin P, Tsangueu P (1985) Incidence de la tuberculose bovin dans le nord ouest du Cameroun. Rev SciTech (Organisation mondiale de la santé animale - OIE) 1:89–93

Muchaal PK (2002) Assessment of Bovine Tuberculosis (*Mycobacterium bovis*) and risk factors of transmission in the peri-urban centres of Bamenda, Northwest Province (Cameroon). Urban Agriculture and Zoonoses in West Africa: an assessment of the potential impact on public health. CFP Report 35. The International Development Research Centre (IDRC), Ottawa

Müller B, Hilty M, Berg S et al (2009) African 1, an epidemiologically important clonal complex of *Mycobacterium bovis* dominant in Mali, Nigeria, Cameroon, and Chad. J Bacteriol 191:1951–1960

Müller B, Dürr S, Alonso SJ et al (2013) Zoonotic *Mycobacterium bovis*-induced tuberculosis in humans. Emerg Infect Dis 19:899–908

Nfi AN, Ndi C (1997) Bovine tuberculosis at the animal research antenna (ARZ) Bangangte, Western province, Cameroon. Bull Anim Prod Health Afr 45:1–3

Niobe-Eyangoh SN, Kuaban C, Sorlin P et al (2003) Genetic biodiversity of *Mycobacterium tuberculosis* complex strains from patients with pulmonary tuberculosis in Cameroon. J Clin Microbiol 41:2547–2553

Njanpop-Lafourcade BM, Inwald J, Ostyn A, Durand B, Hughes S, Thorel MF, Hewinson G, Haddad N (2001) Molecular typing of *Mycobacterium bovis* isolates from Cameroon. J Clin Microbiol 39:222–227

Noeske J, Kuaban C, Cunin P (2004) Are smear-positive pulmonary tuberculosis patients a 'sentinel' population for the HIV epidemic in Cameroon? Int J Tuberc Lung Dis 8:346–351

Oloya J, Opuda-Asibo J, Djønne B et al (2006) Responses to tuberculin among Zebu cattle in the transhumance regions of Karamoja and Nakasongola district of Uganda. Trop Anim Health Prod 38:275–283

Pavlik I (2006) The experience of new European Union Member States concerning the control of bovine tuberculosis. Vet Microbiol 112:221–230

Pešut DP, Gledović ZB, Grgurević AD et al (2008) Tuberculosis incidence in elderly in Serbia: Key trends in socioeconomic transition. Croatian Med J 49:807–812

Regassa A, Medhin G, Ameni G (2008) Bovine tuberculosis is more prevalent in cattle owned by farmers with active TB in central Ethiopia. Vet J 178:119–125

Shitaye JE, Getahun B, Alemayehu T et al (2006) A prevalence study of bovine tuberculosis by using abattoir meat inspection and tuberculin skin testing data, histopathological and IS6110 PCR examination of tissues with tuberculous lesions in cattle in Ethiopia. Vet Medicina 51:512–522

Smith NH, Upton P (2012) Naming spoligotype patterns for the RD9-deleted lineage of the *Mycobacterium tuberculosis* complex; www.Mbovis.org. Infect Genet Evol 2:873–876

Tanya VN (2004) The contribution of animal and fisheries research to poverty alleviation in Cameroon. In: Mbiapo FT, Etoa FX (eds) Proceedings of the 11th Annual Conference of Bioscience: Animal production and poverty alleviation, 10th edn. The Cameroon Bioscience Society, Yaoundé, pp 1–6

Tanya VN, Sallah JNS, Tayou KR (1985) Screening for bovine tuberculosis at Wakwa. Rev Sci Tech OIE Ser Sci Agronomiques et Zootechniques (Cameroon) 1:65–68

Thoen CO, LoBue PA, de Kantor I (2006) The importance of *Mycobacterium bovis* as a zoonosis. Vet Microbiol 112:339–345

Vallat B (2009) One World, One Health. OIE Bull 2:1–2. http://www.oie.int/en/for-the-media/editorials/detail/article/one-world-one-health/

WHO (1994) Zoonotic tuberculosis (*Mycobacterium bovis*) memorandum from a WHO meeting (with the participation of FAO). Bull World Health Org 72:851–857

WHO (2009) Global Tuberculosis Control: Epidemiology, Strategy, Financing. WHO Report: WHO/HTM/TB/2009, vol 411. World Health Organization, Geneva

Zinsstag J, Kazwala RR, Cadmus I et al (2006) *Mycobacterium bovis* in Africa. In: Thoen CO, Steele JH, Gilsdorf MJ (eds) *Mycobacterium bovis* infection in animals and humans, 2nd edn. Blackwell, Iowa, pp 199–210

Chapter 13
Bovine Tuberculosis in Egypt

Aziza Amin

13.1 Introduction

In Egypt, tuberculosis (TB) is an ancient disease, and it is known to have been present from the beginning of this civilization, over 5000 years ago, and was relatively common during the predynastic (ca. 3500–2650 BC) to the late period (ca. 1450–500 BC). The characteristic features of the disease, Pott's deformities in the skeleton of Egyptian mummies, are also vividly depicted in early Egyptian art.

The occurrence of TB throughout ancient Egypt appears to have been surprisingly high due to the dense crowding in cities at times of prosperity. A rough estimate of its prevalence in these populations also indicates an even spatial and temporal distribution in distinct ancient Egyptian populations (Zink et al. 2007). The unexpectedly high frequency of mycobacterial DNA detected in ancient human bone samples suggests the systemic spread of the mycobacteria in infected people and is indicative of the presence of generalized TB in those populations.

According to the estimates of the WHO (2012), 95% of all human TB cases and 99% of deaths due to TB occur in developing countries. Tuberculosis also has a major socioeconomic impact since 75% of people with TB are in the economically productive age group of 15–54 years. Currently, after hepatitis C and bilharziasis, human TB is the third largest killer in Egypt. Approximately 19,000 new human TB cases are reported in Egypt annually, of which, an estimated 9500 suffer from the infectious, smear-positive pulmonary form.

Bovine tuberculosis (BTB), primarily caused by *M. bovis*, is a chronic infectious disease of cattle that inflicts significant losses in infected herds. The bacterium can also infect a large number of other animal species including a wide variety of wildlife

A. Amin (✉)
Faculty of Veterinary Medicine, Pathology Department, Benha University, Benha, Al Qalyubia Governorate, Egypt
e-mail: aziza.amin@fvtm.bu.edu.eg

A. B. Dibaba et al. (eds.), *Tuberculosis in Animals: An African Perspective*,
https://doi.org/10.1007/978-3-030-18690-6_13

species, and its transmission to humans constitutes a significant public health problem (Ameni et al. 2007). Bovine TB is now regarded as one of the most serious animal health problems of cattle in Egypt (Hassanain et al. 2009).

According to a recent report of the General Organization of Veterinary Services (GOVS), Egypt, the annual number of BTB-infected cattle in Egypt is increasing, probably as a consequence of the importation of live animals from countries where BTB is prevalent. It is well known that zoonotic TB caused by *M. bovis* following ingestion of milk and other dairy products from *M. bovis*-infected cattle is a serious public health risk. Human infection due to the inhalation of infected droplets released by *M. bovis* infected animals is usually found in groups of people that are in close contact with animals, such as slaughterhouse workers and farmers (O'Reilly and Daborn 1995), and in developing countries meat handlers are also at risk to contract BTB (Hambolu et al. 2013). Transmission of *M. bovis* between animals, and from animals to humans, represents a major health risk given the complexity of animal husbandry and animal trade systems prevailing in Egypt.

In this chapter an overview of the significance of BTB in Egypt is presented, with emphasis on the common diagnostic tools used, prevalence of BTB and its geographical distribution, and existing BTB eradication schemes.

13.2 Diagnosis of Bovine Tuberculosis in Egypt

Accurate detection of infected animals, particularly during the early phases of the disease, is critical for the effective control of BTB. Prompt diagnosis of BTB is important to identify and remove infectious cases, and failure to do so allows ongoing transmission of *M. bovis* within herds with consequent increased public health risk to humans because of the increasing prevalence of BTB in these herds.

Several diagnostic tests are used for BTB surveillance in Egypt; some are solely used for research purposes. As reported elsewhere (Bezos et al. 2014), all current BTB diagnostic tests have several limitations that must be taken into account when dealing with the control of the disease.

13.2.1 Tuberculin Skin Test in Cattle

The tuberculin skin test suffers from a lack of sensitivity and specificity, but it is universally accepted as the preferred diagnostic method for BTB in live cattle, and it forms the basis for conducting national test-and-slaughter BTB control programs (Ayele et al. 2004). The test is labor- and time-intensive, as it requires a second inspection of the test animals after 72 h to evaluate the reaction caused by the intradermal injection of PPD.

In Egypt, both the single (SIT), more commonly, and the comparative cervical (CCT) intradermal tuberculin tests are used. For the SIT, 0.1 ml of approved bovine PPD is inoculated intradermally in the mid-neck skin, and after 72 h, the skinfold thickness is measured again. According to the regulations of GOVS, a reaction is considered positive if there is an increase in skin thickness of >4 mm at the injection site; while an increase of <3 mm in skin thickness is considered as negative, and an increase of between 3 and 4 mm is interpreted as suspicious. The SIT has practical drawbacks such as the lack of specificity because of false-positive reactions due to sensitization by other mycobacteria, and the development of non-specific, local inflammatory reactions (Bezos et al. 2014; Humblet et al. 2011).

The CCT is not routinely used by the GOVS due to financial constraints but researchers and herd owners often fund it privately. The interpretation of the CCT is based on the method of Ovdiennkop et al. (1987): the test is considered positive, negative, or doubtful, respectively, if the difference in increase in skinfold thickness at the bovine PPD injections site is greater than the corresponding increase at the avian PPD injection site by >4 mm, <3 mm, or between 3 and 4 mm. Even the CCT lacks sensitivity and specificity. For example, of tissues collected from 32 CCT-positive cattle, only 20 (62.5%) were positive for *M. bovis* when cultured (Ramadan et al. 2012). Similarly, of 36 tuberculin-positive cases (9 cattle and 27 buffaloes), 1 of the cows (2.1%) and 23 of the buffaloes (85.2%) had no visible lesions when examined postmortally (Zahran et al. 2014).

The major problem that Egypt faces in its fight against BTB, and the reason for its inability to adequately control the disease, is not the diagnostic limitations of the tests mentioned above but the fact that only about a quarter of the cattle and buffalo populations in Egypt are tuberculin-tested annually, mainly because of a lack of funding.

13.2.2 Meat Inspection and the Detection of Tuberculous Lesions

Postmortal examination of slaughtered animals is considered as an acceptable alternative for detecting BTB in cattle herds although it too lacks sensitivity. Only about 50% of tuberculous cattle are generally detected during routine meat inspection in abattoirs. Lesions may also be present in organs that are not routinely inspected. In addition, the presence of no-visible-lesion (NVL) reactors, particularly during the early stages of the infection when only microscopic lesions are present (Ramadan et al. 2012; Danbirni et al. 2013; Zahran et al. 2014), further reduces the number of positive animals that are detected at slaughter. Not all specimens of tuberculous-like lesions yield *M. bovis* on culture, and it is known that other bacterial and parasitic infections can cause lesions resembling those caused by *M. bovis* (Mohamed et al. 2009). Despite these limitations, and in the absence of other diagnostic techniques, proper meat inspection may be used as a method to detect tuberculous cattle in resource-poor countries (Cosivi et al. 1998).

Other organisms also cause tuberculous-like lesions, and the misinterpretation of a cause may result in tremendous economic losses resulting from the unnecessary condemnation of carcasses containing lesions considered to be due to *M. bovis* infections.

13.2.3 Direct Smear, Culture, and Molecular Techniques

The microscopic detection of acid-fast bacilli (AFBs) in tissue smears is a rapid and cost-effective diagnostic method; however, it too lacks the required sensitivity. In one study acid-fast bacilli were detected in only 49 (73.1%) of smears made from 67 culture-positive specimens (Mohamed et al. 2009).

In general, culture is considered the gold standard for the diagnosis of BTB, but it has disadvantages in that the growth of the tubercle bacilli is slow and it may take 6–12 weeks to detect growth on pyruvate-containing media required for primary culture of *M. bovis* (Zahran et al. 2014). For successful culturing viable organisms must be present in sufficient numbers, and the specimens must also be appropriately handled to ensure survival of the bacteria.

Alternative methods using DNA technology, such as the polymerase chain reaction (PCR) have also been assessed for their suitability under Egyptian conditions (Mohamed et al. 2009; Ramadan et al. 2012; Zahran et al. 2014). Molecular diagnostic techniques are known to have rapid turnaround, they also allow identification of species of mycobacteria other than tubercle bacilli (MOTT) and related genera based on amplification of species-specific targets, but when compared to culture and conventional biochemical tests, they have not shown any superiority in detecting *M. bovis* from clinical specimens in Egypt.

13.3 Prevalence of BTB in Egypt

In developing countries, especially in Africa, where *M. bovis* infection has been reported in a wide range of animal species, there is generally insufficient reliable information about the current distribution, epidemiology, and zoonotic impact of this important zoonosis. Generally, to estimate the prevalence of BTB in Egypt, data are collected from national TB case notification records, vital registration statistics, skin-test-based surveys, and treatment outcomes.

Tuberculosis was first reported in an Egyptian camel in 1888, and the causal organism was demonstrated by inoculation experiments in 1911 (Mason 1912). Reported BTB cases in different species in Egypt over the years are listed in

Table 13.1 Bovine TB prevalence in Egypt over the years

Date	Location	Species	No. examined	Diagnostic method	% Positive	Reference
1910	Cairo abattoir	Camels	1786	PM	2.8	Mason (1917)
1911	Cairo abattoir	Camels	2695	PM	1.63	
1915	Cairo abattoir	Camels	1351	PM	5.4	
1916	Cairo abattoir	Camels	1579	PM	3.2	
1963	Cairo abattoir	Sheep	500	PM	0.4	Moustafa et al. (1964)
–	Monufia	Cattle	1003	CCT	0.3	El-Olemy et al. (1985)
		Buffaloes	180		1.1	
2004/5	Basatin abattoir	Pigs	745	PM	14.8	Mohamed et al. (2009)
2011	Cairo abattoir	Cattle	3347	CCT	1.0	Ramadan et al. (2012)
	Cairo abattoir	Cattle	7235	PM	0.2	
–	Gharbia and Monufia governorates	Cattle	422	CCT	2.14	Zahran et al. (2014)
		Buffaloes	480	CCT	5.6	

Table 13.1. It is evident that BTB is still prevalent in Egypt, and it creates a significant economic burden and constitutes an important zoonotic threat to its population.

13.4 Epidemiology of BTB in Egypt

13.4.1 Biosecurity and Other Herd-Level Risk Factors

According to official reports of the GOVS (1992) of Egypt, the number of BTB-infected cattle in Egypt increased following the importation of animals from countries where BTB is highly prevalent, which constitutes a potential source from which BTB may spread to new herds and/or territories in Egypt. During 2012–2015, the prevalence of BTB in five regions in the Nile Delta was 7.3%. However, on farms with a known history of risky management (such as introduction of animals without prior testing for BTB), the prevalence was as high as 45%.

Abou-Eisha et al. (2002) attributed the high prevalence of BTB infection to additional herd-level risk factors such as ever-increasing herd sizes accompanied by an increasing animal density, unhygienic local husbandry practices, and stress induced by intercurrent diseases and by mass vaccination campaigns. Similarly, the high prevalence of mycobacterial infections in pigs was attributed to the poor hygienic standards of pig farming practices in Egypt (Mohamed et al. 2009). Feeding pigs on household and hospital waste as well as refuge from large animal and chicken farms is a common, but risky, practice in Egypt.

13.4.2 Human Intervention Efforts

In Egypt, intradermal tuberculin testing has been performed since 1920, when the prevalence of BTB in the cattle and water buffaloes (*Bubalus bubalis*) populations was estimated to vary between 2% and 9%, respectively. The prevalence of BTB in cattle and water buffaloes during the 1980s was 6.9% and 26.2%, respectively, although the overall proportion of positive reactors of tested cattle and water buffaloes in 1981, respectively, was 6.2% and 9.4%. However, following the implementation of the national BTB control program in 1981, the prevalence declined to 2.6% during the 1990s (Cosivi et al. 1998) (Table 13.1). Similarly, recent studies demonstrated a declining trend of BTB in certain governorates, including Gharbia and Monufia, where 2.1% and 5.6% tuberculin reactor prevalence was recorded, respectively, in cattle and water buffaloes (Zahran et al. 2014). Meanwhile, in Qena Governorate, the prevalence was 0.4% and 0.3% for cattle and water buffaloes, respectively, in 2014 (Mahmoud et al. 2015).

13.4.3 Geographical Distribution of BTB in Egypt

The extent of BTB, based on postmortal inspection in abattoirs, varies substantially between regions in Egypt (Table 13.1). The data are often unreliable as most of the abattoirs in Egypt lack diagnostic facilities to rapidly confirm the presence of *M. bovis* in exudates collected from tuberculous-like lesions detected during routine meat inspection. The available data indicate that BTB is much more prevalent in the Egyptian Nile Delta and Valley compared to the rest of the country, probably because these regions where rural communities consist of smallholder families that traditionally own a few domestic stock, are the most densely populated areas in Egypt. Additionally, some of the people living in these areas keep animals inside their houses, which creates a major zoonotic BTB health risk. The prevalence of BTB in cattle was also high in other governorates such as Alexandria (6%), and in Dakahlia and Behera (9.6% and 14.1%, respectively) (GOVS 1992). In contrast, low prevalences were recorded at El-Basateen Abattoir, Cairo (0.21%), and the Ismailia Abattoir (0.6%). The underlying causes of the marked wide geographic variation in the prevalence of BTB in Egypt are poorly understood.

In recent years, the proportion of positive reactor cattle in some governorates increased sharply: in Port Said, the prevalence increased from 11.8% in 1990, to 15.9% in 1991; in Behera from 2.7% in 1990, to 13.4% in 1991; and in Dakahlia from 0.4% in 1989, to 23.4% in 1991. Fluctuations in the number of tuberculosis-like lesions detected in cattle slaughtered in Egyptian abattoirs could be ascribed to circumstances governing the volumes of slaughter such as religious feasts and other socio-cultural events when more cattle are slaughtered than normally.

13.4.3.1 Animal-Level BTB Risk Factors in Egypt

In Egypt, BTB affects cattle, camels, water buffaloes, sheep, and pigs (Table 13.1). Cattle are considered the primary source of the infection to other livestock species (Mason 1917) although the direction of transmission can vary in mixed-species production systems. Based on tuberculin testing (El-Olemy et al. 1985) and post-mortem inspection (Mason 1917), BTB was more common in older animals. Reports of the VOGS revealed that, on farms where BTB is endemic, the highest percentage of positive animals (14%) was in the 6–12-month-old age group, while, on newly tested farms, most of the reactors (60%) were older than 24 months. The prevalence of BTB also varied between breeds; it was low or absent in Friesian cows, and 2.7% in local cattle breeds El-Olemy et al. (1985). The periodic application of the test-and-slaughter policy on government farms might have led to the low prevalence in the Friesian cows. The low prevalence in water buffaloes and native cows is attributed to the prevailing farming practice, where individual farmers usually keep only one or two animals.

Other strains of the MTC (including *M. tuberculosis),* and *Mycobacterium spp.* belonging to the MOTT group (including *M. avium*, *M. chelonae*, *M. porcinum*, *M. gallinarum*, *M. simiae*, and *M. scrofulaceum)*, and other genera (such as *Rhodococcus equi)* have been isolated from tuberculous lesions in Egypt. This situation complicates epidemiological surveillance and dealing with the complexity of zoonotic TB as a public health risk, and it emphasizes the need to accurately identify the specific pathogenic mycobacteria from clinical specimens.

13.5 *Mycobacterium bovis* Infection in Humans in Egypt

Four percent of the TB cases in the WHO's Eastern Mediterranean Region comprising 22 countries occurs in Egypt. Generally, countries of the Eastern Mediterranean Region fall into three distinct epidemiological categories, Egypt being a middle-burden country, clustered with other countries with an estimated prevalence of 25–49 TB cases (all forms) per 100,000 of the population. This cluster recorded a 45–57% increase in the number of smear-positive TB cases during 2000–2003. With a prevalence and mortality rate of 24 and 3 per 100,000 people, respectively, TB remains a matter of concern, because 66% of TB cases occur in the socially and economically productive age groups of 15–54 years. According to estimates, 17,200 (new) people contract TB in Egypt every year.

Generally, the proportion of TB caused by *M. bovis* in humans is relatively low compared to *M. tuberculosis*, but it has become increasingly more prevalent in human populations subject to poverty, malnutrition, infection with the HIV, and inadequate health care (Michel et al. 2010). In the developing world, *M. bovis* infection is responsible for 5–10% of human TB cases, but this varies between countries (Haddad et al. 2004). The relatively low rate of *M. bovis* infection in humans in developing countries (in Africa and elsewhere) could probably be attributed to the lack of adequate diagnostic facilities and tests, as those that are used do not allow differentiation between the different species of mycobacteria. In addition, accurate diagnosis is difficult even when culture facilities are available, as *M. bovis* grows poorly on the standard Löwenstein-Jensen medium containing glycerol, which is one of the most widely used culture media for *M. tuberculosis* (Collins and Grange 1983).

In Egypt, the estimate of the number of human TB cases caused by *M. bovis* is inaccurate, mainly because of the difficulties in acquiring data from various Egyptian governorates, the lack of computerized patient record systems, and the vastness of some of the governorates. The available records often also lack important information such as the follow-up of smear-positive patients, while the socioeconomic status and the cause of death when patients die are not recorded.

It appears that the risk of humans contracting *M. bovis* in Egypt has decreased during the last several decades (El Ibiary et al. 1999). Human *M. bovis* infection in Egypt, of which approximately 63% of these patients were from rural areas (FAO 1993), accounted for 12.2% of the total number of human TB cases in 1953. The number declined to 10.0% in 1969, and to 5.4% in 1980 (Cosivi et al. 1998). In a recent investigation, 9 of 20 randomly selected samples from patients with abdominal TB were found to contain *M. bovis* (Nafeh et al. 1992). In another study 5% of 300 mycobacteria cultured from human sputum were *M. bovis* (El-Sabban et al. 1995). Most of these patients lived in the vicinity of the abattoir in Cairo, and some were workers at the abattoir. In addition, *M. bovis* has been identified in the cerebrospinal fluid of a patient suffering from tuberculous meningitis (Cooksey et al. 2002).

The isolation of *M. bovis* from milk and sputum samples collected from tuberculin test-positive cows (El-Olemy et al. 1985) indicates that cattle with BTB excrete *M. bovis* in milk or sputum and that they play a critical role in the transmission of zoonotic TB. Humans are also exposed to pigs with BTB; *M. bovis* was isolated from 12 of 66 swine carcasses containing TB-like lesions (Mohammed et al. 2009). The lack of awareness about the risk of infection and the modes of transmission of *M. bovis* predispose abattoir workers and the general public to the risk of contracting zoonotic TB (Ibrahim et al. 2012).

Given the high prevalence of BTB in cattle and its presence in other species, there is a need for comprehensive disease surveillance with an emphasis on the high-risk populations, to adequately determine the extent of zoonotic TB in Egypt. Due to the known existing high risk, however, effective measures should be instituted immediately to control the disease in farm animals and to advocate the application of heat treatment of milk and other animal products. Regular health inspection of abattoir workers and implementation of public awareness programs are further important steps that must be taken to curtail the risk of occupational zoonotic TB infections.

13.6 Bovine TB Eradication Scheme

To date, the BTB eradication scheme has failed to significantly reduce the prevalence of BTB and the number of TB-infected herds in Egypt. Bovine TB still constitutes a serious public health hazard, and every effort should thus be made to control the disease in Egypt.

Collectively, in Egypt, the use of the single and comparative intradermal tuberculin tests for TB screening, abattoir meat inspection, laboratory monitoring of tissue samples, and improved animal husbandry should lead to an improvement in animal health and a reduction of the zoonotic risk for humans of contracting BTB. The availability of limited laboratory facilities in Egypt, as is the situation in most developing countries, is an additional impediment, and it means that the diagnosis of TB infections is usually only based on the detection of AFBs on smear examination, and misdiagnosis of *M. bovis* infection can be expected to occur frequently (Thoen and Steele 1995).

The national BTB control program in Egypt was based on compulsory periodic testing of females (aged six months and above) and bulls (used for breeding purposes), slaughter of reactors, and monetary compensation for the animals slaughtered. The program was initially implemented in a few Egyptian governorates in 1981, and following the decline in the prevalence of BTB from 6.1% in 1981 to 2.6% during the 1990s (Cosivi et al. 1998), the control program was expanded to cover the entire country in 1986. Currently, the Egyptian Veterinary Service monitors all known BTB-positive farms and subject cattle on the farms to tuberculin testing, and slaughter of reactors. The Veterinary Service does, however, experience difficulties in dealing with household cattle that constitute a large proportion of farmed cattle in Egypt. The control program is also hampered by an unreliable traceability system and the consequent inability of tracing back cattle to their herd of origin whenever a BTB-positive case was found at slaughter.

In summary, frequent testing of animals using the CCT and prompt diagnosis of BTB, especially before restocking a dairy herd, are of great importance and should be used to strengthen control measures. Adopting and applying strict and uniform control measures to animals and workers can reduce the incidence of BTB in animals and humans in Egypt.

References

Abou-Eisha AM, El-Attar AA, El-Sheary MN (2002) Bovine and atypical mycobacterial infections of cattle and buffaloes in Port Said Province, Egypt. Assiut Vet Med J 47:152–162

Ameni G, Aseffa A, Sirak A et al (2007) Effect of skin testing and segregation on the incidence of bovine tuberculosis, and molecular typing of *Mycobacterium bovis* in Ethiopia. Vet Rec 161:782–786

Ayele WY, Neill SD, Zinsstag J et al (2004) Bovine tuberculosis: an old disease but a new threat to Africa. Int J Tuberc Lung Dis 8:924–937

Bezos J, Casal C, Romero B et al (2014) Current ante-mortem techniques for diagnosis of bovine tuberculosis. Res Vet Sci 97:S44–S52
Collins CH, Grange JM (1983) The bovine tubercle bacillus. J Appl Bacteriol 55:13–29
Cooksey RC, Abbadi SH, Woodley CL (2002) Characterization of *Mycobacterium tuberculosis* complex isolates from the cerebrospinal fluid of meningitis patients at six fever hospitals in Egypt. J Clin Microbiol 40:1651–1655
Cosivi O, Grange JM, Daborn CJ et al (1998) Zoonotic tuberculosis due to *Mycobacterium bovis* in developing countries. Emerg Infect Dis 4:59–70
Danbirni S, Okaiyeto SO, Bature C et al (2013) Field determination of tuberculosis prevalence in a herd of cattle using tuberculin and Quicking® bovine tuberculosis antibody rapid tests in Jalingo, Nigeria. J Vet Adv 3:20–23
El Ibiary S, De Coster EJM, Tolba FM et al (1999) Trend in the annual risk of tuberculous infection in Egypt, 1950–1996. Int J Tuberc Lung Dis 3:294–299
El-Olemy GM, El-Bassiouni AA, Negm S (1985) Tuberculosis in Toukh-Tanbisha, Menufia, Egypt. In: Proceeding of the 4th international symposium on veterinary epidemiology and economics. www.sciquest.org.nz/elibrary/download/61287/Tuberculosis+in+Toukh
El-Sabban M.S, Lotfy O, Hammam HM et al (1995) Bovine tuberculosis and its extent of spread as a source of infection to man and animals in Arab Republic of Egypt. In: The international conference on animal tuberculosis in Africa and Middle East, Cairo (Egypt), 28–30 Apr 1992
Food and Agriculture Organization of the United Nations (FAO) (1993) Zoonotic diseases in the near east region. FAO Regional Office for the Near East, Cairo, p 95
GOVS (1992) General Organization for the veterinary services. Annual report. Ministry of Agriculture, Egypt
Haddad N, Masselot M, Durant B (2004) Molecular differentiation of *Mycobacterium bovis* isolates: review of techniques and applications. Res Vet Sci 76:1–18
Hambolu D, Freeman J, Taddese HB et al (2013) Predictors of bovine TB risk behaviour amongst meat handlers in Nigeria: a cross-sectional study guided by the health belief model. PLoS One 8: e56091
Hassanain NA, Hassanain MA, Soliman YA et al (2009) Bovine tuberculosis in a dairy cattle farm as a threat to public health. Afr J Microbiol Res 3:446–450
Humblet MF, Walravens K, Salandre O et al (2011) Monitoring of the intra-dermal tuberculosis skin test performed by Belgian field practitioners. Res Vet Sci 91:199–207
Ibrahim S, Cadmus SIB, Umoh JU et al (2012) Tuberculosis in humans and cattle in Jigawa state, Nigeria: risk factors analysis. Vet Med Int 2012:Article ID 865924:5. https://doi.org/10.1155/2012/865924
Mahmoud YAH, Ali AO, Abdel Alim AA (2015) Prevalence of tuberculosis in cattle and buffalo in Qena, Egypt. World Vet J 5:42–45
Mason FE (1912) Some observations on tuberculosis in camels in Egypt. J Comp Pathol Therap 25:109–111
Mason FE (1917) Tuberculosis in camels. J Comp Path Therap 30:80–84
Michel AL, Müller B, Van Helden PD (2010) *Mycobacterium bovis* at the animal–human interface: a problem, or not? Vet Microbiol 140:371–381
Mohamed AM, El-Ella GAA, Nasr EA (2009) Phenotypic and molecular typing of tuberculous and nontuberculous *Mycobacterium* species from slaughtered pigs in Egypt. J Vet Diagn Investig 21:48–52
Moustafa H, Mostafa AMB, Zayed I (1964) Some observations on the incidence and histopathology of pneumonia in Egyptian sheep with special reference to two cases with tuberculosis. Zentralbl Veterinarmed B 11:231–239
Nafeh MA, Medhat A, Abdul-Hameed AG et al (1992) Tuberculous peritonitis in Egypt: the value of laparoscopy in diagnosis. Am J Trop Med Hyg 47:470–477
O'Reilly LM, Daborn CJ (1995) The epidemiology of *Mycobacterium bovis* infections in animals and man: a review. Tuber Lung Dis 76:1–46

Ovdiennkop NP, Shchur EVE, Naimanor AK et al (1987) Frequency of tuberculin injection in cattle. Vet Muscow USSR 8:29–33

Ramadan HH, El-Gohary AHN, Mohamed AA et al (2012) Detection of *Mycobacterium bovis* and *Mycobacterium tuberculosis* from clinical samples by conventional and molecular techniques in Egypt. Glob Vet 9:648–654

Thoen CO, Steele JH (eds) (1995) Regional and country status reports, part 2. Iowa State University Press, Ames, Iowa, pp 169–345

WHO (2012) Global tuberculosis report. http://www.stop.org/wg/news-vaccines/

Zahran NR, El Behiry A, Marzouk E et al (2014) Comparison of LCD array and IS6110-PCR with conventional techniques for detection of *Mycobacterium bovis* isolated from Egyptian cattle and buffalo. Int J Microbiol 3:197–204

Zink AR, Molnár E, Motamedi N et al (2007) Molecular history of tuberculosis from ancient mummies and skeletons. Int J Osteoarchaeol 17:380–391

Chapter 14
Status of Bovine Tuberculosis in Ethiopia: Challenges and Opportunities for Future Control and Prevention

Demelash B. Areda, Adrian Muwonge, and Asseged B. Dibaba

14.1 Introduction

Ethiopia has the largest cattle population of all the African countries, and it comprises 80% of all livestock units in the country (FAOSTAT 2003). Over 85% of the Ethiopian population lives in rural areas, and more than 80% of them depend on livestock production for their livelihood and survival (CSA 2012). Their livestock serve as a source of food, household cash income, transportation, traction power, and accumulation of wealth. The livestock sector contributes 16.5% to the national and 35.6% to agricultural GDP (Metaferia et al. 2011), generates up to 15% of Ethiopia's export earnings, and provides 30% of agricultural employment opportunities (Behnke 2010). The value of livestock, in terms of their contribution to the national economic development, poverty alleviation, and supporting the livelihood of a significant segment of the population, is often underrated (Kimball 2011), and a mere 0.3% of the annual Ethiopian governmental expenditure is allotted for livestock development (FAO 2004).

Because livestock, especially cattle, are seen as indicators of wealth and social standing in Ethiopia, livestock keepers endeavor to maximize the number of their

D. B. Areda (✉)
College of Science Engineering and Technology (CSET), Grand Canyon University, Phoenix, AZ, USA

A. Muwonge
The Roslin Institute, College of Medicine and Veterinary Medicine, University of Edinburgh, Midlothian, Edinburgh, UK
e-mail: Adrian.muwonge@roslin.ed.ac.uk

A. B. Dibaba
Department of Pathobiology, College of Veterinary Medicine, Tuskegee University, Tuskegee, AL, USA
e-mail: adibaba@tuskegee.edu

A. B. Dibaba et al. (eds.), *Tuberculosis in Animals: An African Perspective*,
https://doi.org/10.1007/978-3-030-18690-6_14

animals that they own, rather than to focus on the quality of products such as milk and meat that they produce, and their market value.

The productivity of the livestock sector is low, and its contribution to the regional and national economic development is suboptimal. However, traditional smallholder dairy farms produce 97% of all the milk in the country, and over 75% of milk is delivered to commercial processors. Most of the cattle are of the indigenous zebu breeds, and they are generally poor milk producers with a yield of about 400–680 kg of milk/cow over a lactation period of 150–180 days, compared to the average of about 9500 kg/annum of the American national herd (Yilma et al. 2011). Poor food, both in quality and quantity, poor management and health care, and the low genetic potential of indigenous breeds all contribute to their low productivity (MoA 1998).

The impact of infectious diseases imposes major constraints on livestock production in Ethiopia. In addition to the direct economic losses associated with mortality and morbidity, these diseases cause retarded growth, reduced fertility, and reduced traction power (Haftu et al. 2014). They also prevent the country from participating in the lucrative international livestock and meat trade and with strategies to improve indigenous stock through crossbreeding programs.

Bovine tuberculosis (TB) is an endemic disease of high economic significance in Ethiopia where it affects cattle, sheep, goats, and camels. This situation which poses a zoonotic risk in the Ethiopian population, enhanced by their impaired immunity caused by HIV/AIDS and malnutrition, should be a major concern. However, despite the high prevalence of BTB in animals and the risk of zoonotic infection, control and preventative measures for the disease in both human and animal populations are virtually nonexistent in the country.

This chapter aims to provide information about the status and outline the challenges and opportunities for the future control and prevention of BTB in Ethiopia.

14.2 Livestock Resources and Husbandry

In Africa, Ethiopia has the largest number of livestock, comprising 52.1 million cattle, 24.2 million sheep, 22.6 million goats, and about 1.0 million camels (Fig. 14.1) (CSA 2012). These numbers are expected to increase substantially during the course of the next 10 years (Leta and Mesele 2014). This creates a major problem since the parallel increase in the numbers of the human population will put more pressure on the available grazing and arable lands, thus further limiting its availability and the ability of farmers to sustain their livestock.

Livestock are unevenly distributed throughout the country, and there are also regional variations in the distribution of the various livestock species. Larger livestock concentrations (per km^2) occur in the central and northern highland regions because of its more favorable climatic and ecological characteristics (Fig. 14.2). Depending on the form of land use, its geographic location and features,

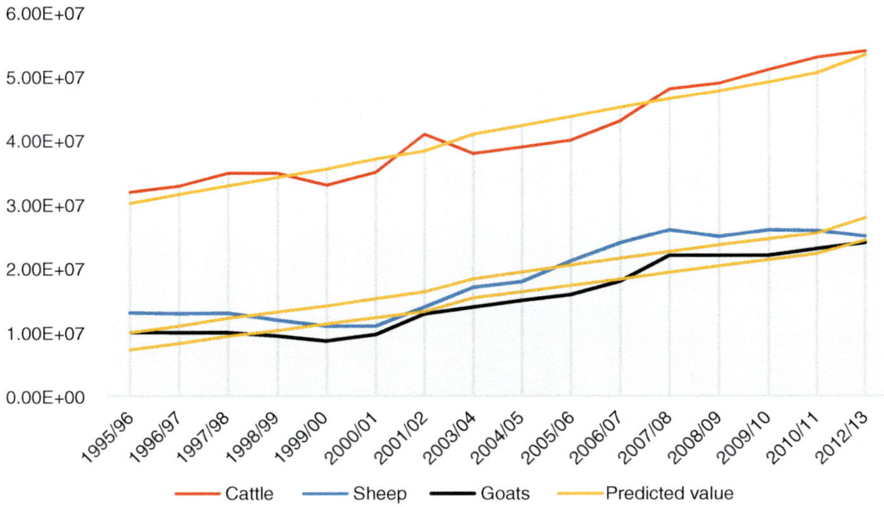

Fig. 14.1 Population growth of cattle, sheep, and goats in Ethiopia (Leta and Mesele 2014)

and the lifestyle of the population, livestock husbandry practices in Ethiopia are categorized as:

Mixed farming which is a subsistence, mixed crop-livestock system, commonly practiced in the central highland regions. Cattle are the predominant species, with relatively large numbers of oxen used for traction on crop fields.

Intensive farming practices such as in dairy (pure or cross-bred) farms in urban and peri-urban areas, or feedlots (fattening zebu cattle for local and export markets), and state owned breeding centers where selected indigenous breeds are maintained for breed evaluation and for cross-breeding purposes.

Pastoral/agropastoral farming is common in marginal, lowland areas such as in Borana, Somali, and Afar regions. Livestock kept in communally owned natural grazing ranges are at times driven far from settlements in search of good pastures and surface water. Cattle are the dominant species in this system in association with small ruminants and camels.

14.3 Bovine TB in Ethiopia

Little is known about the status of BTB in Ethiopia prior to the 1960s when the first cases were documented in the annual abattoir meat inspection report of the Ministry of Agriculture (Hailemariam 1975). It is assumed that the brief Italian occupation from 1895 to 1896, and the importation of cattle as part of their dairy cattle improvement program during the second occupation in the twentieth century (1935–1941), created opportunities for the introduction of BTB into the country.

During the course of recent decades, the situation improved, and more extensive work determining the prevalence and distribution of BTB was done. In the late 1980s,

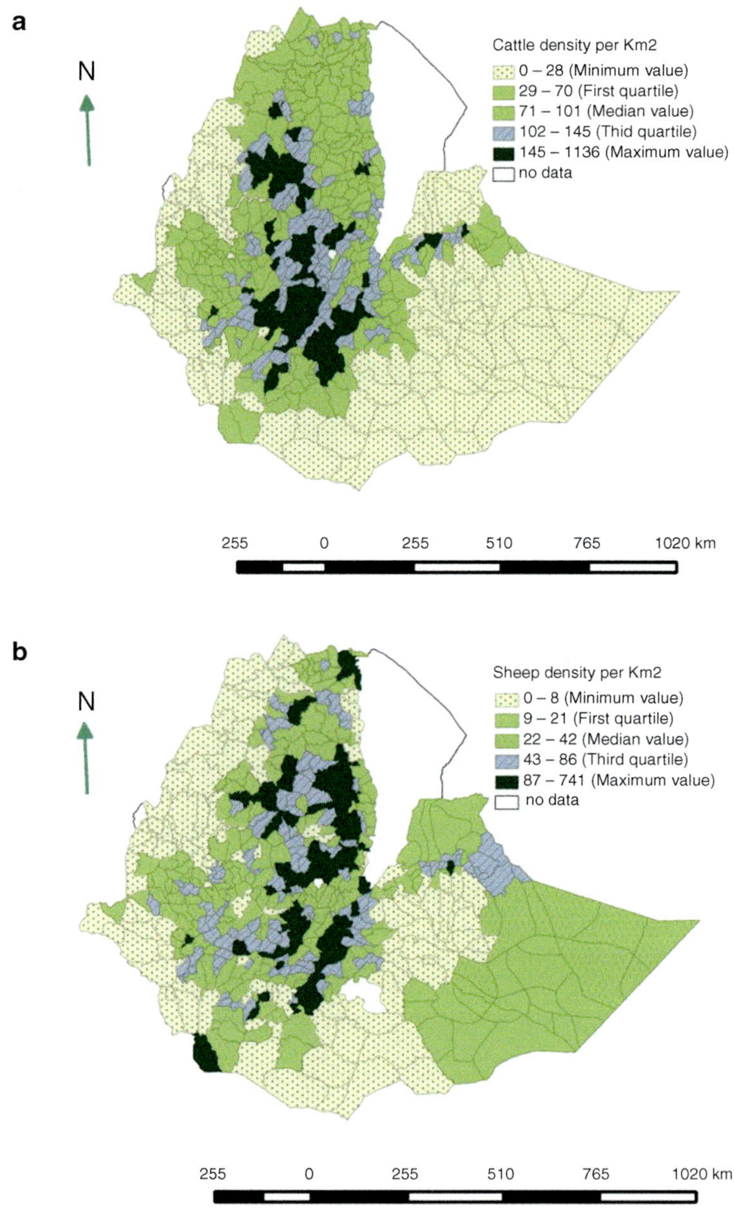

Fig. 14.2 Geospatial distribution of cattle (**a**), sheep (**b**), and goat (**c**) populations in Ethiopia showing higher density (per km^2) in the central region (Leta and Mesele 2014)

c

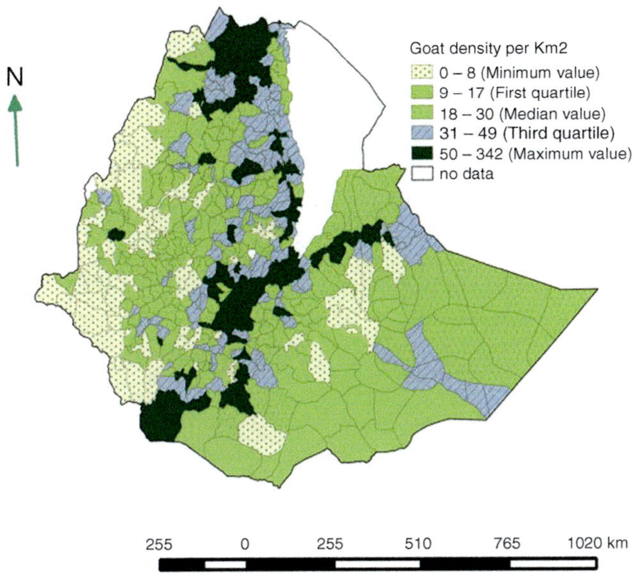

Fig. 14.2 (continued)

the Institute of Agricultural Research (IAR) conducted blanket screening of all state-owned dairy farms in the country, and in 2000 the Armauer Hansen Research Institute (AHRI), a biomedical research institute based in Addis Ababa, expanded its TB research program to include zoonotic TB. The establishment of new veterinary schools at the Jima, Mekele, Hawassa, and Haramaya universities, and expansion of the graduate programs at the College of Veterinary Medicine, Addis Ababa University, enabled extensive surveying of BTB in various regions and agroecological zones in the country. These activities were further enhanced by the establishment of academic and research collaboration between Ethiopian and European universities in mid-2000.

Bovine TB is prevalent in all agroecological zones of the country (Tables 14.1 and 14.2), mainly affecting cattle, sheep, goats, and camels (Table 14.3). The disease

Table 14.1 Prevalence of BTB in various agroecological zones of Ethiopia based on tuberculin skin testing

Agroecological zones[a]	District/area	Number tested	Prevalence (%)	Reference
Dega (highland) (2500–3500 masl)	Fiche	1041	16.2	Ameni et al. (2007)
	Salaale/Holeta	5424	13.5	
Woyina-Dega (mid-highland) (1500–2500 masl)	Arsi	625	12.2	Dinka and Duressa (2011)
	Addis Ababa	1869	23.7	Elias et al. (2008)
	Gurage zone	1214	6.8	Tschopp et al. (2011)
Kolla (lowland) (500–1500 masl)	Hamar	499	0.8	Tschopp et al. (2010a)
	Filtu	421	2.0	Gumi et al. (2012)

[a]Based on traditional Ethiopian agroecological zonation criteria; masl-meter above sea level

Table 14.2 Prevalence of BTB in Ethiopia based on abattoir surveys conducted in various regions

Abattoir	Location	No of carcasses	Prevalence (%)	Reference
Addis Ababa	Central	600	15.2	Biffa et al. (2010)
Adama	Southeast	522	24.7	Biffa et al. (2010)
Melge-Wondo	South	1325	4.5	Biffa et al. (2010)
Hawassa		442	8.8	Biffa et al. (2010)
Yabello		433	4.2	Biffa et al. (2010)
Butajira	Southwest	446	9.0	Nemomsa et al. (2014)
		4606	16.4	Berg et al. (2009)
Jinka	South	3471	8.0	Berg et al. (2009)
Gimbi	West	3250	12.7	Berg et al. (2009)
Gonder	North	14,314	1.8	Berg et al. (2009)
Weldiya		4338	5.53	Berg et al. (2009)
Addis Ababa	Central	500	5.0	Mekibeb et al. (2013)
		1350	3.0	Asseged et al. (2004)
		984	3.5	Shitaye et al. (2006)
Jimma	West	780	2.7	Tigre et al. (2011)
		1102	5.4	Bekele and Belay (2011)
Hossana	Southwest	751	4.5	Teklu et al. (2004)

Table 14.3 Prevalence estimates of BTB in different livestock species in Ethiopia

Species	Diagnostic method	Number examined	Prevalence	Reference
Cattle	CCT	1214	6.8	Tschopp et al. (2011)
		499	0.8	Tschopp et al. (2010a)
		421	2.0	Gumi et al. (2012)
		110	13.6	Ashenafi et al. (2013)
Sheep		77	1.3	Ashenafi et al. (2013)
		203	0.5	Amenu et al. (2010)
		347	1.4	Mamo et al. (2012)
Goats	PME	1744	4.4	Gumi et al. (2012)
		1536	4.2	Hiko and Agga (2011)
	CCT	186	0	Tschopp et al. (2010a)
		518	0.2	Gumi et al. (2012)
		320	6.3	Ashenafi et al. (2013)
		1884	4.3	Mamo et al. (2012)
	SIT	630	7.6	Tafess et al. (2011)
Camels	PME	276	5.1	Ethiopie et al. (2009)
		906	10.0	Mamo et al. (2011)
	CCT	479	0.41	Gumi et al. (2012)
		480	6.0	Beyi et al. (2014)
	PME	398	8.3	Beyi et al. (2014)

CCT comparative cervical intradermal test, *SIT* single cervical intradermal test, *PME* postmortem examination

prevalence is particularly high in crossbred dairy cows in intensive husbandry systems. For example, in a recent study in central Ethiopia where commercial dairy farming is widely practiced, 90% of the herds were positive for BTB, and a prevalence of as high as 41.3% was recorded in some of the large dairy herds (Firdessa et al. 2012).

14.3.1 BTB Epidemiology in Ethiopia

14.3.1.1 Molecular Epidemiology of *M. bovis* in Ethiopia

Spoligotyping and Deletion Analysis

Three features characterize the East African *M. bovis* clonal complex (Berg et al. 2011):

1. The majority of spoligotypes isolated in eastern Africa characteristically lack spacers 3–7 in their spoligopattern (Fig. 14.3).
2. All isolates with this feature also lack RDAf2 chromosomal DNA.
3. Parsimonious phylogenetic inferences, based on 43 spacers of spacer oligotyping, suggest that the most recent ancestor of this group of isolates has a spoligopattern similar to that of SB0133 (Berg et al. 2011).

Only 26 unique spoligotypes have been detected in the country to date. Molecular data (Firdessa et al. 2012; Biffa et al. 2010; Berg et al. 2011; Kidane et al. 2002) indicate that unlike in most Western African regions, SB0133 is not the dominant spoligotype but that SB1176 is the most frequently isolated strain. In Ethiopia, the vast majority of the unique spoligotypes belong to the Af2 clonal complex, and it is plausible that the low genetic diversity in this endemic setting reflects an ongoing transmission event (Firdessa et al. 2012). The vast majority of the unique spoligotypes belong to the Af2 clonal complex, and SB1176 is the most frequently isolated strain. SB1476, the second-most-frequently isolated spoligotype in Ethiopia, is also the most frequently isolated non-Af2 spoligotype (Kidane et al. 2002). The BCG-like spoligopattern, SB0120, has only been isolated in Addis Ababa.

MIRU-VNTR Typing

Allelic Diversity of the MIRU-VNTR Typing Tool

Although spoligotyping has been widely used in molecular epidemiological studies in Ethiopia, its discriminatory power does not exceed 90% (0.90) (Biffa et al. 2010), and a considerable amount of genetic diversity is left un-decoded when spoligotyping only is used for typing purposes. MIRU-VNTR is more sensitive

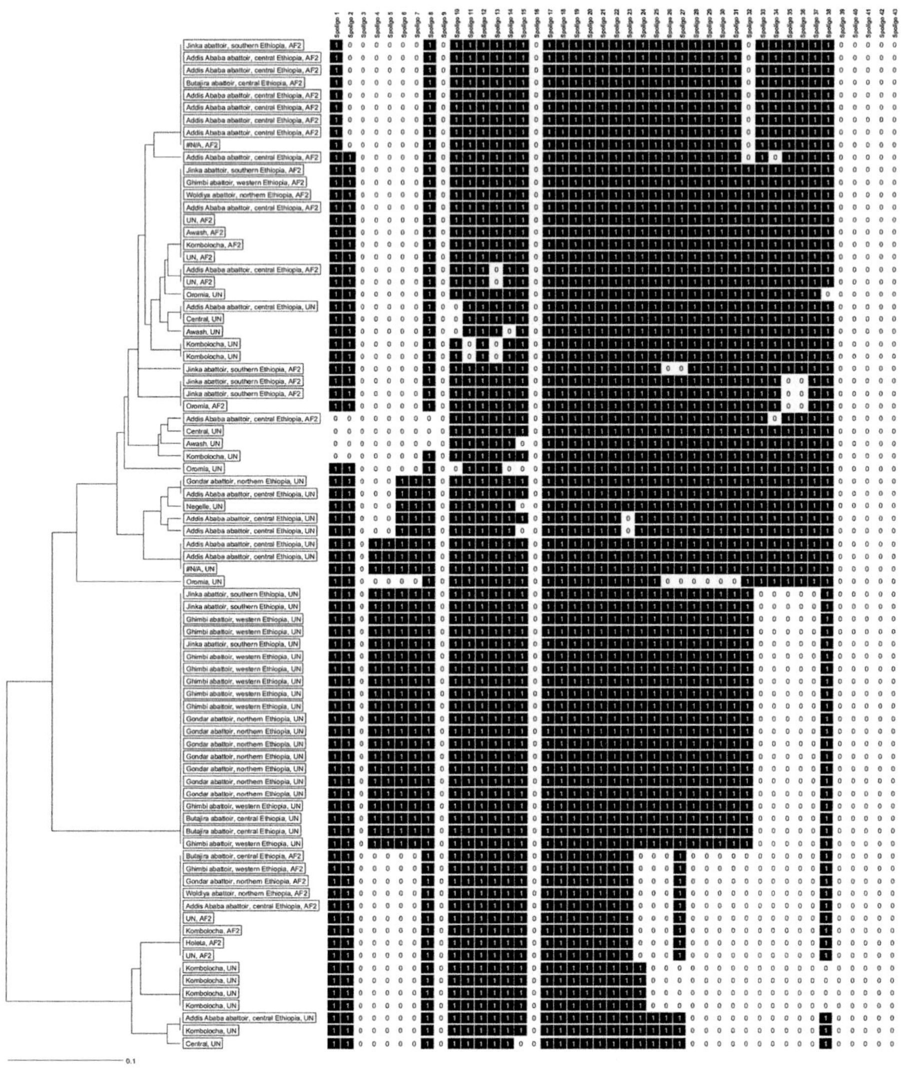

Fig. 14.3 Unweighted pair-group method with arithmetic mean (UPGMA) tree generated based on 64 isolates with 26 unique spoligotypes recovered in specific woredas (districts) of Ethiopia. Note that UN (white color band) represents spoligotypes whose district of origin was not documented

and allows assessment of variable allelic diversity depending on the settings and ecological niches from which the isolates are recovered (Hilty et al. 2005; Muwonge et al. 2014). Of the 55 previously described isolates (Biffa et al. 2010), loci 4052, 2461, 4156, 2165, 2163b, and 2059 (Table 14.4), have the highest allelic

Table 14.4 Allelic diversity of 55 Ethiopian *M. bovis* isolates on a 24 MIRU-VNTR panel. The diversity was calculated using the method of Selander et al. (1986)

Loci	Allelic diversity
2165	0.639
2461	0.678
577	0.551
580	0.442
3192	0.053
154	0.586
960	0.266
1644	0.593
2059	0.605
2531	0.053
2687	0.119
2996	0.587
3007	0.539
4348	0.546
802	0.59
424	0.131
1955	0.564
2163b	0.604
2347	0.489
2401	0.477
3171	0.369
3690	0.482
4052	0.727
4156	0.649

diversity and should be included in MIRU-VNTR panels of Ethiopian *M. bovis* isolates.

Genetic Diversity of the Spoligotypes

A previously published dataset (Firdessa et al. 2012; Biffa et al. 2010) was reanalyzed using MIRU-VNTR, allowing analysis of the genetic diversity within a given spoligotype (Fig. 14.4). For this, two spoligotypes (SB1176 and SB0133) belonging to the Af2 clonal complex, and another spoligotype signature (SB0134) that does not belong to the Af2 clonal complex, were chosen because of their high frequency of isolation and also because of the availability of data on the 24 MIRU-VNTR loci. It transpired that the non-Af2 spoligotype is almost exclusively monomorphic, an attribute that is shared by isolates from all geographic regions regardless of how distant they were from each other. The two spoligotypes belonging to Af2 clonal complex, however, exhibited polymorphism that tended to be clustered according to location.

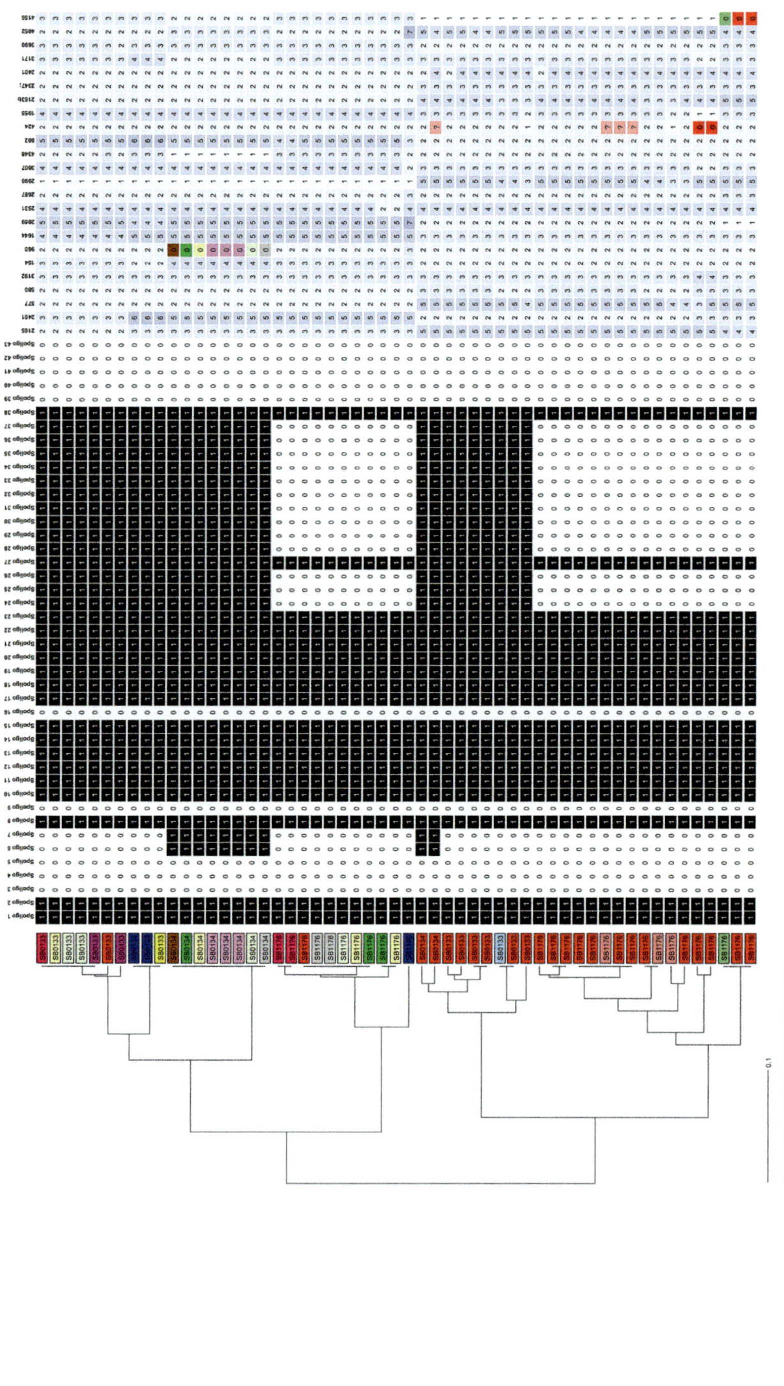

Fig. 14.4 Unweighted pair-group method with arithmetic mean (UPGMA) phylogenetic tree based on 24 MIRU-VNTR loci of three spoligotypes (SB0134, SB1176, and SB0133). The figure was generated using RIDOM MLVA and was based on 55 isolates that produced 26 unique genotypes, with a combined discriminatory power of 0.95% (0.93–0.97), which translates into a cluster rate of 0.52

14.3.2 Cattle Movement in Ethiopia

Ethiopia's vast cattle population (FAOSTAT 2005) is farmed in different production systems that provide different opportunities for the transmission of *M. bovis*. The dissemination of the disease appears to be dependent on two main practices: pastoralism and the search for and the location of markets. Pastoralism is the predominant farming system, and it requires extensive cattle movement over long distances, mainly in search of water and grazing. The location of cattle markets is another important determinant of livestock movement in Ethiopia (Jabbar et al. 2007). The search for markets is driven by price, causing a centripetal movement of cattle to urban areas that offer better prices for livestock. From various rural areas, cattle are moved to Addis Ababa and Holeta, for export to Bati, Tagochale, or Sudanese, Eritrean, and Kenyan markets (Fig. 14.5), thus acting as a conduit for the dissemination of BTB in the country.

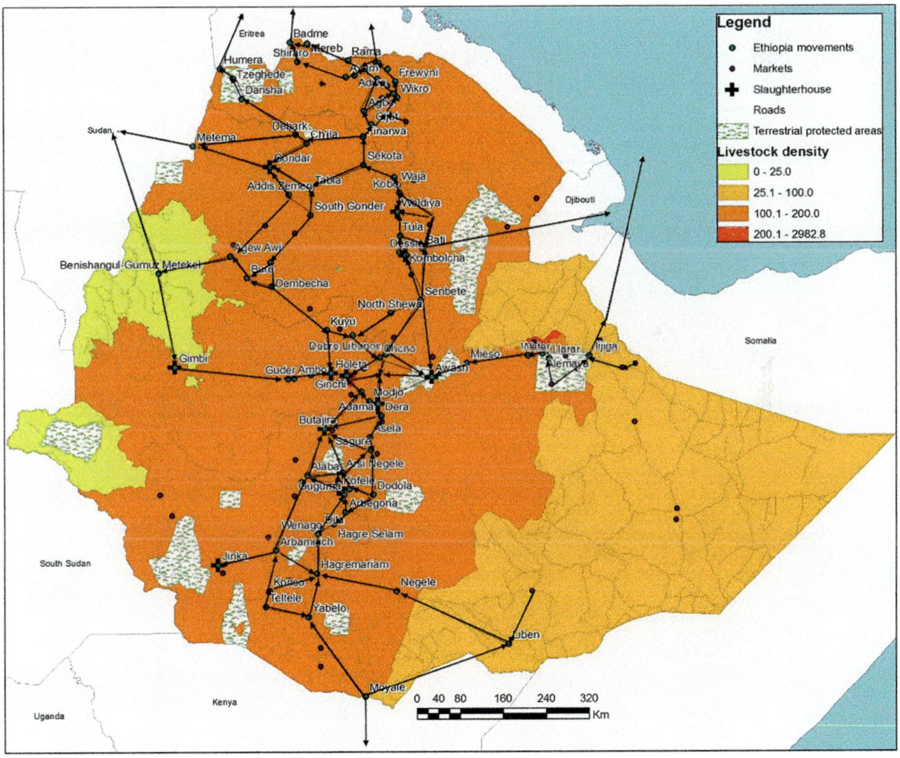

Fig. 14.5 Cattle movement network in Ethiopia (adapted from empirical data collated by Jabbar et al. (2007). Color gradient shows cattle density based on data from FAOSTAT (2005). The map was generated using Arch GIS

14.3.3 *Molecular Epidemiology of* Mycobacterium bovis *in Ethiopia*

To understand the distribution of the different *M. bovis* types in Ethiopia, it is essential to understand the movement of cattle in Ethiopia, as they are the most important maintenance hosts of *M. bovis.* There appears to be a bi-directional movement of cattle in Ethiopia that influences the spread and distribution of *M. bovis* types in the country. Cattle from rural areas are brought to and marketed and slaughtered in the urban centers, and new breeding stock is distributed throughout the country from the urban areas.

The distribution of the five most frequently isolated spoligotypes in Ethiopia appears to reflect the cattle movement network from rural areas to the abattoirs (Fig. 14.5). The largest diversity of spoligotypes, which includes almost all the spoligotypes that are present in the country, has also been isolated from cattle in Addis Ababa and its surroundings. The spoligotype diversity observed in the major slaughterhouses in Addis Ababa, and in Debre Zeit, Jinka, and Kombolcha (Figs. 14.6 and 14.7), appears to be a cross-sectional reflection of the molecular diversity of *M. bovis* isolates from the entire country, rather than that of the *M. bovis* types present in the greater Addis Ababa area and the other urban centers.

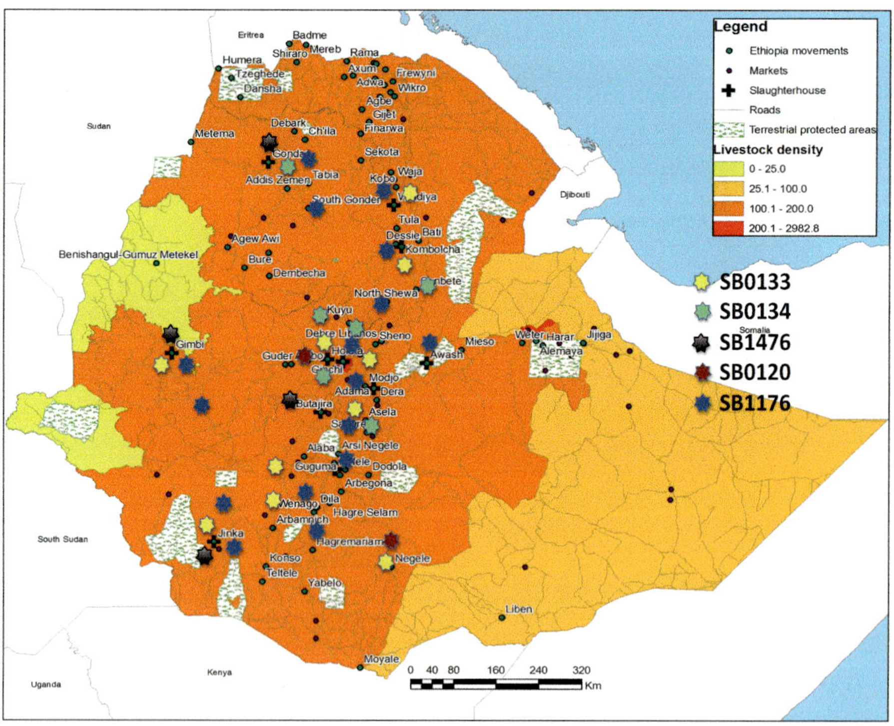

Fig. 14.6 Spatial distribution of the five most frequently isolated spoligotypes in Ethiopia

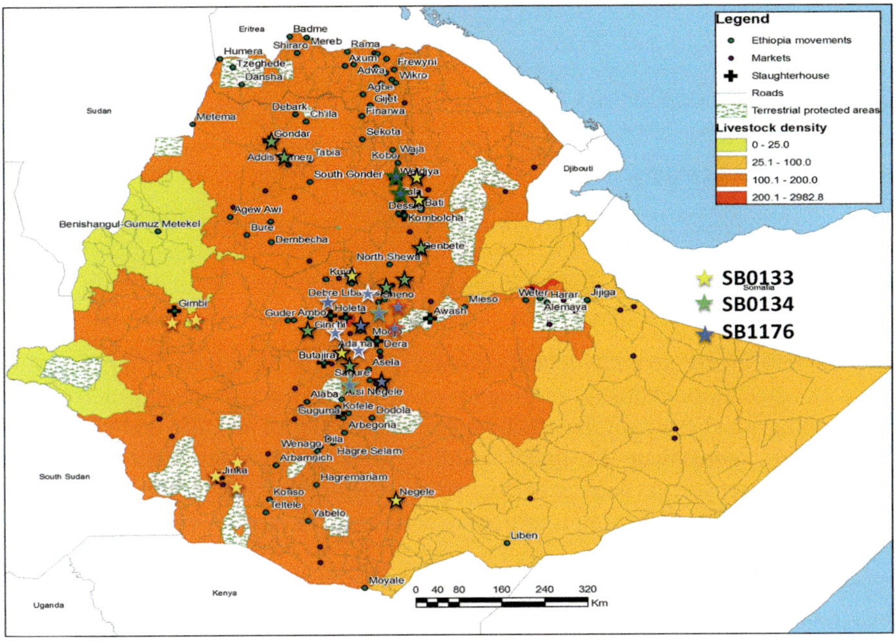

Fig. 14.7 Spatial distribution of the genotypes of the three most frequently isolated spoligotypes in Ethiopia. The color in the middle of the star corresponds to the spoligotype, while the color of the margin of the star corresponds to the specific MIRU-VNTR variant/clone

The centrifugal movement of cattle is the consequence of the recent shift to dairy farming and attempts to improve the milk production of the local breeds, also in rural areas. For this purpose, purebred cattle are imported into the country, mostly through Addis Ababa. The intensive husbandry systems in urban areas into which they are incorporated provide a perfect environment for the transmission of BTB, which is then spread by pure and crossbred bulls when they are distributed to rural breeding centers. It is conceivable that this centrifugal movement of potentially BTB-infected cattle is the reason for the presence of certain identical genotypes, such as SB1176 (the most frequently isolated genotype) throughout the country (Fig. 14.6).

At a granular level, when spoligotyping is combined with MIRU-VNTR, centripetal cattle movement is still discernible. Another dynamic, however, is revealed (Figs. 14.6 and 14.7). Certain clones of the SB0133 and SB1176 circulate exclusively in Addis Ababa and the surrounding towns, Sululta, Senbete, Sendafa, and Debre Zeit. This localization of strains likely reflects the effects of the shift to intensive dairy farming in Addis Ababa and the surrounding towns in response to an increasing demand for milk and, to a lesser extent, meat (Firdessa et al. 2012).

The isolation of identical SB0133 clones from districts (Woredas) as far apart as Nagelle in the south and Weldiya in the north suggests that cattle movement is extensive and linked to the road network. The presence though of two SB0133 genotypes unique to Jinka and Gimbi implies that a mode of animal movement may exist between localities other than that revealed by the empirical data reflecting cattle movement in the country (Fig. 14.7).

14.3.4 Tuberculosis in Ethiopian Wildlife

Ethiopia has diverse geographic and climatic features characterized by an extensive range of rugged mountains (averaging above 13,000 ft.), a vast savanna and rivers, and suitable weather conditions that support a diversity of wildlife and birds. The country has more than 14 major wildlife reserves that represent the entire sub-Saharan ecosystem (Amare 2015), and it is regarded as one of the top 25 countries in the world with a rich endemic collection of higher vertebrate species (Groombridge 1992).

Although diseases threaten wildlife survival worldwide, little is known about their effect on the Ethiopian wildlife populations. Tuberculosis can have serious consequences for wildlife conservation by endangering species, thus affecting biodiversity and national tourism (Tschopp 2015). Bovine TB is known to occur in a wide range of wildlife species, some of which may also serve as maintenance hosts. Many of these wildlife species, particularly larger mammals, known to be natural hosts of, or susceptible to BTB, also occur in Ethiopia (Amare 2015). To our knowledge, only two attempts have been made to investigate BTB in Ethiopian wildlife. None of the members of the *M. tuberculosis* complex (MTC) was detected in 28 different mammalian species (Tschopp et al. 2010a, b), but environmental mycobacteria, predominantly *M. terrae*, were abundant. Failure to detect *M. bovis* was unexpected given the extensive wildlife-livestock interface, which undoubtedly creates many opportunities for the interspecies transmission of *M. bovis*.

Because of the known presence and spread of BTB in various African wildlife species, it is essential that a carefully designed epidemiological study, employing more sensitive tests and using larger sample sizes, be conducted to determine the status of the disease in Ethiopia wildlife and the role that they can play in sustaining the infection in the country.

14.4 Zoonotic Tuberculosis

14.4.1 Policy

As is the case in many other African countries, occupational health and safety issues are largely disregarded in Ethiopia. The ministerial occupational health and safety directives make no provision for dealing with BTB, and the Ethiopian Occupational Health Team under the Ministry of Labor and Social Affairs (MOLSA) focuses on more immediate health crises such as childhood mortality, malaria, waterborne illnesses, and HIV/AIDS. They do not recognize *M. bovis* infections as an occupational risk, and its role in the human TB epidemic is disregarded (Pal et al. 2013). Although the Ethiopian Ministries of Health and Agriculture attempt to document the occurrence of zoonotic TB (along with other zoonotic diseases), the information is inadequate and is insufficient to serve as a basis for the development of control programs.

14.4.2 High-Risk Cultural Practices

The government's attitude about the control of TB is surprising given that *M. bovis* was isolated from 17 to 30% of human TB patients in rural Tanzania, Uganda, and Ethiopia (Kidane et al. 2002; Kazwala et al. 2001) Because pastoralists in Ethiopia drink large amounts of raw milk (Regassa et al. 2009), consume undercooked meat (Kazwala et al. 2001; Neill et al. 1989), and live in very close contact with their animals, they are more likely then to contract zoonotic TB. Abattoir and animal health workers, herdsmen, dairy farmers, and dairy farm workers are similarly at a higher risk of contracting the disease because of its high prevalence in the national herd and their intimate contact with tuberculous cattle and their products (Biet et al. 2005). Zoonotic TB is often overlooked or misdiagnosed because of poor healthcare and treatment services, inadequate diagnostic facilities, health monitoring and surveillance, and public awareness programs.

14.4.3 Public Awareness of the Risk of Contracting BTB

The high prevalence of BTB in slaughtered cattle in Ethiopia should be of concern given the lack of awareness of its importance by the population at risk. In central Ethiopia, only 32% of the population interviewed was aware that humans could become infected by *M. bovis* (Ameni and Erkihun 2007). In southwestern Ethiopia where 66.8% of the population reported that they consumed raw milk, only 29.1% of the study population was aware of zoonotic TB. In general, the dairy farmers appear to be better informed; 50% of them knew about the diseases, while 34.7% of the general public, 20.6% of butchers, and only 10% herdsmen were aware of it (Tesfaye et al. 2013). The lack of knowledge of abattoir and butchery workers in the Tigray region is disturbing since about 61.5% of the abattoir workers there never received training in the practice of meat hygiene, and 53.8% of them remarked on the absence of a legal framework for reporting tuberculous lesions detected during meat inspection (Haileselassie et al. 2013). These practices will markedly increase the amount of *M. bovis*-infected milk and meat consumed by the population at risk.

14.4.4 The Role of HIV-AIDS and Malnutrition in Zoonotic TB

The large number of immune-deficient inhabitants, attributable to chronic malnutrition and HIV/AIDS, makes the Ethiopian population highly susceptible to *M. bovis* infection. Elsewhere, *M. bovis* in HIV-infected patients according to some estimates may account for 20–50% of all forms of TB in immune-deficient (including HIV-positive) patients (Hlavsa et al. 2008; Cicero et al. 2009).

The number of *M. bovis* infections in humans in Ethiopia has been reported to be low. The actual numbers of zoonotic *M. bovis* cases in Ethiopia, however, are likely to be more substantial because of the high prevalence of the diseases in livestock, the lack of BTB control and prevention, and the application of inadequate food safety measures.

14.5 Control of BTB in Ethiopia

14.5.1 Current State of Affairs and Future Direction

Despite the substantial economic burden of BTB on the livestock industry and its considerable threat to public health in Ethiopia, efforts to control or prevent it are limited.

Currently, BTB control is applied only on large dairy farms (both state and privately owned) in and around Addis Ababa and in the export beef cattle sector to meet the requirements of the Middle Eastern and North African importing countries. For the beef sector, a mandatory test-and-isolation policy is applied, but in the dairy sector, only cases suspected clinically to suffer from BTB are tested. Reactors are usually quarantined, treated, and then returned to the herd of origin or are moved to a different farm. The authorities are also not notified of positive reactors, and the current processes appear to enhance the dissemination of the disease.

Due to socioeconomic and logistical constraints in Ethiopia, abattoir meat inspection remains the only way to monitor the occurrence of BTB in livestock. It has the added advantage of allowing an estimation of economic losses caused by the condemnation of tuberculous organs and carcasses. Data collected from routine meat inspection can also be used to make evidence-based decisions when developing control and prevention strategies for BTB in Ethiopia.

14.5.2 *Challenges*

Currently, there is not a national BTB control or prevention program in Ethiopia primarily because of inadequate financial support and the lack of infrastructure and human resources. A lack of awareness, by farmers and traders of a risk-based trading approach allowing buyers to minimize or avoid the risk of infection with BTB, is also prevalent and should be addressed when implementing future control programs (DEFRA 2013).

Another matter that should be rectified is the current unregulated movement of infected livestock, since contact with other livestock or wildlife constitutes a major way in which the infection is disseminated.

The lack of a mandatory test-and-slaughter policy, the absence of livestock movement control, animal identification and a tracking system, and the general lack of knowledge and awareness about isolation and quarantine practices are some of the challenges that make the future prevention and control of BTB in Ethiopia a daunting task.

14.5.3 *Opportunities*

14.5.3.1 Policy

Recognition and inclusion of livestock health as a major component of the Ethiopian livestock master plan (LMP) would be an important step forward for initiating control and prevention strategies for BTB (Shapiro et al. 2015).

14.5.3.2 Human Resources

Establishing more and strengthening the 15 current regional veterinary diagnostic laboratories and an upgrade and expansion of the National Animal Health Diagnostic and Investigation Center (NAHDIC) would create better opportunities for designing and implementing evidence-based disease control and prevention policies and strategies. In the past two decades, Ethiopia has expanded the number of higher educational facilities including the establishment of nine new veterinary schools. Additionally, five more colleges (two public and three private) have been established to train animal health assistants. The training of thousands of veterinary professionals at various levels (undergraduate, graduate, and technicians) may assist to reduce the technical and administrative challenges that the country faces when implementing BTB control and preventive measures.

14.5.3.3 Policy Development and Implementation

The development of its export trade in livestock and livestock products is one of Ethiopia's major objectives. This venture requires adherence to the international policies for the management of trade limiting, trans-boundary livestock diseases such as BTB. Keeping the sanitary and phytosanitary standards of the World Trade Organization (WTO) that govern international trade in livestock in mind, the Ethiopian government will have to review their policies and practices, to allow them to participate in this trade. These measures include:

- Applying stringent sanitary regulations to livestock production and livestock products, including their auditing and certification requirements
- Introducing an animal identification and traceability system
- Creating a national livestock registry and traceability databank

Implementation of these processes is likely to create a better opportunity for initiating sustainable BTB control and prevention programs in Ethiopia. The opportunity to do it already exists as the Ethiopian authorities have already enacted and applied the mandatory test-and-isolation policy in its export beef cattle sector to meet the requirements of importing countries (mainly the Middle East and North Africa), and it merely needs an extension of the existing measures to deal with BTB on a national scale.

References

Amare A (2015) Wildlife resources of Ethiopia: opportunities, challenges and future directions from ecotourism perspective: a review paper. Nat Res 6:405–422

Ameni G, Erkihun A (2007) Bovine tuberculosis on small-scale dairy farms in Adama town, Central Ethiopia, and farmer awareness of the disease. Rev Sci Tech (OIE) 26:711–719

Ameni G, Aseffa A, Engers H et al (2007) High prevalence and increased severity of pathology of bovine tuberculosis in Holsteins compared to zebu breeds under field cattle husbandry in Central Ethiopia. Clin Vac Immunol 14:1356–1361

Amenu K, Thys E, Regassa A et al (2010) Brucellosis and tuberculosis in Arsi-Negele District, Ethiopia: prevalence in ruminants and people's behaviour towards zoonoses. Tropicultura 28:205–210

Ashenafi D, Mamo G, Ameni G et al (2013) Epidemiology and molecular characterization of causative agents of BTB in ruminants. J Bacteriol Parasitol 4:161

Asseged B, Woldesenbet Z, Yimer E et al (2004) Evaluation of abattoir inspection for the diagnosis of *M. bovis* infection in cattle slaughtered at Addis Ababa abattoir. Trop Anim Health Prod 36:537–546

Behnke R (2010) The contribution of livestock to the economies of IGAD Member States. IGAD Livestock Policy Initiative (IGAD LPI) working paper 02-10. www.fao.org/fileadmin/user_upload/drought/docs/IGAD LPI WP 02-10.pdf

Bekele M, Belay I (2011) Evaluation of routine meat inspection procedure to detect bovine tuberculosis suggestive lesions in Jimma municipal abattoir, south West Ethiopia. Glob Veterinaria 6:172–179

Berg S, Firdessa R, Habtamu M et al (2009) The burden of mycobacterial disease in Ethiopian cattle: implications for public health. PLoS One 4:e5068
Berg S, Garcia-Pelayo MC, Müller B et al (2011) African 2, a clonal complex of *M. bovis* epidemiologically important in East Africa. J Bacteriol 193:670–678
Beyi AF, Gezahegne KZ, Mussa A et al (2014) Prevalence of BTB in dromedary camels and awareness of pastoralists about its zoonotic importance in eastern Ethiopia. J Vet Med Anim Health 6(4):109–115
Biet F, Boschiroli ML, Thorel MF et al (2005) Zoonotic aspects of *Mycobacterium bovis* and *Mycobacterium avium-intracellulare* complex (MAC). Vet Res 36:411–436
Biffa D, Skjerve E, Oloya J et al (2010) Molecular characterization of *Mycobacterium bovis* isolates from Ethiopian cattle. BMC Vet Res 6:28
Cicero R, Olivera H, Hernandez-Solis A et al (2009) Frequency of *Mycobacterium bovis* as an etiologic agent in extrapulmonary tuberculosis in HIV-positive and negative Mexican patients. Eur J Clin Microbiol Infect Dis 28:455–460
CSA (2012) Report on livestock and livestock characteristics (private peasant holdings). Agricultural sample survey, vol II. Central Statistical Authority (CSA), Addis Ababa, Ethiopia
DEFRA (2013) Bovine TB risk-based trading: empowering farmers to manage TB trading risks. Department of Environment, Food and Rural Affairs (DEFRA), London
Dinka H, Duressa A (2011) Prevalence of bovine tuberculosis in Arsi zone of Oromia, Ethiopia. Afr J Agric Res 6:3853–3858
Elias K, Hussein D, Asseged B et al (2008) Status of bovine tuberculosis in Addis Ababa dairy farms. Rev Sci Tech 27:915–923
Ethiopie CE (2009) A cross sectional study of camel tuberculosis in Ethiopia. Bull Anim Health Prod Afr 57:13–20
FAO (2004) Livestock sector brief: Ethiopia. Livestock sector analysis and policy branch. FAO, Rome
FAOSTAT (2003) FAO statistics database on world website. http://apps.fao.org/.3
FAOSTAT (2005) Cattle population. Food and agricultural Organization of the United Nations, Rome. http://faostat3.fao.org/search/CATTLE/E; http://faostat.fao.org
Firdessa R, Tschopp R, Wubete A et al (2012) High prevalence of BTB in dairy cattle in Central Ethiopia: implications for the dairy industry and public health. PLoS One 7:e52851
Groombridge B (1992) Global biodiversity: status of the earth's living resources. Chapman and Hall, London, p 465
Gumi B, Schelling E, Firdessa R et al (2012) Low prevalence of bovine tuberculosis in Somali pastoral livestock, south East Ethiopia. Trop Anim Health Prod 44:1445–1450
Haftu B, Asresie A, Haylom M (2014) Assessment on major health constraints of livestock development in eastern zone of Tigray: the case of "Gantaafeshum Woreda", northern Ethiopia. J Vet Sci Technol 5:174
Hailemariam S (1975) A brief analysis of activity of the meat inspection and quarantine division. Ministry of Agriculture (MoA), Department of Veterinary Services, Addis Ababa, Ethiopia
Haileselassie M, Taddele H, Adhana K et al (2013) Food safety knowledge and practices of abattoir and butchery shops and the microbial profile of meat in Mekelle City, Ethiopia. Asian Pac J Trop Biomed 3:407–412
Hiko A, Agga GE (2011) First-time detection of *Mycobacterium* species from goats in Ethiopia. Trop Anim Health Prod 43(1):133–139
Hilty M, Diguimbaye C, Schelling E et al (2005) Evaluation of the discriminatory power of variable number tandem repeat (VNTR) typing of *Mycobacterium bovis* strains. Vet Microbiol 109:217–222
Hlavsa M, Moonan P, Cowan LS et al (2008) Human tuberculosis due to *M. bovis* in the United States, 1995-2005. Clin Infect Dis 47:168–175
Jabbar M, Negassa A, Gidyelew T (2007) Geographic distribution of cattle and shoats populations and their market supply sheds in Ethiopia. ILRI improving market opportunities discussion paper no 2, Nairobi, Kenya, p 54

Kazwala RR, Daborn CJ, Sharp JM et al (2001) Isolation of *M. bovis* from human cases of cervical adenitis in Tanzania: a cause for concern? Int J Tuberc Lung Dis 5:87–91

Kidane D, Olobo JO, Habte A et al (2002) Identification of the causative organism of tuberculous lymphadenitis in Ethiopia by PCR. J Clin Microbiol 40:4230–4234

Kimball T (2011) Environmental policy review: livestock production systems and their environmental implications in Ethiopia. Environmental Policy Group. Environmental Studies Program. Colby College, Maine, USA

Leta S, Mesele F (2014) Spatial analysis of cattle and shoat population in Ethiopia: growth trend, distribution and market access. Springerplus 3:310

Mamo G, Bayleyegn G, Sisay T et al (2011) Pathology of camel tuberculosis and molecular characterization of its causative agents in pastoral regions of Ethiopia. PLoS One 6:e15862

Mamo KG, Abebe F, Worku Y et al (2012) Tuberculosis in goats and sheep in Afar pastoral region of Ethiopia and isolation of *Mycobacterium tuberculosis* from goat. Vet Med Int 2012:869146

Mekibeb A, Fulasa TT, Firdessa R et al (2013) Prevalence study on BTB and molecular characterization of its causative agents in cattle slaughtered at Addis Ababa municipal abattoir, Central Ethiopia. Trop Anim Health Prod 45:763–769

Metaferia F, Cherenet T, Gelan A et al (2011) A review to improve estimation of livestock contribution to the national GDP. Ministry of Finance and economic development and Ministry of Agriculture. Addia Ababa, Ethiopia. http://hdl.handle.net/10568/24987

MoA (1998) National livestock development project (NLDP) working paper 1–4. Ministry of Agriculture, Addis Ababa, Ethiopia

Muwonge A, Oloya J, Kankya C et al (2014) Molecular characterization of *Mycobacterium avium* subspecies *hominissuis* isolated from humans, cattle and pigs in the Uganda cattle corridor using VNTR analysis. Infect Genet and Evol 21:184–191

Neill SD, Hanna J, O'Brien JJ et al (1989) Transmission of tuberculosis from experimentally infected cattle to in-contact calves. Vet Rec 124:269–271

Nemomsa B, Gebrezgabeher G, Birhanu T et al (2014) Epidemiology of BTB in Butajira, southern Ethiopia. A cross sectional abattoir based study. Afr J Microbiol Biotechnol Res 33:3112–3117

Pal M, Tesfaye S, Dav P (2013) Zoonoses occupationally acquired by abattoir workers. J Environ Occup Sci 2:155–162

Regassa G, Mekonnen D, Yamuah L et al (2009) Human brucellosis in traditional pastoral communities in Ethiopia. Int J Trop Med 4:59–64

Selander RK, Caugant DA, Ochman H (1986) Methods of multilocus enzyme electrophoresis for bacterial population genetics and systematics. Appl Environ Microbiol 51:873–884

Shapiro BI, Gebru G, Desta S et al (2015) Ethiopia livestock master plan. International Livestock Research Institute (ILRI) Project Report. Nairobi, Kenya

Shitaye JE, Getahun B, Alemayehu T et al (2006) A prevalence study of BTB by using abattoir meat inspection and tuberculin skin testing data, histopathological and IS6110 PCR examination of tissues with tuberculous lesions in cattle in Ethiopia. Veterinarni Medicina 51:512–522

Tafess K, Dawo F, Sori T et al (2011) Prevalence of caprine tuberculosis in mid-rift valley area of Oromia, Ethiopia. Afr J Microbiol Res 5:1473–1478

Teklu A, Asseged B, Yimer E et al (2004) Tuberculous lesions not detected by routine abattoir inspection: the experience of the hosanna municipal abattoir, southern Ethiopia. Rev Sci Tech 23:957–964

Tesfaye D, Fekede D, Tigre W et al (2013) Perception of the public on the common zoonotic diseases in Jimma, southwestern Ethiopia. Int J Med Sci 5:279–285

Tigre W, Alemayehu G, Abetu T et al (2011) Preliminary study on public health implication of bovine tuberculosis in Jima town, south western Ethiopia. Glob Veterinaria 4:369–373

Tschopp R (2015) Bovine tuberculosis at human-wildlife-livestock interface in sub-Saharan Africa. In: Zinsstag J et al (eds) One health: the theory and practice of integrated health approaches. CAB International, London

Tschopp R, Aseffa A, Schelling E et al (2010a) Bovine tuberculosis at the wildlife-livestock-human interface in Hamer Woreda, south Omo, southern Ethiopia. PLoS One 5:e12205

Tschopp R, Berg S, Argaw K et al (2010b) Bovine tuberculosis in Ethiopian wildlife. J Wild Dis 46:753–762
Tschopp R, Bobosha K, Aseffa A et al (2011) Bovine tuberculosis at a cattle-small ruminant-human interface in Meskan, Gurage region, Central Ethiopia. BMC Infect Dis 11:318
Yilma Z, Emannuelle GB, Ameha S (2011) A review of the Ethiopian dairy sector. In: Fombad R (ed) FAO sub regional Office for Eastern Africa (FAO/SFE). Addis Ababa, Ethiopia, p 81

Chapter 15
Bovine Tuberculosis in Ghana

Dorothy Yeboah-Manu and Adwoa Asante-Poku

15.1 Introduction

Geographically, Ghana (within latitude 4°44′ N and 11°11′ N and 3°11′ W and 1°11′ E) is bordered on the east by the Republic of Togo, to the north by Burkina Faso (formerly Upper Volta), and to the west by Côte d'Ivoiré. The Gulf of Guinea lies south of the country, forming a coastline of 550 km (Fig. 15.1). According to the 2012 population census report, Ghana harbors 24,658,823 humans (Ghana Statistical Service 2012), while reports from the Ministry of Agriculture indicate that the cattle, sheep, and goat populations in that same year were 1,543,000, 4,019,000, and 5,435,000, respectively (Ministry of Agriculture 2012). The most common cattle breed in the country is the West African Shorthorn (WASH) (Teye and Sunkwa 2010) that includes all small, non-humped cattle, generally black and white but sometimes fawn and white (Karbo et al. 2004). The WASH is an indigenous, tough breed of cattle and thickset with short, fine-boned limbs. This breed accounts for about half of the cattle in the country, and is followed numerically by the Sanga, which is a cross between the WASH and large-humped Zebu cattle, and then the Zebu breed (Hutchinson 1962). There are currently no pure, exotic dairy cattle in the country. Most of the about 500 crossbred dairy cattle in the country were bred by artificial insemination with imported semen (Gushiegu District Report 2006).

In the majority of cases, cattle are owned by ethnic groups of farmers or families or even by a whole village. Traditionally, most farmers in Northern Ghana rear cattle for socioreligious reasons, and they are used in ritual events. Two additional factors based on tradition and customs, namely, the prestige of the herd and its value as a self-sustaining investment, are of great importance in sustaining cattle farming in the region (Veterinary Services Directorate 2010).

D. Yeboah-Manu (✉) · A. Asante-Poku
Noguchi Memorial Institute for Medical Research, University of Ghana, Legon, Ghana
e-mail: dyeboah-manu@noguchi.ug.edu.gh; aasante-poku@noguchi.ug.edu.gh

A. B. Dibaba et al. (eds.), *Tuberculosis in Animals: An African Perspective*,
https://doi.org/10.1007/978-3-030-18690-6_15

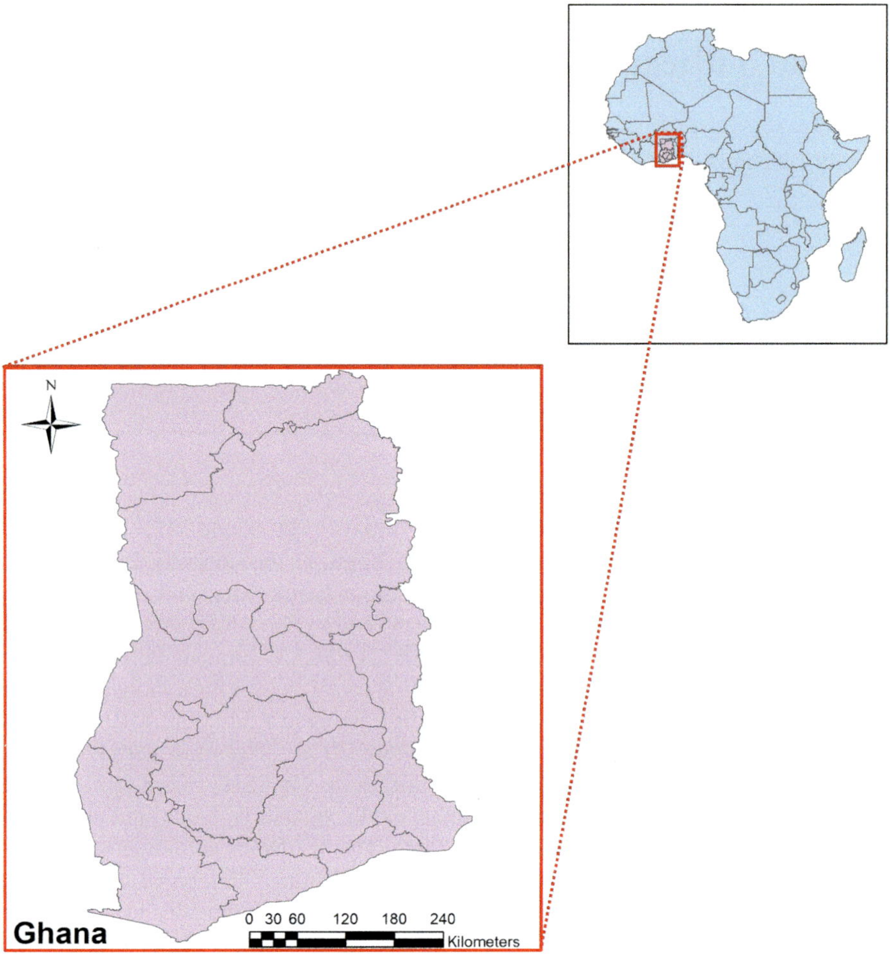

Fig. 15.1 Map of Ghana zoomed from its location on West Africa

Similar to other countries in Africa, human tuberculosis is an important public health problem in Ghana. With 9 million recorded cases and 1.5 million deaths globally in 2013, tuberculosis (TB) in humans is the leading cause of adult death by a single infectious disease. In 1995, the number of cases of human tuberculosis in Africa accounted for 16% of the global burden, and it has now increased to about 30% even though the number of people in Africa constitutes only 11% of the global population (WHO 2014). Furthermore, about 70% of individuals co-infected with both HIV and TB, live in sub-Saharan Africa (Corbett et al. 2006).

A group of Gram-positive bacteria, the *Mycobacterium tuberculosis* complex (MTC), causes tuberculosis in various species. The MTC comprises various species and sub-species (Comas and Gagneux 2011; Corbett et al. 2003; Brosch et al. 2002;

Cagneux and Small 2008; Garnier et al. 2003; Frota et al. 2004; Cousins et al. 2003). In Africa, *Mycobacterium tuberculosis* and *M. africanum* are the main causative agents of TB in humans (hereafter referred to as "human MTC"). Of the other members of the MTC, *M. bovis* is primarily a pathogen of cattle (bovine MTC); *M. microti* is a pathogen of voles (Frota et al. 2004; Wells 1937, 1946; Cavanagh et al. 2002); and *M. caprae* is a pathogen of goats (Aranaz et al. 1999). Other members of the group include *M. mungi* (the mongoose pathogen) (Alexander et al. 2010), *M. orygis*, a pathogen of antelopes (van Ingen et al. 2012), and *M. pinnipedii*, a pathogen of seals and sea lions (Cousins et al. 2003). Contrary to many of the species contained in the group, *M. bovis* is not host-specific and displays a wide host range (Gibson et al. 2004) and can infect a range of mammals including humans. With the semi-high HIV prevalence of 1.9% in Ghana, which translates to close to 500,000 PLHIV (Corbett et al. 2006), BTB represents a significant zoonotic risk.

15.2 Diagnosis of Bovine TB in Ghana

The isolation of viable mycobacteria by culture and their identification is the laboratory diagnostic method accepted globally as the gold standard for confirming the diagnosis of tuberculosis in various species (Asante-Poku et al. 2014). Culture also has the advantage of providing isolates that can be used for in-depth studies aiming to understand the mode of transmission, characterize the mycobacterial species, and determine drug resistance/susceptibility.

Diagnostic specimens for culture are often collected from lesions in lymph nodes and parenchymatous organs such as the lungs, liver, and spleen (Dungworth 1993). However, this is not routinely done in Ghana due to the lack of the required diagnostic infrastructure of the Ghanaian Veterinary Services Department (VSD), though such facilities are available in some research centers in the country. Thus, to the best of our knowledge, only few research studies have tried to isolate mycobacteria for identification from suspected lesions to confirm BTB. Direct microscopic smear examination for acid-fast bacilli (AFB) after staining with carbol fuchsin is simple and routinely used for the diagnosis of human TB (Forrellad et al. 2013). However, it is rarely used for BTB due to its low sensitivity, as it requires a concentration of 10^4 bacilli per milliliter suspension to give a positive smear test (Asante-Poku et al. 2014; Allen and Mitchison 1992; Pearson et al. 2008). In addition, this method lacks specificity, as it cannot differentiate between the various mycobacterial species. The standard Single Intradermal Cervical Test (SIT) with purified protein derivative (PPD) tuberculin for detection of infection in live cattle (Pearson et al. 2008) is available, but it is not used routinely due to logistical constraints. The main routine method for diagnosing BTB in Ghana is a macroscopic inspection done at various abattoirs to detect lesions in carcasses (Menzies 2000). This method, though it may not be sensitive, offers the opportunity for targeted screening of tuberculous carcasses to protect the consumers' health. It also remains the only viable surveillance regimen for the VSD to establish the geographic extent of the disease within different geographical areas in the country. The Ghanaian VSD has not yet developed the capacity to do the interferon gamma assays that are available (Rothel et al. 1990; Neill and Pollock 2000). Molecular detection methods such as polymerase chain reaction (PCR) and other genotyping methods are also not routinely used due to the cost and lack of infrastructure. Nevertheless, several in-country research studies have used PCR-based methods to characterize mycobacterial isolates from both humans and animal sources. In Ghana, these molecular techniques are increasingly being applied primarily to enhance detection, and for the typing of mycobacteria (Bonsu et al. 2000).

15.3 Prevalence of BTB in Ghana

The true prevalence of BTB in Ghana is not known as data on its occurrence in the different districts and administrative regions of the Veterinary Services Department (VSD) are minimal and are usually gathered from sporadic inspection and research activities. However, all findings from these activities point to BTB being endemic in animals and probably also in the human population.

Most sporadic studies were conducted in the Greater Accra Region (GAR). Between 2005 and 2010, 516 BTB cases were diagnosed at abattoirs and slaughterhouses in Ghana (Amemor 2012). Bonsu et al. (2000) tested a total of 1200 cattle, comprising 747 cows and 400 heifers, all 2–4 years old, and 53 bulls from the GAR using the SIT. Of the 1200 cattle tested, 166 reacted positively to bovine tuberculin, representing an average prevalence of 13.8% in cattle in the district, although the prevalence was as high as 50% in some kraals. All ages of animals and both sexes were affected, but the prevalence in cows was twice as high as that in heifers and in bulls. In another study conducted on four farms from the GAR, a crude prevalence of about 17% was reported (Thoen and Ebel 2006). Additionally, 685 cattle from two dairy farms in the same region were tested; of these, 17 were positive by the tuberculin test, giving a crude BTB prevalence of 2.5% (Asante-Poku et al. 2014). The main difference in the animals used in the later and previous two studies was the husbandry systems used: animals from a dairy farm were included in the later study; while the other two studies involved animals from free-range farms scattered in rural communities. These findings suggest that the control of BTB is more difficult in a free-range grazing system. Within the Accra Metropolis, 145 out of 2886 (5%) cattle slaughtered over a period of 4 months were found to contain BTB-like lesions, most of which were confirmed as being infected by *M. bovis* by further microbiological analysis (Atiadeve et al. 2014).

Additional information exists for the Volta Region where Ankugah (2000) reported a BTB prevalence of 3.1% in the Ho district even though it was as high as 5.9% in one cluster. Amemor (2012), using a rapid antigen capture assay, analyzed 200 cattle consisting of 14 bulls and 186 cows, and found a prevalence of 19% (38/200) (all females) in the North Tongu district of the same region.

15.4 Mycobacterial Species Isolated from Cattle

To the best of our knowledge, mycobacteria were isolated from cattle in only two of the reported studies (Bonsu et al. 2000; Asante-Poku et al. 2014). Our group cultured 17 samples from suspected lesions and only 6 (35.3%) yielded AFBs. Three of the isolates were identified as MTC, which yielded two *M. tuberculosis* strains and one *M. africanum* strain. The remaining three isolates belonged to *Mycobacterium* other-than-tuberculosis (MOTT). Spoligotyping further characterized the two *M. tuberculosis* isolates as Ghana (spoligotype Data Base 4 number 53) and Latin American Mediterranean (LAM), while spoligotyping and single nucleotide polymorphism (SNP) analysis typed the *M. africanum* as West African 1. Microseq 500 analysis identified two of the MOTT isolates as *M. flavescens* and *M. moriokaense*, while the remaining one could not be identified (Asante-Poku et al. 2014). Furthermore, Atiadeve et al. (2014) examined 93 AFB isolates: Of these 11 (11.8%) were members of the MTC and 82 (88.2%) were non-tuberculous mycobacteria. All three main MTC strains, that is, *M. bovis*, *M. tuberculosis*, and *M. africanum*, were identified. These findings from Ghana and elsewhere probably indicate that other mycobacterial species are also important as a cause of BTB (Skuce et al. 2011; Asante-Poku et al. 2014).

15.5 *Mycobacterium bovis* Infection in Humans

The occurrence and prevalence of human TB due to *M. bovis* are difficult to determine accurately and probably remain under-reported owing to the diagnostic limitations of the main method used for diagnosing human TB in Ghana. In a previous work in which 70 pulmonary TB patients attending a health-care facility in Accra were sampled, 3% were identified as infected with *M. bovis* using biochemical methods for characterization of the mycobacteria (Addo et al. 2007). Ongoing population-based prospective studies of human TB cases in the Accra Metropolis and a rural district of northern Ghana, using conventional culture followed by molecular characterization, identified 15 *M. bovis* and 1 *M. caprae* strains from 1794 isolates (from sputum) giving an *M. bovis* prevalence of 0.8% in the study population. The spoligotype patterns of the identified animal strains are shown in Fig. 15.2. The prevalence that was reported, however, seems lower when compared to the prevalence of up to 10% found in neighboring countries (Mawak et al. 2006), but it is within the estimated range for Africa based on the review of Müller et al. (2013).

The National Tuberculosis Control Program of the Ghana Health Service guides TB control in humans in Ghana. In the year 2013, close to 16,000 new TB cases were notified making it the 19th most burdened country with TB in Africa. It was found that in that year, 9% of the over 7000 patients registered died before completing treatment (Kurbatova et al. 2012). This figure, however, may not be a true reflection

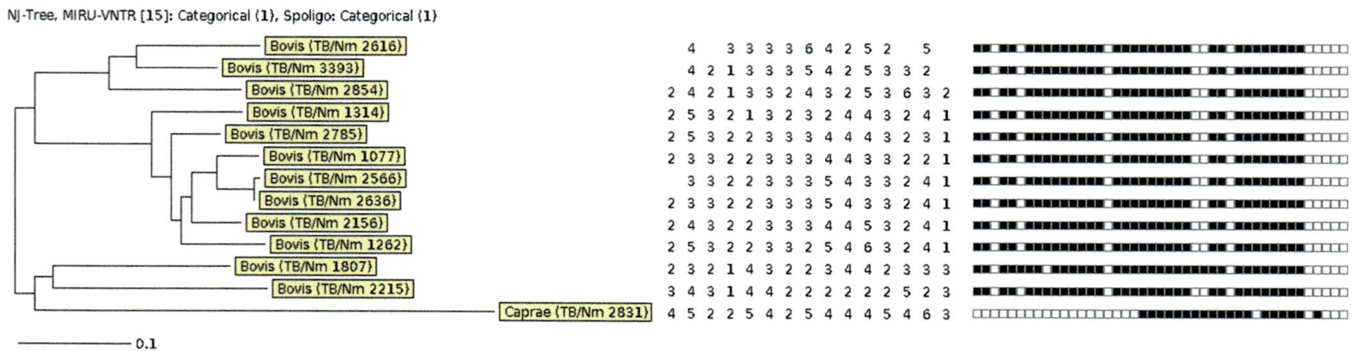

Fig. 15.2 Dendrogram constructed using the Neighbour-Joining (NJ) algorithm for 12 MIRU-VNTR typed *M. bovis* (identified by spoligotyping) isolates and an additional *M. caprae* isolate. Figure was plotted using the MIRU-VNTRplus on-line tool. The respective spoligotype patterns are displayed alongside the dendrogram

of the actual prevalence since there are many patients that do not report to health facilities for treatment. Thus, the real TB burden in Ghana is likely to be substantially higher than the official WHO estimates indicate. This notion is confirmed by preliminary data of the just-ended national TB prevalence survey, which shows a prevalence of about 264 cases per 100,000 human populations (National Tuberculosis Report 2014: unpublished data). In addition, anecdotal evidence in Korle-Bu Teaching Hospital (KBTH), Accra (the highest referral health facility in Ghana), indicates that TB is the cause of death in one out of seven post-mortems conducted (MOH/GHS 2003). Reports from the National Tuberculosis Control Program (NTP) in Ghana indicate that about 23% of TB cases are co-infected with HIV and that as many as 50% of patients with chronic cough could be HIV-positive (MOH/GHS 2007).

Due to the lack of appropriate laboratory infrastructure, drug susceptibility testing is not routinely performed in Ghana; therefore, the actual extent of drug-resistant TB is unknown. Current TB control measures in Ghana (like in most other developing countries) are primarily based on sputum smear microscopy, which has a diagnostic sensitivity of only 50% (Muvunyi et al. 2010). However, TB control activities in the past years have improved greatly due to activities being supported by funds from the Global Fund Initiative that include early case detection, detection among vulnerable groups, increasing access to WHO-recommended rapid and more sensitive diagnostics such as XpertMTB/RIF (Cepheid), and the line probe assay, MTBDRplus (Hain Lifescience).

All forms of human TB are dealt with by the National Tuberculosis Control Program (NTP). The main control strategy of human TB by the NTP is case detection mainly by direct smear microscopy and treatment by the directly-observed-treatment-short-course (DOTs) strategy (MOH/GHS 2003). Thus, determining drug resistance is not routinely done and most efforts are focused on diagnosing pulmonary TB by sputum analysis. Nevertheless, current efforts have been maximized to include recent WHO-recommended diagnostics such as XpertMTB/Rif (Kurbatova et al. 2012) and line probe assays (Asante-Poku et al. 2015) in some regional hospitals. In addition, the NTP receives support from the main biomedical research institute in the country for culture, phenotypic drug-resistance, and molecular genotyping expertise to support case management of retreatment cases and epidemiological studies. The main challenge impeding human TB control is late case reporting and low detection rates due to stigmatization (Weiss et al. 2006).

Control of BTB and Challenges Facing the Veterinary Services in Ghana

In Ghana, the control of BTB in animals is the responsibility of the Veterinary Services Department (VSD) of the Ministry of Agriculture. The VSD is also responsible for public health through controlling diseases communicable from animals to human beings by:

- Ensuring that meat and other products of animal origin are safe for human consumption
- Controlling animal movement by ensuring that only healthy animals are permitted to be moved from one area to another to prevent disease spread in the country

- Ensuring that all animals entering the country are quarantined and only those found healthy allowed passage into the country

The above measures are assumed to be in place to control the transmission of zoonotic diseases, which includes BTB. However, there are a number of challenges for the VSD to carry out its mandate effectively. For example, lack of logistics including reagents and well-equipped laboratories, and the situation that routine surveillance of BTB by the Surveillance System Component (SSC) is not done. Likewise, movement of animals in accordance with the Economic Community of West-African States protocol that allows free movement of animals and humans across borders limits cross-border control and restriction of movement of cattle across borders.

The acclaimed and effective test-and-slaughter protocol, however, is capital-intensive and cannot be supported financially by the Ghanaian Government. The VSD, therefore, has a policy of examining cattle slaughtered at various abattoirs for the presence of granulomatous lesions suggestive of bovine tuberculosis infection in an effort to control the spread of the disease from cattle to humans. Cattle with localized infections at post-mortem, that is, infection in one organ and/or its associated lymph node only, has the affected part trimmed off and the carcasses passed for human consumption. However, in animals with generalized infection, that is, infection throughout the entire carcass, the whole carcass is deemed unwholesome for human consumption and is therefore condemned.

BTB is a potential zoonotic disease that can infect a variety of wild animals including African buffaloes (*Syncerus caffer*) (Katale et al. 2012) that are a known BTB maintenance host and are present in both the Mole and Digya National Parks in Ghana. It is currently not known whether any of the wildlife species are infected with *M. bovis* in Ghana. The presence of multiple hosts including wild animals (Addo et al. 2007) and inefficient diagnostic techniques and the absence of defined national control and eradication programs impede the control of bovine TB in Ghana (Katale et al. 2012).

Finally, not much has been done regarding public education of the zoonotic importance of BTB and raising awareness within communities to prevent the disease. For most rural herdsmen in Ghana, consumption of raw milk and milk products and close association between animals and farmers are common practices, which all enhance exposure to both the alimentary and respiratory routes of infection with *M. bovis*.

15.6 Conclusion

An unknown proportion of cattle in the country are likely to be infected with BTB because the test-and-slaughter policy for the control of the disease has not been implemented, and many of the country's cattle remain untested for BTB. The disease in cattle is currently only detected at slaughter during meat inspection and occasionally by sporadic screening for the disease by skin testing. People who attend to these animals and come into close contact with them, such as herdsmen and veterinarians, and the general public who consume fresh milk or infected meat, are exposed to *M. bovis* from cattle. Humans are exposed especially when animals are slaughtered at home without veterinary supervision and proper meat inspection. Preliminary findings from our ongoing population-based studies indicate a strong association between humans such as farmers or butchers infected with *M. bovis* and contact with animals. For improved control, intensified public education and application of the One Health approach for the control of zoonotic tuberculosis are essential. Furthermore, the capacity of the VSD needs to be enhanced to allow BTB to be dealt with adequately in Ghana.

References

Addo K, Owusu-Darko K, Yeboah-Manu D et al (2007) Mycobacterial species causing pulmonary tuberculosis at the Korle Bu Teaching Hospital, Accra, Ghana. Ghana Med J 41:52–57

Alexander KA, Laver PN, Michel AL et al (2010) Novel *Mycobacterium tuberculosis* complex pathogen, *M. mungi*. Emerg Infect Dis 16:1296–1299

Allen BW, Mitchison DA (1992) Counts of viable tubercle bacilli in sputum related to smear and culture gradings. Med Lab Sci 49:94–96

Amemor EA (2012) Bovine tuberculosis among herdsmen, North Tongu district, Volta Region, Ghana. Doctoral dissertation, University of Ghana. http://hdl.handle.net/123456789/5573

Ankugah DK (2000) Prevalence of bovine tuberculosis in Ho District of Ghana. A potential for human infection. In: Proceedings of the 10th conference of the association of Institutions for Tropical Veterinary Medicine, 20–23 Aug, Copenhagen, Denmark

Aranaz A, Liebana E, Gomez-Mampaso E et al (1999) *Mycobacterium tuberculosis subsp. caprae* subsp. nov: a taxonomic study of a new member of the *Mycobacterium tuberculosis* complex isolated from goats in Spain. Int J Syst Bacteriol 49:1263–1273

Asante-Poku A, Aning KG, Boi-Kikimoto B et al (2014) Prevalence of bovine tuberculosis in a dairy cattle farm and research farm in Ghana. Onderstepoort J Vet Res 81:1–6

Asante-Poku A, Otchere ID, Danso E et al (2015) Establishment and evaluation of the genotype MTBDRplus for rapid detection of drug resistant tuberculosis in Ghana. Int J Tuberc Lung Dis 19:954–959

Atiadeve SK, Gyamfi OK, Mak-Mensah E et al (2014) Slaughter surveillance for tuberculosis among cattle in three metropolitan abattoirs in Ghana. J Vet Med Anim Health 6:198–207

Bonsu OA, Laing E, Akanmori BD (2000) Prevalence of tuberculosis in cattle in the Dangme-West district of Ghana, public health implications. Acta Trop 76:9–14

Brosch R, Gordon SV, Marmiesse M et al (2002) A new evolutionary scenario for the *Mycobacterium tuberculosis* complex. Proc Natl Acad Sci U S A 99:3684–3689

Cagneux S, Small PM (2008) Molecular evolution of mycobacteria. In: Rubin E, Kaufmann SHE (eds) Handbook of tuberculosis: molecular biology and biochemistry, vol 1. Wiley-VCH, Weinheim, pp 393–416

Cavanagh R, Begon M, Bennett M et al (2002) *Mycobacterium microti* infection (vole tuberculosis) in wild rodent populations. J Clin Microbiol 40:3281–3285

Comas I, Gagneux S (2011) A role for systems epidemiology in tuberculosis research. Trends Microbiol 19:492–500

Corbett EL, Watt CJ, Walker N et al (2003) The growing burden of tuberculosis: global trends and interactions with the HIV epidemic. Arch Intern Med 163(9):1009–1021

Corbett EL, Marston B, Churchyard GJ et al (2006) Tuberculosis in sub-Saharan Africa: opportunities, challenges, and change in the era of antiretroviral treatment. Lancet 367:926–937
Cousins DV, Bastida R, Cataldi A et al (2003) Tuberculosis in seals caused by a novel member of the *Mycobacterium tuberculosis* complex: *Mycobacterium pinnipedii* sp. nov. Int J Syst Evol Microbiol 53:1305–1314
Dungworth DL (1993) The respiratory system. In: Jubb KVH, Kennedy PC, Palmer N (eds) Pathology of domestic animals, 4th edn. Academic, San Diego, CA, pp 641–652
Forrellad MA, Klepp LI, Gioffré A et al (2013) Virulence factors of the *Mycobacterium tuberculosis* complex. Virulence 4:3–66
Frota CC, Hunt DM, Buxton RS et al (2004) Genome structure in the vole bacillus, *Mycobacterium microti*, a member of the *Mycobacterium tuberculosis* complex with a low virulence for humans. Microbiology 150:1519–1527
Garnier T, Eiglmeier K, Camus J-C et al (2003) The complete genome sequence of *Mycobacterium bovis*. Proc Natl Acad Sci U S A 100:7877–7882
Ghana Statistical Service (2012) Population and housing census
Gibson AL, Hewinson G, Goodchild T et al (2004) Molecular epidemiology of disease due to *Mycobacterium bovis* in humans in the United Kingdom. J Clin Microbiol 42:431–434
Gushiegu District Report (2006) Ministry of Local Government and Rural Development. Ghana
Hutchinson RA (1962) Stock methods of animal husbandry. In: Wills JB (ed) Agriculture and land use in Ghana. Oxford University Press, London, pp 425–436
Karbo N, Otchere EO, Millar D et al (2004) West African shorthorn breeders association in the Saboba-Chereponi District of Ghana: formation, expectations and perceptions of members in participatory breed development. Ghana J Dev Stud 1:50–59
Katale BZ, Mbugi EV, Kendal S et al (2012) Bovine tuberculosis at the human-livestock-wildlife interface: is it a public health problem in Tanzania? A review. Onderstepoort J Vet Res 79:84–97
Kurbatova EV, Kaminski DA, Erokhin VV et al (2012) Performance of Cepheid® Xpert MTB/RIF® and TB-Biochip® MDR in two regions of Russia with a high prevalence of drug-resistant tuberculosis. Eur J Clin Microbiol Infect Dis 32:735–743
Mawak J, Gomwalk N, Bello C et al (2006) Human pulmonary infections with bovine and environment (atypical) mycobacteria in Jos, Nigeria. Ghana Med J 40:132–136
Menzies D (2000) Tuberculin skin testing. In: Reichman LB, Hershfield ES (eds) Tuberculosis: a comprehensive international approach. Marcel Dekker, New York, pp 279–322
Ministry of Agriculture (2012) Annual report of the ministry of agriculture. Ghana
MOH/GHS (2003) National tuberculosis annual report. Ghana Health Services, Ministry of Health
MOH/GHS (2007) Guidelines for the clinical management of TB and HIV co-infection in Ghana
Müller B, Dürr S, Alonso S et al (2013) Zoonotic *Mycobacterium bovis*-induced tuberculosis in humans. Emerg Infect Dis 19:899–908
Muvunyi CM, Masaisa F, Bayingana C et al (2010) Prevalence and diagnostic aspects of sputum smear positive tuberculosis cases at a tertiary care institution in Rwanda. Afr J Microbiol Res 4:88–91
Neill SD, Pollock JM (2000) Testing for bovine tuberculosis – more than skin deep. Vet J 160:3–5
Pearson JE, Schmitt B, Le Blanc SP et al (2008) Manual of diagnostic tests and vaccines for terrestrial animals (mammals, birds and bees), vol Vol 1, 6th edn. OIE, Paris, p 727
Rothel JS, Jones SL, Corner LA et al (1990) A sandwich enzyme immunoassay for bovine interferon-gamma and its use for the detection of tuberculosis in cattle. Aust Vet J 67:134–137
Skuce RA, Allen AR, McDowell SWJ (2011) Bovine tuberculosis (TB): a review of cattle-to-cattle transmission, risk factors and susceptibility. Thai Agricultural Commodity and Food Standard to Thai Agricultural Standard in accordance with the enforcement of the Agricultural Standards. London, UK, p 2551
Teye GA, Sunkwa WK (2010) Carcass characteristics of tropical beef cattle breeds (West African shorthorn, Sanga and zebu) in Ghana. Afr J Food Agric Nutr Dev 10:2866–2886

Thoen CO, Ebel ED (2006) Diagnostic tests for bovine tuberculosis. In: Thoen CO, Steele JH, Gilsdorf MJ (eds) *Mycobacterium bovis* infection in animals and humans, 2nd edn. Blackwell, Des Moines, IA, pp 49–53

van Ingen J, Rahim Z, Mulder A et al (2012) Characterization of *Mycobacterium orygis* as *M. tuberculosis* complex subspecies. Emerg Infect Dis 18:653–655

Veterinary Services Directorate (VSD) (2010) Ministry of food and agriculture, Ghana

Weiss MG, Ramakrishna J, Somma D (2006) Health-related stigma: rethinking concepts and interventions. Psychol Health Med 11:277–287

Wells AQ (1937) Tuberculosis in wild voles. Lancet 229(5934):1221

Wells AQ (1946) The murine type of tubercle bacillus (the vole acid-fast bacillus). Spec Rep Ser Med Res Counc (G B) 259:1482–1483

WHO (2014) Global tuberculosis report. World Health Organization, Geneva

Chapter 16
The Status of Bovine Tuberculosis in Malawi

Poya E. C. Njoka and Asseged B. Dibaba

16.1 Introduction

Malawi, with a population of 15 million people, shares borders with three southern Africa countries: Tanzania to the north, Mozambique to the south, and Zambia to the west. It is divided into eight agro-ecological zones referred to as agricultural development divisions (ADDs). There are 28 agricultural administrative offices (District Agriculture Offices) distributed throughout Malawi with a number of sub-district offices known as extension planning areas (EPAs) located in each of the districts (Fig. 16.1).

Malawi has an agriculture-based economy that provides employment to about 80% of the national workforce and generates over 80% of the country's foreign exchange earnings. The livestock sector generates about 11% of the Gross Domestic Product (GDP) and about 36% of the value of all agricultural products. The dairy sector consists of a small, commercial, market-oriented segment owning 3.4% of the cattle in the country, and a smallholder, subsistence-farming segment that mainly provides produce for home consumption, and a small surplus of products that are sold on the open market. Currently, there are too few commercial milk processors in Malawi, and those present operate below their capacity due to different factors, including poor road networks, and the lack of cooling and storage facilities.

Because of their low milk yield (2–3 l per day), most Malawian cattle (zebus) are not kept solely for dairying (Tebug et al. 2012) but they also provide an additional income through: provision of manure, provision of animal traction, and by serving as

P. E. C. Njoka (✉)
Ministry of Agriculture, Irrigation and Water Development, Department of Animal Health and Livestock Development, Lilongwe, Malawi

A. B. Dibaba
Department of Pathobiology, College of Veterinary Medicine, Tuskegee University, Tuskegee, AL, USA
e-mail: adibaba@tuskegee.edu

A. B. Dibaba et al. (eds.), *Tuberculosis in Animals: An African Perspective*,
https://doi.org/10.1007/978-3-030-18690-6_16

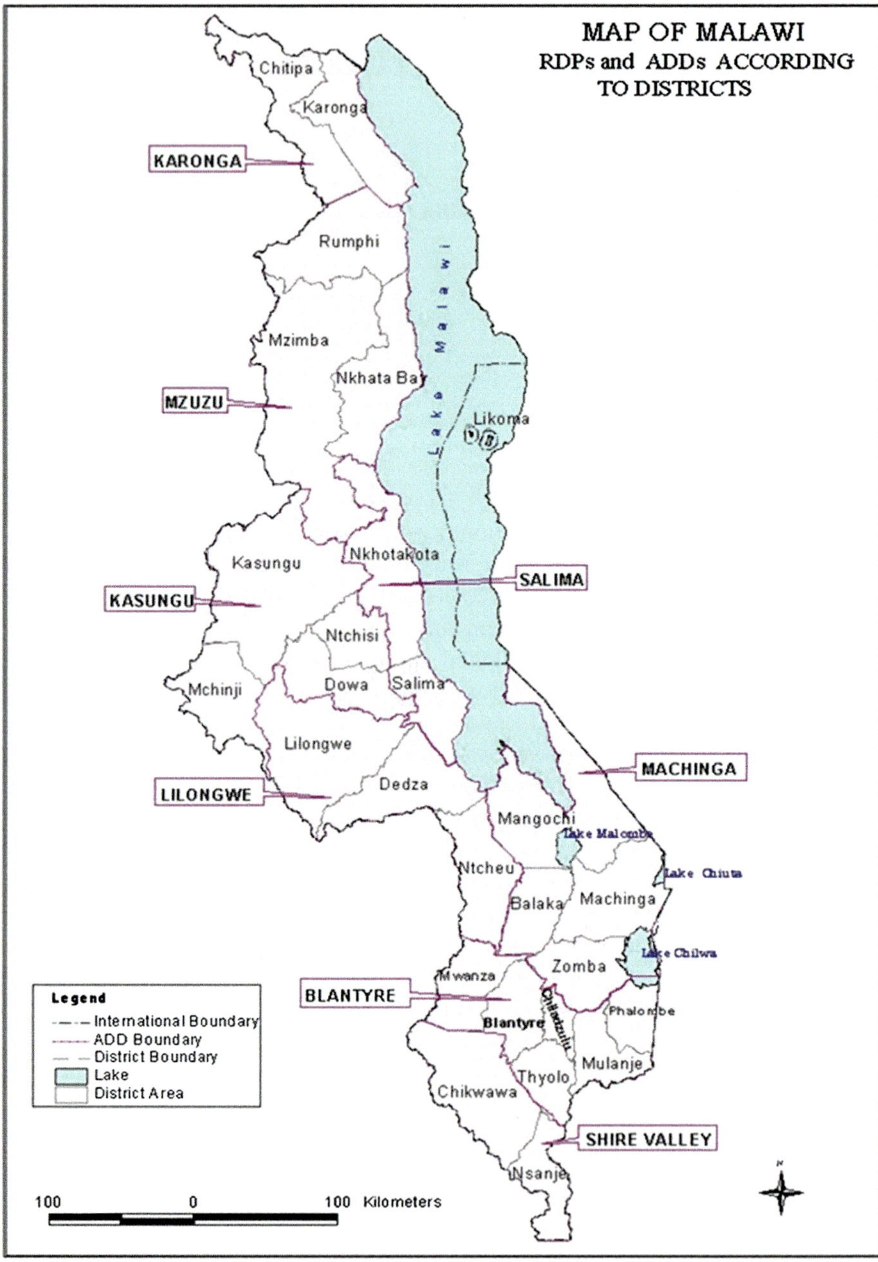

Fig. 16.1 Map of Malawi showing ADDs, District Agriculture Offices, and Extension Planning Areas

social security. In addition to the estimated 1.4 million cattle, there are also 6.3 million goats, 3.5 million pigs, 270,000 sheep, and 7.4 million chickens in Malawi.

16.2 Animal Husbandry

Different dairy farming systems exist in Malawi. Traditionally, and particularly in the Northern Region, where milk constitutes an integral part of the daily human diet, cattle rearing also involves milk production using indigenous zebu breeds. These herds graze on natural pastures shared by communities belonging to a particular dipping group. There are about 300 cattle-dipping stations evenly distributed throughout the country to which farmers used to bring their herds of cattle fortnightly for ectoparasite control during the rainy season from November to April. This practice caused intermingling of herds, and it was conducive to the spread of diseases between herds.

The colonial settlers who started importing and rearing dairy cattle well before the country's independence in 1964, initially, identified the need for the development of smallholder dairy farming around the country's major cities to satisfy the increasing needs for dairy products because of the population increase and urbanization. These farms are predominantly located close to Blantyre in the Southern Region, Lilongwe in the Central Region, and Mzuzu in the Northern Region. Because of the increasing demand for milk in the country's major cities and the heightened awareness of the importance of dairy production, the smallholder dairy industry became a focus for improvement by the Malawian Government in 1969. This culminated in the initiation of a government-supported crossbreeding program in 1973, involving zebu and Holstein-Friesians (Munthali et al. 1992). Currently, several international agencies, corporations, and NGOs are partnering with the Government of Malawi and other stakeholders such as the Malawi Milk Marketing Project, the Malawi Dairy Business Development Program, and the Small Scale Livestock and Livelihood Promotion (SSLLP) to create programs to further expand the dairy industry. As a result of these efforts, there was an increase of 65% in the dairy cattle population between 2004 and 2010 (Banda et al. 2012) although the number of cattle in the sector still comprises less than 5% of the total Malawian cattle population (Kaneene et al. 2016).

There are over 8000 smallholder farmers in Malawi, organized into about 50 Milk Bulking Groups (MBGs). These MBGs are managed by farmers and are tasked with collecting milk from members within an 8–10 km radius and for testing milk (for specific gravity and acidity) before it is pooled in a cooling tank from where it is collected by the commercial milk processors (Banda et al. 2011). Majority of the smallholder farmers (78%) own one to two animals and only 7% own more than five animals. The average animal holding per farmer is only 2.2 heads of cattle (Banda et al. 2011).

Commercial dairy farms are few and far between. Pure Holstein-Friesians and Jerseys and their crosses represent, respectively, 0.6 and 2.7% of dairy cattle

Table 16.1 Summary of BTB surveys conducted in Malawi

Study period	No. examined	Diagnostic method	No. positive (%)	References
1973	5809	PME	1036 (17.8)	Berggren (1977)
1974	5516		915 (16.6)	
	2349		453 (19.3)	
	210		44 (21)	Berggren (1981)
		CCT	73 (34.8)	
		Culture	82 (39.1)	
–	513	CCT	7 (1.4)	Moodie (1977)
1986	3481		134 (3.9)	Bedard et al. (1993)
2009	95		1 (1.0)	Tebug (2012)
2011–2015	35[a]	Direct microscopy[b]	21 (60.0)	[c]

[a]Tissue samples
[b]Ziehl–Neelsen staining
[c]Annual Reports of Department of Animal Health and Livestock Development 2005–2015

Chagunda et al. (1998), but they produce, respectively, 10 and 27% of the annual milk produced (Kaneene et al. 2016).

The abundance of animal diseases is one of the major impediments for livestock production in Malawi. A number of these diseases have a multi-dimensional impact, because they affect both humans and animals. In the following sections, bovine tuberculosis (BTB) in Malawi is discussed.

16.3 Bovine Tuberculosis in Malawi

Information about BTB in Malawi is poorly documented. Its prevalence, however, was such that during the mid-1950s and 1960s the authorities instituted an experimental BCG vaccination program in an attempt to control the disease (see Sect. 16.6.1). The current available information about the prevalence and distribution of BTB is scanty and insufficient to serve as a basis for planning a comprehensive control program in the country (Tables 16.1 and 16.2). There appears to be a marked variation in the prevalence and extent of the distribution of the disease throughout the country (Berggren 1977, 1981; Bedard et al. 1993; Moodie 1977; Tebug 2012). More recent reports covering an 11-year period between 2005 and 2015 indicate that BTB is present in all the ADDs throughout the country, but that certain districts recorded no evidence of the infection over the years (Table 16.2). That no cases were recorded is not necessarily an indication that the disease does not occur in those districts, but it is rather a reflection of the lack of adequate routine abattoir inspection and the shortage of adequately trained human resources to cover all the districts to detect the disease by tuberculin skin testing or at abattoirs doing routine meat inspection (Table 16.3). To deal with these issues, the Department of Animal Health and Livestock Development (DAHLD) is sending officials for training programs in

Table 16.2 BTB in Malawi, by district, during 2005–2015

District	Number of BTB outbreaks	Number of BTB cases	Mean number of BTB cases/ outbreak
Balaka	11	69	6
Blantyre	9	100	11
Chikwawa	26	6384	246
Chitipa	1	1	1
Dedza	16	43	3
Dowa	16	68	4
Karonga	11	107	10
Kasungu	27	363	13
Lilongwe	31	197	6
Machinga	5	10	2
Mchinji	8	33	4
Mulanje	1	1	1
Mwanza	3	3	1
Mzimba	36	320	9
Nkhotakota	1	1	1
Nsanje	3	5	2
Ntcheu	7	16	2
Ntchisi	7	14	2
Phalombe	1	1	1
Salima	15	76	5
Thyolo	5	9	2
Chiradzulu	0	0	0
Likoma	0	0	0
Mangochi	0	0	0
Neno	0	0	0
Nkhatabay	0	0	0
Rumphi	0	0	0
Zomba	0	0	0

Table 16.3 The number of veterinarians in different sectors in Malawi

Institution	Number of veterinarians	Percentage of total veterinarians
Public Sector (Government)	18	51.4
Academic or Training Institutions	4	11.4
Private Practitioners	2	5.7
Expatriates in NGOs	5	14.3
Retired	6	17.1
Total	35	100.0

Source: Annual Reports: DAHLD (2011–2015)

countries with adequate competencies, such as Botswana, in addition to offering in-house refresher courses.

16.4 The Epidemiology of BTB in Malawi

Limited data are available for the distribution, spread, and impact of BTB in Malawi. The number of animals and individual herds infected, sources of infection, the modes of spread, and other drivers of the epidemiology of the disease are poorly investigated. Some information is available for the following parameters:

Age There are contradictory data indicating that in Malawi the prevalence of BTB and age are not linked (Bedard et al. 1993). Generally, however, very few calves react to the CCT (Berggren 1981). These contradictory findings may be related to a number of factors that were not further investigated.

Breed Although there are differences of opinion as to whether breed does, in fact, determine susceptibility, a number of studies seem to support this notion, and many of the differences in the prevalence of BTB in Malawi can be attributed to the location of a specific breed of cattle (Bedard et al. 1993). In southern Malawi 54.5% of zebu cattle carcasses were totally condemned compared to the 66.7% of the Sussex breed (Ellwood and Waddington 1972). Of cattle tested at dip tank sites (representing village zebu cattle), 4.8% reacted to the CCT, while in the bulking-group animals (dairy farms with introduced European breeds), 2.6% reacted. Taking breed into consideration, only 2.7% of the pure- and crossbred dairy cows were positive, compared to 4.7% of the zebu cattle. The data are contradictory and the reactors are often found in clusters within dip tank or bulking groups. The type of farming practice appears more likely to be the cause of the perceived breed predisposition to the disease.

Sex Only one study in Malawi assessed the relationship between the prevalence of BTB and the sex of the animal. The number of reactors was significantly higher in oxen (9.6%) and bulls (8.4%) than in females (3.3%) (Bedard et al. 1993). This has been attributed to the strenuous work to which the males are subjected and their longer life expectancy, compared to cows, that facilitate them to more easily contract BTB. This pattern is difficult to explain since dairy cows are kept in a semi-intensive type of management system, and a higher prevalence rate would thus have been expected. The high prevalence in bulls could be the consequence of the local bull-hiring practice, in which bulls are used for natural mating as an alternative for the often-inadequate artificial insemination (AI) service (Chagunda et al. 1998). Since very few Malawian farmers (9.2%) own bulls (Banda et al. 2011), they are shared by several farms, which increase their risk of acquiring BTB.

Geographic Location There was a general belief that BTB was unevenly spread in Malawi and that higher prevalence rates were recorded in the northern region (Bedard et al. 1993). Other studies (Bedard et al. 1993; Ellwood and Waddington

1972) similarly indicated a clustering of BTB in certain geographic areas and herds, but methodological differences might account for these discrepancies. The reports of the Central Veterinary Laboratory for the period 2011–2015 (Table 16.2) provided more comprehensive data. The largest number of BTB outbreaks occurred in Lilongwe and Kasungu (central) and Mzuzu (north) ADDs that is likely to be due to the concentration of smallholder dairy farms that are subject to compulsory screening for BTB because of public health requirements. The ADDs with the highest number of BTB outbreaks (Shire Valley and Blantyre) are in the South with its concentration of commercial feedlots that are able to pay for BTB screening services. The existing information remains fragmented and is mostly gathered by passive surveillance and opportunistic tuberculin skin testing operations.

Spatial clustering of BTB may be associated with other unknown locality-specific determinants. While the role of wildlife in the epidemiology of BTB in Malawi is unknown, disease hotspots, as is seen in Zambia, may be associated with the presence of wildlife reservoirs (Munyeme et al. 2010). These matters need to be resolved to better understand the epidemiology of the BTB in Malawi and to plan future control strategies.

Herd Management Livestock management practices in Malawi are linked to their geographic location, and the type of system seems to play a pivotal role in determining the prevalence of the disease. Dip tank cattle have a higher BTB prevalence than cattle from bulking groups probably because of their large herd sizes and the intermingling of animals around the dipping tanks (Bedard et al. 1993). In terms of determining priorities for future BTB control programs, it would probably be more appropriate to focus on either bulking or dipping tank groups, rather than to use the data based on geographic location, breed, or sex to determine priority areas in which to initiate control programs.

Several factors are likely to contribute to the increased risk of BTB in smallholder dairy farms in Malawi. The importation of pure-bred dairy cattle, mostly from Zambia and South Africa, may further contribute to the problem because of the occurrence of BTB in cattle and wildlife populations in those countries (Michel et al. 2006; Munyeme et al. 2010).

16.5 Human TB Due to *M. bovis*

It is unknown what the contribution of zoonotic TB is to the total number of human TB cases in Malawi since laboratory facilities that allows differentiation between *M. bovis* and *M. tuberculosis* are not readily available. It is alarming though that a prevalence of as high as 42.8% *M. bovis*-positive specimens have been reported in TB patients in some sections in Malawi (Bedard et al. 1993). These hotspots appear to occur in areas where European cattle breeds had initially been introduced to establish a dairy industry.

A more recent epidemiologic study of human TB in Malawi (Nyirenda 2006) revealed a 45% increase in notified cases between 1994 and 2003; about half of these were reported from the urban districts of Blantyre, Zomba, Lilongwe, and Mzuzu. Overcrowding and the HIV epidemic appear to be the main driving forces of this increase. Of the new TB cases reported, 23% (range = 20–27%) during the period were extra-pulmonary, and many of them are likely to be zoonotic *M. bovis* infections (Kazwala et al. 2001). Owing to the ongoing population growth and urbanization, smallholder dairy farming in Malawi is expected to expand to meet the increasing demand for milk and milk products (Tebug et al. 2012), and unless BTB is adequately controlled, it should be expected that zoonotic TB would remain a major risk factor for the general population.

It is concerning that there is a varying and sometimes very limited awareness by the general population about zoonotic tuberculosis and the persistence of unhygienic practices that increase the risk of contracting zoonotic TB. Although in one survey 74.3% ($n = 140$) of respondents knew that *M. bovis* can be transmitted to humans and 67.0% knew that it is transmitted by infected milk, 96.4% of the respondents reported that they still engage in risky farm practices, including the sale (67.0%) and consumption (34.0%) of unpasteurized milk (Tebug et al. 2014). The practice of souring milk into "*Chambiko*" (instead of boiling or pasteurization) too does not fully eliminate the risk of exposure to milk-borne *M. bovis* (Hailemariam 2014). It is difficult to explain the mismatch between farmers' awareness levels about zoonotic TB and their lack of application of preventive measures (Tebug 2013). There thus remains an urgent need to raise the awareness of all stakeholders about the risk and the control measures of zoonotic TB given the extent of their exposure to *M. bovis*-infected milk.

16.6 Control of Bovine Tuberculosis in Malawi

16.6.1 Vaccination

BCG vaccination was used from the early 1960s to the 1970s in an attempt to control BTB in Malawi following a decision by the Department of Veterinary Services that the test-and-slaughter policy was impracticable, and that it would not allow eradication of BTB from the northern part of Malawi. Early results suggested that vaccination with BCG reduced the spread of BTB in the herds and the number of the condemned carcasses at the abattoirs (Ellwood and Waddington 1972; Moodie 1977), but it later transpired that BCG vaccination had no protective effect against infection with *M. bovis* (Berggren 1981). The situation remains the same today as no effective vaccine has yet been developed for the control of the disease in livestock or in wildlife.

16.6.2 Test-and-Slaughter Policy

Currently, due to cost and the lack of awareness, surveillance of BTB is not regularly conducted in Malawi. In one report only 19.0% of the farms reported ever testing their animals for BTB; the remaining 81% have never been tested (Tebug et al. 2014). It is clear too that the test-and-slaughter approach to control BTB is not easy to implement unless it makes provision for compensation for financial losses following slaughtering of the positive reactors. Because of financial and other constraints, and other priorities, the Malawian government is currently not in a position to fund such a campaign.

An approach that is practical and that will be embraced by all stakeholders, including the government allowing the control of the disease is outlined below:

1. Screening for BTB in Malawi should be based on immunological reactions, specifically the tuberculin skin test, because the few and non-specific clinical signs of BTB do not allow a clinical diagnosis of the disease ante-mortally (Berggren 1981).
2. As is the case in many of the neighboring countries (Kankya et al. 2011), the abundance of atypical mycobacteria (Chilima et al. 2006) and logistical constraints do not allow the use of the SIT for screening, and the CCT should be used in future BTB control programs.
3. There are two broad categories of cattle husbandry in Malawi: the smallholder dairy sector (MBGs) and the dipping tank group. The two management groups carry different burdens of the infection, and it is essential that priority be given to control programs directed at one or the other based on an objective assessment of the public health risks that they pose in Malawi. Control programs can then be expanded to other sectors when capacity allows.

In the interim, there are several measures that could be applied immediately in each sector to halt the spread of BTB in Malawi:

Control Activities in MBGs Small-scale dairy farming is still nascent in Malawi. Farmers, mainly women and those from underprivileged communities, usually acquire dairy cows through loans or as gifts from aid organizations in an attempt to bolster food security. Several of these donors, particularly the NGOs, also provide basic training in dairy management and health as part of the loan process, coined as the pass-on program, that can easily be extended into BTB awareness programs. Furthermore, the Department of Animal Health and Livestock Development should partner with aid organizations to make sure that heifers/cows originate only from BTB-free farms, and that they are tuberculin-tested before relocation to ensure that the newly established farms are BTB-free. It is also important to ensure that these cattle are housed in well-ventilated, clean premises and that they are isolated from other animals that are potentially infected with *M. bovis*.

The existing MBGs, composed on average of 42 members, are registered locally and have fairly adequate access to services and markets (Tebug et al. 2012; Kaneene et al. 2016). For the purpose of BTB control, each MBG should be treated as an epidemiological unit. As few animals are kept on each participating farm, tuberculin testing can be conducted in this sector with relative ease. If all the stakeholders, including the government, aid organizations, and farmers could be convinced to contribute towards the costs, BTB can be eradicated from this sector with a minimum of sacrifice and without adversely affecting other governmental financial commitments. It is also important to raise farmers' awareness about the value of personal and general farm hygiene, the association of BTB with dairy operations, and the value of pasteurizing milk, particularly for children.

Control Activities Involving Dipping Tank Groups These herds are rounded-up periodically, and it is important to consider each dipping tank (area) as an epidemiological unit when devising BTB control strategies (Bedard et al. 1993). The lack of farmer cooperation when undertaking BTB surveys at village level, which requires farmers to present their animals at a dipping station (often located 10–15 km from a village) on two successive occasions at a specific time interval, previously resulted in the number of cattle presented for tuberculin testing being too small (Berggren 1981). Designing a BTB control program for village cattle in Malawi therefore needs a lot of thinking, and a solution to the problem should be devised.

In the interim, two activities can be implemented:

- Although it is true that only intensive strategies such as national testing programs are effective to control BTB (Brooks-Pollock et al. 2014), this approach is too costly and impractical to apply to village cattle in Malawi. As an alternative, the use of abattoir inspection, if systematically applied, can provide information about the area-based prevalence of BTB,
- Irrespective of what is done, public education is pivotal in these operations and any campaign to reduce the prevalence of BTB and zoonotic TB should be supported by comprehensive educational programs, including encouraging boiling milk before consumption.

16.7 Conclusion

In summary, implementing and sustaining a BTB control program is a difficult and costly task. However, it is always important to begin with the end in mind:

- Therefore, the objective of the program has to be stated clearly from the outset.
- The national coordinating office, led by the Department of Animal Health and Livestock Development, should be given a clear mandate for developing an activity roadmap.
- It is necessary to develop policy documents (legal, administrative) and technical guidelines, including standard operating procedures.
- It is equally essential to establish standards for training, surveillance, and diagnosis at the national level.

- Epidemiologic research is an integral part of any BTB control program. Therefore, apart from enhancing the diagnostic capability, it is also important to enhance the computing ability of researchers at various levels.
- Because of the volume and cost of the work, Malawi needs financial and technical assistance from international donors to strengthen its personnel, diagnostic, and institutional capacity.

References

Banda LJ, Gondwe TN, Gausi W et al (2011) Challenges and opportunities of smallholder dairy production systems: a case study of selected districts in Malawi. Livest Res Rural Dev 23 (11):226. http://www.lrrd.org/lrrd23/11/band23226.htm

Banda LJ, Kamwanja LA, Chagunda MGG et al (2012) Status of dairy cow management and fertility in smallholder farms in Malawi. Trop Anim Health Prod 44:715–727

Bedard BD, Martin SW, Chinombo D (1993) A prevalence study of bovine tuberculosis and brucellosis in Malawi. Prev Vet Med 16(3):193–205

Berggren SA (1977) Incidence of tuberculosis in BCG vaccinated and control cattle in relation to age distribution in Malawi. Br Vet J 133:490–494

Berggren SA (1981) Field experiment with BCG vaccine in Malawi. Br Vet J 137:88–94

Brooks-Pollock E, Roberts GO, Keeling MJ (2014) A dynamic model of bovine tuberculosis spread and control in Great Britain. Nature 511:228–231

Chagunda MGG, Clemens BA, Wollny CBA (1998) Evaluation of artificial insemination program for small scale dairy farms in Malawi. Arch Tierzucht 41(1/2):45–51

Chilima BZ, Clark IM, Floyd S (2006) Distribution of environmental mycobacteria in Karonga District, Northern Malawi. Appl Environ Microbiol 72(4):2343–2350

Ellwood DC, Waddington FC (1972) A second experiment to challenge the resistance to tuberculosis in BCG vaccinated cattle in Malawi. Br Vet J 128:619–626

Hailemariam S (2014) Identification and survival studies of *Mycobacterium tuberculosis* within laboratory-fermented bovine milk. BMC Res Notes 7:175

Kaneene JB, Thiagarajan D, Chigwa F et al (2016) A tri-lateral capacity building approach to strengthen the dairy value chain in Malawi: Overview of the design and implementation. Livest Res Rural Dev 28:154. http://www.lrrd.org/lrrd28/9/kane28154.html

Kankya C, Muwonge A, Djønne B et al (2011) Isolation of non-tuberculous mycobacteria from pastoral ecosystems of Uganda: Public Health significance. BMC Public Health 11:320

Kazwala RR, Daborn CJ, Sharp JM et al (2001) Isolation of *Mycobacterium bovis* from human cases of cervical adenitis in Tanzania: a cause for concern? Int J Tuberc Lung Dis 5(1):87–91

Michel AL, Bengis RG, Keet DF et al (2006) Wildlife tuberculosis in South African conservation areas: implications and challenges. Vet Microbiol 112:91–100

Moodie PA (1977) Tuberculin reaction in BCG vaccinated cattle. Br Vet J 133:642–645

Munthali JTK, Musa FA, Chiwayula CLK (1992) Smallholder dairy development in Malawi. In: Kategile JA and Mubi S (eds) Proceedings of a workshop on future of livestock industries in East and Southern Africa. International Livestock Centre for Africa (ILCA), Addis Ababa

Munyeme M, Muma JB, Munang'andu HM et al (2010) Cattle owners' awareness of bovine tuberculosis in high and low prevalence settings of the wildlife-livestock interface areas in Zambia. BMC Vet Res 6:21. https://doi.org/10.1186/1746-6148-6-21

Nyirenda T (2006) Epidemiology of Tuberculosis in Malawi. Malawi Med J 18(3):147–159

Tebug SF (2012) Smallholder dairy farming in Northern Malawi: husbandry practices, constraints and prevalence of major production and zoonotic diseases. Ph.D. thesis, Christian-Albrechts-Universität zu Kiel, Germany

Tebug SF (2013) Factors associated with milk producer's awareness and practices in relation to zoonoses in northern Malawi. Vet World 6:249–253

Tebug SF, Kasulo V, Chikagwa-Malunga S et al (2012) Smallholder dairy production in Northern Malawi: production practices and constraints. Trop Anim Health Prod 44:55–62

Tebug S, Njunga GR, Mizeck GG et al (2014) Risk, knowledge and preventive measures of smallholder dairy farmers in northern Malawi with regard to zoonotic brucellosis and bovine tuberculosis. Onderstepoort J Vet Res 81(1):594

Chapter 17
Bovine Tuberculosis in Nigeria: Historical Perspective, Burden, Risk Factors, and Challenges for Its Diagnosis and Control

Simeon Idowu Babalola Cadmus

17.1 Introduction

Mycobacterium tuberculosis, known to be responsible for most tuberculosis (TB) infections in humans, is inadvertently considered to be the only cause of the disease in humans in Nigeria. Consequently, very little is known in the country about zoonotic TB caused by *M. bovis*, and other members of the *M. tuberculosis* complex (MTC) including *M. africanum*, *M. caprae*, *M. bovis* BCG, *M. canetti*, *M. microti*, and *M. pinnipedii*. Though human TB and bovine TB (BTB) are highly prevalent, respectively, in the human and cattle populations in Africa (including Nigeria), limited actions, particularly as they relate to BTB, are taken to reduce the prevalence of these dangerous diseases (Ayele et al. 2004).

With over 170 million people, Nigeria has the largest human population in Africa, making it the most populous black nation in the world (WHO 2015). Additionally, Nigeria has a rapidly growing livestock population, estimated at 20.5 million cattle, 23.1 million sheep, and 28.1 million goats (FAO 2014). Despite this huge livestock resource, Nigeria has a large proportion of the world's poor livestock keepers, making it a potential hotspot of neglected zoonoses, and a reservoir of endemic tropical diseases.

Nigeria remains one of the countries with the highest human TB burdens in the world, and it is currently ranked 4th globally, and 1st in Africa (WHO 2015). To reverse this situation, the Federal Government of Nigeria has been collaborating with the World Health Organization (WHO) and other international donors (e.g. Global Fund) over the past decade to improve the infrastructure to control TB (an initiative that is currently directed exclusively at controlling human TB) by increasing the number of directly-observed-treatment-short-course (DOTS) centers in the 774 local government areas (LGAs) in Nigeria. Very little consideration is, however, still

S. I. B. Cadmus (✉)
Tuberculosis and Brucellosis Research Laboratories, Department of Veterinary Public Health & Preventive Medicine, University of Ibadan, Ibadan, Nigeria

A. B. Dibaba et al. (eds.), *Tuberculosis in Animals: An African Perspective*,
https://doi.org/10.1007/978-3-030-18690-6_17

given to the fact that most Nigerians live in settings that facilitate the transmission of zoonotic BTB at the human–animal interface. For example, in most rural areas, people live in close contact with their animals, and the current TB control initiatives do not make any provision for the protection of livestock handlers who are at a greater risk of exposure to zoonotic TB due to their unhygienic working conditions in livestock markets, and in urban and peri-urban abattoirs. Disruption of public health services, poor social amenities, lack of infrastructure such as piped water, and high levels of illiteracy, promote the persistence of poor hygienic conditions and low sanitation levels. These have led to the continuing spread of neglected tropical and other zoonotic diseases, including BTB in Nigeria. Other local risky practices that enhance the occurrence of zoonotic BTB infections in humans specifically occur when people drink unpasteurized milk and milk products, and consume meat and meat products contaminated with *M. bovis* (Cadmus et al. 2006).

Knowing that human and animal health are intricately linked in the Nigerian socio-economic setting and geographical space, it is imperative that we take a closer look at BTB in Nigeria. The objectives of this chapter are to review the burden and risk factors of BTB in Nigeria, and the challenges facing its diagnosis and control. The contribution of BTB to the overall human TB burden in Nigeria and factors contributing to the zoonotic transmission of *M. bovis* are also assessed. Finally, it is hoped that this information will assist governmental decision-makers to design a comprehensive strategy based on the One Health approach to reduce the impact of zoonotic BTB on Nigeria's diverse human population.

17.2 Historical Perspective of BTB in Nigeria

Manley (1929) first reported the presence of BTB in Nigeria in 1929, when it was first diagnosed in purebred and half-bred cattle imported from Germany based on a tuberculin test, post-mortem examination, and laboratory confirmation. The Nigerian cattle herds were assumed to also have been infected by being in contact with Cameroonian herds, where BTB in West Africa was first reported in 1913, due to their historical cultural links, and the extensive transhumant movement of pastoralists from Nigeria and other West African countries, including Cameroon (Alhaji 1976). That cattle from these countries were the source of the infection was further strengthened by the presence of the Af1 clonal complex, a major strain of *M. bovis* in Cameroon, Chad, Mali, and Nigeria. Further epidemiological evidence showed that most TB-infected cattle slaughtered in Southern Nigeria originate from Northern Nigeria and the neighboring Francophone countries, Cameroon, Chad, and Niger (Müller et al. 2009). Strains of *M. bovis* prevalent in Northern Nigeria and neighboring countries (Cadmus et al. 2011) were also isolated from cattle in South-western Nigeria (Cadmus et al. 2006).

Due to the limited availability of pasture and water in Northern Nigeria, particularly during the dry season, some of the Fulani pastoralists from the north settled in South-western Nigeria about five decades ago. The Nigerian government, in an attempt to sustain their livelihood, established grazing reserves in Oyo State, Southwestern Nigeria, which subsequently became the convergent hub for animals from Northern Nigeria, the neighboring West African countries, and resident herders, creating a melting pot of diverse livestock diseases, including BTB.

The absence of an active BTB surveillance program in Nigeria further compounds the problem. Disease reporting is not mandatory, and consequently there is a gap in the information needed for its control (Cadmus and Ayanwale 2014). Additionally, because of the very few adequately trained human resources involved in BTB research, and inadequate diagnostic facilities to monitor the disease status, BTB remains highly under-diagnosed and poorly investigated in Nigeria. The actual economic loss associated with BTB and its impact on livestock production in the country cannot be calculated accurately, while its public health implications also remain in the realm of conjecture.

17.3 Prevalence of BTB in Cattle Herds in Nigeria

Tuberculin skin testing (TST) of cattle using the single intradermal comparative cervical test (SICCT) was first carried out in 1939 on 507 animals on a government farm in Vom, Northern Nigeria. In total, a prevalence of 1.4% was recorded using post-mortem inspection for the confirmation of the presence of lesions caused by *M. bovis* (Alhaji 1976). Several studies, though mostly limited to government-owned farms (Alhaji 1976; Cadmus and Ayanwale 2014; Cadmus et al. 2010), have been conducted since then. To date, the largest number of herds and animals screened in a coordinated manner was carried out in 26 local government areas in four states in the former Northern Nigeria (Alhaji 1976). In total, 13,487 cattle from 346 herds (representing about 0.13% of the national cattle herd at the time, and including nomadic, semi-nomadic, resident, commercial, and government-owned herds) were subjected to the single caudal fold test (CFT) using bovine-purified protein derivative (PPD). In all, 2.8% positive and 3.2% suspect reactors were identified. The positive reactors per herd ranged from 0% (only two herds) to 25%. Of the six major breeds of cattle examined (i.e. Bunaji, Rahaji, Sokoto, Adamawa Gudales, Wadara, and Azwak), the Rahaji, which accounted for 14.7% of all cattle screened, was the most affected (31.2%). The investigation further revealed that BTB was widespread in all the breeds of cattle, in different husbandry systems, and in all states in the region. Overall, based on nation-wide tuberculin tests of cattle herds from 1929 to date, the average BTB prevalence varies from 1.8 to 24%.

Between 1980 and the late 1990s, several governmental and public activities were disrupted or downscaled, and infrastructure destroyed by military incursions. The livestock sector suffered a tremendous downturn following these events, and from then the attention given to livestock only focused on a few serious diseases of economic importance; surveillance for BTB was not carried out in pastoral herds in Nigeria until 1999. Even today, tuberculin skin tests are not used for routine BTB surveillance in the whole of Nigeria. It is usually only used by researchers working

Table 17.1 Surveys of BTB using the tuberculin skin test in Nigerian cattle

Location/region	Sample size	Test	Prevalence	Year	References
North	56	STC	24%	1929	Manley (1929)
North	507	SICCT	1.38%	1939	Mettam (1939)
North	13,497	CFT	2.5%	1975/1976	Alhaji (1976)
Ibadan, South-west	171	SICCT	10.5%	2004	Cadmus et al. (2004)
Kaduna and Federal Capital Territory, Abuja, North-central	967	SICCT	14.6%	2007	Abubakar (2007)
North-west, North-central, and South-west	1360	SICCT	5.7% (45.5% herd prevalence)	2010	Cadmus et al. (2010)
Jigawa, North-west	885	SICCT	2%	2012	Ibrahim et al. (2012)
Taraba, North-east	17	SICCT	11.8%	2012	Ayele et al. (2004) and Danbirni et al. (2012)
North-east	20,000	SICCT	12.2%	2011–2014	Salisu I (unpublished data)
Akure, South-west	36	SICCT	1.1%	2014	Adefegha et al. (2014)

SICCT single intradermal comparative cervical test, *CFT* caudal fold test, *STC* short thermal technique

on BTB (either locally or in collaboration with foreign institutions), in segments of areas that are readily accessible (Table 17.1).

Between 2004 and 2005, 33 selected herds in Northern and Southern Nigeria subjected to tuberculin skin testing, had a BTB prevalence of 45.5%. At the animal level, the infection prevalence ranged from 2.9 to 16.6% in the two regions. Poor management and the absence of disease control were identified as key factors causing these pockets of infections (Cadmus et al. 2010). However, in a coordinated effort to rekindle surveillance of BTB in Nigeria, the Federal Government of Nigeria funded a regional study between 2011 and 2014 to determine the status of BTB in cattle herds in North-eastern Nigeria. Through this initiative, the about 20,000 cattle screened had a reactor prevalence of about 14% (I. Salisu, unpublished data). However, considering the unrestricted movement of animals, poor management, and unhygienic practices including poor sanitary measures, and huge under-funding and limited governmental support of the livestock sector, the data presented so far on BTB in Nigeria are limited, and what is known may just be the tip of the iceberg and a total underestimation of the actual current state of affairs.

Based on the results of the exploratory tuberculin tests carried out to date, there is sufficient evidence to suggest that BTB is endemic in pastoral herds in all the regions of Nigeria. It is also clear that over the years, BTB has spread from neighboring countries and Northern Nigeria to the southern parts of the country due to unrestricted cattle movement within and across different states, and across international borders (Cadmus et al. 2006; Abubakar 2007; Müller et al. 2009). The situation becomes more alarming when the herd prevalence is considered, and the finding that at least one animal is positive in the majority of the herds screened, should be a red flag given the absence of BTB control in Nigeria. However, since very few herds/animals have been screened to date, the epidemiological information obtained so far is inadequate to provide an insight into the true scale of BTB in Nigerian cattle herds and to allow the compilation of an adequate strategy to control the disease (Cadmus and Ayanwale 2014).

If tuberculin screening of cattle in Nigeria is going to be of any value, the challenge posed by concurrent helminth infections must also be resolved. Adedipe et al. (2014) and Adedipe (2014) reported 41.6 and 2.8% for gastrointestinal helminth and helminth-BTB co-infection, respectively, in slaughtered cattle in Ibadan, southwestern Nigeria. It is known that helminth infections adversely complicate estimation of the true status of BTB in cattle herds (Claridge et al. 2012). Until the way helminth-BTB co-morbidity influences the tuberculin skin test is resolved, determining the true extent of BTB in Nigeria may not be possible.

17.4 Prevalence of BTB in Cattle Slaughtered in Nigerian Abattoirs

In the absence of meaningful control and eradication measures against BTB, abattoir meat inspection currently remains the only hope for protecting the public from zoonotic TB in Nigeria. The abattoir regulations thus provide a key control point in the meat and meat product processing chain. Hence, data collected by veterinary meat inspection officers are a useful source of information to determine the extent of BTB in Nigeria. Over the years, study reports in Nigeria were mostly based on abattoir meat inspection (Table 17.2). Other procedures, including Ziehl–Neelsen (ZN) staining and microscopy to detect acid-fast bacilli (AFB), culture, biochemical tests, and molecular techniques, in a few instances (Table 17.3; Fig. 17.1), were used to confirm the diagnoses in abattoirs.

Early reports of BTB from Nigerian abattoirs were chronicled for Northern Nigeria (between the 1930s through 1958) followed by those from the southern parts (between 1937 and 1947) (Alhaji 1976). An overall prevalence of 0.02–15.9% was recorded in the respective abattoirs during this period. From 1970 to date, few data have been reported to the various State Veterinary Departments, the former Federal Livestock Department, and individual research initiatives (Tables 17.2 and 17.3). In general, studies carried out within the last two decades in most abattoirs showed a prevalence of BTB ranging from 0.34 to 26.6% (although the diagnosis was rarely confirmed by culture).

Table 17.2 Tuberculous lesions detected during post-mortem meat inspection in Nigerian abattoirs

Location/region	Prevalence (*n*)	Date	References
South-west	28.9% (568,694)	1976–1979	Antia and Alonge (1982)
Southern	8.2% (5407)	–	
North-east	9.0% (376,500)	1981–1990	Alaku and Moruppa (1993)
North-west	0.49% (104,416)	1989–1993	Dusai and Abdullahi (1994)
Southern	2.25% (66,680)	1991–2000	Bikom and Oboegbulem (2000)
North-east	Overall—28% (47,544) Maiduguri—2.5% Damboa—20% Gashua—2.6% Damaturu—1.9%	1994–1998	Igbokwe et al. (2001)
North-east	Overall—4.05% (302,700) Adamawa—0.3% Bauchi—0.9% Borno—1.7% Gombe—12.3% Taraba—5.0% Yobe—8.7%	2000–2004	Aliyu et al. (2009)
North-east	1.1% (2902)	May–June, 2008	Raufu and Ameh (2010)
South-east	1.4% (748)	2004–2008	Nwanta et al. (2011)
South-east	3.4% (247)	March–November, 2010	Opara et al. (2012)
South-west	1.78% (52,273)	2011–2012	Oluwasile et al. (2013)
North-east	1.9% (1172)	2008–2012	Ejeh et al. (2014b)
North-east	26.6% (85)	March–July, 2015	Adang et al. (2015)

17.5 Post-mortal Diagnosis of BTB in Nigerian Abattoirs

Bovine-TB has been reported from several abattoirs; hence, regular and effective meat inspection remains the cornerstone for BTB prevention and control in Nigeria. However, due to poor infrastructure, the shortage of and inadequately trained personnel, coupled with illicit slaughtering, effective meat inspection is seldom conducted in Nigeria. Thus, more often than not, BTB data are non-existent or incomplete. Again, because of the lack of knowledge about BTB, the high levels of illiteracy, and poverty, Nigerian butchers continue to buy obviously diseased animals for slaughter, which worsens the zoonotic risk and transmission of the disease. Another key hindrance is the butchers' general unwillingness, because they are not compensated for condemned carcasses and, hence, perceive condemnation as a huge economic loss, to cooperate with meat inspectors, as they are likely to hide BTB-infected carcasses to avoid them from being confiscated.

Table 17.3 Reports of other ancillary/molecular methods for the confirmation of BTB in Nigerian abattoirs (Alhaji 1976; Idrisu and Schnurrenberger 1977; Cadmus et al. 2008, 2009, 2011; Ofukwu et al. 2008; Damina et al. 2011; Ejeh et al. 2014c; Okeke et al. 2014)

Location/ region	Type of specimen	Sample size	Diagnostic method	No. of isolates
Northern states	Milk	10	Culture/biochemical	2 (*M. tuberculosis* and *M. africanum*)
Northern states	Sputum	15	Culture/biochemical	10 (8 *M. tuberculosis*, 1 *M. bovis*, 1 *M. africanum*)
Northern states	Milk	10	Culture/biochemical	10 mycobacteria
South-western state	Milk and lesion	130	Culture/biochemical	26 (9 *M. tuberculosis*, 15 *M. bovis*, and 2 *M. africanum*)
Northern state	Milk	34	Culture/biochemical	7 (6 *M. bovis* and 1 *M. africanum*)
South-western state	Lesion	1387	PM, deletion, and spoligotyping	5 (4 *M. bovis* and 1 *M. tuberculosis*)
Northern state	Lesion	3381	PM, ZN, and deletion analysis	115 (107 *M. bovis*, 6 *M. tuberculosis*, and 2 *M. africanum*
South-western state	Lesion and milk	180	Deletion analysis, spoligotyping, and VNTR	180 (178 *M. bovis*, 1 *M. tuberculosis*, and 1 *M. africanum*)
Northern state	Lesion	248	PM and ZN	–
Northern state	Lesion	485	ZN and PCR	–

17.6 Molecular Epidemiology of BTB in Nigeria

Despite the evidence of BTB being present in Nigeria from early 1929 (Alhaji 1976), some insight into the molecular epidemiology of the disease was reported only in 2004 (Cadmus et al. 2006, 2004). Prior epidemiologic investigations were based on tuberculin tests, post-mortem examination at the abattoirs, laboratory analyses (mostly direct microscopy to detect AFBs), and on a few occasions, isolation and identification of mycobacteria using culture and biochemical tests (Alhaji 1976; Dusai and Abdullahi 1994). However, in view of the obvious existing information gap about the extent of zoonotic TB in the country, a scientific initiative initiated in 2003, by the author in collaboration with scientists from the TB Diagnostic Section, Veterinary Laboratories Agency (VLA), UK (now known as Animal and Plant Health Agency, Surrey, UK), to undertake molecular analyses of bovine and human tubercle bacilli in Nigeria (Cadmus et al. 2006). Using spoligotyping and variable-number tandem-repeat (VNTR) analyses, this work unequivocally showed that the majority of human TB in Ibadan, South-western Nigeria, was caused by a single, closely

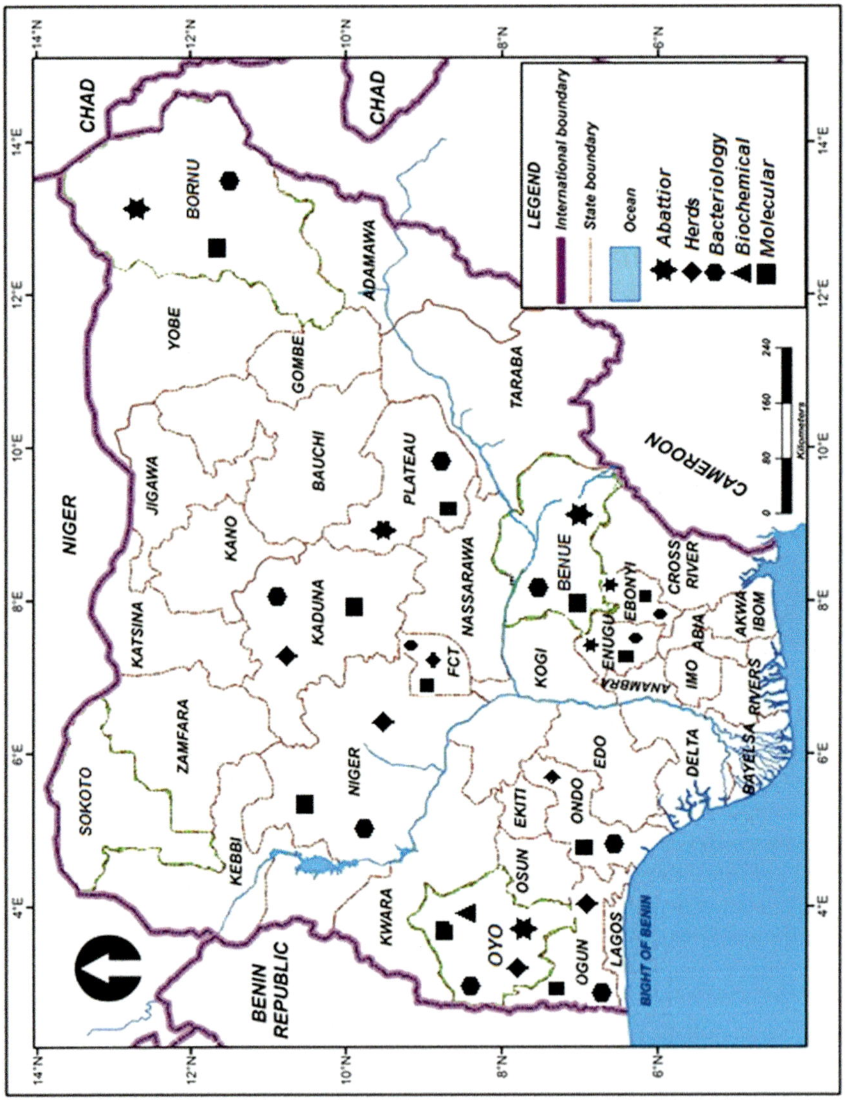

Fig. 17.1 Map of Nigeria showing the different states where BTB investigations using different diagnostic methods have been conducted

related group of *M. tuberculosis* strains. It was further shown by deletion typing that approximately 10% of the human TB cases were caused by strains of *M. africanum* and 5% by *M. bovis*.

Earlier preliminary data had shown the circulation of similar strains of *M. bovis* in a private cattle herd in Ibadan, South-western Nigeria (Cadmus et al. 2004), re-affirming the usefulness of molecular epidemiology as a tool to track specific strains of pathogens to understand the transmission of disease in different populations. Recent studies conducted in different parts of Nigeria (Cadmus et al. 2006, 2011; Abubakar 2007) have shown that diverse strains of *M. bovis* circulate in the country (Table 17.3). These strains, also isolated from neighboring countries of West Africa, have been genotypically classified as the Af1 clonal complex (Müller et al. 2009).

To date, the majority of the MTC species isolated from cattle in Nigeria have been confirmed by spoligotyping to be *M. bovis* (Cadmus et al. 2006, 2011; Abubakar 2007), with the SB0944 strain being the most frequent spoligotype circulating in the country. Several strains of *M. africanum* (lacking spacers 8, 9, and 39) have also been isolated from tuberculous cattle (Cadmus et al. 2006, 2011; Abubakar 2007). Cadmus et al. (2006) and Abubakar (2007) also detected *M. tuberculosis* (intact spacers 39 through 43) in slaughtered cattle in Nigeria using molecular tools like spoligotyping, VNTR (Streicher et al. 2007) and deletion typing (Cadmus et al. 2006).

Molecular epidemiology, using VNTR and MIRU-VNTR, further showed a large diversity in the strains of *M. bovis* in Nigeria (Cadmus et al. 2006, 2011, 2008), although most *M. bovis* strains isolated in Nigeria belong to the African 1 (Af1) clonal complex. The Af1 clonal complex is a group defined by a specific chromosomal deletion (RDAf1), and lacking spacer 30 using the standard spoligotype pattern (Müller et al. 2009), and its occurrence is limited to the West-Central African region compared to strains circulating in Eastern Africa and Europe (Müller et al. 2009). Though there have been several efforts to decipher the epidemiology of *M. bovis* in cattle and other animals in Nigeria using different molecular techniques (Cadmus et al. 2006, 2011; Abubakar 2007; Jenkins et al. 2011), there is still a huge information gap. Inadequate veterinary infrastructure, insufficient and inadequately trained personnel, and most importantly, the lack of funding to support animal health delivery are some of the factors militating against proper investigations into the molecular epidemiology of BTB in Nigeria.

17.7 Risk Factors of BTB in Nigeria

A host of risk factors influence the occurrence and spread of the disease in the Nigerian cattle population. They do not all equally apply to the different management systems, but they do play a substantial role in each of the specific situations.

Peri-Urban and Rural (Pastoral) Animal Husbandry Reports about the cases of BTB in private herds in urban centers in Nigeria (Manley 1929; Ibrahim et al. 2012; Cadmus 2010) support the contention that intensive husbandry is a major risk factor for BTB. Cases of BTB, however, have also been reported in pastoral herds (Cadmus 2010). Over the years, with a fast-growing population, Nigeria has been faced with an increasing demand for animal protein. In the same vein, because of the huge socio-economic pressure on the middle class, there has been a quest for increased personal wealth. This has led to a gradual, but steady shift of animal husbandry from the traditional pastoral system to peri-urban and urban dairy production (mostly operated by civil servants with animals that are reared in living quarters in a semi-intensive management system, and graze within the confines of the owners' properties).

Poor Animal Health Infrastructure and Services Due to limited financial resources and other higher priorities, the livestock sector has received very little attention from the Nigerian government, and livestock production in Nigeria remains underdeveloped. Consequently, almost 95% of the cattle population remains in the extensive management system practiced by the Fulani pastoralists, a group of people that is poorly exposed to modern animal health services, and disease prevention and control strategies. Consequently, routine chemo-prophylactic measures and vaccination efforts are not often embraced. Therefore, BTB surveillance, prevention, and control activities involving the TST are seldom performed and viewed with serious mistrust by the pastoralists, and in several cases prohibited. Furthermore, due to poor farm hygiene and a lack of herd biosecurity measures, healthy herds and those infected with different diseases comingle with the resultant spread of diseases (including BTB) within and across the different states of Nigeria. Additionally, individuals setting up new farms unknowingly acquire foundation stock, which may be infected, from open cattle markets. These and other issues related to the lack of knowledge and poor sanitary practices by farmers ultimately result in the spread of BTB in both dairy and conventional cattle herds in Nigeria (Cadmus et al. 2004; Ibrahim et al. 2012; Adefegha et al. 2014).

Laboratory Infrastructure Currently, in Nigeria, only about two to three laboratories have the infrastructure required to carry out laboratory procedures such as culture and molecular identification of *M. bovis*. These facilities also have limited capabilities to carry out modern techniques like the MIRU-VNTR, single nucleotide polymorphisms (SNPs) and whole genome sequencing (WGS). Therefore, laboratory facilities to carry out prompt and correct confirmation of BTB in infected animals are highly limited and still in their rudimentary stages of development.

Personnel Due to limited governmental support and the lack of funding for BTB surveillance in Nigeria, very little research work is carried out. Additionally, we do not have adequately trained human resources to effectively carry out TSTs of Nigerian cattle herds. In the same vein, there are very few meat inspectors who are employed to carry out the necessary meat inspection services. Hence, tuberculous carcasses and contaminated meat from slaughter slabs and abattoirs are sold to the unsuspecting public.

Non-tuberculous Mycobacteria (NTM) Being unable to accurately differentiate between cases of BTB and those infected with NTMs during meat inspection compounds the problem of BTB in Nigeria. Cases of NTM infections have been reported following preliminary investigations in Nigerian abattoirs (Ejeh et al. 2014a, b, c), thus clouding the true picture of BTB and posing an emerging public health threat. This issue is becoming more significant by the day as NTMs are increasingly isolated from Nigerian patients (Cadmus et al. 2016).

Infection of Cattle with Other Strains of the *M. tuberculosis* Complex Of particular importance is the emerging recognition that cattle across Nigeria are susceptible to infection by other members of the MTC (Cadmus et al. 2006; Abubakar 2007; Raufu and Ameh 2010). Cases of *M. tuberculosis* (Cadmus et al. 2006; Raufu and Ameh 2010) and *M. africanum* infection (Abubakar 2007; Raufu and Ameh 2010) in different parts of the country have been reported. These findings make the epidemiology of BTB in Nigeria more intriguing and complex given the public health implications created by these infections.

Lack of BTB Control Policies In general, there are limited funding and policy guidelines from the government to support the livestock sector. There are no formal operational guidelines or codes of conduct dealing with the registration of cattle herds, their physical location, bio-security, or minimum health standards. Thus, issues relating to surveillance of trans-boundary animal diseases of economic and public health importance, compulsory vaccination to control major epidemics, and a test-and-slaughter policy to eradicate diseases are given little or no attention. Consequently, surveillance measures to monitor BTB in the Nigerian cattle herds and trade animals (especially those imported into the country), and prevention of its introduction and spread within the country have not been established. This has led to several illegal/unwholesome activities and practices in the Nigerian cattle industry. Consequently, animal movement in and out of Nigeria, across and within different states, and between healthy and infected herds is unrestricted, and there is practically no policy or coordinated effort to control BTB (Cadmus 2010). Likewise, dairy animals are also imported into the country without prior screening for or providing a valid record of their BTB status.

Additionally, there is no national TST program to assess the overall burden of BTB in Nigerian cattle herds. To worsen the situation, the numbers of trained veterinarians and meat inspection officers available in most abattoirs/slaughter slabs are insufficient to conduct adequate meat inspection. Therefore, the data generated on BTB are not comprehensive, and often poorly analyzed, and are inadequate to inform government on the burden of the disease. This makes it difficult to formulate a long-term control policy, as currently it would be based mostly on anecdotal information. Consequently, poor control and policy measures against BTB, including the lack of compensation to butchers, affect all stakeholders

(Ejeh et al. 2014a) and result in the ongoing exposure of the public to *M. bovis*-infected carcasses.

International and Intra-country Movement of Animals There is no official BTB control policy in Nigeria or in its neighboring African countries. Because of the unrestricted and uncontrolled continuous in-flow of animals, Nigeria has become a large receptacle of *M. bovis*-infected cattle from neighboring countries, and local cattle herds are at an increased risk of being infected with *M. bovis*. As a result, diverse strains of *M. bovis* are isolated from tuberculous cattle originating from Niger, Mali, Chad, and Cameroon that are slaughtered at abattoirs around the country (Cadmus et al. 2006; Müller et al. 2009).

Cultural Values and Practices The traditional Fulani pastoralists, who own the bulk of the cattle in Nigeria, engage in practices that expose their animals to adverse health conditions due, largely, to their transhumant lifestyle. Each year, the animals are driven over several hundred to thousands of kilometers in search of pasture and water. During this migration, in most instances, the animals are poorly fed, they lack veterinary care, and they are often fatigued, making them more vulnerable to contracting various diseases including BTB.

Lack of Awareness and Illiteracy Due largely to the poor knowledge by the general public about the health risks associated with BTB, and the high level of illiteracy among major stakeholders in the livestock industry, several risky practices that foster the spread of BTB persist in the livestock production chain in Nigeria (Cadmus 2010):

- For example, the uncontrolled movement of cattle over long distances and their comingling around the few watering and salt lick points without any biosafety measures constitute some of the major drivers of the spread of BTB in those herds.
- Another very important risk factor is the intensification of cattle production by individuals in urban areas (mostly government employees) who rear them in their backyards for fattening and sales to generate an extra income. Sadly, some traders in the livestock market keep their animals within their living areas, oblivious of the risk posed by their close association with cattle infected with *M. bovis*.

Cumulatively, these practices lead to cross-contamination and the spread of infection within the cattle populations, and in some instances, to zoonotic transmission (Cadmus et al. 2006).

Eating Habits Consumption of unpasteurized milk and milk products has been identified as the major route of human infection with *M. bovis*. In Nigeria, some of the milk consumed is skimmed, soured, and consumed unpasteurized (Alhaji 1976). There are several reports about the presence of *M. bovis* in cows' milk being drunk directly after it was expressed from the udder (Cadmus et al. 2008), and from milk sold on the open markets in Northern (Abubakar 2007) and South-western Nigeria (Cadmus et al. 2008).

17.8 Economic Losses Associated with BTB in Nigeria

Bovine TB globally remains a disease of major economic and zoonotic importance. This is particularly so in sub-Saharan Africa, where its presence causes a large but unknown number of deaths and economic loss in animals. Data on the number of human infections due to *M. bovis* in Nigeria are very limited, and the human losses attributable to BTB are difficult to calculate.

The economic losses associated with BTB in animals are anticipated to be enormous in Nigeria but there is insufficient documented information to allow the calculation of the actual loss. A rough estimate of the loss can be made by using the number of BTB cases in slaughtered cattle in abattoirs in the country (Alhaji 1976). In this context, in a recent retrospective study conducted between 2008 and 2012, 1172 (1.90%) of 61,654 cattle slaughtered in Makurdi, North-central Nigeria, were positive for TB (Streicher et al. 2007). Overall, 1935 tuberculous organs, weighing 3046.50 kg, were condemned. Given that abattoir meat inspection in Nigeria underestimates the full extent of BTB by far, it is safe to conclude that several millions of US$ are lost annually due to BTB should all the condemned carcasses and tissues be included in the calculation.

The presence of BTB in cattle herds will also aggravate the erosive effects associated with unthriftiness, weight loss, and reduced milk production caused by the poor animal husbandry and health systems in Nigeria, and particularly by the transhumant lifestyle of the Fulani herdsmen.

17.9 Possible Control Measures and the Way Forward

In Nigeria, as in most sub-Saharan African countries, the consumption of raw milk and raw meat remain the major sources of human infection with *M. bovis* (Müller et al. 2009; Abubakar 2007; Cadmus et al. 2008; Ofukwu et al. 2008). It is thus critical that milk processing and meat inspection be conducted at an optimal level of efficiency. It is fundamental to reduce zoonotic tuberculosis by increasing pasteurization of milk in the country, and to encourage improved milk hygiene practices in household and pastoralist herds to ensure the production of wholesome and pathogen-free milk.

Similarly, routine meat inspection (where dedicated veterinarians are present), involving detailed and comprehensive ante-mortem and post-mortem inspections, should be carried out on all animals slaughtered. Importantly, where possible, the entire carcass should be condemned when generalized miliary TB is detected, while entire organs (or their parts) should be condemned when large tuberculous lesions are found in their parenchyma or in the regional lymph nodes. Again, it is very important that efforts should be directed along the entire meat production chain (i.e. from the farm to the abattoir), and to ensure that adequate trace-back is possible to the source of the infected carcasses.

Since the inspection of carcasses in abattoirs remains the critical mitigation point in the food processing chain where BTB can be detected and controlled in Nigeria, it is imperative that the meat inspection services be improved by implementing the following measures:

- Create an optimal working environment and infrastructure (sufficient light source, adequate water) in abattoirs.
- Increase the number of adequately trained and disciplined meat inspectors and support staff.
- Implement regular competency evaluation of meat inspectors to assess their ability to identify lesions consistent with those of BTB in meat and related products.
- Educate butchers and cattle owners to improve cooperation between them and abattoir veterinarians and meat inspectors.
- It is critical that conventional and modern approaches are introduced to allow tracking BTB in Nigeria. Therefore, more personnel and infrastructure must be made available both in the field and in veterinary diagnostic laboratories to properly monitor the disease in animal and human populations.
- Finally, there is a need for improved government policies dealing with the control of BTB and funding of the livestock sector to allow meaningful progress in reducing the menace of BTB in Nigerian cattle herds.

References

Abubakar A (2007) Epidemiology of human and bovine tuberculosis in the Federal Capital Territory and Kaduna State of Nigeria. Ph.D. thesis, University of Plymouth, p 184

Adang KL, Kela SL, Sale S (2015) Prevalence of bovine tuberculosis in cattle slaughtered at Gombe township abattoir, Gombe State, Nigeria. J Vet Med Anim Health 7(7):265–270

Adedipe OD (2014) Prevalence of bovine tuberculosis and helminth co-infection among slaughtered cattle at Bodija municipal abattoir: economic and public health implication. M.Sc. thesis, University of Ibadan, Nigeria, p 52

Adedipe OD, Uwalaka EC, Akinseye VO et al (2014) Gastrointestinal helminths in slaughtered cattle in Ibadan, South-Western Nigeria. J Vet Med 2014:923561, 6. https://doi.org/10.1155/2014/923561

Adefegha OM, Adesokan HK, Cadmus SIB (2014) Investigation of bovine tuberculosis in a private cattle herd in South-Western Nigeria: potential risks for human infection. Afr J Epidemiol 2(1): 16–20

Alaku SO, Moruppa SM (1993) Tuberculosis condemnation in livestock slaughtered for meat in North-Eastern Nigeria. Prev Vet Med 15:67–72

Alhaji I (1976) Bovine tuberculosis in four northern states of Nigeria. Ph.D. thesis, Ahmadu Bello University, Zaria, Nigeria, p 236

Aliyu MM, Adamu JY, Bilyaminu YA (2009) Current prevalence of tuberculous lesions among slaughtered cattle in North-Eastern States of Nigeria. Rev Elev Med Vet Pays Trop 62(1):13–16

Antia RE, Alonge DO (1982) Survey of abattoir data in southern Nigeria. Trop Anim Health Prod 14(2):119–120

Ayele WY, SD Neill J, Zinsstag MG et al (2004) Bovine tuberculosis: an old disease but a new threat to Africa. Int J Tuberc Lung Dis 8:924–937

Bikom PM, Oboegbulem SI (2000) Incidence of bovine tuberculosis in Cross River State: a retrospective abattoir study. Sahel J Vet Sci 6(1):39–43

Cadmus SIB (2010) The transmission chain of *Mycobacterium bovis* from cattle to humans and within the human settings in Nigeria: facts, realities and the way forward. Trop Vet 28(1):1–12

Cadmus SIB, Ayanwale FO (2014) Bovine tuberculosis: epidemiology, zoonotic transmission, activities, and challenges toward its control in Nigeria. In: Thoen CO, Steele JH, Kaneene JB (eds) Zoonotic tuberculosis: *Mycobacterium bovis* and other pathogenic mycobacteria, 3rd edn. Wiley, Ames, IA, p 397

Cadmus SIB, Atsanda AA, Oni SO et al (2004) Bovine tuberculosis in one cattle herd in Ibadan in Nigeria. Vet Med (Praha) 49(11):406–412

Cadmus SIB, Palmer S, Okker M et al (2006) Molecular analysis of human and bovine tubercle bacilli from a local setting in Nigeria. J Clin Microbiol 44(1):29–34

Cadmus SIB, Adesokan HK, Adepoju AF et al (2008) Zoonotic risks and transmission of mycobacteria species from cows' milk and slaughtered cattle to man in Ibadan: role of butchers. Niger Vet J 29(1):30–39

Cadmus SI, Adesokan HK, Jenkins AO et al (2009) *Mycobacterium bovis* and *M. tuberculosis* in goats, Nigeria. Emerg Infect Dis 15(12):2066

Cadmus SI, Agada CA, Onoja II et al (2010) Risk factors associated with bovine tuberculosis in some selected herds in Nigeria. Trop Anim Health Prod 42:547–549

Cadmus SIB, Gordon SV, Hewinson RG et al (2011) Exploring the use of molecular epidemiology to track bovine tuberculosis in Nigeria: an overview from 2002 to 2004. Vet Microbiol 151: 133–138

Cadmus SIB, Diarra B, Traore B et al (2016) Nontuberculous mycobacteria isolated from tuberculosis suspects in Ibadan, Nigeria. J Pathogens 2016:6547363, 5

Claridge J, Diggle P, McCann CM et al (2012) *Fasciola hepatica* is associated with the failure to detect bovine tuberculosis in dairy cattle. Nat Commun 3:853. https://doi.org/10.1038/ncomms1840

Damina MS, Owoludun OA, Chukwukere S et al (2011) The use of deletion analysis in the detection of *Mycobacterium bovis*, *Mycobacterium tuberculosis* and *Mycobacterium africanum* among slaughtered cattle in Plateau State, North Central Nigeria. Niger Vet J 32(1):9–15

Danbirni S, Okaiyeto SO, Joshua IA et al (2012) Prevalence of tuberculosis in a herd of cattle of a tuberculosis herdsman following trace back information from a hospital in Taraba State, Nigeria. J Anim Prod Adv 2(7):325–328

Dusai DHM, Abdullahi DA (1994) Current status of bovine tuberculosis at Sokoto abattoir. Trop Vet 12:134–137

Ejeh EF, Akinseye VO, Igwe D et al (2014a) Molecular characterization of *Mycobacterium bovis* in slaughtered cattle in North-Central Nigeria and the public health implications. Afr J Med Med Sci 43(Suppl):97–103

Ejeh EF, Raji MA, Bello M et al (2014b) Prevalence and direct economic losses from bovine tuberculosis in Makurdi, Nigeria. Vet Med Int 2014:904861, 6

Ejeh EF, Adeshokan HK, Raji MA et al (2014c) Current status of bovine tuberculosis in Otukpo, Nigeria. J Anim Prod Adv 4(8):501–507

FAO (2014) Livestock densities (Gridded livestock of the world), Agriculture and Consumer Production Department, Animal Production and Health Division, FAO, Switzerland. http://www.fao.org/ag/againfo/resources/en/glw/GLW_dens.html. Accessed 10 April 2016

Ibrahim S, Cadmus SIB, Umoh JU et al (2012) Tuberculosis in humans and cattle in Jigawa State, Nigeria: risk factors analysis. Vet Med Int 2012:865924. https://doi.org/10.1155/2012/865924

Idrisu A, Schnurrenberger P (1977) Public-health significance of bovine tuberculosis in 4 northern states of Nigeria-myco-bacteriologic study. Niger Med J 7(4):384–387

Igbokwe IO, Madaki IY, Danburam S et al (2001) Prevalence of pulmonary tuberculous lesions in cattle slaughtered in abattoirs in North-Eastern Nigeria. Rev Élev Méd Vét Pays Trop 54(3–4): 191–195

Jenkins AO, Cadmus SIB, Venter EH et al (2011) Molecular epidemiology of human and animal tuberculosis in Ibadan, South-Western Nigeria. Vet Microbiol 151:139–147

Manley FH (1929) Bovine tuberculosis. Annual report, Veterinary Department, Northern Provinces, Nigeria

Mettam H (1939) Annual report. Federal Veterinary Research Laboratory, Vom, Nigeria

Müller B, Hilty M, Berg S et al (2009) African 1, an epidemiologically important clonal complex of *Mycobacterium bovis* dominant in Mali, Nigeria, Cameroon, and Chad. J Bacteriol 191: 1951–1960

Nwanta JA, Umeononigwe CN, Abonyi GE et al (2011) Retrospective study of bovine and human tuberculosis in abattoirs and hospitals in Enugu State, Southeast Nigeria. J Public Health Epidemiol 3(7):329–336

Ofukwu RA, Oboegbulem SI, Akwuobu CA (2008) Zoonotic *Mycobacterium* species in fresh cow milk and fresh skimmed, unpasteurised market milk (nono) in Makurdi, Nigeria: implications for public health. J Anim Plant Sci 1:21–25

Okeke LA, Cadmus S, Okeke IOM et al (2014) Prevalence and risk factors of *Mycobacterium tuberculosis* complex infection in slaughtered cattle at Jos South Abattoir, Plateau State, Nigeria. Pan Afr Med J 18(Suppl. 1):7

Oluwasile BB, Awoyomi OJ, Kehinde OO (2013) Prevalence and economic loss of bovine tuberculosis in a municipal abattoir, Abeokuta, South-Western Nigeria. Niger J Anim Prod 40(2): 216–223

Opara MN, Nwaeze CN, Olaifa AK et al (2012) Prevalence of bovine tuberculosis in Imo State, South-Eastern Nigeria. J Trop Med Parasitol 35:14–21

Raufu IA, Ameh JA (2010) Prevalence of bovine tuberculosis in Maiduguri Nigeria – an abattoir study. Bull Anim Health Prod Afr 58(2):119–123

Streicher EM, Victor TC, Vander SG et al (2007) Spoligotype signatures in the *Mycobacterium tuberculosis* complex. J Clin Microbiol 45:237–240

WHO (2015) Global tuberculosis report 2015, 20th edn. World Health Organization. http://apps.who.int/iris/bitstream/10665/191102/1/9789241565059_eng.pdf?ua=1. Accessed 27 Aug 2015

Chapter 18
Bovine Tuberculosis in Rwanda

Gervais Habarugira, Joseph Rukelibuga, and Manassé Nzayirambaho

18.1 Introduction

Rwanda, located in central Eastern Africa, is a home to a variety of domestic and wild animal species (Plumptre 1994; Kading et al. 2013). Livestock in Rwanda consists mainly of cattle, small stock, and poultry, and only recently a few exotic animals such as horses were imported into the country (Rawlins et al. 2014). The number of cattle increased significantly since the launch of the Girinka program (one-cow-per-poor-family) in 2006, and more than 150,000 head of cattle have since been distributed to poor families. The program also significantly contributed to the country's economic growth, poverty alleviation, and the reduction of child malnutrition (Argent et al. 2014; de Valk 2010). Although efforts have been made to increase the number of livestock in Rwanda, livestock production still faces a number of challenges. These include animal diseases caused by parasitic, bacterial, and viral causes, and non-infectious diseases. Bovine TB (BTB) is one of the infectious diseases that constitute a threat to animal production in Rwanda (Nshimiyima et al. 2013).

Tuberculosis, a disease caused by bacteria of the genus *Mycobacterium*, is one of the oldest, chronic, communicable, zoonotic infectious diseases affecting both

G. Habarugira (✉)
School of Animal Sciences and Veterinary Medicine, University of Rwanda, Kigali, Rwanda
e-mail: g.habarugira@uq.net.au

J. Rukelibuga
The United States Centers for Disease Control and Prevention, Influenza Program, Kigali, Rwanda
e-mail: JRukelibuga@cdc.gov

M. Nzayirambaho
School of Public Health, College of Medicine and Health Sciences, University of Rwanda, Kigali, Rwanda
e-mail: mnzayira@nursph.org

A. B. Dibaba et al. (eds.), *Tuberculosis in Animals: An African Perspective*,
https://doi.org/10.1007/978-3-030-18690-6_18

humans and animals (Thoen et al. 2008, 2014). Bovine TB is contagious and classified by the World Health Organization (WHO) and the World Organization for Animal Health (OIE) as a major zoonosis, and of public health concern (Boukary et al. 2011; Katigbak et al. 2005). Decreased beef and milk production and carcass condemnation at abattoirs of cattle infected with BTB cause substantial economic losses and have a significant socio-economic impact (Iredia et al. 2011).

Although the disease in cattle and in humans has been considered to be of declining importance because of its successful eradication in many developed countries, its re-emergence during the course of the last three decades in both developing and some of the developed countries is alarming (Baggett 2011; Bynum 2011; Cosivi et al. 1998). Despite its public health and economic significance, its control is neglected in many developing countries in the sub-Saharan region, including Rwanda (Nshimiyima et al. 2013; Habarugira et al. 2014), where it is poorly investigated, and data on its prevalence and impact are scarce and/or poorly reported (Boukary et al. 2011; Vordermeier et al. 2012). Nevertheless, the limited available information generated by post-mortem meat inspection, intradermal tuberculin testing using bovine and avian tuberculin, and laboratory analyses (Zeru et al. 2014) shows that BTB is prevalent in both cattle and humans in these countries (Cosivi et al. 1998; Boukary et al. 2011; Chadha 1997).

This chapter aims to provide an overview and perspectives on the current status of BTB in Rwanda.

18.2 Prevalence and Epidemiology of BTB in Rwanda

Bovine TB in Rwanda is listed as an endemic disease but without data estimates Cosivi et al. (1998). Contrary to the situation in human medicine where the diagnosis of TB is mainly based on microscopy, bacteriology, and molecular biological techniques, the diagnosis of BTB in Rwanda, like in many other developing countries, is reliant on meat inspection, both for diagnosing and generating disease surveillance data. The information about BTB in Rwanda, unlike for human tuberculosis, is very limited and often anecdotal and based on piecemeal prevalence and distribution data obtained by detecting BTB-like lesions during meat inspection at abattoirs.

An abattoir-based study conducted in 2009 at SABAN Nyabugogo Abattoir in Kigali, the capital city, revealed a prevalence of lesions consistent with those of BTB of 0.9% ($n = 16{,}753$). *Mycobacterium bovis* was confirmed to be present in about 60% of these cases (Habarugira et al. 2014). Additional data collected from the same abattoir indicated that the prevalence of BTB-like lesions fluctuates over time: in 2010, the occurrence of BTB lesions was 0.7%, it was slightly more than double in the following year (1.7%), and then varied between 0.5 and 1.5% in 2012 and 2013, respectively.

Although the prevalence of BTB in Rwanda appears to be low, it must be kept in mind that these data are based on the detection of macroscopic lesions during post-mortem examination of carcasses at abattoirs where commonly more than 50% of the BTB cases are missed during routine inspection (Teklu et al. 2004; Shitaye et al. 2006; Awah Ndukum et al. 2010; Biffa et al. 2010; Bekele and Belay 2011).

Bovine TB is an endemic disease in some neighboring countries such as Uganda and Tanzania (Ayele et al. 2004). Given the higher prevalence of BTB in Tanzania where it varies from 0.2 to 13.3% (Katale et al. 2012; Kazwala et al. 2001; Shirima et al. 2003), and in the western parts of Uganda that directly border on Rwanda and where the prevalence is up to 14.5% (Kazoora et al. 2014), it is likely that BTB could have a higher prevalence in Rwanda bordering those areas. Movement of BTB-infected cattle from these areas into Rwanda is highly likely considering the extent of animal trade between the countries, and particularly, the ongoing illegal movement of livestock across these international borders (Fig. 18.1).

As with other infectious diseases (Chatikobo et al. 2009; Juvenal and Edward 2010), it is possible that BTB is also prevalent in some of the wildlife species in the country. It is highly likely that it occurs in some of the wildlife species in the Akagera National Park in the Eastern Province of Rwanda where neighboring farmers graze their animals illegally in the Park allowing intermingling of livestock and wildlife that increases the likelihood of transmission of disease between them.

18.3 Public Health Significance of BTB

The public health significance of *M. bovis* is poorly defined in Rwanda because of the very limited available data. However, from the little information that is available, it appears that the disease poses a definite public health risk (Habarugira et al. 2014; Gafirita et al. 2012).

A number of risk factors are of particular importance:

- The traditional consumption of raw milk and raw milk products by Rwandans enhances the likelihood of transmission of *M. bovis* from cattle to humans. Some Rwandans still believe that it is taboo or culturally improper to boil (pasteurize) milk before it is consumed. They prefer to drink raw, either fresh or a traditionally fermented milk known as *ikivuguto*. Although it has been shown that *M. bovis* has a short half-life of about 12 h in souring raw milk (Michel et al. 2015), it is still unsafe to consume such milk during the early stages of fermentation without pasteurization, as viable *M. bovis* can be present in fermented milk despite its low pH.
- In some of the remote areas of Rwanda, particularly in the Eastern Province where the farming is mostly extensive, cow's milk is consumed on farms where there are no or only rudimentary pasteurization facilities (Karenzi et al. 2013). Although studies have not been carried out to ascertain it, the increased prevalence of extra-pulmonary tuberculosis in humans in Rwanda would suggest the likelihood of *M. bovis* being the causative agent, following ingestion of infected milk (Lorent et al. 2008, 2011).

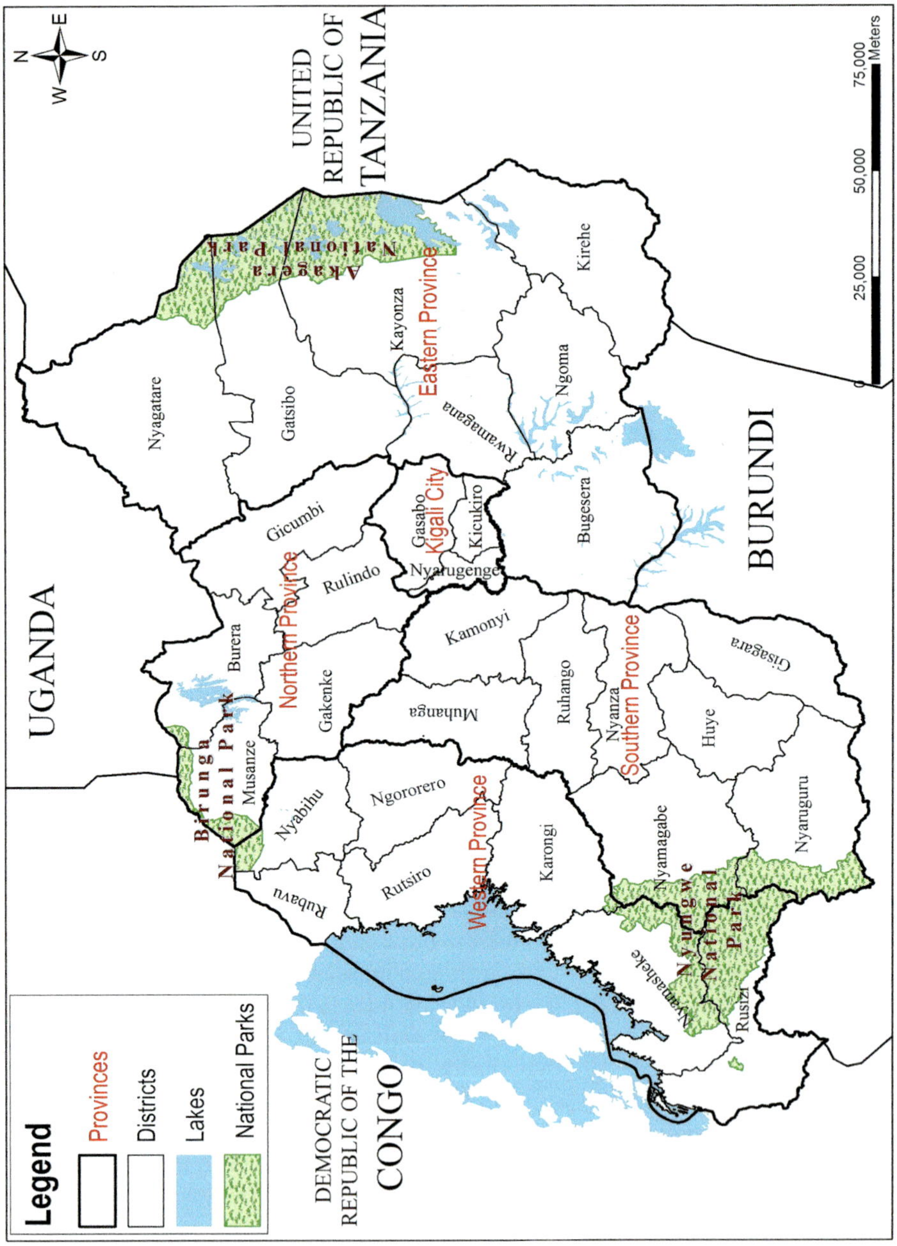

Fig. 18.1 Map of Rwanda showing its borders with Tanzania and Uganda, and the location of the national parks (Map by courtesy of Itegere Basile)

- Considering the occurrence of BTB lesions in meat (Awah Ndukum et al. 2010; Bekele and Belay 2011; Teklu et al. 2004; Youssef and Ahmed 2014), there is potential for acquiring the infection through the consumption of raw and/or undercooked meat.
- With the increasing trend of zero grazing, animals are now kept in backyards in many households in Rwanda. This practice contributes to an increased closer contact between owners and their cattle (Nzeyimana et al. 2015) that may facilitate the transmission of zoonotic infectious diseases such as BTB.

In Rwanda, the presence of BTB has direct and indirect socio-economic consequences. Meat condemnation not only has a direct economic impact, but it also contributes to the loss of animal protein that is a scarce commodity in the country. According to meat inspection data collected from 2009 to 2014 at Nyabugogo Abattoir located in the Capital City, 11,174.5 kg of tissues and offal were condemned because of the presence of BTB-like lesions. Considering the protein content of 22.3% of raw beef (FAO 2007) and an animal protein requirement of 56 g/day for a sedentary average man of 70 kg body weight (Trumbo et al. 2002), the condemned meat would be enough to provide the required protein for 122 adult men for a year.

18.4 Prevention and Control of BTB

In Rwanda, the control of animal diseases, including infectious diseases such as BTB, is the responsibility of the Ministry of Agriculture and Animal Resources (MINAGRI), and the Rwanda Agriculture Board (RAB).

In contrast to developed countries where BTB eradication programs, based on the test-and-slaughter principle, and milk pasteurization, have been successful (Cosivi et al. 1998; Biet et al. 2005), in Rwanda, there is no specific program to prevent and control BTB. In addition to the limited veterinary diagnostic ability, the situation is further complicated by BTB not being listed as one of the diseases for which animals must be tested before importation from neighboring countries that often have a high prevalence of BTB.

The test-and-slaughter policy for the control of BTB as applied in the developed countries, remains prohibitively expensive for Rwanda (Smith et al. 2014; Bezos et al. 2014), and less expensive but effective preventive measures such as testing animals before importation can initially be applied to reduce and subsequently control the disease in the long term.

A pilot study to investigate the prevalence and economic impact of BTB based on intradermal tuberculin tests and meat inspection is currently being conducted nationally. This study is anticipated to shed light on the true burden of the disease on which then to formulate recommendations for control measures. The Government should also consider DNA fingerprinting of strains of *M. bovis* that are circulating in

Rwanda to assess the extent of cross-infection between different animal species and humans.

18.5 Conclusion

Based on the available baseline data, we can confirm that BTB is present in Rwanda. Unfortunately, it remains a neglected disease despite its known socio-economic and public health impact. Although it is too expensive for developing countries to eradicate BTB, at least surveillance programs should be put in place, focusing especially on high-risk animals in areas bordering the national parks, and animals originating from neighboring countries where BTB is present and its prevalence high. Imported animals from neighboring countries should be tested for BTB in addition to other diseases like brucellosis.

Effective control of animal movement in the country can also be used as a way of successfully managing the disease by limiting its further spread into uninfected areas. The effective implementation of the One Health concept would also enhance attempts to effectively prevent, control, and even eradicate the disease in the long term.

Acknowledgment The authors express their appreciation to Mr. Ali Ahsan Bajwa of The University of Queensland (UQ), and Dr. Jared Omolo of the United States Centers for Disease Control and Prevention (CDC), Rwanda, for critically reviewing this book chapter. They are also grateful to Ms. Namubiru Halima of The University of Queensland (UQ) for her assistance during the writing of this book chapter.

References

Argent J, Augsburg B, Rasul I (2014) Livestock asset transfers with and without training: evidence from Rwanda. J Econ Behav Organ 108:19–39

Awah Ndukum J, Kudi AC, Bradley G (2010) Prevalence of bovine tuberculosis in abattoirs of the Littoral and Western highland regions of Cameroon: a cause for public health concern. Vet Med Int 2010:495015. https://doi.org/10.4061/2010/495015

Ayele WY, Neill SD, Zinsstag J et al (2004) Bovine tuberculosis: an old disease but a new threat to Africa. Int J Tuberc Lung Dis 8:924–937

Baggett A (2011) Tuberculosis then and now: perspectives on the history of an infectious disease. J Hist Med Allied Sci 66:578–558

Bekele M, Belay I (2011) Evaluation of routine meat inspection procedure to detect bovine tuberculosis suggestive lesions in Jimma Municipal Abattoir, South West Ethiopia. Glob Vet 6:172–179

Bezos J, Casal C, Romero B et al (2014) Current ante-mortem techniques for diagnosis of bovine tuberculosis. Res Vet Sci 97:S44–S52

Biet F, Boschiroli ML, Thorel MF et al (2005) Zoonotic aspects of *Mycobacterium bovis* and *Mycobacterium avium-intracellulare* complex (MAC). Vet Res 36:411–436

Biffa D, Bogale A, Skjerve E (2010) Diagnostic efficiency of abattoir meat inspection service in Ethiopia to detect carcasses infected with *Mycobacterium bovis*: implications for public health. BMC Public Health 10:462

Boukary A, Thys E, Mamadou S et al (2011) La tuberculose à *Mycobacterium bovis* en Afrique subsaharienne. Ann Méd Vét 155:23–37

Bynum H (2011) Tuberculosis then and now: perspectives on the history of an infectious disease. JAMA 305(14):1491–1492

Chadha V (1997) Global trends of tuberculosis – an epidemiological review. NTI Bull 33:11–18

Chatikobo P, Manzi M, Kagarama J et al (2009) Benchmark study on husbandry factors affecting reproductive performance of smallholder dairy cows in the Eastern Province of Rwanda. Livest Res Rural Dev 21(6):83

Cosivi O, Grange J, Daborn C et al (1998) Zoonotic tuberculosis due to *Mycobacterium bovis* in developing countries. Emerg Infect Dis 4:59

de Valk P (2010) The effect of livestock production on poor and smallholder farmers' income in Rwanda. Case of 'one cow one family program'. Dissertation, Institute of Social Studies, The Hague, The Netherlands, p 75

FAO (2007) Composition of meat. http://www.fao.org/ag/againfo/themes/en/meat/backgr_composition.html. Accessed 26 Aug 2016

Gafirita J, Umubyeyi AN, Asiimwe BB (2012) A first insight into the genotypic diversity of *Mycobacterium tuberculosis* from Rwanda. BMC Clin Pathol 12:20

Habarugira G, Rukelibuga J, Nanyingi MO et al (2014) Bovine tuberculosis in Rwanda: prevalence and economic impact evaluation by meat inspection at Société des Abattoirs de Nyabugogo-Nyabugogo Abattoir, Kigali. J S Afr Vet Assoc 85:1–5

Iredia C, Oguntibeju O, Lewis H et al (2011) Trends and characteristics of patients admitted with musculoskeletal tuberculosis to a referral hospital from 2003 to 2008. Afr J Microbiol Res 5:532–540

Juvenal N, Edward M (2010) Seasonal dynamics and distribution of ticks in Rwanda: implications for tick control strategy in Rwanda. Int J Anim Vet Adv 2:21–25

Kading RC, Borland EM, Cranfield M et al (2013) Prevalence of antibodies to alphaviruses and flaviviruses in free-ranging game animals and nonhuman primates in the greater Congo basin. J Wildl Dis 49:587–599

Karenzi E, Mashaku A, Nshimiyimana AM (2013) Kivuguto traditional fermented milk and the dairy industry in Rwanda. A review. Biotechnol Agron Soc Environ 17(2):383

Katale BZ, Mbugi EV, Kendal S (2012) Bovine tuberculosis at the human-livestock-wildlife interface: is it a public health problem in Tanzania? A review. Onderstepoort J Vet Res 79:84–97

Katigbak MW, Shlasko E, Klein SM (2005) Peritoneal tuberculosis in a 15-month-old male: surgical diagnosis of an insidious disease. Surg Infect 6:255–258

Kazoora HB, Majalija S, Kiwanuka N et al (2014) Prevalence of *Mycobacterium bovis* skin positivity and associated risk factors in cattle from Western Uganda. Trop Anim Health Prod 46:1383–1390

Kazwala R, Kambarage D, Daborn C et al (2001) Risk factors associated with the occurrence of bovine tuberculosis in cattle in the Southern Highlands of Tanzania. Vet Res Commun 25:609–614

Lorent N, Mugwaneza P, Mugabekazi J et al (2008) Risk factors for delay in the diagnosis and treatment of tuberculosis at a referral hospital in Rwanda. Int J Tuberc Lung Dis 12:392–396

Lorent N, Sebatunzi O, Mukeshimana G et al (2011) Incidence and risk factors of serious adverse events during antituberculous treatment in Rwanda: a prospective cohort study. PLoS One 6(5): e19566

Michel AL, Geoghegan C, Hlokwe T et al (2015) Longevity of *Mycobacterium bovis* in raw and traditional souring milk as a function of storage temperature and dose. PLoS One 10(6): e0129926

Nshimiyima J, Nahayo A, Mukamisha P (2013) Prevalence of bovine tuberculosis in Rwanda from 2006 to 2010. Rwanda J Soc Appl Sci 1:43–55

Nzeyimana P, Habarugira G, Udahemuka JC et al (2015) Prevalence of bovine cysticercosis and age relationship at post-mortem in Nyagatare Slaughterhouse. World J Agric Res 3:4–8

Plumptre A (1994) The effects of trampling damage by herbivores on the vegetation of the Pare National des Volcans, Rwanda. Afr J Ecol 32:115–129

Rawlins R, Pimkina S, Barrett CB et al (2014) Got milk? The impact of Heifer International's livestock donation programs in Rwanda on nutritional outcomes. Food Policy 44:202–213

Shirima GM, Kazwala RR, Kambarage DM (2003) Prevalence of bovine tuberculosis in cattle in different farming systems in the eastern zone of Tanzania. Prev Vet Med 57:167–172

Shitaye J, Getahun B, Alemayehu T (2006) A prevalence study of bovine tuberculosis by using abattoir meat inspection and tuberculin skin testing data, histopathological and IS6110 PCR examination of tissues with tuberculous lesions in cattle in Ethiopia. Vet Med (Praha) 51 (11):512

Smith RL, Tauer LW, Sanderson MW et al (2014) Minimum cost to control bovine tuberculosis in cow-calf herds. Prev Vet Med 115:18–28

Teklu A, Asseged B, Yimer E et al (2004) Tuberculous lesions not detected by routine abattoir inspection: the experience of the Hossana municipal abattoir, southern Ethiopia. Rev Sci Tech 23:957–964

Thoen CO et al (2008) *Mycobacterium bovis* infection in animals and humans, 2nd edn. Wiley, Hoboken, NJ

Thoen CO, Steele JH, Kaneene JB (2014) Zoonotic tuberculosis: *Mycobacterium bovis* and other pathogenic mycobacteria, 3rd edn. Wiley-Blackwell, Ames, IA

Trumbo P, Schlicker S, Yates AA et al (2002) Dietary reference intakes for energy, carbohydrate, fiber, fat, fatty acids, cholesterol, protein and amino acids. J Am Diet Assoc 102:1621–1630

Vordermeier M, Ameni G, Berg S et al (2012) The influence of cattle breed on susceptibility to bovine tuberculosis in Ethiopia. Comp Immunol Microbiol Infect Dis 35:227–232

Youssef A, Ahmed A (2014) Bovine tuberculosis survey based on meat inspection and microscopic examination in Central City Abattoir in Ismailia, Egypt and its hazards to the abattoir workers. IFRJ 21(2):77–582

Zeru F, Romha G, Berhe G et al (2014) Prevalence of bovine tuberculosis and assessment of cattle owners' awareness on its public health implication in and around Mekelle, Northern Ethiopia. J Vet Med Anim Health 6:159–167

Chapter 19
BTB Control Strategies in Livestock and Wildlife in South Africa

Anita L. Michel, Donald R. Sibanda, and Lin-Mari de Klerk-Lorist

19.1 Background

Historically in South Africa livestock numbers were comparatively small and managed by transhumance, but they played an important role in the communities of the indigenous Africans and the early settlers (Beinart 2007). During the colonial era, livestock farming in South Africa expanded substantially and gradually transformed into a vibrant commercial agricultural industry with an increasing need for veterinary interventions to control the increasing number of endemic livestock diseases. By the 1870s, governmental veterinary officials in the Cape Colony, deployed by the newly established Veterinary Services, reported the spread of various livestock diseases as a likely consequence of the rapid increase in livestock numbers and movement (Beinart 2007; Department of Agriculture, Forestry and Fisheries 2016a, b). During the first half of the twentieth century, knowledge of livestock diseases increased and new remedies and control measures were researched and implemented.

Before the arrival of the first European settlers, wildlife was very abundant throughout South Africa, but uncontrolled hunting for economic gain decimated the populations of many wildlife species by the late nineteenth century (Mossman and Mossman 1976). The regulatory authorities were also of the opinion that for

A. L. Michel (✉)
Faculty of Veterinary Science, Department of Veterinary Tropical Diseases, University of Pretoria, Onderstepoort, South Africa
e-mail: anita.michel@up.ac.za

D. R. Sibanda
Department of Agriculture and Animal Health, University of South Africa (UNISA), Pretoria, South Africa

L.-M. de Klerk-Lorist
Directorate of Animal Health, Department of Agriculture, Forestry and Fisheries, Pretoria, South Africa
e-mail: linmariedk@daff.gov.za

A. B. Dibaba et al. (eds.), *Tuberculosis in Animals: An African Perspective*,
https://doi.org/10.1007/978-3-030-18690-6_19

livestock production to prosper, wildlife had to make way for livestock because of the devastating diseases that they sustain, and because they competed for the limited available grazing. This approach was strengthened by the impact of serious epidemic diseases such as rinderpest that caused the disappearance of wildlife from large parts of the country.

Because of governmental policy throughout the twentieth century, the white commercial livestock sector enjoyed governmental support and investment while the rural black communities were restricted to communal land where they practiced small-scale or subsistence livestock farming without the benefit of governmental financial or veterinary support (Beinart 2007). Currently, the national cattle herd of 14 million consists of communal cattle (40%), commercial beef and dual-purpose herds (mixed beef and dairy production) (50%), and dairy herds (10%) (Department of Agriculture Forestry and Fisheries 2002). The commercial farmers thus received most of the benefits, and the marginalized communities were, and still remain, insufficiently supported to cope with the effects of livestock diseases and the transmission of indigenous and introduced zoonotic diseases by consuming unpasteurized milk and uninspected meat.

Livestock farming is the largest agricultural sector in South Africa. The wildlife industry is part of this sector, and it recently became the fastest growing agricultural sector with 2–2.5% of farms annually converting from livestock to wildlife farming (Department of Environmental Affairs and Tourism 2005). Wildlife farming is a diverse enterprise, and it comprises a number of subsectors including intensive game farming for harvesting marketable products (e.g., meat, hides) and sought-after blood lines and extensive game ranching focusing on the sustainable utilization of wildlife through hunting and ecotourism (Taylor et al. 2015). Of the approximately 50,000 registered commercial farms in South Africa, the 5000 game farms and 4000 mixed game and livestock farms occupy a larger land surface than the combined South African national and provincial parks and other officially declared wildlife conservation areas. This development led to an unprecedented expansion of a dispersed and fragmented wildlife/livestock interface creating an increasing risk of bidirectional transmission of disease between wildlife, commercially and communally farmed domestic animals, and humans. The presence of these multi-host diseases, such as BTB, at this interface has been ignored for a long time, particularly within the context of the risk that they pose, and not incorporating them in the strategies to control animal diseases in South Africa.

It is considered that (BTB) was introduced into South Africa with the importation of dairy cattle from Europe and from other continents by the early European settlers in the Cape Colony (Hutcheon 1880; Cousins et al. 2004). Hutcheon (1880) first detected BTB in cattle in South Africa in 1880. During the twentieth century, the disease became widespread in South Africa with herd infection rates of 60% or more in dairy cattle herds on the Cape Peninsula, and it was declared a notifiable disease with the promulgation of the Diseases of Stock Act (Act no. 14) of the Union of South Africa in 1911.

More specific information about the occurrence and dynamics of BTB in wildlife is provided in Chap. 5.

19.2 BTB Control in Cattle in South Africa

Tuberculin skin testing of cattle for BTB was instituted in the Cape Colony in 1905, until which time the disease was believed to have been absent from South Africa. The Cape Act (No. 16 of 1906) that replaced the Animal Diseases Act of 1893 made provision for extensive tuberculin testing and the payment of partial compensation for cattle destroyed because of BTB and was incorporated in the subsequent Union of South Africa's Disease of Stock Act in 1911. The control of BTB at that time was voluntary, and because of the high herd infection rates and the poor compensation paid for infected animals that had to be destroyed, dairymen cooperated poorly with the national veterinary authorities to control the disease. This situation prevailed until 1969 when a comprehensive Bovine Tuberculosis Control and Eradication Scheme focusing on commercial cattle was introduced under the Animal Diseases and Parasites Act (Act 13 of 1956). The regulations dealing with BTB were later incorporated in the Animal Diseases Act (Act 35 of 1984) (South African Government 1984; Department of Agriculture Forestry and Fisheries 2013) with the aim of eradicating BTB from the national cattle herd. These control measures provided for compulsory notification of a diagnosis of BTB in herds, compulsory testing of herds in which the disease was suspected to be present, and application of control measures such as quarantine, slaughter of infected animals, INH treatment, and the disinfection of premises to contain and eradicate the disease.

The Animal Disease Act (Act 35 of 1984), in conjunction with the Animal Disease Regulations (R2026 of 1986 as amended), and the Bovine Tuberculosis Scheme Regulations (R1953 of 1988) (South African Government 1984; Department of Agricultural Economics and Marketing 1986; Department of Agricultural Economics and Marketing 1988), regulated the control of BTB in cattle in South Africa from 1984 until 2002 when Act 7 of 2002 (the Animal Health Act) was promulgated and replaced Act 35 of 1984.

Under the various Acts, a number of testing programs were introduced in an attempt to eradicate BTB. These included:

- An accreditation scheme that aimed to create a growing nucleus of BTB-free cattle
- A maintenance scheme in which accredited herds were retested every 2 years
- A diagnostic herd scheme
- Ordinary diagnostics
- Infected herds
- Import
- Export

The testing schemes were voluntary, and only approximately 10–20% of the country's cattle population was part of the program. It was thus not possible to determine the prevalence of BTB in the national cattle herd and to assess the effectiveness of the BTB control schemes. Nevertheless, the herd prevalence in the participating commercial herds was reduced from an average of 11.9% in 1971 to

0.4% in 1995, at which time BTB in cattle was on the brink of being eradicated from the commercial herds.

Following the election of a democratic government in South Africa in 1994 and the subsequent fragmentation caused by the provincialization of the South African state veterinary services, the Accreditation Program was revoked as compliance was considered to be too low. The existing "Annual Diagnostic Program" was amended to address the needs of controlling BTB in the country. Once-off diagnostic testing of herds or individual animals, including those for the purpose of importing and exporting cattle, was implemented.

Following provincialization of the veterinary services, the number of tuberculin tests performed declined dramatically from the early 1990s onward because of various reasons including a shortage of human resources, lack of funding, changes in the farming systems that caused a reduction in the number of dairy herds, and an increase in the number of game farms. In 1991, 1.1 million cattle were tested using the comparative cervical tuberculin skin test (CCT) reflecting a BTB prevalence of 0.04% (Department of Agriculture, Forestry and Fisheries 2016a, b), but unfortunately, from 1986 to 1996, the number of tuberculin tests performed decreased by 75%.

Currently, testing for BTB in South Africa is not compulsory and the national prevalence of the disease in cattle is unknown. Sporadic outbreaks continue to occur in all the country's nine provinces (Michel et al. 2008). From 2000 to 2014, 103 outbreaks were reported to the National Department of Agriculture, Forestry and Fisheries, Directorate of Animal Health, and the OIE (Figs. 19.1 and 19.2)

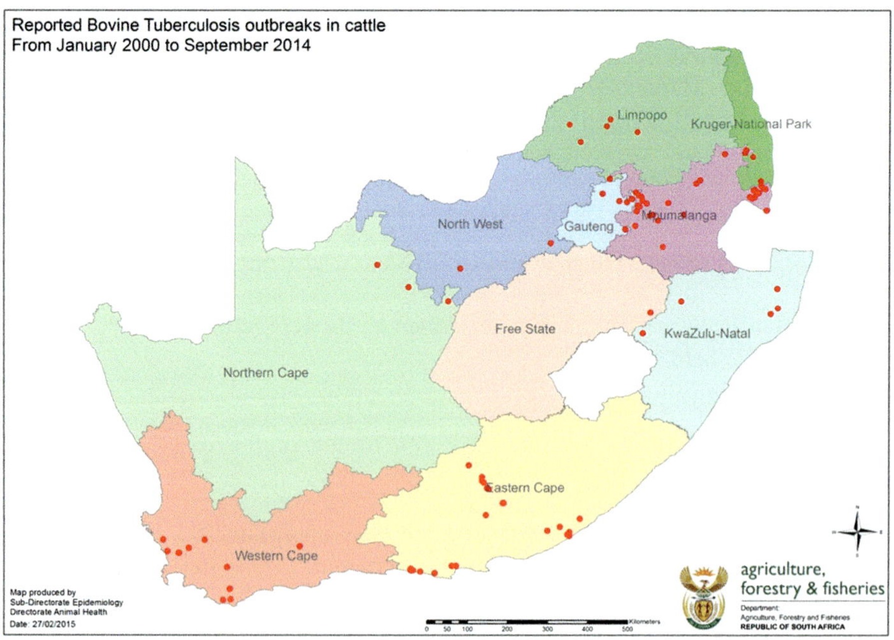

Fig. 19.1 Bovine tuberculosis outbreaks in cattle in South Africa between 2000 and 2014

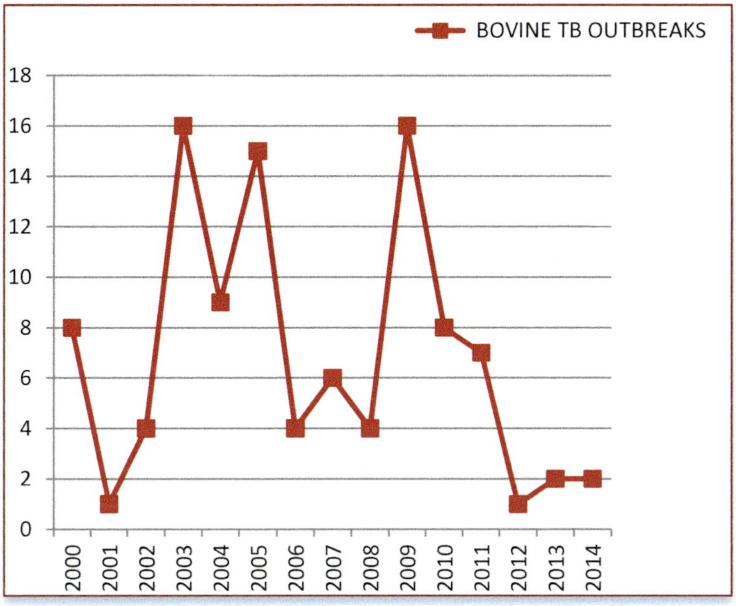

Fig. 19.2 Number of BTB outbreaks in South Africa from 2000 to 2014 (DAFF 2015)

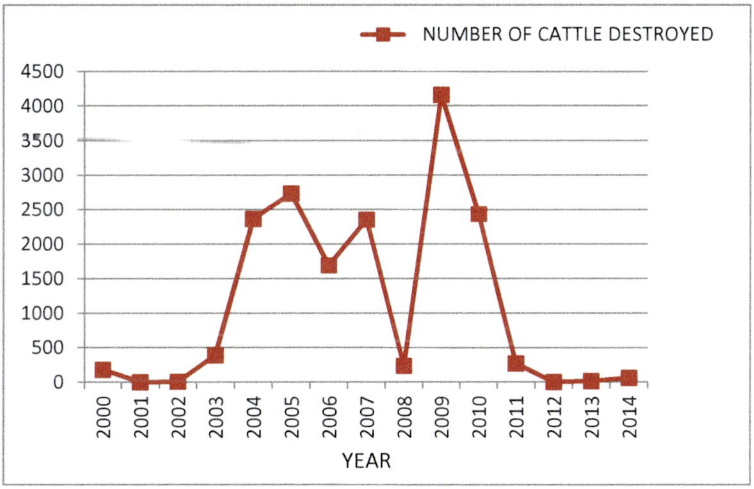

Fig. 19.3 Number of BTB-infected cattle destroyed per annum from 2000 to 2014

(Department of Agriculture, Forestry and Fisheries 2015). During this time, 16,881 head of cattle with a market value of approximately US$ 14 million were culled as part of the National Bovine Tuberculosis Scheme (Fig.19.3). The decline in the recorded number of culled animals from 2010 onward cannot be explained but does not appear to be the result of the control program and is more likely the result of insufficient funding to sustain the surveillance programs. The number of cases given in Figs. 19.1 and 19.2 may therefore be an underestimation of the actual prevalence of the disease in South Africa, and it may be substantially higher than anticipated.

19.3 Bovine Tuberculosis Control Programs in South Africa

Bovine tuberculosis in South Africa is currently controlled according to the guidelines contained in the Interim Bovine Tuberculosis Manual, dated October 6, 2016. This manual will remain in use until such time that the Tuberculosis Advisory Group constituted to deal with this matter formulates a new control strategy for BTB in South Africa.

The objectives of tuberculin testing are multifold and include the detection, control, and eventual eradication of BTB from a country. In South Africa, the single cervical tuberculin test (SIT) is used to screen herds for the presence of BTB. With this test, if any animal shows a reaction of more than 6 mm increase in the skin thickness at the test site, the entire herd is retested after 3 months using the CCT for increased specificity.

The following programs are currently employed to control BTB in South Africa:

- Maintenance program
- Diagnostic program
- Infected herd program

Participation in the maintenance and diagnostic programs is voluntary, while the infected herd program is compulsory. That means, if a BTB-infected herd is detected in any of the test programs, the herd is automatically incorporated in the "infected herd" program. Further tests and other actions required for eradication then become compulsory, and these are enforced in conformity with Act 7 of 2002 (the Animal Health Act).

In South Africa, the SIT is currently used for screening or diagnostic purposes in BTB-free herds. In all herds with a confirmed BTB-positive or suspect BTB status, the CCT is used. Briefly, the CCT involves the preparation of two injection sites about 12–15 cm apart, on either side of the mid-neck by shaving the skin with a clipper, and measuring of the skinfold thickness with a calibrated caliper, followed by the intradermal injection of 0.1 ml of avian tuberculin PPD containing 2500 IU/ml at the cranial site and 0.1 ml of bovine tuberculin PPD containing 3000 IU/ml at the distant site using preset automatic McLintock syringes. The skin reactions are evaluated after 72 h, and the skinfold thickness is remeasured by the same operator. For interpretation of the skin test result, the standard OIE method (World Organisation for Animal Health 2008) is used according to which an animal is considered positive if the increase in skinfold thickness at the bovine injection site is at least

4 mm and is at least 4 mm greater than at the avian site, inconclusive when the reaction difference is from 1 to 4 mm, and negative when the bovine reaction is less than or equal to the reaction to the avian PPD. In addition to measuring the extent of the reaction, changes in the skin at the site of the reaction such as the presence of necrosis, edema, and pain are assessed since the physical characteristics of this reaction also influence the interpretation of the test. At the discretion of the responsible state veterinarian, the commercial IFN-γ assay (Bovigam®) may be used as an ancillary test for diagnostic purposes.

The South African BTB control schemes are described in the following sections, with an emphasis on the requirements for being admitted into the various test programs, the required activities in each program, the expected measures of compliance, and how to deal with BTB-positive herds.

19.3.1 Maintenance Program

The maintenance program includes all the BTB-free herds including those herds that were infected and cleaned. The program does not require individual animal identification. All animals of all sexes above the age of 12 months are retested every 2 years. Cattle that develop suspect reactions to the tuberculin skin test are tagged with official ear tags to identify them permanently for further tests. If any of the cattle react positive to the skin test, the herd is transferred to the infected program and all the animals in the herd are then identified with official ear tags.

New herds that enter the program must undergo two consecutive negative tests with an interval of not less than 3 months before being awarded a BTB-negative status. The initial tests can be done by state officials and are free of charge.

The declaration of BTB negative only certifies that the animals were tested negative for BTB on the date of the second test, but it does not certify freedom from the disease. This declaration is valid for 2 years, and animals must be retested every 2 years. It is the owner's responsibility to keep the herd free from BTB.

19.3.2 Diagnostic Program

The diagnostic program is designed for farmers who want to determine the tuberculosis status of their herds or that of individual animals. State officials, to determine the prevalence of BTB in a herd, district, or municipality, also use this program. For this purpose, a state veterinary officer subjects all animals in a herd, older than 18 months, to one CCT at the state's expense. Where selected individual animals are tested, a private veterinarian, at the owner's expense, must preferably do the tests. Animal identification is preferred but not required for these tests, but official ear tags are attached to BTB-suspect or -positive animals following the specific test. When suspect reactors are detected, the entire herd is retested after 3 months, and if positive reactors are then found, the herd is incorporated in the infected herd program. These herds may be transferred to the maintenance program once the whole herd has had two consecutive negative CCTs, at least 3 months apart.

Imports Cattle that are imported and are kept at one of the registered quarantine stations must undergo a compulsory CCT done by the officer-in-charge of the quarantine station.

Exports Most importing countries require a negative BTB test before issuing import permits, or as a requirement to allow cattle into their country. Cattle that are destined for export are subjected to a CCT at the owner's expense.

19.3.3 Infected Herd Program

A herd is regarded as infected if any of the following procedures indicate infection with *M. bovis*: meat inspection, milk examination, postmortem examination, clinical examination, a positive tuberculin skin test, or the introduction of animals from a confirmed BTB-infected herd. The infected herd is placed under official supervision and quarantined. Necessary steps are then taken to confirm and eradicate the infection according to Act 7 of 2002 (the Animal Health Act), the Regulations enacted under the Act, the Bovine Tuberculosis Scheme regulations (R1953 of 1988 as amended), and the BTB Scheme Manual (South African Government 1984; Department of Agriculture, Forestry and Fisheries 1988; Department of Agriculture, Forestry and Fisheries 2013). Trial slaughter of tuberculin test-positive reactor animals to collect tissue samples for mycobacterial culture is recommended, but not mandatory, to confirm *M. bovis* infection in a herd.

Procedures for Handling BTB-Positive Herds The state veterinarian in charge of the process has the responsibility of implementing quarantine measures, supervising all testing and diagnoses, communication with the farmer, record-keeping, and implementing all applicable control measures in the infected herd.

According to existing legislation, every effort should be made to deal with positive reactor animals in such a way that there is no further spread of the disease within and beyond the herd. Immediate slaughter of the reactors is mandatory. Should a large number of reactor animals be involved that will seriously affect the owner financially should they all be slaughtered immediately, disrupt the supply of milk to the community, or threaten the family's livelihood, slaughtering may be postponed, but all reactors must be slaughtered within 12 months. Extension may also be considered in situations in which valuable animals such as high milk producers are involved, if more than 20% of the animals are infected, if the reactors are valuable pregnant cows, or those with small calves at foot. During the interim, milk produced by positive reactors must be boiled, pasteurized, or sterilized before it can be used for human or animal consumption. Cows that are known to produce infected milk are slaughtered immediately. Should the animal be condemned at the abattoir because of generalized tuberculosis, farmers may apply for compensation that, at present, would be the slaughter value of the animal that he/she would have been paid at the abattoir.

Following the removal of all known reactors, quarantine can be lifted after two additional consecutive negative CCTs, 3 months apart.

19.4 BTB Control in Wildlife

19.4.1 Background

Mycobacterium bovis is a multi-host pathogen with an extremely broad host range, and although susceptibility varies greatly between species, it is able to infect many mammalian species, whether domesticated, wildlife, or humans. The potential for causing widespread dissemination of the infection and the development of clinical disease is probably greater in animals in managed, multi-species systems with high population densities, than in natural, free-ranging ecosystems (Renwick et al. 2007).

When BTB was first detected in free-ranging wildlife populations in different geographical settings around the globe, including South Africa, it was generally not anticipated that these species would become significant role players in the epidemiology of BTB in cattle. Infected wildlife populations are now known sources of spillback of *M. bovis* infection for cattle (Musoke et al. 2015), and this situation puts additional pressure on BTB control programs in countries in which BTB-infected wildlife reservoirs are present.

The establishment of wildlife reservoirs that allow the maintenance and spread of BTB independent of cattle is a major problem in BTB-infected developed countries such as the UK and New Zealand attempting to eradicate BTB, and they have great difficulty in dealing with the problem because of the presence of wildlife BTB reservoirs. The implications for South Africa with its large commercial game farming enterprise and extensive conservation areas are unparalleled in the world, and they create extreme challenges when attempting to control the spread of BTB and its eventual eradication.

To date, *M. bovis* infection has been diagnosed in 21 free- or semi-free-ranging wild mammal species in South Africa (Michel et al. 2015). Here only African buffaloes (*Syncerus caffer*) are regarded as true and greater kudus (*Tragelaphus strepsiceros*) as highly likely maintenance hosts of BTB. The remaining 19 species lack the attributes to sustain a persistent intraspecific infection (De Lisle et al. 2001) and are dead-end, spillover hosts. They should, however, not be ignored as a source of infection as they could still play a role in disease transmission especially during times of drought when animals cluster around water points and exposure to stress factors, such as overstocking. The epidemiology of BTB in wildlife is complex and largely unknown, and unexpected sources of infection should be anticipated. One such example is the recent event of a BTB-free African buffalo population in the

Madikwe Game Reserve that became infected with BTB, most probably by infected greater kudus entering the reserve (Hlokwe et al. 2016).

Private ownership of wildlife in South Africa was sanctioned by the state in 1991, which led to a boom in activities relating to hunting, breeding, sale, ecotourism, and game meat production. The wildlife industry has become one of the fastest growing economic sectors in South Africa and generated US$ 1,000,000,000 and created 120,000 job opportunities in 2012. During that time, the number of commercial game farms increased from 3500 in 1992 to 5000 in 2014. Additionally, there are over 4000 mixed livestock/game farms. Sixty percent of the country's game animals are privately owned compared to the 40% (6 million) in protected areas. Over the past 15 years, the financial turnover from game ranching grew at an annual rate of 20%, and it now outstrips that of the dairy sector (Lewis 2015).

In addition to the risk of severe economic losses by the growing game industry when BTB-free animals are infected, the disease is also a potential threat to wildlife conservation, particularly when endangered species such as the African wild dog (*Lycaon pictus*), black rhinoceros (*Diceros bicornis*), and others are affected.

It is imperative that future BTB control programs in South Africa, and the other African countries with large wildlife populations, take wildlife maintenance species of BTB into account when designing and executing their national BTB control programs. Unless this is done, attempts to control and eradicate the disease will be a futile waste of financial and other resources.

19.4.2 BTB Control in African Buffaloes

African buffaloes play a major role in the maintenance and transmission of several important livestock diseases including foot-and-mouth disease (FMD), Corridor disease (CD), bovine tuberculosis (BTB), and bovine brucellosis (BR) in South Africa (Michel and Bengis 2012). Buffalo herds in the Kruger National Park (KNP) and the Hluhluwe/iMfolozi Park (HIP) are endemically infected with BTB, and they serve as sources of infection for other wild animal species and as a source of spillback transmission to cattle (Musoke et al. 2015; Michel et al. 2015; Michel 2002; Keet et al. 1996). In some of the larger multi-host conservation areas such as these, it is considered to be virtually impossible to eradicate the disease.

Breeding and the sale of African buffaloes are a particularly lucrative business with increasing numbers of prime animals being sold at auctions, sometimes at exorbitant prices (Anon 2013; Wildlife Ranching 2015). To control BTB, and the other important diseases that they carry, the Directorate of Animal Health of the Department of Agriculture, Forestry and Fisheries maintains strict control over all buffalo movements in the country (Department of Agriculture, Forestry and Fisheries 2014). When contemplating keeping buffaloes, landowners must apply to the National Director of Animal Health for approval, and they must also comply with the fencing regulations as stipulated by in the Buffalo Veterinary Procedural Note (VPN) regulating the keeping of African buffaloes in South Africa. Irrespective of their disease status, buffaloes may under no circumstances be kept on the same land as cattle, and they may only be moved between farms or conservation areas following the issuing of a veterinary red-cross movement permit after having been tested for the presence of specified diseases, and after the results have been communicated to the responsible state veterinarian (Department of Agriculture, Forestry and Fisheries 2014).

The comparative cervical tuberculin skin test (CCT) for the diagnosis of BTB is the required test when moving African buffaloes. The OIE guidelines for interpreting skin reactions in cattle for BTB testing have been adapted for buffaloes, and these guidelines are included in the Buffalo VPN. Although the IFN-γ test has great value as a screening tool in free-ranging buffalo populations (Michel et al. 2011), it is currently not approved by DAFF as a sole test for movement purposes. The IFN-γ test may only be used as an ancillary test at the discretion of the state veterinarian or the provincial Director of Veterinary Services. Because BTB is a controlled disease, the CCT must be carried out under the supervision of a state veterinarian. All the buffaloes presented for testing at any time should be visibly identifiable and microchipped to provide an unambiguous record of the test results.

To facilitate the uniform application of the diagnostic tests and standardized recording of the results, the National Director of Animal Health developed a guideline defining the BTB status of buffaloes and buffalo herds on private buffalo farms.

19.4.2.1 BTB Status of a Buffalo or a Buffalo Herd

BTB-Free Buffalo Herd or BTB-Free Facility This is a herd or a facility in which all animals have undergone three consecutive negative CCT tests if sourced from a herd of unknown status, or five consecutive negative tests if sourced from a known BTB-infected herd.

Unknown Status Herd This is a buffalo herd that has never been tested, or in which a proportion of the herd may have been sampled or examined through either testing or necropsies, and no positive cases have been identified.

Infected Herd This is a herd in which BTB has been detected and confirmed by culture in, at least, one animal.

BTB-Positive Animal This is an animal that has tested positive on the appropriate blood-based or intradermal tuberculin test, or in which the disease has been confirmed by culture.

19.4.2.2 Standard Procedures for the Eradication of BTB from a Buffalo Herd

The Buffalo VPN determines that any suspect or positive BTB test result must be reported to the Provincial Director of Veterinary Services and that the entire portion of land or a farm must then be placed under quarantine until the diagnosis can be either confirmed or dismissed. In such a case, no further movement of buffaloes, or any other susceptible species, will be allowed onto, off, or through the land registered for keeping buffaloes.

Should the diagnosis of BTB be confirmed by culture, the land/farm will remain under quarantine until such time that all the buffaloes on the property have undergone five consecutive negative CCT tests. However, if the population is too big and the testing scheme cannot be applied, then the manager or owner of the land may decide to remain under permanent quarantine with no movement of buffaloes allowed from the land to any BTB-free area. If the owner/manager decides to cull all the positive animals with the intention of regaining BTB disease-free status, a suitable, science-based action plan must be developed in collaboration with the relevant Provincial Director of Veterinary Services. Action plans must always also be submitted to the Directorate of Animal Health for approval before the final approval can be given for the implementation of the plan. A complete, accurate, and digitally accessible record of all test results for each buffalo is required to allow forward and backward tracing of the disease status of all the buffaloes on the property.

19.4.2.3 BTB Control in Other Free-Ranging Wildlife Species

Although the BTB risk posed by the South African buffalo population for the transmission of BTB is very well monitored and controlled, the same is not true for most of the other susceptible wildlife, especially the cloven-hoofed species. Thousands of head of game are captured and translocated annually across the country, yet unless they are clinically ill, none of these animals is ever screened for BTB or any other disease.

A major obstacle to screening the various wildlife species for the presence of BTB is that, with the exception of tests for African buffaloes, the accepted diagnostic tests have been validated for use in only a few wildlife species. The CCT has been validated for use in lions (Keet et al. 2010), and it also holds some promise for use in warthogs (de Klerk pers. comm.). The IFN-γ test and the various serological tests used to detect BTB infection in lions and rhinoceros are also being evaluated (Duncan et al. 2009; Miller et al. 2015; Morar et al. 2013; Maas et al. 2012). The intra-palpebral tuberculin test is the test of choice for the diagnosis of BTB in baboons in Africa (Keet et al. 2000).

An additional impediment to BTB risk management in wildlife is that owners of these species, other than African buffaloes, are not required by law to have any of their animals tested prior to movement from one area to another. It does, however, sometimes so happen that owners might request that an animal be tested to prove freedom from BTB prior to an auction or for specific breeding purposes. This is problematic because no validated tests are available for most antelope species, and there are no validated guidelines for the interpretation of the SIT or CCT in those species. A "positive" or "negative" test result in these cases is unreliable and may create a false sense of security (Maas et al. 2013). This situation emphasizes the need for future research and the development of validated diagnostic assays for the detection of BTB in a range of game species, most of all, those that are in high demand for sale at auctions and in which BTB has been detected, i.e., greater kudu, nyala (*Tragelaphus angasii*), and other medium-sized and small antelopes. Equally important is the development of suitable and practical, long-term surveillance programs for the early detection of the disease at herd level in wildlife. It is important to understand that the practice of translocating wildlife of which the BTB status is unknown may have serious consequences for other wildlife species in the same environment. Most of these game species easily contract BTB, and they constitute sources of infection that are mostly ignored by owners of BTB-free buffalo herds and their veterinarians.

References

Anon (2013) It's a buffalo's life. Destiny. http://www.destinyconnect.com/2013/09/21/its-a-buffalos-life/

Beinart W (2007) Transhumance, animal diseases and environment in the Cape, South Africa. S Afr Hist J 58(1):17–41. https://doi.org/10.1080/02582470709464743

Cousins D, Huchzermeyer HF, Griffin JF et al (2004) Tuberculosis. In: Coetzer JAW, Tustin RC (eds) Infectious diseases of livestock, 2nd edn. Oxford University Press, Cape Town, pp 1973–1993

De Lisle GW, Mackintosh CG, Bengis RG (2001) *Mycobacterium bovis* in free-living and captive wildlife, including farmed deer. Rev – Off Int Epizoot 20:86–111

Department Agriculture, Forestry and Fisheries (2002) Animal production. http://www.nda.agric.za/doaDev/sideMenu/animalAndAquacultureProduction/animalProduction.htm. Accessed 14 Jan 2017

Department of Agricultural Economics and Marketing (1986) Animal Diseases Regulation R2026. Government Gazette, vol 255. http://www.nda.agric.za/vetweb/Legislation/govgezet/Gov%20Gaz%20%20Act%2035%20of%201984%20-%20No%2010469%20-%20Part%202.pdf. Accessed 21 July 2015

Department of Agricultural Economics and Marketing (1988) Bovine Tuberculosis Scheme: Regulation 1953. Government Gazette, vol 279. Cape Town, South Africa. http://www.nda.agric.za/vetweb/Legislation/govgezet/Gov%20Gaz%20No%2011524%20R1953%20%2030%20Sept%201988.pdf

Department of Agriculture, Forestry and Fisheries (2013) Interim bovine tuberculosis scheme manual. http://www.nda.agric.za/vetweb/Pamphlets&Information/Policy/TB%20Manual%20%206Dec2013%20interim%20signed.pdf

Department of Agriculture, Forestry and Fisheries (2014) Buffalo Veterinary Procedural Notice (VPN) (In Preparation). http://www.daff.gov.za/vetweb/pamphlets&Information/Policy/Government%20Gazette%20Notice%2039424%20Extension%20of%20Buffalo%20VPN%20comment%20period%2011%20December%202015.pdf

Department of Agriculture, Forestry and Fisheries (2015) Animal diseases information. http://www.nda.agric.za/vetweb/epidemiology/Disease%20Database/OIEData/OIE_query_Criteria.asp

Department of Agriculture, Forestry and Fisheries (2016a) The development of veterinary services in South Africa. http://www.nda.agric.za/vetweb/History/H_History_Main_Initual.htm http://www.huntingreport.com/images2/pdf/Final_Draft_Panel_of_Expert_Report_to_DEAT.pdf. Accessed 14 Dec 2016

Department of Agriculture, Forestry and Fisheries (2016b) Annual reports. http://www.nda.agric.za/vetweb/History/H_Annual_reports_Index.htm

Department of Environmental Affairs and Tourism (2005) Report of the panel of experts on professional and recreational hunting in South Africa, Pretoria

Duncan AE, Lyashchenko KR, Greenwald M (2009) Application of elephant TB Stat-Pak assay and MAPIA (multi-antigen print immunoassay) for detection of tuberculosis and monitoring of treatment in black rhinoceros (*Diceros bicornis*). J Zoo Wildl Med 40:781–785

Hlokwe TM, de Klerk-Lorist LM, Michel AL (2016) Wildlife on the move: a hidden tuberculosis threat through introduction of untested species in an ecosystem. J Wildl Dis 52:837–843

Hutcheon DT (1880) Consumption, tables Mesenterica. Annual report, Colonial Veterinary Surgeon, Cape of Good Hope

Keet DF, Kriek NPJ, Penrith ML et al (1996) Tuberculosis in buffaloes (*Syncerus caffer*) in the Kruger National Park: spread of the disease to other species. Onderstepoort J Vet Res 63:239–244

Keet DF, Kriek NPJ, Bengis RG et al (2000) The rise and fall of tuberculosis in a free-ranging Chacma baboon troop in the Kruger National Park. Onderstepoort J Vet Res 67:115–122

Keet DF, Michel AL, Bengis RG (2010) Intradermal tuberculin testing of wild African lions (*Panthera leo*) naturally exposed to infection with *Mycobacterium bovis*. Vet Microbiol 144:384–391

Lewis A (2015) The private wildlife industry. BTB Outreach Day, 02 March 2015, Faculty of Veterinary Science, Onderstepoort, South Africa. https://www.up.ac.za/media/shared/678/btb-outreach-day-booklet-print_03042017.zp114198.pdf

Maas M, van Kooten PJS, Schreuder JD et al (2012) Development of a lion-specific interferon-gamma assay. Vet Immunol Immunopathol 149:292–297

Maas M, Michel AL, Rutten VP (2013) Facts and dilemmas in diagnosis of tuberculosis in wildlife. Comp Immunol Microbiol Infect Dis 36:269–285

Michel AL (2002) Implications of tuberculosis in African wildlife and livestock. Ann N Y Acad Sci 969:251–255

Michel AL, Bengis RG (2012) The African buffalo: a villain for inter-species spread of infectious diseases in southern Africa. Onderstepoort J Vet Res 79:26–30

Michel AL, Hlokwe TM, Coetzee ML et al (2008) High *Mycobacterium bovis* genetic diversity in a low prevalence setting. Vet Microbiol 126:151–159

Michel AL, Cooper D, Jooste J et al (2011) Approaches towards optimising the gamma interferon assay for diagnosing *Mycobacterium bovis* infection in African buffalo (*Syncerus caffer*). Prev Vet Med 98:142–151

Michel AL, de Klerk LM, Buss P et al (2015) Tuberculosis in South African wildlife: lions, African buffalo and other species. In: Mukundan H, Chambers MA, Waters WR et al (eds) Tuberculosis, leprosy and mycobacterial diseases in animals: the many hosts of mycobacteria. CAB International, pp 365–385

Miller M, Buss P, Hofmeyr J et al (2015) Antemortem diagnosis of *Mycobacterium bovis* infection in free-ranging African lions (*Panthera leo*) and implications for transmission. J Wildl Dis 51:493–497

Morar D, Schreuder J, Mény M et al (2013) Towards establishing a rhinoceros-specific interferon-gamma (IFN-γ) assay for diagnosis of tuberculosis. Transbound Emerg Dis 60(Suppl 1):60–66

Mossman SL, Mossman AS (1976) Wildlife utilization and game ranching. IUCN Occasional Paper No. 17. International Union for Conservation of Nature and Natural Resources Morges, Switzerland

Musoke J, Hlokwe T, Marcotty T et al (2015) Spill-over of *Mycobacterium bovis* from wildlife to livestock, South Africa. Emerg Infect Dis 21:448–451

Renwick AR, White PC, Bengis RG (2007) Bovine tuberculosis in southern African wildlife: a multi-species host-pathogen system. Epidemiol Infect 135:529–540

South African Government (1984) Animal Diseases Act (Act 35 of 1984), Pretoria

Taylor WA, Lindsey PA, Davies-Mostert H (2015) An assessment of the economic, social and conservation value of the wildlife ranching industry and its potential to support the green economy in South Africa. The Endangered Wildlife Trust, Johannesburg. https://www.ewt.org.za/scientific%20publications/An%20assessment%20of%20the%20economic,%20social%20and%20conservation%20value%20of%20the%20wildlife%20ranching%20industry%20and%20its%20potential%20to%20support%20the%20green%20economy%20in%20SA.pdf

Wildlife Ranching SA (2015) An industry of the future. http://www.wrsa.co.za/news/item/562-wrsa-an-industry-of-the-future

World Organisation for Animal Health (OIE) (2008) Manual of diagnostic tests and vaccines for terrestrial animals. OIE, 6th edn. OIE, Paris, pp 683–697

Chapter 20
Bovine Tuberculosis in the Republic of Sudan: A Critical Review

Z. A. Ishag, El Tigani Asil, Ali Parsaeimehr, and Guo-Qing Shao

20.1 Introduction

Bovine tuberculosis (BTB) caused by *Mycobacterium bovis* is globally also one of the most important zoonotic diseases, and it is of considerable public health importance (O'Reilly and Daborn 1995). Infection with *M. bovis* has been reported in 69% of the tropical countries around the world and in 80% of African countries (OIE 2000), including Sudan (Awad et al. 1959; Awad 1962).

The Republic of Sudan, situated in the northeastern part of the African continent, is vast (1.9 million km^2), the second largest country in Africa (Fig. 20.1), and a substantial proportion of its inhabitants are nomads (FAO 2015). Sudan's livestock population, estimated at about 105 million, consists of approximately 40 million sheep, 30 million goats, 30 million cattle, and 5 million camels (Fig. 20.1). After the separation of the Republic of South Sudan from the Republic of Sudan, increasing

Z. A. Ishag (✉)
Key Laboratory of Veterinary Biological Engineering and Technology, Ministry of Agriculture, National Research Center for Engineering and Technology of Veterinary Bio-products, Institute of Veterinary Medicine, Jiangsu Academy of Agricultural Sciences, Nanjing, China

College of Veterinary Sciences, University of Nyala, Nyala, Sudan

E. T. Asil
College of Veterinary Sciences, University of Nyala, Nyala, Sudan

Faculty of Agriculture and Veterinary Medicine, Department of Veterinary Medicine, Qassim University, Buraydah, Kingdom of Saudi Arabia

A. Parsaeimehr
Laboratory of Biosystems Engineering, Institute of Biotechnology, Jiangsu Academy of Agricultural Sciences, Nanjing, Jiangsu, China

G.-Q. Shao
Key Laboratory of Veterinary Biological Engineering and Technology, Ministry of Agriculture, National Research Center for Engineering and Technology of Veterinary Bio-products, Institute of Veterinary Medicine, Jiangsu Academy of Agricultural Sciences, Nanjing, China

A. B. Dibaba et al. (eds.), *Tuberculosis in Animals: An African Perspective*,
https://doi.org/10.1007/978-3-030-18690-6_20

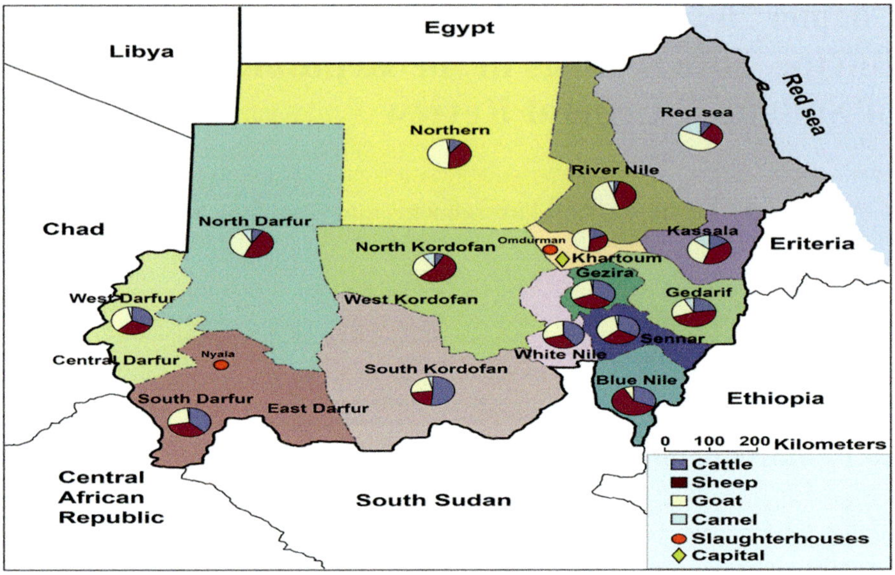

Fig. 20.1 Map of Sudan showing the estimated animal population by state and the location of the abattoirs

the number of livestock in the country as part of a program to stimulate the national economy and to increase exports became a national priority. To this end, the country's livestock exports improved by 96.6% in 2012, generating an income of US$ 408 million, compared to the US$ 333 million earned in 2011 (Sudan Tribune 2013).

Unfortunately, in Sudan, livestock suffer from a range of diseases, many of which are endemic, including BTB, bovine farcy, and Johne's disease (Awad 1962; Hamid et al. 2002; Mohammed and Mohammed 2009). As is the case in other developing countries, information in Sudan about the occurrence, distribution, risk factors, and the economic impact of BTB is insufficient, and because of financial constraints, programs to control tuberculosis in animals and humans have not been implemented.

Financial constraints limit the way in which mycobacterial infections in Sudan can be diagnosed. For this reason, the tuberculin skin test, bacteriology, and molecular biological methods are not routinely used, and information about the occurrence and prevalence of the disease in cattle is primarily deduced from abattoir records (Fig. 20.1) that reflect the presence of BTB-like lesions detected during routine meat inspection. Although flawed, this does provide some information about the presence and distribution of BTB in livestock in Sudan, and it provides the opportunity to control the disease in infected herds (Biffa et al. 2010).

The presence of, and extent of, human infections with *M. bovis* in Sudan is unknown, but the current lack of adequate control measures in livestock implies that BTB in cattle is likely to continue to spread in the country, and as long as poverty and malnutrition prevail, it will remain an important zoonotic threat (Moda et al. 1996). Co-infection of humans with BTB and HIV is an increasingly serious health problem in the developing countries (Narain et al. 1992). This situation is of particular concern given the known presence of HIV and HIV/TB co-infection in Sudan (Cosivi et al. 1998). The close association between the nomads and their animals further increases the risk of human infection by *M. bovis*, in addition to the likelihood of them also contracting other zoonotic diseases. For these reasons, there is an urgent need to step up, devise, and implement strategies to control BTB both in humans and in animals in Sudan.

In this chapter, we review the available information about BTB in cattle in the Republic of Sudan, with a focus on the threat posed by zoonotic TB, and especially on the risk of HIV-infected patients becoming co-infected with *M. bovis*. We also propose a number of strategies that should help to effectively control and eradicate BTB in cattle, humans, and wildlife in Sudan.

20.2 The Lesions and Diagnosis of BTB in Sudan

In Africa, BTB is widespread in cattle, and it impacts negatively on the animal industry causing economic losses due to carcass condemnation and the imposition of international trade restrictions. It also poses a serious public health risk (Cosivi et al. 1998; Thoen et al. 2006; Munyeme et al. 2009; OIE 2009).

Bovine TB is a neglected, transmissible disease caused by *M. bovis*. Humans are infected with *M. bovis* either by inhalation of aerosols or by consuming milk contaminated with the bacteria. Clinically, zoonotic BTB in humans is indistinguishable from that caused by *M. tuberculosis* (Ocepek et al. 2005; Mathews et al. 2006).

In Sudan, there are limitations, because of financial and infrastructural constraints, in the ability to confirm the diagnosis of BTB in animals. Although routine testing for BTB in cattle using the tuberculin skin test is done occasionally for surveillance purposes, the number of cattle that are tested is limited, and the tests generate insufficient data on which to base assessment of the prevalence of the disease in the country.

Most of the data about the occurrence and prevalence of the disease are obtained from abattoirs (Fig. 20.1) where BTB is diagnosed during the course of meat inspection of slaughtered livestock. The physical appearance of the lesions in animals with BTB in Sudan is similar to that reported elsewhere (Manal et al. 2005; El Tigani et al. 2013), and detection of these lesions is used as a screening mechanism for the presence of the disease in the country (Salih et al. 2010). However, the sensitivity and specificity of this procedure are very low because of the difficulty of distinguishing macroscopically between the various tuberculous-like lesions that can be encountered in livestock and the inability to make a final diagnosis of the specific cause of the disease based on the macroscopical appearance of these lesions. In spite of this, under the current circumstances, it is considered that abattoirs offer an affordable source of information to assess the prevalence of BTB in livestock in Sudan.

The lack of trained laboratory personnel with the ability to detect acid-fast bacteria microscopically is a further impediment. Using microscopy for the detection of acid-fast bacilli is also of limited value, as mycobacteria are difficult, if not impossible, to detect when they occur in low numbers in the smears of the exudate obtained from suspect tuberculous lesions. Many of the specimens contain few bacteria, and these paucibacillary specimens require indirect microscopy to detect the pathogen.

Isolation of *M. bovis* by culturing is time-consuming, but also in Sudan, it is considered the best method for diagnosing BTB (Hamid et al. 2002; Vincent et al. 2003; Manal et al. 2005; Zackaria et al. 2008; Salih et al. 2010; El Tigani et al. 2013, 2014). Inoculation of guinea pigs and rabbits with suspected *M. bovis*-infected specimens can also be used to confirm the presence of bovine tubercle bacilli (Awad et al. 1959).

PCR techniques have recently been introduced for diagnosing BTB in Sudan (Taylor et al. 2007), and they have reduced the diagnostic time and increased the sensitivity and specificity of the investigations (El Tigani et al. 2013; Zackaria et al. 2008). Utilization of these techniques should in future generate more reliable data about the presence, distribution, and importance of BTB in the country.

20.3 The Status of Bovine Tuberculosis in Sudan

Little information about the occurrence and prevalence of BTB in Sudan is available. While most of the states in Sudan do not have the means to do research, Khartoum State is unique in having access to facilities allowing them to do research in animal diseases, and most of the diagnostic work that has been done thus focused on the central regions of the country where these research facilities are located. The nature of the rural nomadic farming systems and ongoing border conflicts hamper research in these remote regions, with the result that almost no data are available for them.

Various publications reported a limited number of fragmented studies that were undertaken to determine the presence of BTB in some of the provinces of Sudan. Given the size of the national herd and the small number of animals tested, the data obtained are inadequate to provide a reliable estimate of the prevalence of the disease in any one of the regions. The limited available information suggests though that the distribution and prevalence of the disease are not uniform, in that the northern regions harbor cattle with a low prevalence of BTB, while higher infection rates are present in the central and southern regions.

BTB was first reported in Sudan in 1915 (Anon 1915). Subsequently, screening for the presence of BTB using the single or comparative intradermal tuberculin skin tests, microscopy, and culture for diagnostic purposes also indicated that the disease was more prevalent in Southern Sudan (16%; $n = 631$) than in Northern Sudan where the prevalence was much lower (2.4%; $n = 1143$) (Awad et al. 1959; Awad 1962). Several successive studies during the 1960s and early 1970s confirmed the persistence of the disease (Karib 1962; El Nasri 1966; Mustafa 1970; Tageldin 1971; El-Kareem and Moustafa 1974).

Limited information about its occurrence and rough estimates of the prevalence are available for some of the regions. In Central Sudan (Khartoum) (Fig. 20.1), cattle with BTB were detected in its abattoir (Sulieman and Hamid 2002; Manal et al. 2005; Osman 2007). Additionally, by using the single intradermal comparative tuberculin test (CCT), 1.5% of 587 cattle in 35 randomly selected dairy herds were infected with *M. bovis* (Naglaa 2007). In another study, suspected tuberculous lesions collected in abattoirs and subjected to microscopic examination revealed a BTB infection rate of 7.1% (174 of 1940 cattle) (Salih et al. 2010).

In Southern Sudan, an investigation of the presence of BTB demonstrated the presence of acid-fast rods in 3.3% of smears prepared from lymph node aspirates, while PCR detected *M. bovis* in 26 (65%) of milk and in 12 (40%) of lymph node samples (Zackaria et al. 2008). Given the limited number of specimens examined, it was not possible to determine the prevalence of the infection in this region.

A cross-sectional, abattoir-based study of BTB and other tuberculous diseases (bovine farcy) in South Darfur State (Western Sudan) (Fig. 20.1) during 2007–2009 revealed that 0.2% (12 of 6680) of the slaughtered animals contained lesions consistent with those of BTB and that 3% of the specimens collected from the caseous exudates from these lesions and subjected to m-PCR diagnostics contained mycobacteria. Filamentous, acid-fast mycobacteria consistent with those that cause bovine farcy were present in 0.9% (59 of 6680) of the animals slaughtered (Salih et al. 2010). During 2006 to 2008 at the Nyala abattoirs, in the same region (Fig. 20.1), 1.4% of the 2794 carcasses examined contained tuberculous lesions (Aljameel et al. 2014). Following culture of specimens from these lesions, 17 (10.3%) of the isolates were identified as *M. bovis* (6.7%) and six (3.6%) as *M. farcinogenes*.

Limited data too are available for Eastern Sudan. In a number of small surveys, 64 of 120 (53.3%) caseous lesions, similar to those caused by mycobacterial infections, contained acid-fast bacilli, while 56 (46.7% of 569 cattle) were due to other causes (Sulieman and Hamid 2002). Recently, a BTB prevalence of 1.9%, using the comparative intradermal test (CCT), was detected in an area that shares its border with Eritrea (Fig. 20.1), suggesting that the infection could be the consequence of in-bound cattle movement from the neighboring country (Ayman et al. 2014).

Although BTB has not been detected in them, the presence of various species of wildlife susceptible to BTB in Sudan could play a role in sustaining the infection and in the transmission of BTB to cattle. Wildlife as a potential source of infection should thus be taken into consideration when developing BTB prevention and control measures for this region.

20.4 Human Tuberculosis and its Risk Factors in Sudan

As is the case with citizens of other sub-Saharan countries, Sudanese suffer from various epidemic and communicable diseases. Tuberculosis is one of the leading causes of mortality of humans in developing countries (Corbett et al. 2006; Wrigh et al. 2009), and it is also an important cause of death in Sudan, the other being the consequence of ongoing human conflict, poverty, and population displacement (UNDP 2012).

The incidence of human tuberculosis in Sudan was estimated at 108/100,000 population in 2013. This accounts for 15% of the TB burden in the World Health Organization's Eastern Mediterranean Region (WHO 2014). The average prevalence of zoonotic tuberculosis in African countries is estimated to be 2.8% of all tuberculosis cases (264/100,000 population/year); this translates into an estimated seven zoonotic human BTB cases/100,000 population/year in Sudan (Müller et al. 2013). In addition, HIV and mycobacterial drug resistance are some of the major drivers sustaining endemic tuberculosis in developing countries (Corbett et al. 2006; Wright et al. 2009).

In 2014, the prevalence of HIV in Sudan was estimated to be 0.2% (World Bank 2015). This is of some concern given that in the neighboring Tanzania, *M. bovis* was isolated from seven of 65 (10.8%) HIV-infected patients with cervical adenitis (Cleaveland et al. 2007; Mfinanga et al. 2004). Based on this observation, it is expected that HIV infections will exacerbate the number of zoonotic tuberculosis cases and that it will increase the difficulty of combating TB worldwide (Katale et al. 2012). This situation should be of great concern in Sudan given the lack of information about *M. bovis*, and particularly HIV/*M. bovis* co-infections in humans.

The close association between humans and their animals in rural areas in Sudan may also enhance the risk of them contracting zoonotic BTB. The common practice here of consuming uninspected meat from illegally slaughtered livestock and the lack of awareness of dairy workers about the risk of BTB are additional factors that increase the risk of becoming infected with *M. bovis*. This appears to be the situation in both the Southern and Western Regions of the country where the cattle density is high, and the consumption of raw milk and meat is a normal practice. Here, there is a correlation between the prevalence of BTB and the percentage of human tuberculin reactors (Awad et al. 1959). To what extent this reflects human infection with *M. bovis* is not known. Zoonotic tuberculosis, irrespectively, will probably remain of major concern in parts of Sudan until such time that it is possible to easily and cheaply distinguish between *M. tuberculosis* and *M. bovis*.

Wildlife reservoirs may also serve as a source of human infection at the human/wildlife/livestock interface in large ecosystems (Ocepek et al. 2005). Trans-boundary nomadic movement can similarly contribute to the dissemination of the disease between domestic animals and wildlife that share pastures with livestock owned by the nomads.

In conclusion, it is imperative that a large-scale, population-based, surveillance program is launched to assess the actual prevalence of zoonotic TB in humans in Sudan. The high prevalence of BTB in cattle here supports the likelihood that the disease remains a major health risk for humans in Sudan. Therefore, proper meat inspection, regular health checks of the abattoir workers for occupational zoonotic tuberculosis, and the implementation of awareness-enhancing campaigns are important activities to consider when controlling BTB in Sudan.

20.5 Prevention and Control of BTB in Sudan

The World Organisation for Animal Health (OIE) classifies BTB as a List B disease, i.e., those transmissible diseases with socioeconomic and/or public health implications, that are significant to be controlled. In many African countries, however, there is a lack of understanding of the disease's epidemiology, and an inability to control it principally because of the difficulty of diagnosing the infection, and the lack of adequate reporting systems (Cosivi et al. 1998).

In developed countries, the effective control and eradication of BTB from herds and/or farms are dependent on the identification and isolation of the sources of infection and application of the test-and-slaughter strategy (Cosivi et al. 1998; Ayele et al. 2004). In Sudan, these strategies pose major challenges to decision-makers because of a number of factors. These include the lack of knowledge of the actual prevalence of BTB in the national herd, technical and financial limitations, lack of an adequate veterinary infrastructure, and the negative views held by old-fashioned politicians who fail to understand the need to deal with the problem. Within a financial context, much less is spent in Sudan on maintaining animal and public health, compared to the amount of money spent on defense.

Although a number of current constraints impact negatively on attempts to deal with BTB, there remain a number of opportunities that will allow the authorities to deal with the infection. These include the extensive use of the intradermal tuberculin skin test, bacteriology, and molecular biological techniques for diagnostic purposes, in addition to using the relevant information emanating from abattoirs (Probst et al. 2011). This information can be supplemented by changes in animal husbandry and herd management, inspection of carcasses at the abattoirs for the presence of tuberculous lesions, and condemning tuberculous carcasses and slaughtering BTB-infected cattle, with indemnity to the producers. Tracing-back to the herd of origin of tuberculous cattle will also be useful in combating animal tuberculosis and to limit its spread before it becomes more widespread and thus much more difficult to control (Smith et al. 2014).

Unfortunately, even though considerable efforts have been made to improve the Sudanese economy, there are still major financial constraints limiting efforts to control BTB in humans and animals. These constraints are compounded by the loss of fuel export earnings following the separation from South Sudan. One of the consequences of the ongoing lack of finances is that many of the well-trained Sudanese veterinarians are leaving the country to find employment in the nearby rich Arabian Gulf states (A-Rahman and Jacquet 2014). An enormous effort is needed by all concerned (the government, experts, international donors, and civil society organizations) to create and maintain a high-quality animal health service that will benefit Sudan.

In Sudan, the control of BTB by means of the test-and-slaughter policy has not yet been implemented. In dealing with BTB, initially testing and isolating reactors, combined with the pasteurization of milk, could be beneficial control practices on dairy farms. When dealing with the traditional, extensive production systems in rural

Sudan, implementing control measures will be more difficult and complex due to the large numbers of livestock involved, the free and uncontrolled movement of animals (nomadic and pastoral practices), and the related economic and social challenges associated with the nomadic way of life.

For the successful implementation of a control program, Sudanese authorities will have to address important economic considerations, governmental commitment, active partnerships, and collaboration with international organizations. The Sudanese Government will also have to provide adequate security for the personnel involved in the BTB control program, particularly in those instances when they employ foreigners to assist them with the program (Safari Notes 2010).

As *M. bovis* is mainly transmitted from cattle to humans through the consumption of milk from tuberculous cattle, the control of human infection can be achieved by pasteurization of milk in addition to the general control of bovine tuberculosis in livestock (O'Reilly and Daborn 1995). Additionally, enhancing public awareness about the importance of the disease in humans and animals, the risks involved, the routes of infection, the measures required to control infection, the application of efficient and effective meat inspection practices, and changes in animal management would contribute substantially to the control of tuberculosis in Sudan. It must also be kept in mind that elimination of the disease may be complicated by the possibility that several wildlife reservoirs of *M. bovis* may be present in the country (Flamand et al. 1994).

20.6 Conclusion and Recommendations

Given the limited knowledge regarding the status of BTB in Sudan, it is recommended that a systematic surveillance program for BTB be launched throughout Sudan to estimate the prevalence of the disease in both animals and humans. This should establish a solid knowledge base that would enable the country to control tuberculosis in animals, to increase the productivity of dairy farms, and to decrease the risk for humans to contract this zoonosis. Used alongside existing BTB control measures, vaccination of domestic animals in Sudan against BTB could in future decrease the spread in cattle. However, there is currently no effective BTB vaccine available for use in cattle.

The lack of application of quarantine measures in most African countries in addition to the free movement of cattle from one region to another across international borders facilitates the transmission and spread of diseases between and within countries. It would thus be necessary for the Sudanese Government to trace the sources of infection based on molecular epidemiological data to assist with controlling the further introduction and spread of BTB from neighboring countries. Generating epidemiological data, including strain characterization by using molecular epidemiological tools, is also critical for the future successful control of the disease in Sudan. However, with the extent of environmental crises in the country, civil wars, and the general levels of poverty and illiteracy prevailing at this stage, the challenges to implement such a program are substantial.

References

Aljameel MA, Fayza MAWA, Abdellatif AETM (2014) Occurrence of bovine tuberculosis at Nyala abattoirs in South Darfur State, Sudan. Rev Elev Med Vet Pays Trop 67:61–65

Anon (1915) Annual report of the Sudan Veterinary Services

A-Rahman NHA, Jacquet GA (2014) The state of emergency care in the Republic of the Sudan. Afr J Emerg Med 4:55–60

Awad FI (1962) Studies on type-determination and epidemiology of tuberculosis among cattle in Sudan. Transbound Emerg Dis 9(9):890–898

Awad FI, Karib AA, Fawi MT (1959) Some observations on tuberculosis among cattle in the Sudan. Transbound Emerg Dis 6(2):180–184

Ayele WY, Neill SD, Zinsstag J et al (2004) Bovine tuberculosis: an old disease but a new threat to Africa. Int J Tuberc Lung Dis 8:924–937

Ayman EA, Salih DA, Gumaa MM (2014) Prevalence of bovine tuberculosis in Kassala State, Eastern Sudan. Int J Adv Sci Tech Res 4:238–249

Biffa D, Bogale A, Skjerve E (2010) Diagnostic efficiency of abattoir meat inspection service in Ethiopia to detect carcasses infected with *Mycobacterium bovis*: implications for public health. BMC Public Health 10(1):462

Cleaveland S, Shaw DJ, Mfinanga SG et al (2007) *Mycobacterium bovis* in rural Tanzania: risk factors for infection in human and cattle populations. Tuberculosis (Edinb) 87:30–43

Corbett EL, Marston B, Churchyard GJ et al (2006) Tuberculosis in sub-Saharan Africa: opportunities, challenges, and change in the era of antiretroviral treatment. Lancet 367(9514):926–937

Cosivi O, Grange JM, Daborn CJ et al (1998) Zoonotic tuberculosis due to *Mycobacterium bovis* in developing countries. Emerg Infect Dis 4:59–70

El Nasri M (1966) Present status of diseases and disease control. Sudan J Vet Sci Anim Husb 7.34–43

El Tigani A, El Sanousi SM, Gameel A et al (2013) Bovine tuberculosis in South Darfur State, Sudan: an abattoir study based on microscopy and molecular detection methods. Trop Anim Health Prod 45:469–472

El Tigani AE, El Sanousi SM, Aljameel MA et al (2014) Molecular identification of nontuberculous mycobacteria isolated from pyogenic bovine tissues in South Darfur State and Alsabalouga slaughterhouse at Omdurman area. Sudan Vet J 4:16–19

El-Kareem MA, Moustafa AA (1974) Bovine nocardiosis, tuberculosis and other caseous infections at Omdurman central abattoir. Sudan J Vet Sci Anim Husb 15:57–60

FAO (2015) Country programming framework for Sudan. Plan of action (2015–2019). Resilient livelihoods for sustainable agriculture, food secutity and nutrition. FAO, Sudan, pp 11–15

Flamand JR, Greth A, Haagsma J et al (1994) An outbreak of tuberculosis in a captive herd of Arabian oryx (*Oryx leucoryx*): diagnosis and monitoring. Vet Rec 134:115–118

Hamid ME, Roth A, Landt O et al (2002) Differentiation between *Mycobacterium farcinogenes* and *Mycobacterium senegalense* strains based on 16S-23S ribosomal DNA internal transcribed spacer sequences. J Clin Microbiol 40:707–711

Karib E (1962) Bovine tuberculosis in the Sudan. Sudan J Vet Sci Anim Husb 3:9–19

Katale BZ, Mbugi EV, Kendal S et al (2012) Bovine tuberculosis at the human-livestock-wildlife interface: is it a public health problem in Tanzania? A review. Onderstepoort J Vet Res 79:84–97

Manal SH, Hamid ME, El Jalii MI et al (2005) Correlation between microscopic examination and culture for detection and differentiation of mycobacterial isolates from cattle in the Sudan. Pakistan J Biol Sci 8:284–286

Mathews F, Macdonald DW, Taylor GM et al (2006) Bovine tuberculosis (*Mycobacterium bovis*) in British farmland wildlife: the importance to agriculture. Proc R Soc Lond B Biol Sci 273 (1584):357–365

Mfinanga SGM, Morkve O, Kazwala RR et al (2004) Mycobacterial adenitis: role of *Mycobacterium bovis*, non-tuberculous mycobacteria, HIV infection, and risk factors in Arusha, Tanzania. East Afr Med J 81(4):171–178

Moda G, Daborn CJ, Grange JM et al (1996) The zoonotic importance of *Mycobacterium bovis*. Tuber Lung Dis 77:103–108

Mohammed KB, Mohammed ZA (2009) Comparison between different techniques for diagnosis of bovine paratuberculosis in dairy cows in Khartoum State, Sudan. Sudan J Vet Res 24:27–34

Müller B, Dürr S, Alonso S et al (2013) Zoonotic *Mycobacterium bovis*-induced tuberculosis in humans. Emerg Infect Dis 19:899–908

Munyeme M, Rigouts L, Shamputa I et al (2009) Isolation and characterization of *Mycobacterium bovis* strains from indigenous Zambian cattle using spacer oligonucleotide typing technique. BMC Microbiol 9(1):144

Mustafa AA (1970) Bovine tuberculosis in the Sudan, vol 8. Al Hakeem Medical Student Association, University of Khartoum, Sudan, p 22

Naglaa A (2007) Epidemiological and zoonotic aspect of bovine tuberculosis in Khartoum State, Sudan. Dissertation, University of Khartoum, Sudan

Narain JP, Raviglione MC, Kochi A (1992) HIV-associated tuberculosis in developing countries: epidemiology and strategies for prevention. Tuber Lung Dis 73:311–321

O'Reilly LM, Daborn CJ (1995) The epidemiology of *Mycobacterium bovis* infections in animals and man: a review. Tuber Lung Dis 76:1–46

Ocepek M, Pate M, Žolnir-Dovč M et al (2005) Transmission of *Mycobacterium tuberculosis* from human to cattle. J Clin Microbiol 43:3555–3557

OIE (2000) World animal health in 1999: part 2, tables. OIE, Paris, p 650. https://web.oie.int/delegateweb/eng/en_reports.php?WAHISPHPSESSID=9740ab7dfe4993d21056b14a5d1d049f

OIE (2009) Bovine tuberculosis. World assembly of delegates of the OIE. Chapter 2.4.7. https://web.oie.int/eng/normes/MMANUAL/2008/pdf/2.04.07_BOVINE_TB.pdf

Osman AB (2007) Caseation in lymph nodes of slaughtered cattle with a special reference to bovine tuberculosis. Vet Med J Giza 55:1101

Probst C, Freuling C, Moser I et al (2011) Bovine tuberculosis: making a case for effective surveillance. Epidemiol Infect 139:105–112

Safari Notes (2010) Dinder National Park, Sudan. http://safarinotes.blogspot.com/2010/05/dinder-national-park-sudan.html. Accessed 22 Jan 2013

Salih MH, Mohammed ZA, El Eragi AM et al (2010) Bovine tuberculosis at Omdurman Central Abattoir, Khartoum State and Wau slaughterhouses (Bahr El-Ghazal) State, Sudan. Sudan J Vet Res 25:1–8

Smith RL, Tauer LW, Sanderson MW et al (2014) Minimum cost to control bovine tuberculosis in cow–calf herds. Prev Vet Med 115(1):18–28

Sudan Tribune (2013) Sudan livestock exports almost double in 2012, minister says. http://www.sudantribune.com/spip.php?article45496. Accessed 6 Dec 2016

Sulieman MS, Hamid ME (2002) Identification of acid fast bacteria from caseous lesions in cattle in Sudan. J Veterinary Med Ser B 49:415–418

Tageldin MHA (1971) Comparative study of the pathology and bacteriology of human and animal tuberculosis in the Sudan. Thesis, University of Khartoum, Sudan

Taylor GM, Worth DR, Palmer S et al (2007) Rapid detection of *Mycobacterium bovis* DNA in cattle lymph nodes with visible lesions using PCR. BMC Vet Res 3:12

Thoen C, LoBue P, De Kantor I (2006) The importance of *Mycobacterium bovis* as a zoonosis. Vet Microbiol 112:339–345

UNDP (2012) Annual report, 2012. The sustainable future we want. http://www.undp.org/content/dam/undp/library/corporate/UNDP-in-action/2012/English/UNDP-AnnualReport_ENGLISH.pdf

Vincent V, Brown-Elliot B, Jost KC et al (2003) *Mycobacterium*: phenotypic and genotypic identification. In: Murray PR, JoBaron E, Jorgensen JH et al (eds) Manual of clinical microbiology, 8th edn. ASM Press, Washington, pp 560–584

WHO (2014) Global tuberculosis report. pp 117. http://apps.who.int/iris/bitstream/10665/137094/1/9789241564809_eng.pdf

World Bank (2015) Prevalence of HIV (% of population ages 15–49). http://data.worldbank.org/indicator/SH.DYN.AIDS.ZS

Wright A, Zignol M, Van Deun A et al (2009) Epidemiology of antituberculosis drug resistance 2002–07: an updated analysis of the global project on anti-tuberculosis drug resistance surveillance. Lancet 373(9678):1861–1873

Zackaria H, Mukhtar MM, Bakheit SM (2008) Molecular detection of bovine tuberculosis in El Rank Area, Northern Upper Nile State, Sudan. Doctoral dissertation, University of Khartoum

Chapter 21
The Changing Landscape of Bovine Tuberculosis in Tanzania

Bugwesa Z. Katale, Hezron E. Nonga, and Rudovick R. Kazwala

21.1 Introduction

Bovine tuberculosis (BTB) is a zoonotic bacterial disease caused by *Mycobacterium bovis*, a member of *Mycobacterium tuberculosis* complex (MTC). The members of the *M. tuberculosis* complex that consists of *M. tuberculosis*, *M. bovis*, *M. canettii*, *M. africanum*, *M. pinnipedii*, *M. microti*, *M. caprae*, *M. orygis*, *M. mungi*, and the dassie bacillus (see Chap. 6) are not very diverse in terms of their DNA sequences, but they differ widely in their host tropism, phenotypic properties, and pathogenicity (Brosch et al. 2002). *Mycobacterium bovis* is primarily a pathogen of cattle (Garnier et al. 2003) but has a wide host range (Brosch et al. 2002), including humans (Cosivi et al. 1998; Mfinanga et al. 2004; Kazwala et al. 2006), and a number of wildlife species (Cleaveland et al. 2005; de Lisle et al. 2001; Katale et al. 2015).

The introduction of *M. bovis* into Africa is linked to the historical livestock trade between the UK and African countries during colonization in the nineteenth and twentieth centuries. For this reason, a number of members of the European 1 (Eu1) *M. bovis* clonal complex present in the UK and Ireland occur in some of the former colonies such as Tanzania, South Africa, and Zambia (Smith et al. 2011). In Africa, *M. bovis*, a known multi-host pathogen, became embedded in a multi-species system following spillover of the disease to various African wildlife species, some of which then became maintenance hosts (Renwick et al. 2007).

In Tanzania, *M. bovis* was first isolated from cattle in 1952 (Markham 1952). Currently, the infection is prevalent in all the regions of the country (Cleaveland et al. 2005; Durnez et al. 2009; Katale et al. 2013; Kazwala 1996; Mwakapuja et al.

B. Z. Katale
Tanzania Wildlife Research Institute (TAWIRI), Arusha, Tanzania

H. E. Nonga · R. R. Kazwala (✉)
Department of Veterinary Medicine and Public Health, Sokoine University of Agriculture (SUA), Morogoro, Tanzania

A. B. Dibaba et al. (eds.), *Tuberculosis in Animals: An African Perspective*,
https://doi.org/10.1007/978-3-030-18690-6_21

2013a, b; Shirima et al. 2003). It has been detected in indigenous cattle (Kazwala et al. 1998; Cleaveland et al. 2007; Mdegela et al. 2004; Durnez et al. 2009; Makondo 2013; Mwakapuja et al. 2013a, b; Katale et al. 2015), humans (Kazwala et al. 2006; Mfinanga et al. 2004), and a number of wildlife species (Cleaveland et al. 2005; Katale et al. 2015) including lions (*Panthera leo*), blue wildebeest (*Connochaetes taurinus*), topi (*Damaliscus lunatus*) (Cleaveland et al. 2005), African buffaloes (*Syncerus caffer*), and the African civet (*Civettictis civetta*) (Katale et al. 2015). Spillover and spillback of *M. bovis* infection between wildlife and livestock are facilitated by anthropogenic factors such as encroachment on wildlife habitat and animal translocation (Palmer et al. 2012). Although spillover and spillback infections in wildlife, livestock, and humans are generally considered to be relatively rare in Africa (de Garine-Wichatitsky et al. 2013), the presence of maintenance hosts, whether wild or domestic, facilitates the persistence of the disease (Palmer et al. 2012). In such complex systems, the transmission rate of the infection between species is largely determined by the spatial distribution of susceptible hosts, the timing of contact between them, resource utilization patterns, and the frequency and intensity of interactions between these hosts at the human/livestock/wildlife interface (Renwick et al. 2007).

There seems to be a large variety of *M. bovis* strains circulating in the various susceptible hosts in Tanzania. In one study, 13 different pTBN12 RFLP types were encountered, of which the RFLP patterns reflected a high degree of relatedness (86%) between the dominant pTBN12 genotypes. There were 13 different spoligotypes found in this study, whose genetic relatedness was also high (79%). Their DNA profiles were also confirmed by IS986 RFLP, which revealed that these strains have 1–13 copies of IS986. Geographically, there is an overlap between pTBN12 RFLP and spoligotypes among strains isolated, and the diversity of the RFLP and spoligotype patterns observed probably reflects the extensive movement of pastoral cattle and contact with the different sources of the infection (Kazwala et al. 2006; Katale et al. 2015; Mwakapuja et al. 2013a, b). Spoligotype SB0133 is widespread and commonly occurs in cattle and wildlife in East Africa (Berg et al. 2011).

Intermingling of livestock and wildlife during grazing and at water sources might have contributed to the spread of *M. bovis* infection between livestock and wildlife. The role of each of these species in spreading the disease depends on the way of transmission, the abundance of each host, and interaction between hosts, but the actual situation is largely unknown (Nugent 2011).

21.2 Risk Factors for the Transmission of BTB in Indigenous Cattle in Tanzania

In Tanzania, the prevalence of BTB in cattle varies from one region to another, and it ranges from 0.2 to 13.2% (Katale et al. 2012). The highest prevalence was found in the large numbers of cattle in the extensive farming systems in the Southern

Highlands region of Tanzania (Kazwala et al. 2001; Durnez et al. 2009). Persistence of *M. bovis* infection here is linked to the traditional, extensive pastoral farming system where cattle are free to move extensively in search of pasture and water (Durnez et al. 2009; Pusic et al. 2009). In other regions of the country, the prevalence of *M. bovis* infection was higher in intensive systems than in the pastoral production system (Shirima et al. 2003), but there is no consistency in this pattern since there is a higher prevalence of *M. bovis* in certain regions in the intensive farming systems (Shirima et al. 2003), while the opposite is true for others (Durnez et al. 2009). The variation in the prevalence of BTB in different regions in Tanzania suggests that there are foci of *M. bovis* (or hotspots) produced by different factors including the presence of mycobacteria in the environment, specific management practices (Shirima et al. 2003), and flooding (Cleaveland et al. 2007).

The prevalence of BTB infection in animals in Tanzania also appears to be dependent on age, sex, breed, lactational status of cows, and variation in climate (Makondo 2013; Kazwala et al. 2001). Older cattle were more infected than younger ones and calves, males were more affected than females, and castrates were significantly more infected than bulls. Cattle on the plains in the middle of the Rift Valley are more infected than those on the highlands (Kazwala et al. 2001).

Other factors that influence the presence and prevalence of BTB infection include proximity to and contact with wildlife, their physiological status, geographical location, and husbandry practices (Kazwala et al. 2001; Ameni et al. 2002; Kaneene et al. 2002; Ameni and Erkihun 2007; Munyeme et al. 2008; Cadmus et al. 2010; Katale et al. 2013). In the coastal region of Tanzania, there is a significant association of the prevalence of BTB with the type of production system, as there is a higher prevalence in cattle in intensive systems there than in the pastoral system (Shirima et al. 2003) (Table 21.1).

21.3 Genetic Diversity of *Mycobacterium bovis* Strains in Tanzania

There is a relatively extensive genetic polymorphism of *M. bovis* isolates in the different geographical regions in Tanzania, and the various strains are not uniformly distributed throughout the country (Mwakapuja et al. 2013a, b). The wide dispersal of the various spoligotypes probably reflects the consequences of the extensive pastoral movement of cattle in certain areas (Kazwala et al. 2006). Some of these spoligotypes, such as SB0133, one of the predominant types, occur in livestock and in wildlife probably due to intermingling of these species at the various interfaces (Katale et al. 2015; Mwakapuja et al. 2013a, b; Berg et al. 2011).

Strain diversity is a characteristic of the isolates from across the country. Katale et al. (2013) and Mwakapuja et al. (2013a, b) reported a set of *M. bovis* strains, some of which were novel strains, in indigenous cattle. Another set of 13 different pTBN12 RFLP and 13 different spoligotypes isolated from animals and humans

Table 21.1 Risk factors associated with BTB in Tanzanian cattle herds (Katale et al. 2013; Shirima et al. 2003; Kazwala et al. 2001)

Study site	Variable	Categories	Sample size (n)	Prevalence (%)
Northern region (Serengeti ecosystem)	Location	Bunda	160	0.63
		Serengeti	569	2.64
		Ngorongoro	374	2.94
	Sex	Male	379	3.17
		Female	724	2.07
	Age	< 2 years	231	1.30
		2–4 years	304	3.6
		> 4 years	568	2.3
	Animals tested	1–20	170	3.5
		21–40	265	1.9
		>41	668	2.4
Coast region	Production system	Intensive	1253	62.9
		Extensive	740	37.1
Southern highlands	Sex	Male	1814	12.6
		Female	4039	14.8
	Age	<18 months calves	1211	9.1
		<1½–3 years cattle	855	9.7
		>3 years cattle	3723	15.2
	Breed	Exotic	244	8.2
		Short horn zebu	5692	13.4
	Pregnancy	Pregnant	493	12.0
		Nonpregnant	3600	12.5
	Lactation	Lactating	1627	14.6
		Non-lactating	2079	12.0
	Climate	Highland	1117	8.6
		Rift Valley	4614	14.0
	Male cattle	Entire bull	922	12.4
		Castrated (oxen)	892	17.3

also confirmed the high degree of relatedness (86%) between the dominant pTBN12 genotypes (Kazwala et al. 2006).

The diversity of *M. bovis* spoligotypes in Tanzania requires further epidemiological studies given the detection of novel strains such as SB2289 and SB2290 (Katale et al. 2015) and SB2190 (de Garine-Wichatitsky et al. 2013) in indigenous cattle (Table 21.2). Some of these isolates from (SB2290) differed from those found in wild animals by the loss of a single spacer, suggesting that BTB infections in wild animals and cattle are epidemiologically linked (Katale et al. 2015). The genetic relatedness of *M. bovis* isolates from indigenous cattle and wild animals further suggests the existence of either evolutionary spillback of *M. bovis* infection from wild animal reservoirs to livestock or micro-evolutionary events of *M. bovis* in the cattle populations in the ecosystem, although some of them may have been

Table 21.2 *Mycobacterium bovis* strains (spoligotypes) isolated from cattle, other animals, and humans in Tanzania

Spoligotype	MTC strain	Host species
SB0133	*M. bovis*	Cattle, African buffalo, African civet
SB2191	*M. bovis*	Cattle, human
SB1467	*M. bovis*	Cattle
SB2190	*M. bovis*	Cattle
SB1467	*M. bovis*	Cattle
SB2289	*M. bovis*	Cattle
SB2290	*M. bovis*	Cattle

introduced from elsewhere into the cattle and wildlife populations (Katale et al. 2015). In the Mikumi-Selous ecosystem, *M. bovis* was isolated only from indigenous cattle, highlighting the importance of pastoral livestock as a reservoir for *M. bovis* in systems in which both livestock and wildlife occur.

21.4 Other Mycobacterial Infections in Livestock and Wildlife in Tanzania

There is a large variety of non-tuberculous mycobacterial (NTM) species, some of which have the potential of causing diseases in animals and humans (Mwikuma et al. 2015; Durnez et al. 2009; Mdegela et al. 2004; Hoefsloot et al. 2013; Asiimwe et al. 2013; Makondo et al. 2014; Mwikuma et al. 2015; Shojaei et al. 2011). In Tanzania, isolation from indigenous cattle and wildlife species rendered a large diversity of NTM species including *M. simiae, M. confluentis, M. neoaurum, M. nonchromogenicum, M. terrae, M. thermoresistibile, M. genavense, M. gilvum, M. intermedium, M. poriferae, M. spaghni, M. kansasii, M. gastri, M. indicus pranii, M. hiberniae, M. engbaekii, M. septicum, M. arupense, M. peregrinum, M. moriokaense, M. palustre, M. flavescens, M. goodii, M. gordonae, M. smegmatis, M. fortuitum, M. phlei*, and *M. avium intracellulare* (Kazwala et al. 1998; Mdegela et al. 2004; Makondo et al. 2014; Katale et al. 2014). The isolation of NTMs from tuberculosis-like lesions in the absence of members of the MTC calls for further research to elucidate their actual role as a cause of disease.

Mycobacterium intracellulare is known to cause disease, not only in immunocompromised but also in immunocompetent humans (Han et al. 2005). *Mycobacterium intracellulare* has been isolated from humans with both pulmonary and extrapulmonary forms of tuberculosis and from cattle and wildlife (Han et al. 2005; Nishiuchi et al. 2007). *Mycobacterium kansasii* is an exceedingly rare pathogen and has been isolated from lesions in the respiratory tract and associated lymph nodes in animals (Waters et al. 2006).

In summary, differences in the distribution of the various NTM species may partly determine the frequency and clinical manifestations of pulmonary disease in humans in various geographical locations (Hoefsloot et al. 2013). In Uganda, *M. fortuitum* was the most common cause of infection by the NTMs in human infants and adolescents in the rural areas (Asiimwe et al. 2013).

21.5 Challenges for the Control of BTB in Tanzania

Implementation of BTB control and eradication programs in most developing countries has been impeded by the lack of application of policy, the lack of financial resources (Katale et al. 2012), and more recently, the presence of wildlife reservoirs (Cosivi et al. 1998). In Africa, only seven countries list BTB as a notifiable disease and apply the global test-and-slaughter policy as part of their control measures (Cosivi et al. 1998). In Tanzania, BTB is not actively controlled, as the disease is not considered to be of sufficient economic importance, and because of a lack of human and financial resources.

The control of BTB in Tanzania is particularly complex given the extensive pastoral husbandry system in the country where cattle intermingle extensively and are moved extensively in search of water and grazing. It is further complicated by the presence of maintenance hosts of *M. bovis* in wildlife (Fitzgerald and Kaneene 2013). Minimizing contact between wildlife and livestock should thus be attempted if successful management of the disease is to be achieved (Katale et al. 2012).

Given the complexity of dealing with BTB in a multi-host system internationally, there is a focus on the development of new *M. bovis* vaccines with the hope of augmenting the current, often inefficient and excessively costly, control measures. Such a vaccine, however, is yet to be developed, and those currently available are not effective for use in animals (Morrison et al. 2004).

Bovine tuberculosis can be eradicated in Tanzania only if mandatory testing of cattle is implemented, animal movement is restricted, farmers are compensated for all positive reactors that are culled, and skilled technical personnel and adequate financial resources are made available (Theon et al. 2006).

21.6 Conclusion

In Tanzania, control of BTB is impeded by the lack of a clear policy to control the disease, traditional pastoral husbandry systems, and more recently, the presence of wildlife reservoirs.

Despite the perceived current low prevalence of BTB in the different regions of Tanzania, continuous monitoring of the disease in both livestock and wildlife is of paramount importance. The presence of the large diversity of *M. bovis* strains in indigenous cattle indicates uncontrolled movement and, possibly, intermingling with wildlife, and it is essential that educational and public awareness programs be given priority in areas where livestock are in close proximity to wildlife. Moreover, high-risk areas should be identified to allow targeted interventions, for better management of the control program, and the rational application of scarce resources. Effective control will need collaboration between government institutions, herdsmen, and other stakeholders, and it is important that all concerned embrace any future action plan to control the disease.

References

Ameni G, Erkihun A (2007) Bovine tuberculosis on small-scale dairy farms in Adama Town, Central Ethiopia, and farmer awareness of the disease. Rev Sci Tech 26(3):711–719

Ameni G, Amenu K, Tibbo M (2002) Prevalence and risk factor assessment in cattle and cattle owners in Wuchale-Jida District, Central Ethiopia. Int J Appl Res Vet Med 1(1):17–26

Asiimwe BB, Bagyenzi GB, Ssengooba W et al (2013) Species and genotypic diversity of non-tuberculous mycobacteria isolated from children investigated for pulmonary tuberculosis in rural Uganda. BMC Infect Dis 13:88

Berg S, Garcia-Pelayo MC, Muller B et al (2011) African 2, a clonal complex of *Mycobacterium bovis* epidemiologically important in East Africa. J Bacteriol 193:670–678

Brosch R, Gordon S, Marmiesse M et al (2002) A new evolutionary scenario for the *Mycobacterium tuberculosis* complex. Proc Natl Acad Sci USA 99:3684–3689

Cadmus S, Agada C, Onoja I et al (2010) Risk factors associated with bovine tuberculosis in some selected herds in Nigeria. Trop Anim Health Prod 42(4):547–549

Cleaveland S, Mlengeya T, Kazwala RR et al (2005) Tuberculosis in Tanzanian wildlife. J Wildl Dis 41:446–453

Cleaveland S, Shaw DJ, Mfinanga SG et al (2007) *Mycobacterium bovis* in rural Tanzania: risk factors for infection in human and cattle populations. Tuberculosis 87:30–43

Cosivi O, Grange J, Daborn C et al (1998) Zoonotic tuberculosis due to *Mycobacterium bovis* in developing countries. Emerg Infect Dis 4:59–70

de Garine-Wichatitsky M, Fritz H, Chaminuka P et al (2013) Consequences of animals crossing the edges of transfrontier parks. In: Andersson JA, de Garine-Wichatitsky M, Cumming DHM et al (eds) Transfrontier conservation areas: people living on the edge. Earthscan, London, pp 137–162

de Lisle GW, Mackintosh CG, Bengis RG (2001) *Mycobacterium bovis* in free-living and captive wildlife, including farmed deer. Rev Sci Tech 20:86–111

Durnez L, Sadiki H, Katakweba A et al (2009) The prevalence of Mycobacterium bovis infection and atypical mycobacterioses in cattle in and around Morogoro, Tanzania. Trop Anim Health Prod 41(8):1653–1659

Fitzgerald SD, Kaneene JB (2013) Wildlife reservoirs of bovine tuberculosis worldwide: hosts, pathology, surveillance, and control. Vet Pathol 50(3):488–499

Garnier T, Eiglmeier K, Camus J-C et al (2003) The complete genome sequence of *Mycobacterium bovis*. Proc Natl Acad Sci USA 100:7877–7882

Han XY, Tarrand JJ, Infante R et al (2005) Clinical significance and epidemiologic analyses of *Mycobacterium avium* and *Mycobacterium intracellulare* among patients without AIDS. J Clin Microbiol 43:4407–4412

Hoefsloot W, van Ingen J, Andrejak C et al (2013) The geographic diversity of nontuberculous mycobacteria isolated from pulmonary samples: an NTM-NET collaborative study. Eur Respir J 42(6):1604–1613

Kaneene JB, Bruning-Fann CS, Granger LM et al (2002) Environmental and farm management factors associated with tuberculosis on cattle farms in northeastern Michigan. J Am Vet Med Assoc 221(6):837–842

Katale BZ, Mbugi EV, Kendal S et al (2012) Bovine tuberculosis at the human-livestock-wildlife interface: is it a public health problem in Tanzania? A review. Onderstepoort J Vet Res 79 (2):463. https://doi.org/10.4102/ojvr.v79i2.463

Katale BZ, Mbugi EV, Karimuribo ED et al (2013) Prevalence and risk factors for infection of bovine tuberculosis in indigenous cattle in the Serengeti ecosystem, Tanzania. BMC Vet Res 9:267

Katale BZ, Mbugi EV, Botha L et al (2014) Species diversity of non-tuberculous mycobacteria isolated from humans, livestock and wildlife in the Serengeti ecosystem, Tanzania. BMC Infect Dis 14:616

Katale BZ, Mbugi EV, Siame K et al (2015) Isolation and potential for transmission of *Mycobacterium bovis* at human-livestock-wildlife interface of the Serengeti ecosystem, northern Tanzania. Transbound Emerg Dis 64(3):815–825. https://doi.org/10.1111/tbed.12445

Kazwala RR (1996) Molecular epidemiology of bovine tuberculosis in Tanzania. PhD thesis, University of Edinburgh, Edinburgh

Kazwala RR, Daborn CJ, Kusiluka LJ et al (1998) Isolation of *Mycobacterium* species from raw milk of pastoral cattle of the Southern Highlands of Tanzania. Trop Anim Health Prod 30:233–239

Kazwala RR, Kambarage DM, Daborn CJ et al (2001) Risk factors associated with the occurrence of bovine tuberculosis in cattle in the Southern Highlands of Tanzania. Vet Res Commun 25:609–614

Kazwala RR, Kusiluka LJM, Sinclair K et al (2006) The molecular epidemiology of *Mycobacterium bovis* infections in Tanzania. Vet Microbiol 112:201–210

Makondo EZ (2013) Mycobacterial infection in the livestock and wildlife interface of Katavi Rukwa ecosystem, Tanzania. International congress on bacteriology and infectious diseases, OMICS Group, Baltimore, USA, November 20–22

Makondo ZE, Kazwala RR, Mwakapuja RS et al (2014) Nontuberculous mycobacterial infections in Katavi Rukwa ecosystems. J Agr Sci Tech B 4:215–223

Markham AEG (1952) Bovine tuberculosis in the Southern Highlands Province of Tanganyika. PhD thesis, University of London

Mdegela RH, Kusiluka LJM, Kapaga AM et al (2004) Prevalence and determinants of mastitis and milk-borne zoonoses in smallholder dairy farming sector in Kibaha and Morogoro districts in eastern Tanzania. Vet J Series B 51:123–128

Mfinanga SGM, Morkve O, Kazwala RR et al (2004) Mycobacterial adenitis: role of *Mycobacterium bovis*, non-tuberculous mycobacteria, HIV infection, and risk factors in Arusha, Tanzania. East Afr Med J 81:171–178

Morrison WI, Bourne FJ, Cox DR et al (2004) Potential use of vaccination in cattle or badgers to control bovine tuberculosis. Dev Biol (Basel) 119:351–359

Munyeme M, Muma JB, Skjerve E et al (2008) Risk factors associated with bovine tuberculosis in traditional cattle of the livestock/wildlife interface areas in the Kafue basin of Zambia. Prev Vet Med 3(4):317–328

Mwakapuja RS, Makondo ZE, Malakalinga J et al (2013a) Prevalence and significant geospatial clusters of bovine tuberculosis infection at livestock–wildlife interface ecosystem in eastern Tanzania. Trop Anim Health Prod 44(8). https://doi.org/10.1007/s11250-11013-10350–11252

Mwakapuja RS, Makondo ZE, Malakalinga J et al (2013b) Molecular characterization of *Mycobacterium bovis* isolates from pastoral livestock at Mikumi-Selous ecosystem in the eastern Tanzania. Tubercle 93:668–674

Mwikuma G, Kwenda G, Hang'ombe BM et al (2015) Molecular identification of non-tuberculous mycobacteria isolated from clinical specimens in Zambia. Ann Clin Microbiol Antimicrob 14 (1). https://doi.org/10.1186/s12941-014-0059-8

Nishiuchi Y, Maekura R, Kitada S et al (2007) The recovery of *Mycobacterium avium-intracellulare* complex (MAC) from the residential bathrooms of patients with pulmonary MAC. Clin Infect Dis 45:347–351

Nugent G (2011) Maintenance, spillover and spillback transmission of bovine tuberculosis in multi-host wildlife complexes: a New Zealand case study. Vet Microbiol 151:34–42

Palmer MV, Thacker TC, Waters RW et al (2012) *Mycobacterium bovis*: a model pathogen at the interface of livestock, wildlife, and humans. Vet Med Int 236205, pp 17

Pusic I, Milićević V, Savić S et al (2009) A preliminary trial to evaluate the gamma interferon assay for the detection of tuberculosis in cattle under local conditions in Serbia. Lucrări Stiinłifice Medicină Veterinară 42:125–130

Renwick A, White P, Bengis R (2007) Bovine tuberculosis in southern African wildlife: a multi-species host-pathogen system. A review. Epidemiol Infect 135:529–540

Shirima GM, Kazwala RR, Kambarage DM (2003) Prevalence of bovine tuberculosis in cattle in different farming systems in Tanzania. Prev Vet Med 57:167–172

Shojaei H, Heidarieh P, Hashemi A et al (2011) Species identification of neglected nontuberculous mycobacteria in a developing country. Jpn J Infect Dis 64(4):265–271

Smith NH, Berg S, Dale J et al (2011) European 1: a globally important clonal complex of *Mycobacterium bovis*. Infec Genet Evol 11:1340–1351

Theon CO, Steele JH, Kaneene JB (2006) Zoonotic tuberculosis: *Mycobacterium bovis* and other pathogenic mycobacteria, 2nd edn. Blackwell, Chichester, p 338

Waters WR, Palmer MV, Thacker TC et al (2006) Immune responses to defined antigens of *Mycobacterium bovis* in cattle experimentally infected with *Mycobacterium kansasii*. Clin Vaccine Immunol 13(6):611–619

Chapter 22
Holes and Patches: An Account of Tuberculosis Caused by *Mycobacterium bovis* in Uganda

A. Muwonge, L. Nyakarahuka, W. Ssengooba, J. Oloya, F. Olea-Popelka, and C. Kankya

22.1 Geography, Demography, and Livestock Systems in Uganda

Uganda is located on an East African plateau, 1100 m above sea level (Bakama 2010). As a land-locked country, it is geopolitically in the center of East Africa, a position that makes it vulnerable to the challenges of livestock and human disease management, and disease surveillance and control. Being neighbors with the Democratic Republic of Congo and South Sudan that have endured years of civil unrest means that it has to contend with the influx of refugees (Ford 2007; Hovil 2007) and their disease-ridden livestock as they cross the increasingly porous international borders. This phenomenon is not limited to the borders with these

A. Muwonge (✉)
The Roslin Institute, The Dick School of Veterinary Medicine, College of Medicine and Veterinary Medicine, University of Edinburgh, Easter Bush, Edinburgh, UK
e-mail: adrian.muwonge@roslin.ed.ac.uk

L. Nyakarahuka
Department of Public Health and Preventive Medicine, College of Veterinary Medicine, Animal Resources & Biosecurity, Makerere University, Kampala, Uganda

W. Ssengooba · C. Kankya
Mycobacteriology Laboratory, Department of Medical Microbiology, Makerere University College of Health Sciences, Kampala, Uganda

J. Oloya
Department of Epidemiology and Biostatistics/Population Health, College of Public Health, 132 Coverdell Center, University of Georgia, Athens, GA, USA
e-mail: joloya@uga.edu

F. Olea-Popelka
Department of Clinical Sciences and Mycobacteria Research Laboratories (MRL), College of Veterinary Medicine and Biomedical Sciences, Colorado State University, Fort Collins, CO, USA
e-mail: francisco.olea-popelka@colostate.edu

A. B. Dibaba et al. (eds.), *Tuberculosis in Animals: An African Perspective*,
https://doi.org/10.1007/978-3-030-18690-6_22

two countries, as reports also indicate the presence of BTB in rustled cattle at the border between Kenya and Uganda (Ford 2007; Hovil 2007).

Uganda has a tropical, two-season climate, characterized by bimodal wet and dry seasons (Huxley 1965), and it has access to abundant sources of fresh water especially from Lake Victoria that is connected to the Mediterranean Sea by the Nile River. The northeastern part of the country though tends to have prolonged annual dry spells. The moderate temperature, adequate precipitation, and fertile loam soils are what make Uganda the food basket of Africa (Huxley 1965; WFO 2012). These attributes not only ensured a steady growth in the agricultural sector, but they are also fueling a human and animal population explosion. The rapid human population growth (~3% annually; currently 37 million) in Uganda, combined with climatic and environmental changes, is increasing the extent of the human/livestock/wildlife interface. This rapid growth has, in turn, also greatly transformed the livestock farming systems in Uganda.

Cattle, mostly reared in the Ugandan cattle corridor (UCC), are an important component of the county's agricultural industry (WFO 2012: Oloya et al. 2007a). The UCC is a diagonal expanse of land spanning the country from its northeastern to its southwestern corners and supports about 45% (5 million) of all the cattle in Uganda (WFO 2012). Most of the large herds of cattle in the UCC are owned by pastoralist communities that keep cattle communally, thus increasing the risk of spreading zoonotic diseases like BTB between herds (Oloya et al. 2006, 2007a).

Subsistence farming is the predominant farming system in Uganda, and it is characterized by farmers keeping a few animals within the confines of a homestead, supplemented by crop cultivation (Ebanyat et al. 2010). This close association with their animals enhances transmission of diseases like BTB (Thoen et al. 2006) and increases the zoonotic risk in general when farmers rear animals infected with zoonotic diseases (Kankya et al. 2010).

22.2 Cattle Movement Networks in Uganda

Disease dynamics are a function of host population characteristics. This effectively means that pathogens are "hitchhikers" on waves of individual and group characteristics and other events that occur in populations. It is therefore imperative to understand the disease dynamics, such as the numbers, direction, and risks associated with animal movements in developing disease control measures (Fèvre et al. 2006). Within this context, there are no empirically documented livestock movement patterns in Uganda. However, Nyakahuma and Kimezire (1995) predicted livestock movements in Uganda based on human and animal census data (Fig. 22.1), and it is clear that Uganda has a centripetal cattle movement pattern, i.e., they filter from the rural areas into urban centers where a unit of animal protein is likely to fetch a higher price (Boysen and Matthews 2012). This phenomenon has also been reported in Ethiopia where the patterns of molecular markers of *M. bovis* reflect a similar centripetal cattle movement (Firdessa et al. 2012).

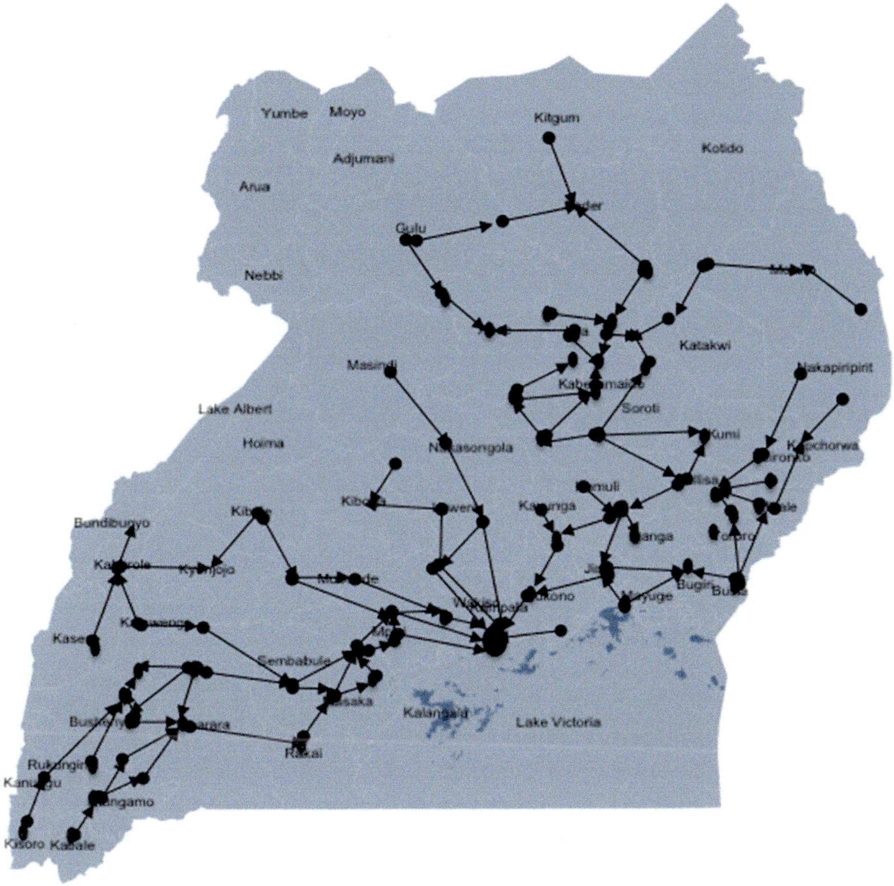

Fig. 22.1 The predicted livestock movement network in Uganda

Animal movement between herds and districts has been linked to the distribution of BTB in two UCC districts: Nakasongola and Nakapiripirit (Oloya et al. 2007a) where cattle mobility is one of the cardinal features of the system of livestock farming (Oloya et al. 2006, 2007a; Muwonge et al. 2012a). Spoligotype profiles in Uganda too reflect this feature in that SB1469 was isolated from humans, cattle, and pigs in both the Karamoja and the Mubende districts. These two districts are approximately 600 km apart, and unless this spoligotype is ubiquitous in Uganda, it is present in these two wide-apart districts because of host movements between these two districts. Other than long-distance movement, there is also evidence of short-range movement, reflecting the situation in the Nakasongola, Mubende, and Luwero districts that share similar spoligotypes (see below). Cattle rustling sporadically occurs at the Uganda borders (Gray et al. 2003; Jabs 2007), and this violent and illegal practice also causes the movement of BTB-infected cattle into the UCC and beyond (Gray et al. 2003). The molecular signature of *M. bovis* from the latter sources can only be examined once molecular data from neighboring countries are made available.

22.3 Bovine Tuberculosis in Uganda

BTB is a chronic infectious disease caused by *M. bovis* that mostly affects ungulates (OIE 2009). Nearly all warm-blooded animals can be infected by *M. bovis*, following which they develop clinical signs of BTB. In Uganda, this disease affects humans, cattle, goats, sheep, pigs, and a number of wildlife species including African buffaloes (*Syncerus caffer*), greater kudus (*Tragelaphus strepsiceros*), warthogs (*Phacochoerus africanus*), and lions (*Panthera leo*) (Muwonge et al. 2012a; Kalema-Zikusoka et al. 2005; Oloya et al. 2008). This section explores our current understanding and the status of BTB in Ugandan livestock, humans, and wildlife and its socioeconomic impact.

22.3.1 Historical Overview

The history of any mycobacterial disease is dependent on how pathogens and different disease conditions evolved over time in response to changing population structures and the environment. Human TB, leprosy, and Buruli ulcer dominated the history of diseases caused by the genus *Mycobacterium* in Uganda (Daniel 1998, 2006). The control of these diseases effectively started in the 1960s after Uganda's independence, but efforts were halted by the civil unrest that plagued the country from 1966 to 1986 (Ministry of Health 1992). Studies conducted during these turbulent times estimated the average prevalence of BTB to be 22% in cattle (Nyakahuma and Kimezire 1995). For a disease known to have been present in Uganda since the early 1900s (Opuda-Asibo 1995), it is only recently that the holes in our knowledge are slowly being patched by comprehensive epidemiological and socio-anthropological studies. The majority of these studies have shown that BTB is most prevalent in the UCC (Oloya et al. 2007a; Inangolet et al. 2008; Muwonge et al. 2012a; Asiimwe et al. 2009; Nasaka 2014).

22.3.2 Diagnostic Challenges

Making an accurate diagnosis is an important component of BTB control and management in animals, but, as is the case with human TB (Mwinga 2005; Sharma et al. 2012), it remains an extremely challenging process. Uganda does not have an official national BTB control program, and it is important to reflect on the following points when interpreting and/or evaluating diagnostics results especially in a resource-poor country like Uganda:

- Currently, not a single test fulfills all the criteria necessary to identify all BTB-infected animals (Strain et al. 2011),
- The effectiveness of a diagnostic test depends on the ability (dependent on the objectives) of the test to correctly identify animals at different stages of the disease (Nielsen et al. 2011), i.e., exposed, infected, infectious (present an immediate threat), and animals with clinical disease.
- Finally, the intervention that will follow and resources available influence the choice of test.

Unlike in developed countries where the life span of animals is shorter due to a younger slaughter age, in Ugandan cattle, the life span is substantially longer. This means that there is a higher likelihood of detecting infected animals at slaughter as a longer life span favors the progression of a chronic disease and the likelihood of becoming infected. In such settings, animals are expected to remain infectious for a longer duration (Strain et al. 2011; Shittu et al. 2013).

If Uganda were to adopt a comprehensive BTB control policy, the choice of tools would have to take these features into consideration. Since Uganda does not have a structured BTB control policy, the diagnostic and prevalence data of BTB presented in this section are results of academic research projects and are not based on statutorily sanctioned BTB surveillance programs and diagnostic techniques.

22.3.3 Ante-mortal and Postmortem Inspection for Bovine Tuberculosis in Livestock

Ante-mortal diagnosis of BTB is difficult due to the lack of specific diagnostic clinical signs. Due to an absence of BTB surveillance programs using the intradermal tuberculin test, most *M. bovis*-infected cattle in Uganda will live their entire life without being tested for BTB. For this reason, the first opportunity for detecting BTB is usually at municipal abattoirs during meat inspection (Oloya et al. 2006; Muwonge et al. 2012a; Asiimwe et al. 2009). Because animals are also slaughtered illegally, an unknown but probably a considerable number of BTB cases are never detected or the presence of the disease reported (Kankya et al. 2010).

At slaughter, BTB presents with nodular, granulomatous tubercles, predominantly localized in anatomical sites reflecting the route of transmission (OIE 2009). In pigs, lesions have predominately been observed in the abdominal lymph nodes (Muwonge et al. 2012a), while in cattle they occur in both pulmonary and extrapulmonary lymph nodes (Muwonge et al. 2012a; Asiimwe et al. 2009). Recent unpublished data also indicated that goats had extrapulmonary tuberculous lesions especially in the mesenteric lymph nodes from which *M. bovis* was isolated (Nalapa et al. 2015, unpublished data).

One of the greatest shortcomings of postmortem inspection is its low sensitivity for the detection of gross BTB (tuberculous) lesions, making it impossible to detect a significant number of animals infected with and containing lesions of BTB (Enøe et al. 2003; Murray 1986). According to similar studies conducted in Uganda, postmortem inspection for BTB had a sensitivity of only 37% (Muwonge et al. 2012a, b).

22.3.4 *Tuberculin Test*

Screening for BTB is both qualitative and quantitative. In Uganda, screening has been based on the comparative intradermal test (CCT) using 0.1 ml of purified bovine and avian tuberculin (purified protein derivative or PPD) injections, at a concentration of 20,000 UCT/ml. After 72 hours, an increase of >4 mm in skin thickness at the *M. bovis* tuberculin injection site, and if in excess of the increase in the thickness of the skin at the site of the *M. avium* tuberculin injection, is considered positive for BTB.

The lack of the ability to control TB due to *M. bovis* in Uganda is a hole caused by the lack of sufficient nation-wide data to provide a reliable estimate of the extent of the problem to guide the decision-making process. In the last decade, only ten reports have been published on the occurrence of tuberculosis in Uganda, most providing information about the prevalence of BTB in districts in the UCC. According to these reports, the prevalence of BTB in cattle, based on the tuberculin skin test, ranges from 1.3% to 51.6% at herd level and 0.6%–13% at slaughter (Table 22.1) (Inangolet et al. 2008). It is critical to quantify the problem countrywide to allow an equitable allocation of resources for its control.

22.4 Molecular Epidemiology of Bovine Tuberculosis in Uganda

Although the use of molecular diagnostic methods has been an integral part of human TB investigation schemes for some time, molecular typing of *M. bovis* is a fairly recent addition to the BTB diagnostic regimen in livestock in Uganda. Funded animal TB research projects collaborate with the National TB Surveillance Program, thus allowing them access to their molecular diagnostic technologies and laboratory facilities.

From the approximately 20,000 animals examined at slaughter in Ugandan abattoirs during the course of the past 10 years, 800 had lesions suggestive of BTB. Specimens from these lesions were cultured for mycobacteria, and the isolates were subsequently subjected to molecular analysis for identification of the strains. These endeavors yielded only 13 unique spoligotypes (Oloya et al. 2007a; Muwonge

Table 22.1 Summary of reports of tuberculosis caused by *M. bovis* in cattle, humans, and pigs in Uganda

Region	Host	Diagnostic tool	Sample size	Positive (%)	References
Karamoja	Cattle	Tuberculin test	1470	1.2	Inangolet et al. (2008)
Kampala	Cattle	Necropsy	3330	0.6	Asiimwe et al. (2009)
Mbarara	Cattle	Necropsy	97	13.4	Nasaka (2014)
Cattle corridor	Cattle and goats	Necropsy	1576	0.8	Nalapa et al. (2015)
Central Uganda	Cattle	Necropsy	61	31.1	Oloya et al. (2007b)
Karamoja and Nakasongola	Cattle	Tuberculin test	1864 (31[a])	45.2	
Moroto	Human	Aspirate exam	43	6.9	
Mbarara	Human	Sputum exam	70	1.4	Byarugaba et al. (2010)
Mubende	Pigs	Necropsy	997	0.3	Muwonge (2012)
Mbarara	Cattle	Tuberculin test	525	2.1	Bernard et al. (2005)

[a]Number of herds

et al. 2012a; Asiimwe et al. 2009). This spoligotype diversity is higher and lower, respectively, than that reported in Ethiopia and Tanzania. The dominant spoligotypes (SB1470 and SB0133) belong to the Af2 clonal complex (Fig. 22.2) reported to be geographically restricted to East Africa (Berg et al. 2011). Based on the currently available data, SB1470 and SB1405 appear to be restricted to the northeastern and southwestern parts of the UCC, respectively, while SB0133 has been isolated from cattle originating from both the northeast of the UCC and western part of Uganda's oil basin. One of the less frequently observed spoligotypes, (SB1469), has been isolated from humans, cattle, and pigs from the northeastern and central parts of the UCC (Oloya et al. 2007a; Asiimwe et al. 2009). It is important to note that with the limited data available, the molecular epidemiological picture of *M. bovis* in Uganda has a lot of holes to be patched. Nonetheless, the presence of identical spoligotypes in districts that are at a great distance from each other suggests the following:

- Some of the *M. bovis* spoligotypes are ubiquitous in the UCC.
- The occurrence of identical spoligotypes in districts separated by large and/or short distances suggests the existence of an uncontrolled animal movement network in Uganda (Fig. 22.3).
- The distribution of spoligotypes, which reflects the inferred cattle movement network (Fig. 22.1), corroborates this assumption.
- If verified, the animal movement network will provide a critical base for targeted infectious disease control measures.

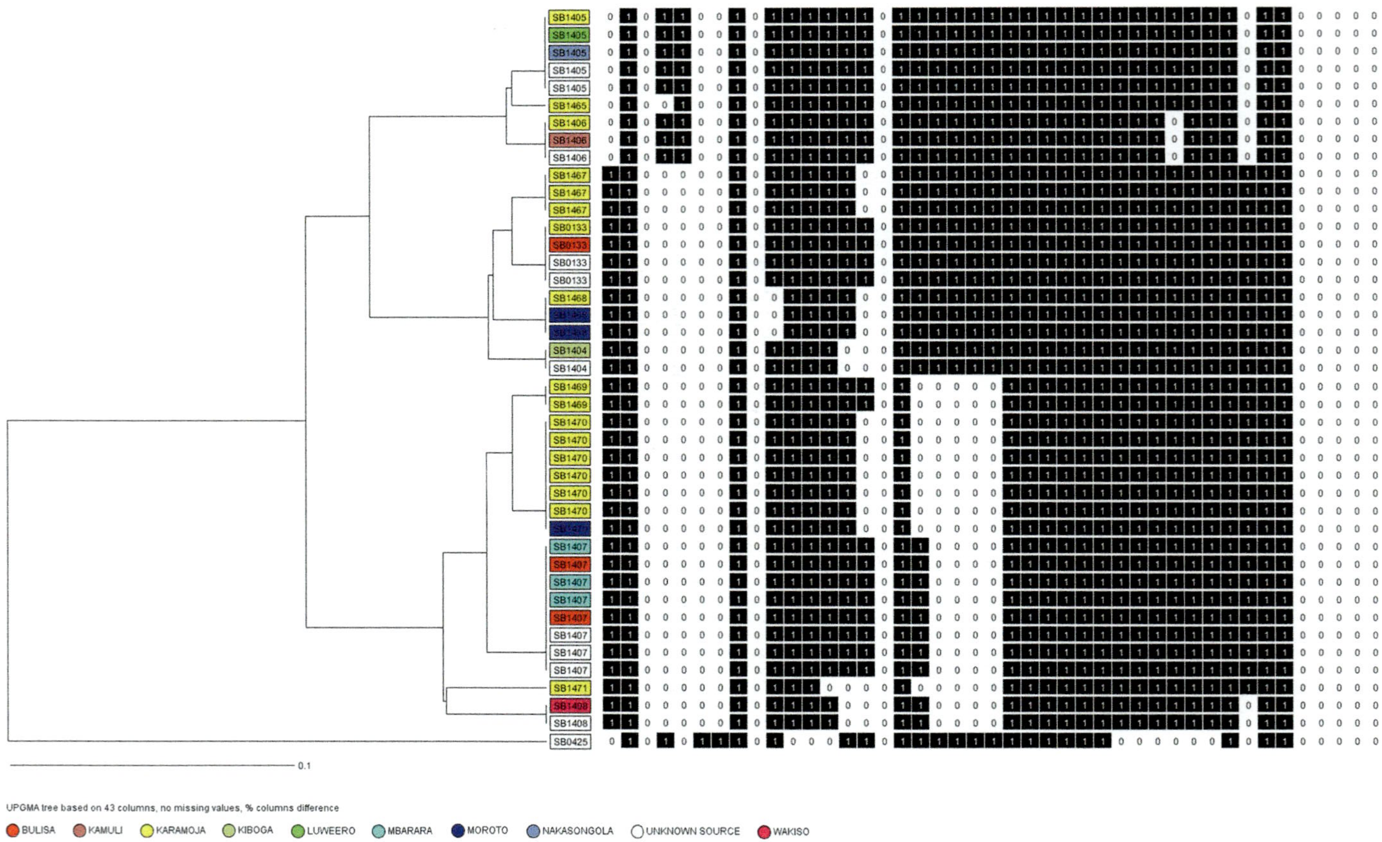

Fig. 22.2 Spoligotypes isolated in Uganda during the past 13 years. The dendrogram was generated using RIDOM ver 1.2. There are two groups of isolates in Uganda: the majority belong to the Af2 clonal complex, characterized by the loss of spacer 3–7; the rest belong to a group yet to be characterized

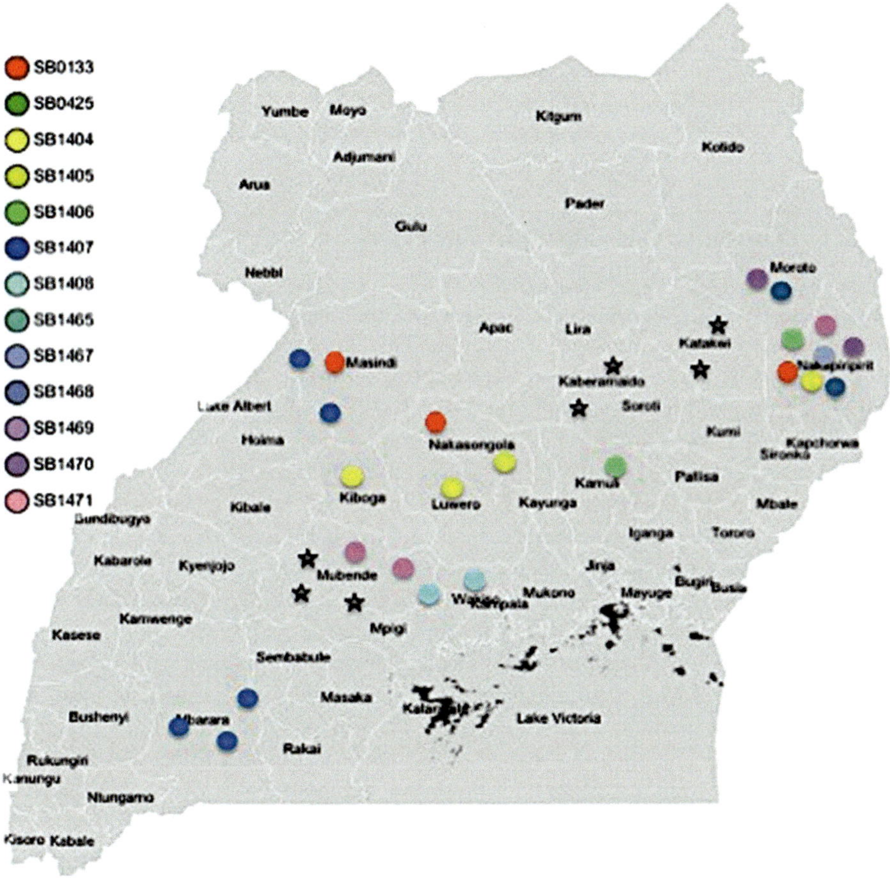

Fig. 22.3 Spatial distribution of the spoligotypes recovered in Uganda in the past 13 years. The black stars represent isolates that are not yet typed (Nalapa et al. 2015, unpublished data)

22.5 Tuberculosis Caused by *M. bovis* in Humans in Uganda

In humans, TB is mostly caused by *Mycobacterium tuberculosis*; however, *M. bovis* remains an important pathogen capable of infecting and causing clinical disease (zoonotic TB) in humans (See Chap. 3). The lack of routine surveillance for *M. bovis* as the causal agent of human TB in most of the developing world (Cosivi et al. 1998; Müller et al. 2013) represents an important challenge to our understanding of the true global burden of zoonotic TB. This is also the case for Uganda, in which the systemic lack of surveillance for zoonotic TB most likely results in underestimating the true burden of this disease. From a public health perspective, defining the role of *M. bovis* (and environmental mycobacteria) as a causal agent of human TB in developing countries such as Uganda is critical, but it has become increasingly difficult because of financial and technical constraints and disregard of its importance to obtain the required information by the governmental regulatory authorities.

22.5.1 Diagnosis of Zoonotic Tuberculosis

Direct microscopic examination of sputum specimen is the most widely used method for TB screening in humans in Uganda. This approach presents an important obstacle because direct microscopy does not differentiate between *M. tuberculosis* and *M. bovis*. Additionally, zoonotic TB mostly presents as extrapulmonary TB (Cosivi et al. 1998; Dürr et al. 2013). As only sputum samples are customarily examined, cases with extrapulmonary lesions are not detected.

The correct identification of *M. bovis* as a causal agent of TB in humans is important especially in rural communities in which BTB is endemic in the local cattle populations. In such areas, sociocultural practices create close interaction between livestock and humans, thus increasing the probability of direct and indirect transmission of *M. bovis* from livestock (and their products) to humans.

From a public health perspective, it is critical to identify patients with TB infected with *M. bovis*, because it differs from that of *M. tuberculosis* by being naturally resistant to pyrazinamide. Thus, zoonotic TB patients require longer treatment regimens and are inadequately treated if *M. bovis* is not identified. Additionally, because of the attrition that occurs during protracted treatment periods, antimicrobial resistance is more likely to develop in *M. bovis* (Dürr et al. 2013).

The epidemiology and transmission dynamics of *M. bovis* to humans differ considerably from *M. tuberculosis*, which is mostly transmitted by spread of aerosol droplets during coughing and sneezing. These differences ought to be reflected in the diagnostic regimen used, but unfortunately in Uganda, sensitive and specific diagnostic tests to correctly identify *M. bovis* as the causal agent of TB in humans are not readily available in rural areas. It is therefore not surprising that most of the reported zoonotic TB cases in Uganda are the results of individual research projects (Oloya et al. 2008; Nasaka 2014) rather than of the national TB surveillance programs. These reports showed that the prevalence of zoonotic TB ranges from 1.4% to 6.9% in various parts of the UCC. There have been efforts to patch these holes (Bretzel et al. 1999), but to acquire accurate and more comprehensive estimates of the true burden of zoonotic TB in Uganda, a holistic One Health approach, involving the Ministries of Health and Agriculture, Animal Industry and Fisheries, the National TB program, National Animal Disease Diagnostics, and the National Epidemiology Centre, is needed to conduct adequate surveillance programs using adequate diagnostic techniques.

22.5.2 *Mycobacterial Culture*

Mycobacterial culture is performed in four laboratories, three of which are located in the capital city, Kampala. These include the National TB Program Laboratory (under the Ministry of Health), the laboratory at the Makerere University Teaching Hospital, and the HIV/AIDS Research-Oriented TB Culture Laboratory (Everett et al. 2010; Ssengooba et al. 2015). This urban localization of the sophisticated diagnostic infrastructure means that it is out of reach of rural communities, mainly because of the cost implications related to accessibility. The fourth TB culture laboratory is in the western part of the country, at the Mbarara University Teaching Hospital, which operates under the same remit as the three laboratories in Kampala.

22.5.3 *Molecular Typing Tools*

Molecular typing is the process by which different types of organisms within a species are identified (Sabat et al. 2013). Spacer Oligonucleotide typing (spoligotyping) and MIRU-VNTR are the two commonly used molecular methods to type mycobacteria in Uganda. The former is reported to be highly reproducible with results that allow inter-laboratory comparison, but it has low discriminatory power, and it can produce homoplasy (Rodriguez-Campos et al. 2011). In Uganda, it is used to differentiate between members of the MTC including *M. bovis* (Asiimwe et al. 2009) and for predicting TB transmission patterns (Lukoye et al. 2014). MIRU-VNTR analysis was introduced in Uganda in 2010 and is used mainly to detect multiple infection and transmission patterns (Dickman et al. 2010). Although rarely used now, IS6110 restriction fragment length polymorphism (IS6110 RFLP) has been used in Uganda for the last 15 years, and there is a considerable amount of molecular data stored in this format (Joloba et al. 2000; Niemann et al. 2002). The National TB Program Laboratory and the Mycobacteriology Diagnostic Laboratory in Kampala have also recently acquired sequencing machines that are likely to change the landscape of molecular typing and epidemiology because of the access to the high-throughput data generated by this technique.

22.6 Environmental Non-tuberculous Mycobacteria (NTMs)

The ubiquitous and opportunistic nature of some mycobacterial species in the Ugandan environment has further complicated the diagnostic and therapeutic picture of *Mycobacterium* as a genus. It is therefore imperative to include a brief account of non-tuberculous mycobacteria (NTM) in this chapter, because NTMs have recently gained attention owing to their increased rates of isolation in livestock and human populations sharing the same environment (Oloya et al. 2007a). Due to their ubiquitous presence in the environment, exposure to NTMs is extremely common. Nevertheless, as they can colonize the respiratory tract without causing disease, the finding of an NTM in respiratory secretions is not an indication of active disease (Falkinham 2009). Identification of an NTM is, however, important because microscopy, the mainstay of TB diagnosis in Uganda, cannot differentiate between the members of the *M. tuberculosis* complex and NTMs, and detecting them may lead to false positives.

The common assumption in Uganda is that most individuals presenting with pulmonary symptoms are infected with *M. tuberculosis*. Therefore, NTMs are not identified during routine TB diagnosis. This is mainly attributed to the limited diagnostic capacities of rural health centers and the lack of awareness along the diagnostic pipeline (Muwonge et al. 2012a, b). In the presence of HIV infections, zoonotic and infections with NTMs continue to present challenges to case management of immune-compromised individuals. This is specifically so because of the following reasons:

- A positive smear result for NTM can easily be mistaken for TB infection.
- The majority of NTMs are naturally resistant to most anti-mycobacterial drugs.
- The extrapulmonary nature of *M. bovis* infection further undermines the diagnostic value of sputum specimens.
- Differentiating a true NTM lung infection from contamination and/or colonization is difficult.

All these attributes make the treatment of TB cases increasingly challenging in Uganda, especially in rural settings.

Recent studies in Uganda have identified the *M. avium* complex (29.0%; $n = 48$), *M. intracellulare* (18.8%; $n = 48$), the *M. fortuitum peregrinum* complex (25.0%; $n = 48$), *M. gordonae* (10.4%; $n = 48$), *M. nonchromogenicum* (10.4%; $n = 48$), *M. engbaekii* (4.2%; $n = 48$), and *M. kubicae* (4.2%; $n = 48$) as the most commonly isolated mycobacterial species in human–animal ecosystems. The isolation of NTMs in the cattle corridor ecosystem of the Mubende and Nakasongola areas showed seasonal variations, with the highest number of isolates originating from the water sources in November and December (Kankya et al. 2011). The isolation of NTMs in the ecosystem was also directly linked to cattle-keeping activities, open or natural water sources such as valley dams, and the presence of wildlife, especially antelopes. Studies further confirmed that these organisms were not only found in the environment but were also infecting communities and their cattle (Table 22.2) (Oloya et al. 2006, 2008). In the Karamoja District, the NTMs recovered from human and cattle shared identical profiles; i.e., *M. avium* complex (41.7%), *M. bovis* (12.5%), *M. intracellulare* (8.3%), and unidentified species of mycobacteria (4.7%) were isolated from humans, whereas *M. avium* sub. sp. *hominissuis* (8.1%; $n = 37$), *M. intracellulare* (2.7%; $n = 37$), and *M. avium* (2.7%; $n = 37$) were isolated from slaughtered cattle. Incidentally, multi-species infections (5.4%; $n = 37$) were also observed in the region. This means that both tuberculous and non-tuberculous mycobacterial species were recovered from the same tuberculous lesion (Oloya et al. 2007a).

Table 22.2 Non-tuberculous mycobacterial species commonly isolated in Uganda

NTM species	Source			References
	Environment	Livestock	Human	
M. intracellulare	9/48	–	1/37	Kankya et al. (2011)
M. fortuitum peregrinum	10/48	2/48	–	Kankya et al. (2011)
M. avium subsp. *avium*	–	6/143	1/37	Oloya et al. (2008) and Muwonge et al. (2014)
M. avium subsp. *hominissuis*	–	21/143	3/37	Oloya et al. (2008) and Muwonge et al. (2014)
M. gordonae	5/48	1/143	–	Muwonge (2012)
M. nonchromogenicum	5/48	–	–	Kankya et al. (2011)
M. senuense	1/48	22/143	–	Muwonge et al. (2012a) and Kankya et al. (2011)
M. terrae	1/48	8/143	–	Muwonge et al. (2012a) and Kankya et al. (2011)
M. asiaticum	–	7/143	–	Muwonge et al. (2012a)
M. chelonae	–	2/143	–	Muwonge et al. (2012a)
M. simiae	1/48	1/143	1/299	Kankya et al. (2011) and Ssali et al. (1998)

Insufficient surveillance data of zoonotic TB in Uganda are attributed mainly to inadequate resource allocation and the minor attempts to identify such patients. To patch these holes, there is a need to review the current protocol of sputum smear mycobacteriology and chest x-ray as diagnostic criteria for TB in rural communities that keep livestock and use open water sources. There is a further need for additional diagnostic protocols for pulmonary NTM and BTB and increasing the awareness of the health workers about the appropriate diagnostic tools and managing NTM and *M. bovis* infections in rural health centers.

22.7 Socioeconomic Consequences of BTB

This is yet another critical aspect of bovine and zoonotic TB in Uganda that has not been fully explored. Socio-anthropological and socio-epidemiological factors, such as rural livelihoods and microenvironments in the context of the local communities, are key factors fuelling the emergence of infectious diseases such as BTB (Farmer 1996; Hudelson 1996).

Since community beliefs, societal norms, values, myths, and perceptions are drivers of the spread of BTB, there is more that needs to be understood about the impact of BTB in terms of income lost due to slaughterhouse condemnation or rejection of cattle or animals during cultural ceremonics such as marriage, funerals, and other social events. It has been emphasized that management of emerging diseases needs to be dynamic, systematic, and critical, with a focus on society, culture, and economy (Farmer 1996). In a report from the UCC, for example, the role of food of animal origin was ranked high, and communities were aware that consuming raw animal products such as meat and milk increases the risk of contracting zoonotic TB and other infections. Nonetheless, they still do it, as it was part of their culture (Kankya et al. 2011). Therefore, bovine and zoonotic TB control efforts focusing exclusively on individuals without involving the communities may not yield the desired results.

Wildlife is one of the major concerns of the farming communities in Uganda, as a number of wild animal species are considered to be potential reservoirs of BTB and other mycobacterial pathogens. A study conducted on mycobacterial infections (Kankya et al. 2011) gave insights into the community's perception of their perceived role in the risk that they pose for the transmission of mycobacterial disease in Uganda. A 50-year-old key informant expressed the opinion that:

> Wild animals feed on fresh food and fruits like mangoes in our garden, leaving mucus and saliva. Animals such as monkeys, chimpanzees, and baboons feed on bananas, maize, or fruit. This therefore is likely to result in transmission of infections to our community members... *Tuli bubi enu Musawu* (we are badly off here, Doctor).

The ever-shifting human–livestock–wildlife interface requires constant vigilance by communities to prevent wildlife from invading and destroying their crops and other produce and introducing zoonotic diseases (Kankya et al. 2010).

22.8 Policy Provisions for Bovine and Zoonotic Tuberculosis in Uganda

In the grand scheme of disease management and control, adequate epidemiological data underpin policy. It is therefore hardly surprising that Uganda does not have a specific policy for BTB control, nor for dealing with zoonotic TB because of the lack of sufficient information about BTB and zoonotic TB.

There are, however, regulations in the Animal Disease Act that determine how infectious diseases like BTB should be prevented and controlled (Ministry of Agriculture 2010). The irony about this act is that it instructs farmers and/or livestock handlers to isolate infected animals, but they do not have access to routine diagnostic tools. The only achievable component of this act is that of performing postmortem inspection to determine the presence of BTB in the Ugandan cattle population. This act has recently been supplemented by the revised Public Health Act of 2000 (Ministry of Agriculture 2010), which determines that only government-appointed inspectors should inspect all slaughtered animals to ensure their freedom from any form of TB. It is unfortunate that although Uganda has these acts in place, they do not contain all the requirements for BTB control as recommended by the OIE (2009). There are also challenges with traceability when a tuberculous carcass is found during abattoir inspection, as most cattle are not individually identifiable and cannot be traced to their herd of origin.

22.9 Future Challenges

The current diagnostic methods for *M. bovis* are too expensive and complicated to be financed and applied by the Ugandan animal industry, and the inability to diagnose the disease is the source of all the holes that we identified if our objective is to control disease caused by *M. bovis* at the human–animal interface in Uganda. We strongly believe that unless these matters are fully addressed, the status quo is unlikely to change in the near future.

Investment in developing less expensive, rapid, and efficient diagnostic techniques is required to improve our knowledge of the socioeconomic impact of BTB on Ugandan people and the country's economy. Such studies will help to quantify the impact of BTB on the animal industry and the Ugandan economy at large. At a micro-economic level, such studies will reveal the health impact of zoonotic TB in Uganda measured by disability-adjusted life years (DALYS) and years of potential life lost (YPLL). Such data will not only highlight the problem and force people to become involved in finding relevant community-tailored solutions, but will also provide an opportunity for African problems to be addressed, and solved, by Africans.

If the WHO's ambitious goal of elimination TB by 2035 (Mario 2014) is to be achieved, there is a dire need to determine the impact of zoonotic TB on the global TB control initiatives. Now, more than ever before, it is imperative to make a definitive diagnosis of TB in humans because, for the first time in 100 years of TB control and management, every single case of TB, regardless of the causative agent, matters. We therefore strongly believe that this requires attention and investment by both local and international stakeholders.

22.10 Conclusions and Recommendations

Tuberculosis due to *M. bovis* is underreported in Uganda, mainly due to the lack of a specific policy on surveillance, prevention, and control of the disease in the animal populations. This is likely compounded by the lack of data on the socioeconomic impact of the disease to justify financial or other forms of investment in its control. This, in turn, stems from the requirement for expensive and complicated diagnostic procedures that render attempts to investigate the status of BTB and its impact on communities and the national economy a complicated endeavor. Nonetheless, the available data show that approximately one in 15 slaughtered cattle originating from the UCC will have BTB and that one in 100 TB patients in Mbarara hospital will have zoonotic TB caused by *M. bovis*. The distribution of *M. bovis* molecular markers mirrors the inferred cattle movement pattern, an indication that uncontrolled animal movement plays an important role in the spread of BTB in Uganda.

We therefore recommend the following to resolve with these issues:

1. There is a need to invest in understanding cattle movement patterns in Uganda. This, will not only enable cost-effective surveillance of infectious diseases like BTB, but will also allow the regulatory authorities to design targeted interventions along the cattle movement network.
2. Regardless of its shortcomings, abattoir meat inspection remains the cheapest method to diagnose BTB in low-income settings (OIE 2009). Its effectiveness is, however, dramatically reduced if personnel are not well trained. It is therefore important to increase the competency of meat inspection personnel across the country with regard to postmortem inspection techniques and the ability to detect lesions consistent with those of BTB. This should decrease the risk to the public of exposure to *M. bovis* and to other zoonotic diseases.
3. To identify zoonotic TB cases, there is a need to revise the current diagnostic algorithm. In particular, history taking that captures and reflects exposure to *M. bovis* should be included in the TB diagnostic regimen. This would identify cases that are likely to be due to *M. bovis* that requires pyruvate-infused media to enhance growth when attempting to culture them. Further definitive molecular diagnostic techniques should then be used to differentiate between *M. bovis* and *M. tuberculosis* to improve the effectiveness of treating cases with zoonotic BTB and of disease management schemes.
4. There is a need to conduct impact assessments and to document the effects of BTB at the human–animal interface and on the economy of the country. Only then can decisions be made to allocate adequate resources to control and eradicate BTB and zoonotic TB in Uganda.

References

Asiimwe BB, Asiimwe J, Kallenius G et al (2009) Molecular characterisation of *Mycobacterium bovis* isolates from cattle carcases at a city slaughterhouse in Uganda. Vet Rec 164:655–658

Bakama B (2010) A contemporary geography of Uganda, 1st edn. Mkuki na Nyota, Dar es Salaam

Berg S, Garcia-Pelayo MC, Müller B et al (2011) African 2, a clonal complex of *Mycobacterium bovis* epidemiologically important in East Africa. J Bacteriol 193:670–678

Bernard F, Vincent C, Matthieu L et al (2005) Tuberculosis and brucellosis prevalence survey on dairy cattle in Mbarara milk basin (Uganda). Prev Vet Med 67:267–681

Boysen O, Matthews A (2012) The differentiated effects of food price spikes on poverty in Uganda. In: 123rd European Association of Agricultural Economists Seminar, Price volatility and farm income stabilization: modeling outcomes and assessing market and policy based responses, Dublin, Ireland, February, pp 23–24

Bretzel G, Aziz M, Wendl-Richter U et al (1999) Anti-tuberculosis drug resistance surveillance in Uganda 1996–1997. Int J Tuberc Lung Dis 3:810–815

Byarugaba F, Grimaud P, Godreuil S (2010) Risk assessment in zoonotic tuberculosis in Mbarara, the main milk basin of Uganda. Bull Anim Health Prod Afr 58:125–132

Cosivi O, Grange JM, Daborn CJ et al (1998) Zoonotic tuberculosis due to *Mycobacterium bovis* in developing countries. Emerg Infect Dis 4:59–70

Daniel TM (1998) The early history of tuberculosis in Central East Africa: insights from the clinical records of the first twenty years of Mengo Hospital and review of relevant literature. Int J Tuberc Lung Dis 2:784–790

Daniel TM (2006) The history of tuberculosis. Respir Med 100:1862–1870

Dickman KR, Nabyonga L, Kateete DP et al (2010) Detection of multiple strains of *Mycobacterium tuberculosis* using MIRU-VNTR in patients with pulmonary tuberculosis in Kampala, Uganda. BMC Infect Dis 10:349

Dürr S, Müller B, Alonso S et al (2013) Differences in primary sites of infection between zoonotic and human tuberculosis: results from a worldwide systematic review. PLoS Negl Trop Dis 7: e2399

Ebanyat P, de Ridder N, de Jager A et al (2010) Drivers of land use change and household determinants of sustainability in smallholder farming systems of eastern Uganda. Popul Environ 31:474–506

Enøe C, Christensen G, Andersen S et al (2003) The need for built-in validation of surveillance data so that changes in diagnostic performance of post-mortem meat inspection can be detected. Prev Vet Med 57:117–125

Everett CA, Davis JL, Worodria W et al (2010) Performance of LED fluorescence microscopy for the diagnosis of pulmonary tuberculosis: preliminary results from the Uganda National Reference Laboratory. Am J Respir Crit Care Med 181:A1769

Falkinham JO (2009) Surrounded by mycobacteria: nontuberculous mycobacteria in the human environment. J Appl Microbiol 107:356–367

Farmer P (1996) Social inequalities and emerging infectious diseases. Emerg Inf Dis 2:256–269

Fèvre EM, Bronsvoort BMDC, Hamilton K et al (2006) Animal movements and the spread of infectious diseases. Trends Microbiol 14:125–131

Firdessa R, Tschopp R, Wubete A et al (2012) High prevalence of bovine tuberculosis in dairy cattle in Central Ethiopia: implications for the dairy industry and public health. PLoS One 7:e52851

Ford LB (2007) Civil conflict and sleeping sickness in Africa in general and Uganda in particular. Confl Heal 1:6

Government of Uganda (2000) Public Health Act, Uganda. pp 1–17 http://www.kcca.go.ug/uploads/acts/Building%20Regulations.pdf

Gray S, Sundal M, Wiebusch B et al (2003) Cattle raiding, cultural survival, and adaptability of east African pastoralists. Curr Anthropol 44(5 Suppl):S3–S30

Hovil L (2007) Self-settled refugees in Uganda: an alternative approach to displacement? J Refug Stud 20:599–620

Hudelson P (1996) Gender differentials in tuberculosis: the role of socio-economic and cultural factors. Tuber Lung Dis 77:391–400

Huxley PA (1965) Climate and agricultural production in Uganda. Exp Agric 1:81–97

Inangolet FO, Demelash B, Oloya J et al (2008) A cross-sectional study of bovine tuberculosis in the transhumant and agro-pastoral cattle herds in the border areas of Katakwi and Moroto districts, Uganda. Trop Anim Health Prod 40:501–508

Jabs L (2007) Where two elephants meet, the grass suffers: a case study of intractable conflict in Karamoja, Uganda. Am Behav Sci 50:1498–1519

Joloba ML, Whalen CC, Cave DM et al (2000) Determination of drug susceptibility and DNA fingerprint patterns of clinical isolates of *Mycobacterium tuberculosis* from Kampala, Uganda. East Afr Med J 77:111–115

Kalema-Zikusoka G, Bengis RG, Michel AL et al (2005) A preliminary investigation of tuberculosis and other diseases in African buffalo (*Syncerus caffer*) in Queen Elizabeth National Park, Uganda. Onderstepoort J Vet Res 72:145–151

Kankya C, Muwonge A, Olet S et al (2010) Factors associated with pastoral community knowledge and occurrence of mycobacterial infections in human-animal interface areas of Nakasongola and Mubende districts, Uganda. BMC Public Health 10:471

Kankya C, Muwonge A, Djønne B et al (2011) Isolation of non-tuberculous mycobacteria from pastoral ecosystems of Uganda: public health significance. BMC Pub Health 11:320

Lukoye D, Katabazi FA, Musisi K et al (2014) The T2 *Mycobacterium tuberculosis* genotype, predominant in Kampala, Uganda, shows negative correlation with antituberculosis drug resistance. Antimicrob Agents Chemother 58:3853–3859

Mario R (2014) The end TB strategy. WHO/HTM/GTB/2015.09

Ministry of Agriculture (2010) Agricultural sector development strategy and investment plan 2010–15. Ministry of agriculture report, pp 11–16

Ministry of Health (1992) Manual of the national tuberculosis and leprosy program in Uganda for district TB/leprosy supervisors, 1st edn. Ministry of Health report, pp 1–9

Müller B, Dürr S, Alonso S et al (2013) Zoonotic *Mycobacterium bovis*-induced tuberculosis in humans. Emerg Infect Dis 19:899–908

Murray G (1986) Ante-mortem and post-mortem meat inspection: an Australian inspection service perspective. Aust Vet J 63:211–215

Muwonge A (2012) Non tuberculous mycobacteria in swine: is it a public health problem? Mycobacteria Dis 20(2):e110. https://doi.org/10.4172/2161-1068.1000e110

Muwonge A, Johansen TB, Vigdis E et al (2012a) *Mycobacterium bovis* infections in slaughter pigs in Mubende district, Uganda: a public health concern. BMC Vet Res 8:168

Muwonge A, Kankya C, Johansen TB et al (2012b) Non-tuberculous mycobacteria isolated from slaughter pigs in Mubende district, Uganda. BMC Vet Res 8:52

Muwonge A, Oloya J, Kankya C et al (2014) Molecular characterization of *Mycobacterium avium* subspecies *hominissuis* isolated from humans, cattle and pigs in the Uganda cattle corridor using VNTR analysis. Infect Genet Evol 21:184–191

Mwinga A (2005) Challenges and hope for the diagnosis of tuberculosis in infants and young children. Lancet 365(9454):97–98

Nasaka J 2014 Occurence of bovine tuberculosis in slaughtered cattle. MSc Dissertation, Department of Environment and Natural Resources Management, Makerere University, Uganda

Nielsen SS, Toft N, Gardner A (2011) Structured approach to design of diagnostic test evaluation studies for chronic progressive infections in animals. Vet Microbiol 150:115–125

Niemann S, Rüsch-Gerdes S, Joloba ML et al (2002) *Mycobacterium africanum* subtype II is associated with two distinct genotypes and is a major cause of human tuberculosis in Kampala, Uganda. J Clin Microbiol 40:3398–3405

Nyakahuma D, Kimezire M (1995) Uganda. In: Thoen C, Steel J (eds) *M. bovis* infection in animals and humans. Iowa State University Press, Ames, pp 303–330

OIE (2009) Bovine tuberculosis: The OIE terrestrial manual. http://www.oie.int/fileadmin/Home/fr/Health_standards/tahm/2.04.07_BOVINE_TB.pdf

Oloya J, Opuda-Asibo J, Djønne B et al (2006) Responses to tuberculin among Zebu cattle in the transhumance regions of Karamoja and Nakasongola district of Uganda. Trop Anim Health Prod 38:275–283

Oloya J, Kazwala R, Lund A et al (2007a) Characterisation of mycobacteria isolated from slaughter cattle in pastoral regions of Uganda. BMC Microbiol 7:95

Oloya J, Muma JB, Opuda-Asibo J et al (2007b) Risk factors for herd-level bovine-tuberculosis seropositivity in transhumant cattle in Uganda. Prev Vet Med 80:318–329

Oloya J, Opuda-Asibo J, Kazwala R et al (2008) Mycobacteria causing human cervical lymphadenitis in pastoral communities in the Karamoja region of Uganda. Epidemiol Infect 136:636–643

Opuda-Asibo J (1995) Regional and country status reports, Uganda. In: Thoen C, Steel J (eds) *Mycobacterium bovis* infection in animals and humans. Iowa State University Press, Ames, pp 299–303

Rodriguez-Campos S, Aranaz A, de Juan L et al (2011) Limitations of spoligotyping and variable-number tandem-repeat typing for molecular tracing of *Mycobacterium bovis* in a high-diversity setting. J Clin Microbiol 49:3361–3364

Sabat AJ, Budimir A, Nashev D et al (2013) Overview of molecular typing methods for outbreak detection and epidemiological surveillance. Euro Surveill 18(4):15. http://www.eurosurveillance.org/ViewArticle.aspx

Sharma SK, Mohan A, Sharma A (2012) Challenges in the diagnosis and treatment of miliary tuberculosis. Indian J Med Res 135:703–730

Shittu A, Clifton-Hadley RS, Ely ER et al (2013) Factors associated with bovine tuberculosis confirmation rates in suspect lesions found in cattle at routine slaughter in Great Britain, 2003–2008. Prev Vet Med 110:395–404

Ssali FN, Kamya MR, Wabwire-Mangen F et al (1998) A prospective study of community-acquired bloodstream infections among febrile adults admitted to Mulago Hospital in Kampala, Uganda. J Acquir Immune Defic Syndr 19:484–489

Ssengooba W, Gelderbloem SJ, Mboowa G et al (2015) Feasibility of establishing a biosafety level 3 tuberculosis culture laboratory of acceptable quality standards in a resource-limited setting: an experience from Uganda. Health Res Policy Syst 13:1–10

Strain SAJ, Mcnair J, Mcdowell SWJ (2011) Bovine tuberculosis: a review of diagnostic tests for *M. bovis* infection in badgers. Bacteriology Branch Veterinary Sciences Division Agri-Food and Biosciences Institute. http://www.bovinetb.info/docs/bovine-tuberculosis-a-review-of-diagnostic-tests-for-m-bovis-infection-in-badgers.pdf. p 45

Thoen C, Lobue P, de Kantor I (2006) The importance of *Mycobacterium bovis* as a zoonosis. Vet Microbiol 112:339–345

WFO (2012) Re-greening the Ugandan Cattle Corridor. Water Food Organisation. http://waterandfood.org/2011/10/21/re-greening-the-ugandan-cattle-corridor/. Accessed 22 Jul 2015

Chapter 23
Bovine Tuberculosis in Zambia

Sydney Malama, Musso Munyeme, and John B. Muma

23.1 Bovine Tuberculosis (BTB) in Cattle in Zambia

There are about 3 million cattle in Zambia of which the traditional farmers own approximately 80% (UKaid 2011). Livestock farming is mainly concentrated in three provinces: the Southern, Western, and Eastern Provinces (Fig. 23.1).

In rural areas, cattle are typically grazed communally on land held in trust by the local chiefs. In certain areas, farmers practice transhumant grazing, where there is a seasonal migration of livestock from the highlands to suitable grazing and watering areas. One such area is the Kafue flats, or Butwa, one of the few lacustrine wetlands in Zambia, which supports about 300,000 cattle (Munyeme et al. 2010a, b). Here, cattle are moved to the flood plains immediately after the harvest season (March to May) and returned to the highlands with the onset of the rainy season (November to December). Some herds though remain permanently on the flood plains. Three types of herding systems exist on the Kafue flats (Munyeme et al. 2008, 2009a, b):

- Village-resident herding, where herds are kept and graze in the villages
- Transhumant herding with cattle moving between villages and flood plains depending on the water levels in the plains
- Interface herding, in which herds are permanently kept on the flood plains where they are in ongoing contact with BTB-infected wildlife on the floodplains

Bovine TB is endemic in the traditional Zambian cattle sector, but its prevalence and distribution across the country is unknown. There are indications that the disease

S. Malama (✉)
Biological Sciences Department, School of Natural Sciences, University of Zambia, Lusaka, Zambia
e-mail: Sydney.malama@unza.zm

M. Munyeme · J. B. Muma
Department of Disease Control, School of Veterinary Medicine, University of Zambia, Lusaka, Zambia

A. B. Dibaba et al. (eds.), *Tuberculosis in Animals: An African Perspective*,
https://doi.org/10.1007/978-3-030-18690-6_23

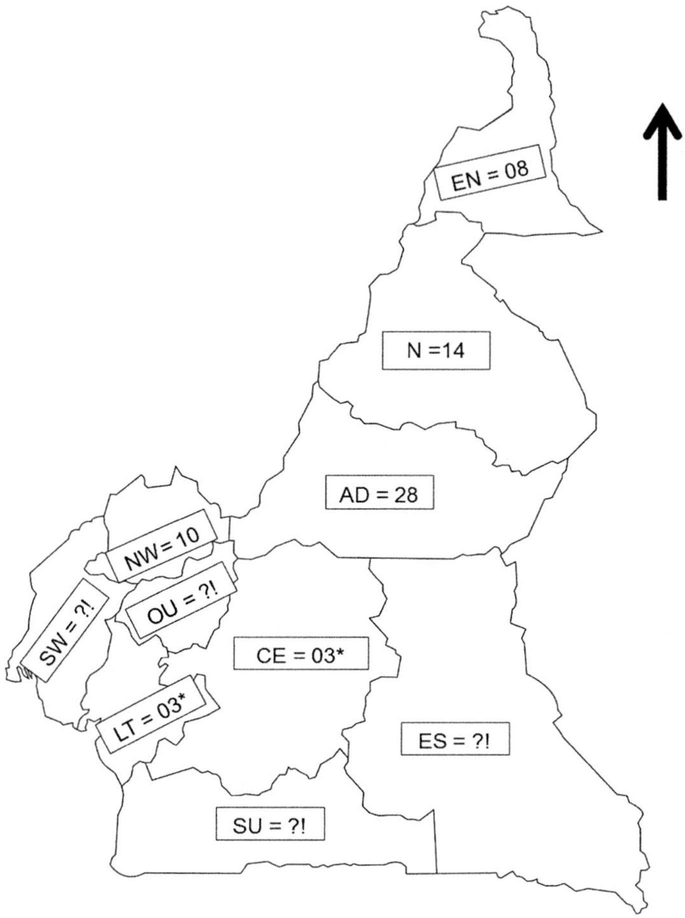

Fig. 23.1 Map of Zambia showing the Kafue basin

is widespread and that it may have high herd prevalences. As far back as 1947, herd prevalences of up to 49.8% were recorded in areas within and adjacent to the Kafue flats (Cook et al. 1996; Munyeme et al. 2009a, b). Abattoir reports from Namwala, a district located within the Kafue Basin, indicate that, based on the presence of typical tubercles detected at slaughter, 16.8% of cattle in this area suffered from BTB (Munyeme and Munang'andu 2011), and in the Western Province of Zambia, BTB was reported as one of the most common causes of beef carcass condemnation following postmortem inspection at abattoirs, signifying its economic importance (Phiri 2006). This situation is likely to be similar in most of the high cattle-producing areas in the country where similar management systems are practiced.

23.2 Bovine Tuberculosis in Zambian Wildlife

Bovine tuberculosis is known to occur in wildlife species in Zambia. Of those infected, the Kafue lechwe (*Kobus leche kafuensis*) is a maintenance host, thus sustaining the infection with spread back to cattle. The Kafue lechwe is a medium-sized, semi-aquatic antelope with a population of 44,000 and is endemic to the Kafue flats (Munyeme et al. 2010a, b) where there is a close association between livestock and wildlife as over 300,000 head of cattle are annually moved from the highlands to the flats during the dry season.

Bovine TB was diagnosed in Kafue lechwe for the first time in 1954 in the Lochinvar National Park, which is located within the Kafue Basin. It was estimated that on the southern bank of the Kafue flats, BTB was annually responsible for the death of at least 20% of the lechwe (Gallagher et al. 1972). A recent study has shown a BTB prevalence of 27.7% in the Kafue lechwe (Munyeme et al. 2010a, b). This high prevalence poses a potentially high risk of transmission of *M. bovis* to livestock, to other wildlife, and to members of the local communities (Malama et al. 2013a, b, c).

The resident population of African buffaloes (*Syncerus caffer*) in the Kafue basin appears to be free from BTB (Munang'andu et al. 2011). However, the results of the single study done to detect BTB in them are unreliable because of the small number of animals tested and also because primarily young buffaloes that were less likely to be infected with BTB and intended for translocation were tested.

The difference in the occurrence of BTB in lechwe and buffaloes sharing the same ecosystem is probably because of the differences in their interaction with cattle. Buffaloes normally do not come into direct contact with cattle as do lechwe. Within this context, cattle keepers and herdsmen in the Kafue basin report that cattle are never seen to go close to buffaloes, and vice versa, yet lechwe and cattle are regularly observed grazing together (Fig. 23.2).

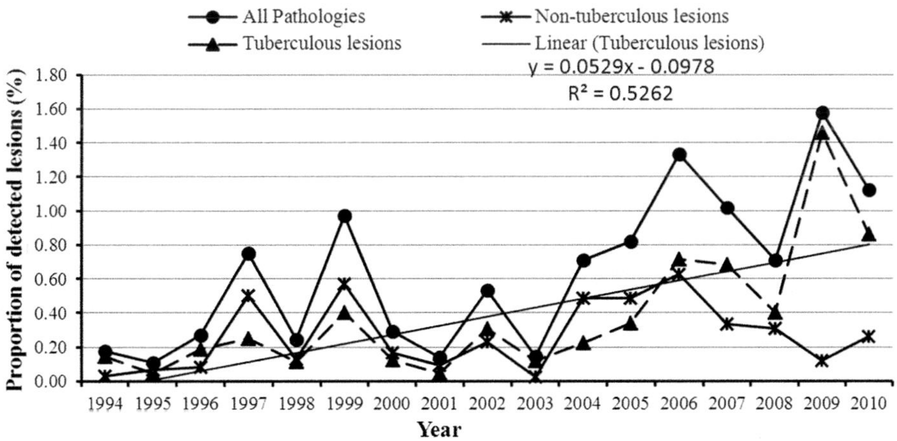

Fig. 23.2 Cattle and Kafue lechwe grazing side by side in the Kafue basin, Zambia

DNA molecular typing supports the occurrence of bidirectional transmission of the infection between cattle and Kafue lechwe in this area, as similar *M. bovis* isolates, containing spoligotype SB0120 (SIT 482), occur in Kafue lechwe and in cattle. This is the only spoligotype detected so far in Kafue lechwe in the Kafue Basin indicating that it is highly conserved in Kafue lechwe and in livestock in this area.

23.3 Zoonotic Tuberculosis in Zambia

Zambia has an estimated human population of 13 million (Malama et al. 2014), and the World Health Organization (WHO) estimates that the incidence of all forms of tuberculosis here is 444/100,000 people (Phiri 2006; UKaid 2011). The WHO reported in 1998 that 3.1% of human cases of TB worldwide are attributable to infection with *M. bovis* and that 0.4–10% of isolates from the sputum of patients in African countries could be *M. bovis*.

The situation in respect of zoonotic BTB in Zambia is unknown, as data on the prevalence of human disease due to *M. bovis* are limited, owing to the technical problems limiting the ability to identify the organism, such as the lack of trained personnel, and inadequate microbiological diagnostic laboratory facilities in the country (Cook et al. 1996; Moda et al. 1996). In Zambia, generally only Ziehl-Neelsen staining as a diagnostic technique is performed on sputum samples to detect acid-fast bacilli (AFB), and this technique cannot distinguish between the different *Mycobacterium* species (Grange et al. 1996) that may cause tuberculosis in humans. In Zambia, *M. bovis* has been isolated from human sputum samples, indicating that the infection, once thought to mainly cause extrapulmonary tuberculosis in humans, has the potential of being transmitted by droplet infection between humans (Michel et al. 2010; Malama et al. 2013a, b, c). Given this situation, zoonotic tuberculosis is now receiving increasing attention in Zambia as a public health issue that is enhanced by the risky practice by pastoral communities of consuming raw, unpasteurized milk. In the Kafue basin, consumption of raw and soured milk is a common practice (Fig. 23.3), and it poses a health risk when milk obtained from *M. bovis*-infected animals is consumed (Grange and Yates 1994).

Zambia's dairy sector consists of three types of producers: commercial farmers, emergent farmers (smallholder), and traditional/small-scale dairy farmers owning, respectively, 5%, 15%, and 80% of the cattle (UKaid 2011). Of the about 300 million liters of milk produced in the country, 25% is supplied by commercial farmers, 13% by smallholders, and 62% by traditional cattle farmers (Pandey 2014). Of this, only 95 million liters are processed and pasteurized, and the rest, mainly produced by the traditional farmers, remain in the informal market, or are retained for home consumption. This situation emphasizes the extent of the zoonotic risk posed by BTB and the importance of controlling milk-borne zoonoses such as BTB by encouraging pasteurization. In addition, physical contact with cattle and sharing shelter/space are

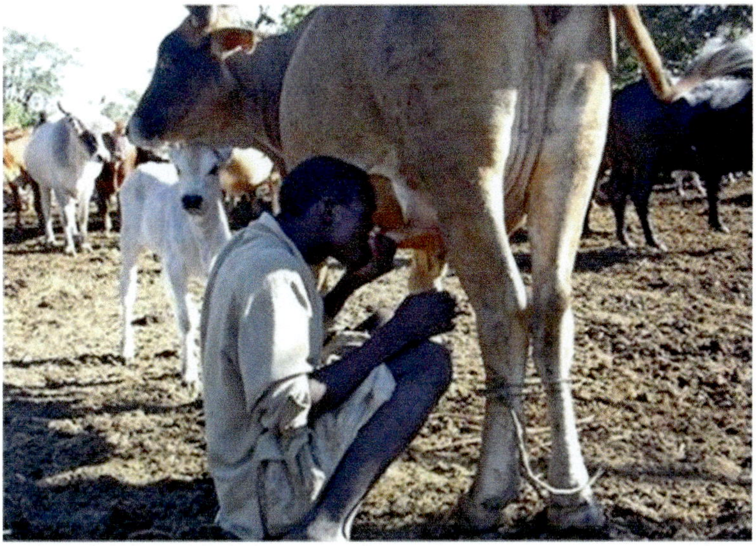

Fig. 23.3 A boy drinking milk directly from a cow's udder

other common practices of the people in this area, and this also constitutes a substantial risk of contracting the disease.

Kafue lechwe, a known maintenance host of BTB, are hunted for meat, trophies, and hides. Lechwe are the most hunted antelope for human consumption in Zambia (Siamudaala et al. 2005a), and it is estimated that about 80% of lechwe carcasses hunted for meat could be infected with BTB. Additionally, the number of lechwe poached is about 50% of the official annual hunting quota (Siamudaala et al. 2005b), and this large number of potentially BTB-infected carcasses increases the risk for humans to contract zoonotic tuberculosis. Molecular epidemiological studies conducted in the Namwala district, located in the Kafue Basin, have shown a close relationship between the *M. bovis* strains isolated from humans and from cattle. Based on spoligotyping and the 9-loci MIRU-VNTR results, *M. bovis* isolated from humans and cattle are closely clustered and highly homogenous and are circulating in both cattle and humans in the district. These findings confirm the zoonotic significance of *M. bovis* in this area (Malama et al. 2014). Isolates from other districts of Zambia are similar to those isolated in the Kafue Basin, indicating that the SB120/SIT 420 spoligotype is highly conserved in Zambia (Malama et al. 2014).

23.4 Diagnosis of Bovine Tuberculosis in Zambia

In Zambia, the diagnosis of BTB in cattle and wildlife is constrained by numerous challenges. The comparative intradermal tuberculin test (CCT) (Table 23.1) (Munyeme et al. 2009a, b), as recommended by the World Organisation for Animal

Table 23.1 Herd prevalence of BTB in cattle determined by cross-sectional studies around the Kafue basin based on comparative intradermal tuberculin testing (CCT)

Study site	Prevalence (%)	Method	Reference
Blue lagoon	48	CCT	Munyeme et al. (2009a, b)
Lochinvar	43	CCT	Munyeme et al. (2009a, b)
Kazungula	4	CCT	Munyeme et al. (2009a, b)
Monze	33	CCT	Cook et al. (1996)
Livingstone	1	Necropsy	Dept of Veterinary and Tsetse Control (1957)
Mazabuka	5	Necropsy	Dept of Veterinary and Tsetse Control (1957)
Lusaka	2	Necropsy	Dept of Veterinary and Tsetse Control (1957)
Namwala	17	Necropsy	Dept of Veterinary and Tsetse Control (1957)

Table 23.2 Area-based prevalence of BTB using postmortem examination, Ziehl-Neelsen staining, and culture results from Kafue lechwe tissue samples ($n = 119$) (Munyeme et al. 2010a, b and Malama et al. 2013a, b, c)

Study area	Test method	Prevalence (%)
Lochinvar/blue lagoon	Necropsy	24.34
Lochinvar/blue lagoon	Ziehl-Neelsen	17.6
Lochinvar/blue lagoon	Culture	27.7

Health (OIE), is recommended for making an antemortal diagnosis of tuberculosis, but it is not often used in Zambia. Normally, when it is done, employees of the Veterinary Services perform the test. The limited use of the test is due to the reluctance of farmers to allow it to be done since they are not compensated for positive reactors that must be culled. Implementation of a test-and-slaughter policy is not feasible because of this reluctance, and the difficulties of controlling BTB in rural, nomadic cattle pose additional challenges.

Farmers supplying milk to commercial processing companies must screen their animals annually for BTB by using the tuberculin test if they are to have their milk accepted by these companies and they are thus forced to comply. In cattle owned by traditional farmers, BTB is mainly diagnosed by postmortem exanimation to detect the presence of gross lesions compatible with those of BTB in the lungs and/or associated lymph nodes. Whole carcass or partial condemnation is regularly applied depending on the distribution of the lesions. When the lesions are localized, carcasses are according to standard regulations declared fit for human consumption once the organs containing gross lesions have been removed (Gracey et al. 1999).

Laboratory diagnostic techniques such as acid-fast staining, culture, biochemical typing, and PCR molecular typing methods (Munyeme et al. 2010a, b; Hang'ombe et al. 2012) are only available at the Veterinary Research Institutes, and the School of Veterinary Medicine at the University of Zambia in Lusaka (Table 23.2), and have limited value in terms of controlling the disease throughout the country.

23.5 National Bovine Tuberculosis Control Policies and Strategies

Although BTB is a notifiable disease, there is no official control program for the disease in Zambia. The Zambian authorities profess that diseases such as BTB do not have a dramatic effect on livestock populations and have little effect on their international trade, and they consequently allocate financial and human resources to more pressing problems. Unlike in developed countries that control the disease in cattle with consequent benefit to public health, most African countries, including Zambia, argue that BTB is not a disease of national economic importance, and its control is left to individual farmers as they classify it as "a management disease." The Zambian government only encourages the screening of cattle herds for the presence of BTB and culling of the positive animals (Munyeme et al. 2010a, b). Currently, BTB control is mainly applied in the private dairy sector as the major milk processing factories do not buy milk from cattle herds that are not certified BTB free. For this reason, commercial cattle farmers privately fund the required BTB screening tests, and those animals that test positive are sold off to recover some of the costs.

This approach in Zambia is considered to be poorly justified given the zoonotic risk posed by BTB, the high prevalence of the infection in sectors of the cattle population, the existence of a wildlife maintenance host, and the large-scale consumption of potentially BTB-infected raw milk by a significant proportion of the rural communities. This situation may change in the foreseeable future as the Zambian Government realizing the public health importance of BTB and the need to control it to limit the risk for the at-risk population, is currently developing a National Bovine Tuberculosis Surveillance and Control Program that will be funded by the World Bank.

A general lack of knowledge about the disease is another critical factor in Zambia impeding the control of BTB (Munyeme et al. 2010a, b). Most farmers are unable to tell if their animals have BTB because of the insidious nature of the disease, especially in the initial stages. This poses a major risk of exposure for those people who milk the cattle. Communities cognizant of the disease usually attempt to control the disease in spite of the lack of governmental support. They also reduce the chances of contracting the disease by slaughtering suspected BTB-positive animals (Kazwala et al. 2001). It is thus critical when establishing future viable, workable control programs, to implement public education and awareness campaigns, which currently are not in place.

Another aspect complicating control of BTB in Zambia is the nature and complexity of the natural reservoir hosts other than cattle. It is known that once BTB establishes itself in wildlife populations, the likelihood of controlling and/or eradicating it because of the presence of sylvatic reservoir hosts is extremely difficult, if not impossible with the currently available means of controlling the disease (Michel et al. 2010). The extensive livestock/wildlife interface further complicates the situation due to bidirectional transmission of the disease across it and the further likelihood of environmental contamination (Cross et al. 2009). This situation prevails in the Kafue basin, where it has been shown that cattle initially introduced the disease to wildlife and that there now is spillback to cattle from wildlife.

A number of key factors that further negatively affect BTB control in Zambia include the different types of cattle farming and pastoral practices (Aranaz et al. 2006). Herd size has an influence on the prevalence of BTB (Ameni et al. 2010), and on the Kafue flats, it is not uncommon to find pastoralists owning over a 1000 head of cattle that allows easy spread of the disease. The spread of the disease is further enhanced by the aggregation of large numbers of cattle at water sources in arid environments and by communal grazing (Oloya et al. 2007). These factors should be kept in mind when formulating a workable control strategy.

The choice of workable control measures and strategies should take all the various factors dealt with into account. At the livestock/wildlife interface, more detailed studies will be needed to provide data on how the disease is maintained, and how it is spread and transmitted between susceptible hosts to allow an adequate control strategy to be designed.

References

Ameni G, Desta F, Firdessa R (2010) Molecular typing of *Mycobacterium bovis* isolated from tuberculosis lesions of cattle in north eastern Ethiopia. Vet Rec 167:138–141

Aranaz A, de Juan L, Bezos J et al (2006) Assessment of diagnostic tools for eradication of bovine tuberculosis in cattle co-infected with *Mycobacterium bovis* and *M. avium* subsp. *paratuberculosis*. Vet Res 37:593–606

Cook AJC, Tuchili LM, Buve A et al (1996) Human and bovine tuberculosis in the Monze District of Zambia – a cross-sectional study. Br Vet J 152:37–46

Cross PC, Heisey DM, Bowers JA et al (2009) Disease, predation and demography: assessing the impacts of bovine tuberculosis on African buffalo by monitoring at individual and population levels. J Appl Ecol 46:467–475

Department of Veterinary and Tsetse Control Services (1957) Annual report, Government Printer, Northern Rhodesia

Gallagher J, Macadam I, Sayer J et al (1972) Pulmonary tuberculosis in free-living lechwe antelope in Zambia. Trop Anim Health Prod 4:204–213

Gracey JF, Collins DS, Huey RJ (1999) Animal slaughter: meat inspection statistics. In: Gracey JF (ed) Meat hygiene. WB Saunders & Co, London, pp 190–202

Grange JM, Yates MD (1994) Zoonotic aspects of *Mycobacterium bovis* infection. Vet Microbiol 40:137–151

Grange JM, Yates DM, de Kantor NI (1996) World Health Organization, 1994. Guidelines for speciation within the *Mycobacterium tuberculosis* complex. WHO/EMC/ZOO/964. Geneva: WHO

Hang'ombe M, Munyeme M, Fukushima Y et al (2012) *Mycobacterium bovis* infection at the interface between domestic and wild animals in Zambia. BMC Vet Res 8(1):1. http://www.biomedcentral.com/1746-6148/8/221

Kazwala RR, Daborn CJ, Sharp JM et al (2001) Isolation of *Mycobacterium bovis* from human cases of cervical adenitis in Tanzania: a cause for concern? Int J Tuberc Lung Dis 5:87–91

Malama S, Muma JB, Godfroid J (2013a) A review of tuberculosis at the wildlife-livestock-human interface in Zambia. Infect Dis Poverty, 2(\) p 1. doi:https://doi.org/10.1186/2049-9957-2–1

Malama S, Johansen TB, Muma JB (2013b) Isolation and molecular characterization of *Mycobacterium bovis* from Kafue lechwe (*Kobus leche kafuensis*) from Zambia. Trop Anim Health Prod 46:153–157

Malama S, Muma JB, Popelka-Olea F et al (2013c) Isolation of *Mycobacterium bovis* from human sputum in Zambia: public health and diagnostic significance. J Infect Dis Ther 1:1–4

Malama S, Johansen TB, Muma JB et al (2014) Characterization of *M. bovis* from humans and cattle in Namwala District, Zambia. Vet Med Int 2014:7. https://doi.org/10.1155/2014/187842

Michel AL, Muller B, van Helden PD (2010) *Mycobacterium bovis* at the animal-human interface: a problem, or not? Vet Microbiol 140:371–381

Moda G, Daborn CJ, Grange JM et al (1996) The zoonotic importance of *Mycobacterium bovis*. Tuber Lung Dis 77:103–108

Munang'andu HM, Siamudaala VM, Matandiko W et al (2011) Comparative intradermal tuberculin testing of free-ranging African buffaloes (*Syncerus caffer*) captured for ex situ conservation in the Kafue ecosystem in Zambia. Vet Med Int 2011:5

Munyeme M, Munang'andu HM (2011) A review of bovine tuberculosis in the Kafue Basin ecosystem. Vet Med Int 2011:9

Munyeme M, Muma JB, Skjerve E et al (2008) Risk factors associated with bovine tuberculosis in traditional cattle of the livestock/wildlife interface areas in the Kafue basin of Zambia. Prev Vet Med 85:317–328

Munyeme M, Muma JB, Skjerve E et al (2009a) Prevalence of bovine tuberculosis and animal level risk factors for indigenous cattle under different grazing strategies in the livestock/wildlife interface areas of Zambia. Trop Anim Health Prod 41:345

Munyeme M, Rigouts L, Shamputa IC et al (2009b) Isolation and characterization of *Mycobacterium bovis* strains from indigenous Zambian cattle using spacer oligonucleotide typing technique. BMC Microbiol 9:144

Munyeme M, Muma JB, Munang'andu HM (2010a) Cattle owners' awareness of bovine tuberculosis in high and low prevalence settings of the wildlife-livestock interface areas in Zambia. BMC Vet Res 6:21. http://www.biomedcentral.com/1746-6148/6/21

Munyeme M, Muma JB, Siamudaala VM et al (2010b) Tuberculosis in Kafue lechwe antelopes (*Kobus leche kafuensis*) of the Kafue Basin in Zambia. Prev Vet Med 95:305–308

Oloya J, Kazwala R, Lund A et al (2007) Characterisation of mycobacteria isolated from slaughter cattle in pastoral regions of Uganda. BMC Microbiol 7:95. https://doi.org/10.1186/1471-2180-7-95

Pandey GS (2014) Food security and poverty mitigation through smallholder dairy — the Zambian case. Int J Rural Dev 4:26–27

Phiri AM (2006) Common conditions leading to cattle carcass and offal condemnations at 3 abattoirs in the Western Province of Zambia and their zoonotic implications to consumers. J S Afr Vet Assoc 77:28–32

Siamudaala VM, Muma J, Munang'andu HM et al (2005a) In: Osofsky SA (2005) Conservation and development interventions at the wildlife/livestock interface: implications for wildlife, livestock and human health. Proceedings of the Southern and East African Experts Panel on designing successful conservation and development interventions at the wildlife/livestock interface: implications for wildlife, livestock and human health, AHEAD (Animal Health for the Environment And Development) Forum, IUCN Vth World Parks Congress, Durban, South Africa, 14th and 15th September 2003 (No. 30). IUCN

Siamudaala VM, Muma JB, Munang'andu HM (2005b) Veterinary challenges regarding the utilization of the Kafue lechwe (*Kobus leche kafuensis*) in Zambia. In: Osofsky SA (2005) Conservation and development interventions at the wildlife/livestock interface: implications for wildlife, livestock and human health. Proceedings of the Southern and East African Experts Panel on designing successful conservation and development interventions at the wildlife/livestock interface: implications for wildlife, livestock and human health, AHEAD (Animal Health for the Environment And Development) Forum, IUCN Vth World Parks Congress, Durban, South Africa, 14th and 15th September 2003 (No. 30). IUCN

UKaid (2011) What would it take for Zambia's beef and dairy industry to achieve their potential? UKaid, The World Bank. www.worldbank.org/zambia

Printed by Printforce, the Netherlands